Advances in Intel
Soft Computing 107

Editor-in-Chief: J. Kacprzyk

Advances in Intelligent and Soft Computing

Editor-in-Chief

Prof. Janusz Kacprzyk
Systems Research Institute
Polish Academy of Sciences
ul. Newelska 6
01-447 Warsaw
Poland
E-mail: kacprzyk@ibspan.waw.pl

Further volumes of this series can be found on our homepage: springer.com

Vol. 93. M.P. Rocha, J.M. Corchado, F. Fernández-Riverola, and A. Valencia (Eds.)
5th International Conference on Practical Applications of Computational Biology & Bioinformatics 6-8th, 2011
ISBN 978-3-642-19913-4

Vol. 94. J.M. Molina, J.R. Casar Corredera, M.F. Cátedra Pérez, J. Ortega-García, and A.M. Bernardos Barbolla (Eds.)
User-Centric Technologies and Applications, 2011
ISBN 978-3-642-19907-3

Vol. 95. R. Burduk, M. Kurzyński, M. Woźniak, and A. Żołnierek (Eds.)
Computer Recognition Systems 4, 2011
ISBN 978-3-642-20319-0

Vol. 96. A. Gaspar-Cunha, R. Takahashi, G. Schaefer, and L. Costa (Eds.)
Soft Computing in Industrial Applications, 2011
ISBN 978-3-642-20504-0

Vol. 97. W. Zamojski, J. Kacprzyk, J. Mazurkiewicz, J. Sugier, and T. Walkowiak (Eds.)
Dependable Computer Systems, 2011
ISBN 978-3-642-21392-2

Vol. 98. Z.S. Hippe, J.L. Kulikowski, and T. Mroczek (Eds.)
Human – Computer Systems Interaction: Backgrounds and Applications 2, 2011
ISBN 978-3-642-23186-5

Vol. 99. Z.S. Hippe, J.L. Kulikowski, and T. Mroczek (Eds.)
Human – Computer Systems Interaction: Backgrounds and Applications 2, 2011
ISBN 978-3-642-23171-1

Vol. 100. S. Li, X. Wang, Y. Okazaki, J. Kawabe, T. Murofushi, and Li Guan (Eds.)
Nonlinear Mathematics for Uncertainty and its Applications, 2011
ISBN 978-3-642-22832-2

Vol. 101. D. Dicheva, Z. Markov, and E. Stefanova (Eds.)
Third International Conference on Software, Services and Semantic Technologies S3T 2011, 2011
ISBN 978-3-642-23162-9

Vol. 102. R.S. Choraś (Ed.)
Image Processing and Communications Challenges 3, 2011
ISBN 978-3-642-23153-7

Vol. 103. T. Czachórski, S. Kozielski, and U. Stańczyk (Eds.)
Man-Machine Interactions 2, 2011
ISBN 978-3-642-23168-1

Vol. 104. D. Jin and S. Lin (Eds.)
Advances in Computer Science, Intelligent System and Environment, 2011
ISBN 978-3-642-23776-8

Vol. 105. D. Jin and S. Lin (Eds.)
Advances in Computer Science, Intelligent System and Environment, 2011
ISBN 978-3-642-23755-3

Vol. 106. D. Jin and S. Lin (Eds.)
Advances in Computer Science, Intelligent System and Environment, 2011
ISBN 978-3-642-23752-2

Vol. 107. P. Melo-Pinto, P. Couto, C. Serôdio, J. Fodor, and B. De Baets (Eds.)
Eurofuse 2011, 2011
ISBN 978-3-642-24000-3

Pedro Melo-Pinto, Pedro Couto,
Carlos Serôdio, János Fodor, and
Bernard De Baets (Eds.)

Eurofuse 2011

Workshop on Fuzzy Methods for Knowledge-Based Systems

Springer

Editors

Prof. Pedro Melo-Pinto
CITAB-UTAD University
Quinta dos Prados
5000-911 Vila Real
Portugal
E-mail: pmelo@utad.pt

Prof. Pedro Couto
CITAB-UTAD University
Quinta dos Prados
5000-911 Vila Real
Portugal
E-mail: pcouto@utad.pt

Prof. Carlos Serôdio
CITAB-UTAD University
Quinta dos Prados
5000-911 Vila Real
Portugal
E-mail: cserodio@utad.pt

Prof. János Fodor
Óbuda University
Bécsi út 96/B
H-1034 Budapest
Hungary
E-mail: fodor@uni-obuda.hu

Prof. Bernard De Baets
KERMIT
Research Unit Knowledge-based Systems
Universiteit Gent
Coupure links 653
9000 Gent
Belgium
E-mail: Bernard.Debaets@UGent.be

ISBN 978-3-642-24000-3 e-ISBN 978-3-642-24001-0

DOI 10.1007/978-3-642-24001-0

Advances in Intelligent and Soft Computing ISSN 1867-5662

Library of Congress Control Number: 2011936640

Typeset & Cover Design: Scientific Publishing Services Pvt. Ltd., Chennai, India

Printed on acid free paper

5 4 3 2 1 0

springer.com

Preface

This book comprises the papers from EUROFUSE 2011 Workshop on Fuzzy Methods for Knowledge-based Systems.

EUROFUSE was established in 1998 as the EURO (the Association of European Operational Research Societies) Working Group on Fuzzy Sets, as a successor of the former European Chapter of IFSA (the International Fuzzy Systems Association). The working group is coordinated by Bernard De Baets (KERMIT, Ghent University, Belgium) and János Fodor (Óbuda University, Hungary).

EUROFUSE 2011 is the next in a series ofEUROFUSE workshops, including the successful EUROFUSE 1999 (Budapest, Hungary), EUROFUSE 2000 (Mons, Belgium), EUROFUSE 2001 (Granada, Spain),EUROFUSE 2002 (Varenna, Italy), EUROFUSE 2004 (Warsaw, Poland), EUROFUSE 2005 (Belgrade, Serbia), EUROFUSE 2007 (Jaen, Spain) and EUROFUSE 2009 (Pamplona, Spain).

The present edition is held in Régua, Portugal, in the World Heritage Site of the Douro Wine Region and is organised by Pedro Melo-Pinto and Pedro Couto. The theme of the workshop is *Fuzzy Methods for Knowledge-based Systems*. As was the case for previous editions, the aim of the workshop is to bring together researchers and practitioners in an informal atmosphere. For this reason, the number of participants is limited and all presentations are plenary.

EUROFUSE 2011 has three distinguished invited speakers: Francisco Herrera (Granada, Spain), Radko Mesiar (Bratislava, Slovakia) and José Luis Garcia Lapresta (Valladolid, Spain). Next to the invited speakers, the three day program consists of 37 lectures. In total, there are 70 participants from 10 countries. This edited volume contains the final revised manuscripts on the basis of which the program was put together.

We are grateful to the invited speakers, the authors and all reviewers for contributing to the success of this workshop. We also want to thank all the people from the organizing committee for the invaluable help, which helped turning this workshop into a successful one.

Bernard De Baets
János Fodor
Pedro Couto
Pedro Melo-Pinto

This conference was partially supported by Program Operacional Regional do Norte (ON.2)

QUADRO DE REFERÊNCIA ESTRATÉGICO NACIONAL
ON.2 O NOVO NORTE
UNIÃO EUROPEIA
Fundo Europeu de Desenvolvimento Regional

Organisation

Organising Commitee

Chairmen	Members
• B. de Baets	• J. Matias
• J. Fodor	• P. Mestre
• P. Melo-Pinto	• R. Ribeiro
• P. Couto	• C. Serodio

Invited Speakears

- José Luis García-Lapresta
- Francisco Herrera
- Radko Mesiar

Scientific Program Committee

E. Barrenechea (Spain)	J.M. Benítez (Spain)	U. Bodenhofer (Austria)
H. Bustince (Spain)	J. Carvalho (Portugal)	J. Chamorro (Spain)
F. Chiclana (UK)	O. Cordón (Spain)	C. Cornelis (Belgium)
I. Couso (Spain)	P. Couto (Portugal)	B. De Baets (Belgium)
S. Diaz (Spain)	J. Dombi (Hungary)	D. Dubois (France)
P. Fazendeiro (Portugal)	M. Fedrizzi (Italy)	J. Fernandez (Spain)
J. Fodor (Hungary)	J.L. Garcia-Lapresta (Spain)	G. Georgescu (Romania)
D. Gómez (Spain)	M. Grabisch (France)	F. Herrera (Spain)
E. Herrera-Viedma (Spain)	E. Hüllermeier (Germany)	J. Kacprzyk (Poland)
E. Kerre (Belgium)	E.P. Klement (Austria)	A. Kolesárová (Slovak Republic)
J. Liu (UK)	V. Loia (Italy)	L. Martínez (Spain)
G. Mayor (Spain)	P. Melo-Pinto (Portugal)	R. Mesiar (Slovak Republic)
J. Montero (Spain)	S. Montes (Spain)	E. Montseny (Spain)
J.A. Olivas (Spain)	V. Oliveira (Portugal)	M. Pagola (Spain)
E. Pap (Serbia)	I. Perfilieva (Czech Republic)	H. Prade (France)
Da Ruan (Belgium)	P. Salgado (Portugal)	D. Sánchez (Spain)
P. Sobrevilla (Spain)	E. Szmidt (Poland)	V. Torra (Spain)
J. Torrens (Spain)		J.L. Verdegay (Spain)

Contents

Invited Talks

On the Usefulness of Interval Valued Fuzzy Sets for Learning Fuzzy Rule Based Classification Systems

Francisco Herrera

Extended Abstract

One of the main advantages of the Fuzzy Rule Based Classification Systems (FRBCSs) is the high interpretability of the model. However, the disadvantage of these systems may be their lack of accuracy when dealing some complex systems, due to the inflexibility of the concept of linguistic variable, which imposes hard restrictions to the fuzzy rule structure. For example, sometimes when the classes are overlapped, we have not exact knowledge about the membership degree of some elements to the fuzzy sets that characterize the attributes defining the class.

This situation suggests the possibility to represent the membership degrees of the objects to the fuzzy set by means of an interval. That is, to employ the Interval-Valued Fuzzy Sets (IVFSs) to characterize the linguistic labels that compound the attributes of the problems. IVFSs allow us to take into account the effect of the ignorance of the experts in the membership function definition.

The aim of this talk is to shown the performance of FRBCSs by extending the Knowledge Base with the application of the concept of IVFSs. The modeling of the linguistic labels by means of IVFSs implied an adaptation of the original fuzzy reasoning method to allow us to handle the uncertainty that is inherent to the definition process of the membership functions. We define new reasoning methods meaning use of the interval-valued restricted equivalence functions to increase the relevance of the rules in which the equivalence of the interval membership degrees of the patterns and the ideal membership degrees is greater, which is a desirable behavior. Furthermore, the parametrized construction of this fuzzy reasoning method allows the choice of the optimal function for each variable to be performed, which could involve a potential improvement of the system behavior. These parameters will be

Francisco Herrera
Dept. of Computer Sciences and Artificial Intelligence, University of Granada, 18071-Granada, Spain
e-mail: Herrera@decsai.ugr.es

B. De Baets et al. (Eds.): Eurofuse 2011, AISC 107, pp. 3–4, 2011.
springerlink.com

tuned using genetic algorithms in order to further improve the performance of the systems in a general framework.

We will show different experimental studies showing the usefulness of the IVFSs for enhancing the FRBCSs performance.

Using Aggregation Functions for Measuring Social Inequality and Poverty

José Luis García-Lapresta

Poverty reduction is without doubt a goal of development policy in most countries. To evaluate the evolution of poverty over time in some particular region, the differences of poverty across different countries or the effect of different policies in the alleviation of poverty, one should be first able to measure poverty.

According to the 1998 Nobel Prize Laureate A.K. Sen [17], any poverty index should be sensitive to the number of people below the poverty line, to the extent of the income shortfall of the poor from the poverty line, and to the exact pattern of the income distribution of the poor. In other words, every poverty measure should be expressed as a function of these three poverty indicators, showing the incidence, the intensity and the inequality of the poor, respectively. Poverty changes can be more meaningful and easily understandable if poverty indices can be decomposed into these underlying contributing factors. In fact, a number of poverty indices and their decompositions have been proposed to explicitly identify these three components[1].

The use of aggregation functions[2] for analyzing poverty and inequality[3] can provide new results within the theory of welfare economics. In fact, the dual decomposition of aggregation functions into a self-dual core and anti-self-dual remainder proposed by García-Lapresta and Marques Pereira [9] has been applied to the design of new poverty measures in García-Lapresta *et al.* [8], and to avoid inconsistencies in some decompositions of the Sen poverty index [17] in Aristondo *et al.* [1]). We note that Maes *et al.* [15] propose a generalization of the dual decomposition framework introduced in García-Lapresta and Marques Pereira [9], based on a

José Luis García-Lapresta
PRESAD Research Group, Departamento de Economía Aplicada,
Universidad Valladolid, Avenida Valle de Esgueva 6, 47011 Valladolid, Spain
e-mail: lapresta@eco.uva.es

[1] Besides Clark *et al.* [6], Osberg and Xu [16], Xu and Osberg [23] and Aristondo *et al.* [2], some of them may be found in Kakwani [13].

[2] See Fodor and Roubens [7], Calvo *et al.* [4], Beliakov *et al.* [3] and Grabisch *et al.* [11].

[3] For comprehensive surveys on poverty and inequality measures see Silber [20] and Chakravarty [5].

B. De Baets et al. (Eds.): Eurofuse 2011, AISC 107, pp. 5–8, 2011.
springerlink.com

family of binary aggregation functions satisfying a form of twisted self-duality condition. However, the dual decomposition framework introduced in García-Lapresta and Marques Pereira [9] remains the only one which preserves stability under translations, a crucial requirement in the present construction of poverty measures.

Poverty Measures Based on Exponential Means

The class of poverty measures introduced in García-Lapresta *et al.* [8] is based on the one-parameter family of exponential means, which are used in order to aggregate the normalized income gaps of the poor individuals. This new class of poverty measures has three interesting features. First, they are invariant under changes in the unit in which income is measured. Secondly, given a poverty line and a set of poor individuals, when all poor individuals receive the same amount of extra income the reduction in the overall poverty level depends only on that extra amount. Finally, all the members of the new family of poverty measures are more sensitive to changes at the lower end of the income profile, that is, to what happens to the poorest of the poor. In addition, the family parameter can be interpreted as a measure of this sensitivity. It is interesting to note that none of the indices introduced in the literature are able to capture these three issues together.

The core and the remainder of the proposed poverty measures can be interpreted as measures of the intensity and of the inequality among the poor, respectively. This decomposition allows us to know if increasing poverty is due to more people becoming poor, or increasing deprivation of the poor, or because income short-fall below the poverty line has become more unequal, or some combination of the above. Therefore, the decomposition proposed can be useful to have a better understanding of the measurement of poverty, particularly in a complex scenario.

The Consistent Measurement of Inequality among the Poor

Two different decompositions of the classical Sen poverty index [17, 18] have been proposed in the literature: Sen [17] and Xu and Osberg [23]. In both cases, the inequality among the poor is captured by the Gini index [10], but applied either to the poor income or to the shortfall of the poor. Recently, Lambert and Zheng [14] show that no relative inequality index offers consistent results. In fact, the choice between income (Sen [17]) and shortfall (Xu and Osberg [23]) inequality is not innocuous and may produce different rankings, so contradictory results.

In order to avoid the mentioned problem, Aristondo *et al.* [1] investigate the structure of the Sen poverty index [17, 18] within the framework of the dual decomposition of aggregation functions proposed by García-Lapresta and Marques Pereira [9]. In [1], the Sen index has been written as a product of the standard headcount ratio and an OWA operator (Yager [24]) applied to the poverty gaps. This OWA operator decomposes into a self-dual core, corresponding to the average poverty gap, and an anti-self-dual remainder which corresponds to the classical absolute Gini index of the normalized incomes of the poor. In this new decomposition of the Sen poverty index, therefore, the self-dual core and the anti-self-dual remainder measure (respectively) the intensity and the inequality of poverty within the given income distribution. The central result of Aristondo *et al.* [1] is that the dual decomposition of the Sen poverty index contains an inequality measure which is naturally achievement/shortfall consistent: The anti-self-duality of the remainder component guarantees that inequality among the poor does not change if one focus either on incomes or on shortfalls. These inequality components will allow policy makers to determine in a consistent way if inequality among the poor has increased or decreased.

It is interesting to note that the new proposal works for a number of poverty indices in which the inequality is captured by the Gini index [10], such as the Thon index [22], the index introduced by Kakwani [12], the Takayama proposal [21], and the Sen index modified by Shorrocks [19].

Acknowledgements. This contribution is based on my research with Ricardo A. Marques Pereira, Casilda Lasso de la Vega, Ana M. Urrutia and Oihana Aristondo. The author gratefully acknowledges the funding support of the Spanish Ministerio de Ciencia e Innovación (Project ECO2009-07332) and ERDF.

References

1. Aristondo, O., García-Lapresta, J.L., Lasso de la Vega, C., Marques Pereira, R.A.: The Gini index and the consistent measurement of inequality among the poor through the dual decomposition of OWA operators (submitted)
2. Aristondo, O., Lasso de la Vega, C., Urrutia, A.M.: A new multiplicative decomposition for the Foster-Greer-Thorbecke poverty indices. Bulletin of Economic Research 62, 259–267 (2010)
3. Beliakov, G., Pradera, A., Calvo, T.: Aggregation Functions: A Guide for Practitioners. Springer, Heidelberg (2007)
4. Calvo, T., Kolesárova, A.: Komorníková, M., Mesiar, R.: Aggregation operators: Properties, classes and construction methods. In: Calvo, T., Mayor, G., Mesiar, R. (eds.) Aggregation Operators: New Trends and Applications, pp. 3–104. Physica-Verlag, Heidelberg (2002)
5. Chakravarty, S.R.: Inequality, Polarization and Poverty: Advances in Distributional Analysis. Springer, New York (2009)
6. Clark, S., Hemming, R., Ulph, D.: On indices for the measurement of poverty. Economic Journal 91, 515–526 (1981)
7. Fodor, J., Roubens, M.: Fuzzy Preference Modelling and Multicriteria Decision Support. Kluwer Academic Publishers, Dordrecht (1994)
8. García-Lapresta, J.L., Lasso de la Vega, C., Marques Pereira, R.A., Urrutia, A.M.: A class of poverty measures induced by the dual decomposition of aggregation functions. International Journal of Uncertainty, Fuzziness and Knowledge-Based Systems 18, 493–511 (2010)
9. García-Lapresta, J.L., Marques Pereira, R.A.: The self-dual core and the anti-self-dual remainder of an aggregation operator. Fuzzy Sets and Systems 159, 47–62 (2008)
10. Gini, C.: Variabilità e Mutabilità. Tipografia di Paolo Cuppini, Bologna (1912)
11. Grabisch, M., Marichal, J.L., Mesiar, R., Pap, E.: Aggregation Functions. Cambridge University Press, Cambridge (2009)

12. Kakwani, N.: On a class of poverty measures. Econometrica 48, 437–446 (1980)
13. Kakwani, N.: Inequality, Welfare and Poverty. In: Silber, J. (ed.) Handbook of Income Inequality Measurement, pp. 599–628. Kluwer Academic Publishers, Dordrecht (1999)
14. Lambert, P., Zheng, B.: On the consistent measurement of achievement and shortfall inequality. Journal of Health Economics 30, 214–219 (2011)
15. Maes, K., Saminger, S., De Baets, B.: Representation and construction of self-dual aggregation operators. European Journal of Operational Research 177, 472–487 (2007)
16. Osberg, L., Xu, K.: International comparisons of poverty intensity: Index decomposition and bootstrap inference. The Journal of Human Resources 35, 1–81 (2000)
17. Sen, A.K.: Poverty: An ordinal approach to measurement. Econometrica 44, 219–231 (1976)
18. Sen, A.K.: Issues in the measurement of poverty. Scandinavian Journal of Economics 81, 285–307 (1979)
19. Shorrocks, A.: Revisiting the Sen poverty index. Econometrica 63, 1225–1230 (1995)
20. Silber, J.: Handbook on Income Inequality. Kluwer Academic Press, Dordrecht (1999)
21. Takayama, N.: Poverty, income inequality, and their measures: Professor Sen's axiomatic approach reconsidered. Econometrica 47, 747–759 (1979)
22. Thon, D.: On measuring poverty. Review of Income and Wealth 25, 429–440 (1979)
23. Xu, K., Osberg, L.: The social welfare implications, decomposability, and geometry of the Sen family of poverty indices. Canadian Journal of Economics 35, 138–152 (2002)
24. Yager, R.R.: Ordered weighted averaging operators in multicriteria decision making. IEEE Transactions on Systems, Man and Cybernetics 8, 183–190 (1988)

Fuzzy Integrals as a Tool for Multicriteria Decision Support

Radko Mesiar

Abstract. We recall several types of discrete fuzzy integrals and their recent generalizations based on level–dependent capacities. In multicriteria decision support these functionals can be seen as special utility functions. We include some examples of axiomatically defined utility functions and show their relationship to fuzzy integrals.

1 Introduction

In a set $\mathscr{A}$ of alternatives, based on n criteria, we aim to find the best alternatives, or to construct a preference structure. For several approaches to these problems we recommend a recent edited volume [3]. One possible approach is based on the utility theory. To make the problem more transparent, in this contribution we consider that each alternative $a \in \mathscr{A}$ is represented by a score vector $\mathbf{x} \in [0,1]^n$, $\mathbf{x} = (x_1, \ldots, x_n)$, where x_i is the degree of satisfaction of i–th criterion by the considered alternative a. Thus $\mathscr{A}$ can be isomorphically embedded into the space $[0,1]^n$, which will be the basic domain for our next considerations.

Utility function $U : [0,1]^n \to \Re$ is a non–decreasing real function (to preserve the Pareto property when constructing a preference structure) which can be transformed into a normed utility function $A : [0,1]^n \to [0,1]$, $A(0,\ldots,0) = 0$, $A(1,\ldots,1) = 1$, preserving the preference structure stated by U, i.e., $A(\mathbf{x}) \leq A(\mathbf{y})$ if and only if $U(\mathbf{x}) \leq U(\mathbf{y})$. Observe that normed utility function is called also (n–ary) aggregation function [4, 11, 12].

Formally, each score vector can be considered as a fuzzy subset of the universe $N = \{1,\ldots,n\}$. Expected value of a fuzzy set was introduced by Zadeh [20] as a

Radko Mesiar
Faculty of Civil Engineering, Slovak University of Technology, 813 68 Bratislava, Slovakia,
Institute of Theory of Information and Automation, Czech Academy of Sciences,
Prague, Czech Republic
e-mail: Radko.Mesiar@stuba.sk

B. De Baets et al. (Eds.): Eurofuse 2011, AISC 107, pp. 9–15, 2011.
springerlink.com

Lebesgue integral $A_P(\mathbf{x}) = \int_N \mathbf{x}\, dP = \sum_{i=1}^n p_i x_i$, $p_i = P(\{i\})$, where P is a probability measure on N. Clearly A_P is just an additive normed utility function. However, there are several other approaches how to define an expected value of a fuzzy set, namely by means of fuzzy integrals with respect to a capacity μ on N. These special utility functions are frequently applied in many multicriteria decision problems. The aim of this contribution is to recall the concept of fuzzy integrals with several special examples, including Choquet integral and Sugeno integral, among others, and their axiomatization. Next, we recall and introduce some new types of fuzzy integrals based on level–dependent capacities.

The paper is organized as follow. In the next section, fuzzy integrals are discussed, including several particular classes of fuzzy integrals. Section 3 considers the level–dependent capacities and some fuzzy integrals. In Section 4, we look on fuzzy integrals as special axiomatically defined classes of normed utility functions. Section 5 brings some examples, and in Section 6, some concluding remarks are added.

2 Fuzzy Integrals

The notion "fuzzy integral" was proposed by Sugeno [17] for an integral he has introduced based on a capacity (fuzzy measure) $\mu : 2^N \to [0,1]$, $\mu(\emptyset) = 0$, $\mu(N) = 1$, $\mu(E) \leq \mu(F)$ whenever $E \subseteq F \subseteq N$, i.e., for a normed utility function $S_\mu : [0,1]^n \to [0,1]$ given by

$$S_\mu(\mathbf{x}) = \bigvee_{i=1}^{n} \left(x_{(i)} \wedge \mu\left(\{j | x_j \geq x_{(i)}\}\right)\right), \tag{1}$$

where $(\cdot)$ is a permutation of N such that $x_{(1)} \leq \cdots \leq x_{(n)}$. Though the Sugeno integral S_μ can be introduced on any measurable space, we will consider only its discrete form (1) acting on N (similarly as for all other fuzzy integrals). Later, all kinds of integrals acting on $[0,1]^n$ as normed utility functions were called fuzzy integrals, and we will keep this convention. Maybe the most applied fuzzy integral goes back to Choquet [7], though its first traces can be found already in Vitali's paper [18].

For a given capacity μ on N, the Choquet integral $C_\mu : [0,1]^n \to [0,1]$ is given by

$$C_\mu(\mathbf{x}) = \sum_{i=1}^{n} x_{(i)} \cdot \left(\mu\left(\{j | x_j \geq x_{(i)}\}\right) - \mu\left(\{j | x_j \geq x_{(i+1)}\}\right)\right), \tag{2}$$

with convention $x_{(n+1)} = \infty$. Observe that both S_μ and C_μ are piecewise linear functionals. In general, any fuzzy integral $I_\mu : [0,1]^n \to [0,1]$ satisfying

(i) I_μ is normed utility function
(ii) $I_\mu(\mathbf{c}) = c$ (unanimity)
(iii) $I_\mu(\mathbf{1}_E) = \mu(E)$ (i.e., I_μ extends μ)
(iv) $I_\mu(c \cdot \mathbf{1}_E)$ depends on c and $\mu(E)$ only

can be seen as a functional determined by a capacity M on $\mathscr{B}(]0,1[^2)$, $M(]0,c[\times]0,1[) = c$, $M(]0,1[\times]0,b[) = b$, in the next form:

$$I_\mu(\mathbf{x}) = A_{\mu,M}(\mathbf{x}) = M\left(\left\{(x,y) \in]0,1[^2 | y < \mu\left(\{j | x_j \geq x\}\right)\right\}\right). \tag{3}$$

For more details see [8].

For example, putting $M(E) = sup\ \{xy | (x,y) \in E\}$, formula (3) yields the Shilkret integral [15] $Sh_\mu : [0,1]^n \to [0,1]$ given by

$$Sh_\mu(\mathbf{x}) = \bigvee_{i=1}^{n} x_{(i)} \cdot \mu\left(\{j | x_j \geq x_{(i)}\}\right). \tag{4}$$

Each fuzzy integral given by (3) can be interpreted also as an instant of some universal integral [9] based on a semicopula $\odot : [0,1]^2 \to [0,1]$, $\odot$ is a binary aggregation function on $[0,1]^2$ with neutral element 1. Then $I_\mu(c \cdot 1_E) = c \odot \mu(E)$. We recall three classes of universal integrals acting on $[0,1]^n$:

- the weakest $\odot$–based universal integral $I_{\mu,\odot} : [0,1]^n \to [0,1]$ given by

$$I_{\mu,\odot}(\mathbf{x}) = \bigvee_{i=1}^{n} x_{(i)} \odot \mu\left(\{j | x_j \geq x_{(i)}\}\right); \tag{5}$$

- the strongest $\odot$–based universal integral $I_\mu^{\odot} : [0,1]^n \to [0,1]$ given by

$$I_\mu^{\odot}(\mathbf{x}) = \mathbf{x}^{(\mu)} \odot \mu\left(supp(\mathbf{x})\right), \tag{6}$$

  where $\mathbf{x}^{(\mu)} = sup\{t \in [0,1] | \mu(\{i | x_i \geq t\}) > 0\}$ and $supp(\mathbf{x}) = \{i | x_i > 0\}$;
- copula–based universal integral $I_{C,\mu} : [0,1]^n \to [0,1]$ given by

$$I_{C,\mu}(\mathbf{x}) = \sum_{i=1}^{n} \left(C\left(x_{(i)}, \mu\left(\{j | x_j \geq x_{(i)}\}\right)\right) - C\left(x_{(i)}, \mu\left(\{j | x_j \geq x_{(i+1)}\}\right)\right)\right), \tag{7}$$

  where $C : [0,1]^2 \to [0,1]$ is a copula, i.e., a semicopula which is 2–increasing, $C(x',y') - C(x,y') - C(x',y) + C(x,y) \geq 0$ for all $0 \leq x \leq x' \leq 1$, $0 \leq y \leq y' \leq 1$.

Note that for the product copula $\Pi : [0,1]^2 \to [0,1]$, $\Pi(x,y) = xy$ we have $I_{\Pi,\mu} = C_\mu$ (Choquet integral) and $I_{\mu,\Pi} = Sh_\mu$ (Shilkret integral). For the strongest copula $Min : [0,1]^2 \to [0,1]$, $Min(x,y) = min\ \{x,y\}$ we have $I_{\mu,Min} = S_\mu = I_{Min,\mu}$ (Sugeno integral).

3 Fuzzy Integrals Based on Level–Dependent Capacities

A level–dependent capacity M on N is a system $M = (\mu_t)_{t\in[0,1]}$ such that each μ_t is a capacity on N, and for each $E \subseteq N$, the function $h_{E,M} : [0,1] \to [0,1]$ given by $h_{E,M}(t) = \mu_t(E)$ is Borel measurable. For more details we recommend [6] and [10].

For any $\mathbf{x} \in [0,1]^n$ and a level–dependent capacity M on N one can define a function $h_{M,\mathbf{x}} : [0,1] \to [0,1]$ given by $h_{M,\mathbf{x}}(t) = \mu_t(\{i|x_i \geq t\})$, which is Borel measurable. Based on $h_{M,\mathbf{x}}$, we can introduce a level–dependent capacity M based fuzzy integrals extending the fuzzy integrals discussed in Section 2 as follows:

- $$I_{M,\odot}(\mathbf{x}) = \bigvee_{t\in[0,1]} t \odot h_{M,\mathbf{x}}(t); \tag{8}$$
- $$I_M^{\odot}(\mathbf{x}) = \lambda\left(\{h_{M,\mathbf{x}} > 0\}\right) \odot sup\left\{h_{M,\mathbf{x}}(t) | t \in]0,1]\right\}, \tag{9}$$
  where λ is the Lebesgue measure on $\mathscr{B}([0,1])$;
- $$I_{C,M}(\mathbf{x}) = P_C\left(\left\{(x,y) \in [0,1]^2 | x \leq h_{M,\mathbf{x}}(y)\right\}\right), \tag{10}$$
  where P_C is a probability measure on $\mathscr{B}([0,1]^2)$ induced by a copula C. If $C = \Pi$ (and then P_Π is the Lebesgue measure on $\mathscr{B}([0,1]^2)$, one gets the level–dependent capacity based Choquet integral as proposed in [6].

Observe that all three $I_{M,\odot}$, $I_M^{\odot}$ and $I_{C,M}$ are normed utility functions, and for a constant level–dependent capacity $M = (\mu)_{t\in[0,1]}$, $I_{M,\odot} = I_{\mu,\odot}$, $I_M^{\odot} = I_\mu^{\odot}$ and $I_{C,M} = I_{C,\mu}$.

4 Fuzzy Integrals as Axiomatically Defined Classes of Normed Utility Functions

Recall that two score vectors $\mathbf{x}, \mathbf{y} \in [0,1]^n$ are comonotone whenever there is a common permutation (.) of N such that $x_{(1)} \leq \cdots \leq x_{(n)}$, and $y_{(1)} \leq \cdots \leq y_{(n)}$. It is well–known that the Choquet integral $C_\mu : [0,1]^n \to [0,1]$ can be characterized as a comonotone additive normed utility function [16, 1] with a capacity μ given by $\mu(E) = C_\mu(\mathbf{1}_E)$, $E \subseteq N$. Similarly, the Sugeno integral $S_\mu : [0,1]^n \to [0,1]$ can be characterized as a comonotone maxitive normed utility function which is $\wedge$–homogeneous ($\wedge = Min$), i.e., $S_\mu(\mathbf{c} \wedge \mathbf{x}) = c \wedge S_\mu(\mathbf{x})$ for any $c \in [0,1]$ and $\mathbf{x} \in [0,1]^n$. Here $\mu(E) = S_\mu(\mathbf{1}_E)$ for any $E \subseteq N$. Maxitive normed utility function A can be seen as a level–dependent capacity M based Sugeno integral, $A = I_{M,Min}$, where $M = (\mu_t)_{t\in[0,1]}$, $\mu_t(E) = A(t \cdot \mathbf{1}_E)$, $E \neq N$ and $t > 0$. Moreover, each maxitive normed utility function A satisfying $A(t\,1_E) = A(t\,1_F)$ for all $t \in [0,1]$ whenever $A(1_E) = A(1_F)$ can be seen as the weakest universal integral $A = I_{\mu,\odot}$, where $\mu(E) = A(\mathbf{1}_E)$, $E \subseteq N$, and $u \odot v = A(u\,\mathbf{1}_E)$ whenever $v = \mathbf{1}_E$.

For anonymous normed utility functions, it is well known that if we require also the comonotone additivity, the resulting normed utility function is an OWA operator [19]. Similarly, imposing the comonotone maxitivity and $\wedge$–homogeneity, we recover the OWMAX operators (Ordered Weighted Maximum) [2]. Moreover, we have the next interesting links to the fuzzy integrals, see [13]:

- each comonotone unanimous modular anonymous normed utility function A can be seen as a copula–based fuzzy integral, $A = I_{C,\mu}$, where $\mu(E) = A(1_E)$, $E \subseteq N$ and the copula C is constrained by $A(t \cdot \mathbf{1}_E) = C(t, \mu(E))$.

5 Examples

First of all, recall that if a capacity μ has minimal range $\{0,1\}$, then each fuzzy integral based on μ from a score vector $\mathbf{x}$ collapses into the essential supremum $\mathbf{x}^{(\mu)}$. In special cases, considering the minimal capacity μ_* and the maximal capacity μ^*, we have $\mathbf{x}^{(\mu_*)} = min\,\{x_1,\cdots,x_n\}$ and $\mathbf{x}^{(\mu^*)} = max\,\{x_1,\cdots,x_n\}$.

For the simplicity, consider $n=2$, i.e., $N=\{1,2\}$. Then each capacity μ on N is determined by $\mu(\{1\}) = a$ and $\mu(\{2\}) = b$. It holds:

- $$I_{\mu,\odot}(x,y) = (min\,\{x,y\}) \vee ((max\,\{x,y\}) \odot c),$$

  where
  $$c = \begin{cases} 1 & \text{if } x=y \\ a & \text{if } x>y \\ b & \text{if } x<y; \end{cases} \tag{11}$$
  if $\odot = Min = \wedge$ (i.e., $I_{\mu,Min}$ is the Sugeno integral), we have the well known formula
  $$I_{\mu,\wedge}(x,y) = (x\wedge y) \vee (x \wedge a) \vee (y \wedge b),$$
  which in the case a = b turns into
  $$I_{\mu,\wedge}(x,y) = med\,(x,y,a);$$

- $$I_\mu^{\odot}(x,y) = \mathbf{x}^{(\mu)} \odot S_D(a,b)$$

  where $S_D : [0,1]^2 \to [0,1]$ is the drastic sum defined by
  $$S_D(u,v) == \begin{cases} u\vee v & \text{if } u\wedge v = 0, \\ 1 & \text{else.} \end{cases}$$
  Note that
  $$\mathbf{x}^{(\mu)} = \begin{cases} x\vee y & \text{if } a>0,\, b>0, \\ x & \text{if } a>0,\, b=0, \\ y & \text{if } a=0,\, b>0, \\ x\wedge y & \text{if } a=0,\, b=0. \end{cases}$$
  Therefore,
  $$I_\mu^{\odot}(x,y) = x\vee y \text{ whenever } a>0,\ b>0,\ x>0,\ y>0;$$

- $$I_{C,\mu}(x,y) = (x\wedge y) - C(x\wedge y, c) + C(x\vee y, c),$$

  where c is given by (11).

For $N=\{1,2\}$, consider level–dependent capacities $M_1 = (\mu_t)_{t\in[0,1]}$ and $M_2 = (\eta_t)_{t\in[0,1]}$, where $\mu_t = \eta_u = \mu^*$ whenever $t \in [0,\frac{1}{2}]$ and $u \in [\frac{1}{2},1]$, and $\mu_t = \eta_u = \mu_*$ whenever $t \in]\frac{1}{2},1]$ and $u \in [0,\frac{1}{2}[$. Then

- $$I_{M_1,\odot}(x,y) = (x \wedge y) \vee ((x \vee y) \wedge \frac{1}{2}) = med(x,y,\frac{1}{2});$$
- $$I_{M_2,\odot}(x,y) = \begin{cases} x \wedge y & \text{if } x \vee y < \frac{1}{2}, \\ x \vee y & \text{if } x \vee y \geq \frac{1}{2}, \end{cases}$$

  (observe the non–continuity of $I_{M_2,\odot}$).

Note that as the range of M_1 is minimal (i.e., it is $\{0,1\}$), all introduced fuzzy integrals based on M_1 coincide. Similar is the case of M_2–based fuzzy integrals.

6 Concluding Remarks

We have introduced and discussed fuzzy integrals based on capacities or on level–dependent capacities. For special cases, an axiomatic approach was given, too. Concerning the possible further generalizations, induced fuzzy integrals, as well as induced level–dependent capacity based integrals can be introduced. For an introductory discussion on this topic we recommend [5].

We believe that some of presented normed utility functions become useful in the multicriteria decision support. Especially, OMA (Ordered Modular Averages) linked to copula–based fuzzy integrals based on symmetric capacities show a big potential. Observe that if $C = \Pi$, popular OWA's are recovered.

Acknowledgements. The work on this contribution was supported by grants LPP–0004–07 and GACR P 402/11/0378.

References

1. Benvenuti, P., Mesiar, R., Vivona, D.: Monotone set functions-based integrals. In: Pap [14], pp. 1329–1379 (2002)
2. Dubois, D., Prade, H.: A review of fuzzy set aggregation connectives. Inform. Sci. 36, 85–121 (1985)
3. Ehrgott, M., Figueira, J.P., Greco, S.: Trends in Multiple Criteria Decision Analysis. Springer, New York (2002)
4. Grabisch, M., Marichal, J.–L., Mesiar, R., Pap, E.: Aggregation functions. Cambridge University Press, Cambridge (2009)
5. Greco, S., Mesiar, R., Merigó, J.M.: Induced level–dependent integrals. In: AGOP 2011 (submitted 2011)
6. Greco, S., Giove, S., Matarazzo, B.: The Choquet integral with respect to a level dependent capacity. Fuzzy Sets and Systems (in press)
7. Choquet, G.: Theory of capacities. Ann. Inst. Fourier (Grenoble) 5, 131–292 (1953)
8. Klement, E.P., Mesiar, R., Pap, E.: Measure–based aggregation operators. Fuzzy Sets and Systems 142(1), 3–14 (2004)
9. Klement, E.P., Mesiar, R., Pap, E.: A universal integral as a common frame for Choquet and Sugeno integral. IEEE Transactions on Fuzzy Systems 18, 178–187 (2010)

10. Klement, E.P., Kolesárová, A., Mesiar, R., Stupňanová, A.: Universal integrals based on level dependent capacities. In: IFSA-EUSFLAT 2009. Proceedings of International Fuzzy Systems Association World Congress IFSA/EUSFLAT, Lisbon, Portugal, July 20-24 (2009)
11. Klir, G.J., Folger, T.A.: Fuzzy Sets. In: Uncertainty and Information. Prentice Hall, Englewood Cliffs (1987)
12. Kolesárová, A., Komorníková, M.: Triangular norm-based iterative compensatory operators. Fuzzy Sets and Systems 104, 109–120 (1999)
13. Mesiar, R., Mesiarová – Zemánková, A.: The ordered modular averages. IEEE Transactions on Fuzzy Systems 19, 42–50 (2011)
14. Pap, E. (ed.): Handbook of measure theory. Elsevier Science, Amsterdam (2002)
15. Shilkret, N.: Maxitive measure and integration. Indag. Math. 33, 109–116 (1971)
16. Schmeidler, D.: Integral representation without additivity. Proc. Amer. Math. Soc. 97, 255–261 (1986)
17. Sugeno, M.: Theory of fuzzy integrals and its applications. Ph.D. thesis, Tokyo Institute of Technology, Tokyo (1974)
18. Vitali, G.: Sulla definizione di integrale delle funzioni di una variabile. Ann. Mat. Pura Appl. 2, 111–121 (1925)
19. Yager, R.R.: On ordered weighted averaging aggregation operators in multicriteria decision making. IEEE Trans. on Systems, Man and Cybernetics 18, 183–190 (1988)
20. Zadeh, L.A.: Probability measures of fuzzy events. J. Math. Anal. Appl. 23, 421–427 (1968)

Part I

Theory

A Possibilistic Logic View of Sugeno Integrals

Didier Dubois, Henri Prade, and Agnès Rico

Abstract. Sugeno integrals are well-known qualitative aggregation functions in multiple criteria decision making. They return a global evaluation between the minimum and the maximum of the input criteria values. They can model sophisticated aggregation schemes through a system of priorities that applies to any subset of criteria and can take into account some kind of synergy inside subsets of criteria. Although a given Sugeno integral specifies a particular way of implicitly describing a set of entities reaching some global satisfaction level, it is hard to figure out what is the underlying explicit meaning of such an integral in practice (even if the priority level associated to each subset of criteria has a precise meaning). The paper proposes an answer to this problem. Any capacity on a finite set can be represented by a special possibilistic logic base containing positive prioritised clauses, and conversely any possibilistic logic base can represent a set-function. Moreover, Sugeno integral can be represented by a possibilistic logic base expressing how it behaves (thanks to a mapping between the scale and a set of logical atoms reflecting the different values for each criterion). Viewing a Sugeno integral as a set of prioritized logically expressed goals has not only the advantage to make the contents of a Sugeno integral more readable, but it also prompts Sugeno integrals into the realm of logic, and makes it possible to define entailment between them.

1 Introduction

Simple multiple criteria aggregation attitudes are easy to grasp. Thus, a strict conjunctive attitude based on the minimum operator amounts to saying that the global

Didier Dubois · Henri Prade
IRIT, CNRS and Université de Toulouse, France
e-mail: {dubois,prade}@irit.fr

Agnès Rico
ERIC, Université de Lyon, France
e-mail: agnes.rico@univ-lyon1.fr

B. De Baets et al. (Eds.): Eurofuse 2011, AISC 107, pp. 19–30, 2011.
springerlink.com

evaluation reflects the worst rating got among the criteria. With an averaging attitude between two criteria, a global medium score corresponds to having either both criteria half satisfied, or one poorly satisfied and the other well satisfied. With more sophisticated aggregation operators, it becomes difficult to figure out what is their exact meaning from a preference representation point of view, and more generally to provide explanations on the reasons why an object receives a global score [11].

In this paper, we consider an important family of qualitative aggregation operators, namely Sugeno integrals [14, 15]. They range between minimum and maximum operators, and behave more generally as the median of the normalized ratings according to the various criteria, together with priorities of groups of best satisfied criteria. In Sugeno integrals, each subset of criteria is indeed associated with a priority level, which enables some synergy between them to be modelled. Sugeno integrals have found many applications in multi-criteria decision aid [9]. In the following we investigate the possibility of providing an equivalent logical representation of Sugeno integrals in the setting of possibilistic logic where classical logic propositions are associated with priorities, in order to articulate the intended meaning of the aggregation, in logical terms. This may be especially of interest when Sugeno integrals are obtained by means of a learning procedure [18, 17] from observed data describing how people associate a global evaluation to elementary ones. Then, providing the logical counterpart of the elicited Sugeno integral may help understand what really matters for people when giving their global judgement.

The paper is structured as follows. The next section provides the necessary background both on Sugeno integrals defined in a discrete setting (valued on a finite scale), and on possibilistic logic. Section 3 shows how a fuzzy measure can be represented by a possibilistic logic base, and conversely how a possibilistic logic base can encode a set-function prioritizing groups of criteria. Section 4 explains how to represent a fuzzy integral by a possibilistic logic base that provides an equivalent logical reading of the aggregation attitude underlying this integral. Section 5 discusses related works and lines for further development.

2 Background

In this part we provide the necessary background on Sugeno integrals and on possibilistic logic bases, since we want to translate one into the other.

2.1 Discrete Sugeno Integral

We consider a set $\mathscr{P} = \{1, \cdots, n\}$ of criteria (properties) and a set X of alternatives evaluated according to each criterion. We suppose that all criteria share the same evaluation scale, a finite totally ordered set $\mathscr{L}$ with $k+1$ levels denoted by $0 = \alpha_1 < \cdots < \alpha_{k+1} = 1$. Each alternative $x \in X$ is encoded as a vector of ratings $x = (x_1, \cdots, x_n) \in \mathscr{L}^n$ where x_i is the rating of x with respect to criterion i.

A *fuzzy measure*, or *capacity* μ is an increasing set function $\mu : 2^{\mathscr{P}} \to \mathscr{L}$ such that $\mu(\emptyset) = 0$ and $\mu(\mathscr{P}) = 1$. The capacity μ defines a priority system on the groups of criteria. In the following, for criteria i, j, k, priority levels $\mu(\{i\})$, $\mu(\{j,k\})$ are respectively denoted by μ_i and $\mu_{j,k}$ for short. Clearly, $\mu_{i,i} = \mu_i$, $\mu_{j,k} = \mu_{k,j}$ and more generally, $\mu_{i,j,\ldots,r} = \mu_{\sigma(i),\sigma(j),\ldots,\sigma(r)}$ for any permutation σ of the indices. The increasingness of the capacity ensures that $\forall i, j, \ldots, r, \mu_i \leq \mu_{i,j} \leq \mu_{i,j,\ldots,r}$. Moreover, if $\max(\mu_i, \mu_j) < \mu_{i,j}$, it expresses that the criteria i, j are not redundant, and together they have a higher priority than any individual one, which can be viewed as a form of *positive synergy*.[1]

Let $x \in \mathscr{L}^n$ be an alternative and μ be a fuzzy measure on the power set of $\mathscr{P}$, the Sugeno integral of x with respect to μ takes various forms [14, 15, 5, 12], e.g.,

$$\begin{aligned} \oint_\mu(x) &= \max_{T \subseteq \mathscr{P}} \min(\mu_T, \min_{i \in T} x_i) = \min_{T \subseteq \mathscr{P}} \max(\mu_{\mathscr{P}\setminus T}, \max_{i \in T} x_i) \\ &= \max_{i=1}^{k+1} \min(\alpha_i, \mu(X_{\alpha_i})) = \min_{i=1}^{k+1} \max(\alpha_i, \mu(X_{\alpha_{i+1}})) \end{aligned}$$

where $\forall \alpha \in \mathscr{L}, X_\alpha = \{i | x_i \geq \alpha\}$ and $X_{\alpha_{k+2}} = \emptyset$, μ_T is short for $\mu(T)$.

When the evaluation scale is the unit interval $[0,1]$, it has been noticed [12, 4] that any lattice polynomial $P(x_1, \ldots, x_n)$ with variables $x_1, \ldots, x_n$ (representing the degrees of fulfillment of n properties), involving any constants $c_1, \ldots, c_m$ in $[0,1]$, and formed in the usual manner with min, max, and parentheses, is a Sugeno integral provided that the limit conditions $P(1, \ldots, 1) = 1$ and $P(0, \ldots, 0) = 0$ hold. Thus, Sugeno integrals include Boolean combination of properties *involving no negation*, as particular cases. This suggests that Sugeno integrals may be viewed not only as an evaluation tool, but one for knowledge representation as well.

Let x be a Boolean alternative ($x_i \in \{0,1\}$) and let $T^x = \{i, x_i = 1\}$ be the set of criteria satisfied by x. It can be checked that $\oint_\mu(x) = \mu(T^x)$, which provides a simple interpretation of the values of the capacity.

The *qualitative Möbius transform* [13, 8] of the fuzzy measure μ is a set-function on $\mathscr{P}$:

$$\forall T \subseteq \mathscr{P}, \mu_\#(T) = \begin{cases} \mu(T) & \text{if } \mu(T) > \max_{S \subset T} \mu(S) \\ 0 & \text{otherwise} \end{cases}$$

The sets T such that $\mu_\#(T) \neq 0$ are named the focal sets of μ and form the set $\mathscr{F}_\mu$. The qualitative Möbius transform contains the minimal amount of information sufficient to reconstruct the fuzzy measure: $\forall T \subseteq N$, $\mu(T) = \max_{S \subseteq T} \mu_\#(S)$. Moreover:

$$\oint_\mu(x) = \oint_{\mu_\#}(x) = \max_{T \in \mathscr{F}_\mu} \min(\mu_\#(T), \min_{i \in T} x_i) = \max_{i=1}^{k+1} \min(\alpha_i, \mu_\#(X_{\alpha_i})).$$

[1] In numerical approaches to criteria aggregation using Choquet integrals, weights can be added, and a positive synergy between two criteria i, j is modeled by a weight $\mu_{i,j} > \mu_i + \mu_j$. Independent criteria are such that $\mu_{i,j} = \mu_i + \mu_j$. A qualitative approach cannot express this kind of independence. However, in the qualitative approach μ_T represents the priority level of the group T of criteria, not a weight. In this context, where priority levels can only be compared, the inequality $\max(\mu_i, \mu_j) < \mu_{i,j}$ can thus be interpreted as a form of positive synergy, while in the numerical setting with weights, the case when $\max(\mu_i, \mu_j) < \mu_{i,j} < \mu_i + \mu_j$ corresponds to a negative synergy.

2.2 Possibilistic Logic

Let $B^N = \{(\varphi_j, \alpha_j) | j = 1, \ldots, m\}$ be a propositional possibilistic logic base, where φ_j denotes a classical logical formula and its priority level $\alpha_j \in \mathscr{L} \subset [0,1]$ [6]. In the following, logical conjunction and disjunction are respectively denoted by $\wedge$ and $\vee$ in the usual way. Each formula (φ_j, α_j) should be understood as $N(\varphi_j) \geq \alpha_j$, where N is a necessity measure obeying the characteristic property $N(\varphi \wedge \psi) = \min(N(\varphi), N(\psi))$. A necessity measure stems from a possibility distribution π via the formula $N(\varphi) = \min_{\omega \notin M(\varphi)}(1 - \pi(\omega)) = 1 - \Pi(\neg\varphi)$, where Π denotes the dual (potential) possibility measure, and $M(\varphi)$ is the set of interpretations for which φ is true.

The possibilistic logic base B^N is semantically associated with the possibility distribution $\pi_{B^N}(\omega) = \min_{i=1,\ldots,m} \pi_{(\varphi_j,\alpha_j)}(\omega)$ with $\pi_{(\varphi_j,\alpha_j)}(\omega) = 1$ if $\omega \in M(\varphi_j)$, and $\pi_{(\varphi_j,\alpha_j)}(\omega) = 1 - \alpha_j$ if $\omega \notin M(\varphi_j)$. It is the minimally specific possibility distribution such that $N(\varphi_j) \geq \alpha_j, \forall \varphi_j$ appearing in B^N. Here an interpretation ω is all the more possible as it does not violate any formula φ_j with a high priority level α_j. The minimally specific possibility distribution induced by B^N can also be expressed as

$$\pi_B(\omega) = \min_{j=1,\ldots,m} \max(1 - \alpha_j, I_{M(\varphi_j)}(\omega))$$

where $I_{M(\varphi_j)}$ denotes the characteristic function of $M(\varphi_j)$. Then, it is easy to express the possibility distribution on interpretations as a fuzzy integral:

Proposition 1. $\pi_B(\omega) = \oint_{\mathscr{N}}((I_{M(\varphi_1)}(\omega), \cdots, I_{M(\varphi_m)}(\omega))) = \mathscr{N}(K(\omega))$, *where* $\mathscr{N}$ *is the necessity measure defined on* $2^{\{\varphi_j\, j=1,\cdots,m\}}$ *by* $\mathscr{N}(A) = \max_{\varphi_j \notin A}(1 - \alpha_j)$, *viewing* B^N *as a possibility distribution on formulas* φ_j, *and* $K(\omega) = \{\varphi_j | \omega \in M(\varphi_j)\}$.

There exists a dual reading of a possibility distribution π, which insists on the minimal degree of satisfaction reached by elements inside a set, rather than on the impossibility of elements outside it. Then, the information represented by weighted formulas denoted by $[\psi, \gamma]$ expresses constraints $\Delta(\psi) \geq \gamma$, where Δ denotes a measure of *actual possibility* defined by $\Delta(\psi) = \min_{\omega \in M(\psi)} \pi(\omega)$ [1]. Measures of actual possibility obey the following characteristic postulate: $\Delta(\varphi \vee \psi) = \min(\Delta(\varphi), \Delta(\psi))$, and are thus decreasing with respect to logical entailment, which contrasts with necessity (or potential possibility) measures. In other words, the piece of information $[\psi, \gamma]$ expresses that any model of ψ is at least possible with degree γ. A Δ-possibilistic logic base is denoted by $B^\Delta = \{[\psi_\ell, \gamma_\ell] | \ell = 1, \ldots, q\}$, where $[\psi_\ell, \gamma_\ell]$ is understood as $\Delta(\psi_\ell) \geq \gamma_\ell$. Its semantics is given by the possibility distribution

$$\pi_{B^\Delta}(\omega) = \max_{\ell=1,\ldots,q} \pi_{[\psi_\ell,\gamma_\ell]}(\omega)$$

with $\pi_{[\psi_\ell,\gamma_\ell]}(\omega) = 0$ if $\omega \notin M(\psi_\ell)$, and $\pi_{[\psi_\ell,\gamma_\ell]}(\omega) = \gamma_\ell$ if $\omega \in M(\psi_\ell)$.

Note that π_{B^Δ} is obtained as the max-based *disjunctive* combination of the representation of each formula in B^Δ, while π_{B^N} was obtained by a min-based *conjunctive* combination of the representation of each formula in B^N.

3 Fuzzy Measures as Possibilistic Logic Bases

In this section, we show how to encode a fuzzy measure by a possibilistic logic base with positive clauses and show that the set-function induced by a general possibilistic logic base induces a non-necessarily monotonic set-function. Suppose a language where propositional variables p_i encode criteria $i \in \mathscr{P}$. A Boolean alternative x is an alternative for which criteria are fully satisfied or not. It can be encoded as a subset $T^x \subseteq \mathscr{P}$ of satisfied criteria and as an interpretation $\omega_x = \wedge_{i \in T^x} p_i \bigwedge \wedge_{i \notin T^x} \neg p_i$. A set of formulas $\varphi_j, i = 1 \ldots m$ can be viewed as requirements $j \in \mathscr{R}$ expressed as logical combinations of elementary criteria. In such a context, an alternative x satisfies requirement j or not, according to whether ω_x is or is not a model of φ_j. So using the characteristic function of $M(\varphi_j)$, an alternative x can be evaluated according to each requirement j, by computing $I_{M(\varphi_j)}(\omega_x)$.

Consider a Boolean capacity μ taking values on $\{0,1\}$. For any focal set $T \in \mathscr{F}_\mu$, it is clear that $\forall A \subset T, \mu(A) = 0$ and $\forall A \supseteq T, \mu(A) = 1$. In the decision context, a Boolean capacity means that for an alternative to be acceptable, it is necessary and sufficient that all criteria in some focal set of μ be jointly satisfied. Let p_i be the propositional variable associated to criterion $i \in \mathscr{P}$. The logical formula expressing this global requirement is

$$\varphi \equiv \bigvee_{T \in \mathscr{F}_\mu} \bigwedge_{i \in T} p_i.$$

Putting this formula in conjunctive normal form, one obtains a logical base of clauses B_μ that exactly describes the priority scheme expressed by the Boolean capacity, in terms of a conjunction of clauses $\varphi_j = \vee_{i \in T} p_i$, each requiring that at least one criterion in $T \in \mathscr{F}_\mu$ be satisfied.[2] Note that it is obvious that no such clause contains a negative literal (just like for the corresponding Boolean polynomial).

Now consider a fuzzy measure μ taking values on $\mathscr{L}$, and let $\mu_\alpha(T) = 1$ if $\mu_T \geq \alpha$ and 0 otherwise, where $\alpha > 0$. Let $\mathscr{F}_\mu^\alpha$ denote the set of focal sets of μ_α. Recalling the fact that $\mu_\#$ is a possibility distribution on $2^{\mathscr{P}}$, and denoting by $\mathscr{N}_\mu$, the associated necessity measure on the power set of $2^{\mathscr{P}}$, it is clear that, by construction, $\mathscr{N}_\mu(\mathscr{F}_\mu^\alpha) = 1 - \alpha^-$, where $\alpha^- = \sup\{\mu_\#(T) < \alpha\}$. Hence to each Boolean capacity μ^α induced by $\mathscr{F}_\mu^\alpha$ can be associated to a possibilistic formula $(\bigvee_{T \in \mathscr{F}_\mu^\alpha} \wedge_{i \in T} p_i, 1 - \alpha^-)$ that can be put into a set B_μ^α of prioritised positive clauses $(\varphi_j, 1 - \alpha^-)$. Hence the possibilistic base equivalent to the fuzzy measure μ is $B_\mu = \bigcup_{\alpha > 0} B_\mu^\alpha$. From Proposition 1, for any Boolean alternative x, corresponding to a set T^x of criteria, and an interpretation ω_x, it is clear that

Proposition 2. $\pi_{B_\mu}(\omega_x) = \mu(T^x) = \oint_\mu(x) = \mathscr{N}_\mu(B_x)$, *where* $B_x = \{\varphi : (\varphi, \beta) \in B_\mu, \omega_x \models \varphi\}$.

Note that here the possibilistic logic base B_μ reflects the conjugate of capacity μ defined as $\mu_c(A) = 1 - \mu(A^c)$, where A^c denotes the complement of A, namely $\mu_c(A) = \max_{T \subseteq A}\{\beta : \varphi = \vee_{i \in T} p_i, (\varphi, \beta) \in B_\mu\}$.

---

[2] By construction, no such clause is redundant.

Example: Let $n=3$, $\mathscr{L}=\{0<\beta<\alpha<1\}$. Let $\mu_1=\alpha$, $\mu_2=\beta$, $\mu_3=0$, $\mu_{1,3}=1$, $\mu_{2,3}=\alpha$, $\mu_{1,2}=\alpha$. It can be seen that $\mathscr{F}^1_\mu=\{\{1,3\}\}$, which yields $B^1_\mu=\{(p_1,1-\alpha),(p_3,1-\alpha)\}$; $\mathscr{F}^\alpha_\mu=\{\{1\},\{2,3\}\}$, which yields the formula $(p_2\wedge p_3)\vee p_1$, hence $B^\alpha_\mu=\{(p_1\vee p_2,1-\beta),(p_1\vee p_3,1-\beta)\}$; finally $\mathscr{F}^\beta_\mu=\{\{1\},\{2\}\}$, which, yields the formula $(p_1\vee p_2,1)$. Then, the corresponding possibilistic logic base B^N_μ is (after subsumption): $\{(p_1,1-\alpha),(p_3,1-\alpha),(p_1\vee p_3,1-\beta),(p_1\vee p_2,1)\}$.

Such a representation of a capacity is unusual in multiple criteria aggregation. However, while the values of the capacity directly reflect the importance of groups of criteria, we obtain here another view of the interplay between criteria; the possibilistic base expresses that one should satisfy at least one among criteria '1' and '2', that it is better to satisfy at least one among criteria '1' and '3', while it is ideal to satisfy *both* criteria '1' and '3' (although it is a bit less crucial), since $0<1-\alpha<1-\beta<1$. This result might look surprising since the capacity value of criterion '3' is 0. But we should observe that, together, criteria '1' and '3' have a capacity value equal to 1.

In fact a fuzzy measure μ can be more directly represented as a Δ-possibilistic base, interpreting its basic possibility assignment $\mu_\#(T)>0$ as the actual possibility degree of the conjunction of atoms corresponding to criteria in T. On the example, one directly gets $B^\Delta=\{[p_1\wedge p_3],[p_1,\alpha],[p_2\wedge p_3,\alpha],[p_2,\beta]\}$. Besides, it could be checked that $\pi_{B^N}=\pi_{B^\Delta}$.

Conversely, any possibilistic logic base $B=\{(\varphi_j,\alpha_j)|j=1,\ldots,m\}$ can be associated to a set-function on $2^{\mathscr{P}}$, viewing the propositional variables as Boolean criteria (with ratings in $\{0,1\}$), interpretations as alternatives, formulas as more general requirements combining criteria. We view B as a possibility distribution on formulas. The importance of sets of requirements is represented by the necessity measure defined on $2^{\{\varphi_j\, j=1,\cdots,m\}}$ by $\mathscr{N}(K)=\bigwedge_{\varphi_j\notin K}(1-\alpha_j)$. Note that for all i we have $\alpha_j=1-\mathscr{N}(\{\varphi_j\}^c)$. Contrary to the previous case, relying on a capacity, B can be put as a conjunction of *general* prioritised clauses. The set-function can be retrieved by applying Proposition 2 and letting $\mu(T)=\pi_B(\omega_T)$ where $\omega_T=\wedge_{i\in T}p_i\bigwedge\wedge_{i\notin T}\neg p_i$. So, this representation may account for a negative synergy between properties, since the clauses φ_j are not necessarily positive.

Example : Suppose $\mathscr{P}=\{1,2,3\}$. Suppose we want to select alternatives that satisfy property 1 and at least one among properties 2 or 3, preferably 2 but not one that satisfies both properties 2 and 3. This can be translated verbatim as a possibilistic base $B=\{(p_1,1),(p_2\vee p_3,1),(p_2,1-\alpha),(\neg p_2\vee\neg p_3,1)\}$. Let $\varphi_1=p_1$, $\varphi_2=p_2\vee p_3$, $\varphi_3=p_2$ and $\varphi_4=\neg p_2\vee\neg p_3$. Let us reconstruct the set-function μ

- $x=(1,1,1)$: $K(\omega_x)=\{\varphi_1,\varphi_2,\varphi_3\}$ and $\mu(\mathscr{P})=\pi_B(\omega_x)=N(K(\omega_x))=0$.
- $x=(1,1,0)$: $K(\omega_x)=\{\varphi_1,\varphi_2,\varphi_3,\varphi_4\}$; $\mu(\{1,2\})=\pi_B(\omega_x)=N(K(\omega_x))=1$.
- $x=(1,0,1)$: $K(\omega_x)=\{\varphi_1,\varphi_2,\varphi_4\}$; $\mu(\{1,3\})=\pi_B(\omega_x)=N(K(\omega_x))=\alpha$.

It can be checked that $\mu(T)=0$ otherwise. Clearly μ is not monotonic (which accounts for a kind of negative synergy between properties). It comes down to saying that acceptable alternatives with respect to $\mathscr{P}$ are either those which have properties 1 and 2, or to a lesser extent those with properties 1 and 3, and none else (all other alternatives will be rated 0).

4 Sugeno Integrals as Possibilistic Logic Bases

Here we consider the case of non-Boolean properties. We first explain the general idea, before establishing the results that are necessary for obtaining a possibilistic logic base that expresses the contents of a Sugeno integral.

4.1 General Principle

Let $\oint_\mu$ be a Sugeno integral based on a fuzzy measure μ which takes its values in $\mathscr{L}$. It rank-orders vectors x according to the values $\oint_\mu(x)$.

Then, a possibilistic logic base can be associated to any Sugeno integral, via the following general procedure:

- *Representation:* Each alternative $x = (x_1, ..., x_n)$ with $x_i \in \mathscr{L}$ can be represented by a grounded formula in First-Order Logic. More precisely, each criterion is viewed as a predicate P_j and each rating α_j with respect this criterion in the scale $\mathscr{L}$ is a grounded atom $P_j(\alpha_i)$. Then each x is now represented by the logical formula $P_1(x_1) \wedge \cdots \wedge P_n(x_n)$.
  Example: Let $\mathscr{L} = \{1, \alpha, \beta, 0\}$ with $1 > \alpha > \beta > 0$. We consider 3 criteria where ratings correspond to linguistic predicates as follows:

  - **Size:** Large (1), Medium (α), Small (β) Tiny (0).
  - **Location:** Good (1), Acceptable (α), Poor (β), Bad (0)
  - **Price:** Cheap (1), Reasonable (α), High (β), Excessive (0)

  A large, reasonably-located house with high price is thus associated with the vector $x = (1, \alpha, \beta)$, and to the formula $Size(1) \wedge Location(\alpha) \wedge Price(\beta)$.
- *Construction principle of the possibilistic logic base:* (justified in the next subsection). The global rating computed by a Sugeno integral is viewed as the degree of possibility of the corresponding formula:

  - a constraint $\oint_\mu(x) \leq \gamma$ is understood as $N(\neg(P_1(x_1) \wedge \cdots \wedge P_n(x_n))) \geq 1 - \gamma$ in possibility theory, and thus corresponds to the possibilistic logic formula $(\neg P_1(x_1) \vee \cdots \vee \neg P_n(x_n), 1 - \gamma)$.
  - a constraint $\oint_\mu(x) \geq \gamma$ is understood as $\Delta(P_1(x_1) \wedge \cdots \wedge P_n(x_n)) \geq \gamma$, and thus corresponds to the possibilistic logic formula $[P_1(x_1) \wedge \cdots \wedge P_n(x_n), \gamma]$.

As recalled in section 2.1, a Sugeno integral may either be viewed as a min-max, or as a max-min expression. The first expression leads to rules of the type "IF the house is such... AND such... THEN its evaluation cannot be less than some level". The other one leads to rules of the type "IF the house is such... AND such... THEN its evaluation cannot be greater than some level".

Example (continued): Consider $x = (1, \alpha, \beta)$. Whatever the capacity μ, we have $\oint_\mu(1,1,1) = 1 \geq \oint_\mu(1, \alpha, \beta) \geq \oint_\mu(1,0,0) = \mu_1$.

- If we write $\oint_\mu(1,\alpha,\beta) = \min(\max(\beta,\mu_{1,2}),\max(\alpha,\mu_1),1)$ we have $\oint_\mu(x) = 1$ if $\mu_{1,2} = \mu_1 = 1$; $\oint_\mu(x) = \beta$ if $\mu_{1,2} \leq \beta$; and $\oint_\mu(x) = \alpha$ otherwise.
  Thus, one can state the rule: *if a house is large, its location is acceptable and its price is high, then it is α satisfactory* (except if $\mu_{1,2} \leq \beta$, or $\mu_{1,2} = \mu_1 = 1$). It can be represented as $[Size(1) \wedge Location(\alpha) \wedge Price(\beta), \alpha]$ in Δ-possibilistic logic.
- If we write $\oint_\mu(x) = \max(\min(x_1,\mu_1),\min(x_2,\mu_2),\min(x_3,\mu_3),\min(x_1,x_2,\mu_{1,2}),$ $\min(x_2,x_3,\mu_{2,3}),\min(x_1,x_3,\mu_{1,3}),\min(x_1,x_2,x_3))$; assume $\mu_i > 0$ for $i = 1,2,3$ and $\mu_1 = \beta$ and $\mu_i > \beta$ for $i = 2,3$, $\mu_{j,k} > \beta$ for $j \neq k$ and $j,k = 1,3$. Then $\oint_\mu(x) \leq \beta$ requires $x_i \leq \beta$ for $i = 2,3$. Several x's satisfy this constraint, for instance $x = (1,\beta,\beta)$ or $x = (\alpha,\beta,0)$ which correspond to $(\neg Size(1) \vee \neg Location(\beta) \vee \neg Price(\beta), 1-\beta)$ and $(\neg Size(\alpha) \vee \neg Location(\beta) \vee \neg Price(0), 1-\beta)$. So for instance, *if the location is poor and the price is high, then even if the size is large the house is at best poorly acceptable.*

Thus, the above example suggests the following general procedure. Given a capacity μ, determine $\{x | \oint_\mu(x) \leq \gamma\}$ for all $\gamma \in \mathscr{L}$, and turn the x's into the corresponding possibilistic logic formulas. Similarly, considering constraints of the form $\{x | \oint_\mu(x) \geq \gamma\}$ leads to Δ-based possibilistic logic formulas.

4.2 Formal Characterization: Upper Bound

The following result characterizes the set $\{x | \oint_\mu(x) \leq \gamma\}$ of alternatives which lay below a minimal prescribed global evaluation.

Proposition 3. $\{x | \oint_\mu(x) \leq \gamma\} = \{x | \forall T \in \mathscr{F}_\mu,$ *s.t.* $\mu_\#(T) > \gamma, \exists i \in T$ *s.t.* $x_i \leq \gamma\}$.

Proof. Let us use the identity $\oint_\mu(x) = \max_{T \in \mathscr{F}_\mu} \min(\mu_\#(T), \min_{i \in T} x_i)$. Then clearly, $\oint_\mu(x) \leq \gamma$ if and only if $\forall T \in \mathscr{F}_\mu, \min(\mu_\#(T), \min_{i \in T} x_i) \leq \gamma$. It induces a constraint on x only if $\mu_\#(T) > \gamma$. And for each focal set T for which $\mu_\#(T) > \gamma$, the condition $\min_{i \in T} x_i \leq \gamma$ must be satisfied. In other words, the alternatives x for which $\oint_\mu(x) \leq \gamma$ are precisely those for which $\forall T \in \mathscr{F}_\mu$ such that $\mu_\#(T) > \gamma$, $\exists i \in T$ s.t. $x_i \leq \gamma$.

Illustrative example
Let us go back to the example. $\mathscr{P} = \{Size, Location, Price\}$ is the set of criteria, $\mathscr{L} = \{1,\alpha,\beta,0\}$ is the evaluating scale with $1 > \alpha > \beta > 0$. $x = (x_S, x_L, x_P)$ denotes an alternative evaluated according to *Size*, *Location* and *Price*. We consider a Sugeno integral with respect to μ defined by

$\mu(Size) = \mu(Location) = \beta$, $\mu(Price) = \alpha$, $\mu(Size,Location) = \beta$,
$\mu(Size,Price) = 1$, $\mu(Location,Price) = \alpha$, $\mu(Size,Location,Price) = 1$.
This fuzzy measure is associated to the following qualitative Möbius transform:
$\mu_\#(Size) = \mu_\#(Location) = \beta$, $\mu_\#(Price) = \alpha$ and $\mu_\#(Size,Price) = 1$.

- We want to find x such that $\oint_\mu(x) \leq \beta$.
  $\{A \subseteq \mathscr{P} | \mu_\#(A) > \beta\} = \{\{Price\},\{Price,Size\}\}$. The alternatives we look for must have a rating not greater than β for at least one criterion in each set. We must thus enforce $x_{Price} \leq \beta$ in order to respect $\oint_\mu(x) \leq \beta$., i.e. all x Pareto-less than or equal to $(1,1,\beta)$ satisfy $\oint_\mu(x) \leq \beta$ and only them.

- We want to find x such that $\oint_\mu(x) \leq \alpha$.
  $\{A \subseteq \mathscr{P} | \mu_\#(A) > \alpha\} = \{\{Size, Price\}\}$ so x must be such that $x_{Price} \leq \alpha$ or $x_{Size} \leq \alpha$, i.e. all x Pareto-less than or equal to $(1,1,\alpha)$ or Pareto-less than or equal to $(\alpha,1,1)$ satisfy $\oint_\mu(x) \leq \alpha$.
- We want to find x such that $\oint_\mu(x) \leq 1$.
  $\{A \subseteq \mathscr{P} | \mu_\#(A) > 1\} = \emptyset$ so all x are solutions (as expected!).
- We want to find x such that $\oint_\mu(x) \leq 0$.
  $\{A \subseteq | \mu_\#(A) > 0\}$ contains $\{Size\}$,$\{Price\}$ and $\{Location\}$, among others, so x has to be equal to 0 everywhere.

So altogether, we get the following N-based possibilistic logic base as a result:

$B^N_\mu = \{(\neg Size(0) \vee \neg Location(0) \vee \neg Price(0), 1), (\neg Size(1) \vee \neg Location(1) \vee \neg Price(\beta), 1-\beta), (\neg Size(1) \vee \neg Location(1) \vee \neg Price(\alpha), 1-\alpha),$
$(\neg Size(\alpha) \vee \neg Location(1) \vee \neg Price(1), 1-\alpha)\}$.

4.3 Formal Characterization: Lower Bound

In order to obtain a possibilistic logical representation of the Δ-type (in the sense of section 2.2), one needs to find x such that $\oint_\mu(x) \geq \gamma$, i.e. alternatives that can achieve a minimal global evaluation.

Proposition 4. $\{x | \oint_\mu(x) \geq \gamma\} = \{x | \exists T \in \mathscr{F}_\mu \text{ s.t } \mu_\#(T) \geq \gamma \text{ and } \forall i \in T, x_i \geq \gamma\}$.

Proof. Let us use again the identity $\oint_\mu(x) = \max_{T \in \mathscr{F}_\mu} \min(\mu_\#(T), \min_{i \in T} x_i)$. Then clearly, $\oint_\mu(x) \geq \gamma$ if and only if $\exists T \in \mathscr{F}_\mu, \min(\mu_\#(T), \min_{i \in T} x_i) \geq \gamma$. For such a focal set T, the condition $\mu_\#(T) \geq \gamma$ must hold, as well as $\min_{i \in T} x_i \geq \gamma$, i.e. $\forall i \in T, x_i \geq \gamma$ must be satisfied.

Example. Let us go back to the previous example.

- We want to find x such that $\oint_\mu(x) \geq \beta$.
  $\{A \subseteq \mathscr{P} | \mu_\#(A) \geq \beta\}$ contains $\{Size\}$, $\{Location\}$ and $\{Price\}$.
  According to the theoretical results at least one of the sets A is such that for all $i \in A$ $x_i \geq \beta$ so x is such that $x_{Size} \geq \beta$ or $x_{Price} \geq \beta$ or $x_{Location} \geq \beta$, i.e. any x greater than $(\beta,0,0)$ or $(0,\beta,0)$ or $(0,0,\beta)$ satisfies $\oint_\mu(x) \geq \beta$.
- We want to find x such that $\oint_\mu(x) \geq \alpha$.
  $\{A \subseteq \mathscr{P} | \mu_\#(A) \geq \alpha\} = \{\{Price\}, \{Size, Price\}\}$ so x has to be greater than α on at least one of the two sets i.e. any x greater than $(0,0,\alpha)$ satisfies $\oint_\mu(x) \geq \alpha$.
- We want to find the alternatives x such that $\oint_\mu(x) \geq 0$.
  $\{A \subseteq \mathscr{P} | \mu_\#(A) \geq 0\}$ contains $\{Size\}$, $\{Location\}$ and $\{Price\}$ so all x are solutions (as expected).
- We want to find the alternatives x such that $\oint_\mu(x) \geq 1$.
  $\{A | \mu_\#(A) \geq 1\} = \{Size, Price\}$ so all x greater than $(1,0,1)$ satisfy $\oint_\mu(x) \geq 1$.

So altogether, we get the following Δ-possibilistic logic base as a result:

$K_\mu^\Delta = \{[Size(1) \wedge Location(0) \wedge Price(1), 1], [Size(0) \wedge Location(0) \wedge Price(\alpha), \alpha],$ $[Size(\beta) \wedge Location(0) \wedge Price(0), \beta], [Size(0) \wedge Location(\beta) \wedge Price(0), \beta],$ $[Size(0) \wedge Location(0) \wedge Price(\beta), \beta]\}.$

Note that one could write still more compactly $[Size(1) \wedge Price(1), 1]$ for the first formula, taking into account that $Location(0)$ is the worst location rating (which is clear for the end-user, even if this piece of information (the ordering in the value scale) is not encoded in the logic).

5 Related Work

There do not exist many works trying to provide a logical reading of decision processes. In decision under uncertainty, let us mention a logical approach [3] where uncertain knowledge and prioritized preference are respectively represented by means of two distinct possibilistic logic bases, and where the pessimistic or optimistic decision criteria that are maximized are particular cases of Sugeno integrals. In [7], in a multiple criteria decision perspective, a qualitative approach, also in the spirit of possibilistic logic, is compared to a counterpart of a Choquet integral-based aggregation process. However, the closest related work is [10] in which Greco *et al.* have proposed a preliminary study to prove that the set of the elements for which a Sugeno integral is less than a given score γ is similar to finding if - then rules associated to a Sugeno integral.

More precisely Greco *et al.* considers a slightly different setting from the one in this paper. We use the same notations. For each $i \in P$, X_i is the set of ratings of alternatives with respect to criteria i. The set of alternatives is thus $X = \Pi_{i=1}^n X_i$. On X there is a comprehensive weak preference relation $\succeq$. For all $x, y \in X$ $x \succeq y$ means x is at least as good as y. Its asymmetric part is denoted by $\succ$.

In such a context, there exists the following link between the Sugeno integral representation and a decision rule (in the sense below):

Proposition 5. *$\succeq$ is representable by a Sugeno integral if and only if there exists*

- *functions $g_i : X_i \rightarrow [0,1]$ for all $i = 1, \cdots n$ called criteria*
- *a function $g : X \rightarrow [0,1]$ called comprehensive evaluation*
- *a set of decision rules of the form "if $g_{i_1}(x_{i_1}) \geq g(y), \cdots, g_{i_k}(x_{i_k}) \geq g(y)$, then $x \succeq y$", where $\{i_1, \cdots, i_k\} \subseteq \{1, \cdots, n\}$, called* single-graded, *such that*
  - *each $z \in X$ such that $z \succeq y$ satisfies the antecedent of at least one rule whose consequent is "then $x \succeq w$" where $w \succeq y$.*
  - *each $z \in X$ such that not $z \succeq y$ does not satisfy the antecedent of any rule whose consequent is "then $x \succeq w$" with $w \succeq y$.*

  If the decision rule: "if $g_{i_1}(x_{i_1}) \geq g(y), \cdots, g_{i_k}(x_{i_k}) \geq g(y)$" holds, then $x \succeq y$", then for any $w \in X$ such that $g(y) \geq g(w)$, the decision rule "if $g_{i_1}(x_{i_1}) \geq g(w), \cdots, g_{i_k}(x_{i_k}) \geq g(w)$" then $x \succeq w$ also holds.

In [10], Greco *et al.* consider both sets $\{x | \oint_\mu(x) \geq \gamma\}$ and $\{x | \oint_\mu(x) \leq \gamma\}$ using the two expressions of the Sugeno integral (min-max and max-min). Considering an alternative x and a real γ, this work entails the following remarks:

- Let $A(x,\gamma)$ be the set of the criteria on which x is better or equal to γ. If $A(x,\gamma)$ is large enough i.e. if $\mu(A(x,\gamma)) \geq \gamma$ then $\oint_\mu(x) \geq \gamma$.
- Let $B(x,\gamma)$ be the set of the criteria on which x is smaller or equal to γ. If $B(x,\gamma)^c$ is small enough i.e $\mu(B(x,\gamma)^c) \leq \gamma$ then $\oint_\mu(x) \leq \gamma$.

Note that these remarks tend to suggest that the decision rules representable by a Sugeno integral are only single-graded rules, i.e., that involve a single threshold $g(y)$ (see pages 284-286 in [10]). However, the following example indicates that other rules are representable by a Sugeno integral as well.

Example. We consider two criteria: mathematics, French; students evaluated according to these criteria and the following rules involving two thresholds:

- if mathematics ≥ 0.2, French ≥ 0.3 then student ≥ 0.2.
- if mathematics ≥ 0.3, French ≥ 0.2 then student ≥ 0.3.

A Sugeno integral representation appears to agree with the previous rules.
$A(x,0.2) = \{mathematics, French\}$ so $\mu(\{mathematics, French\}) \geq 0.2$
$A(y,0.3) = \{mathematics\}$ so $\mu(mathematics) \geq 0.3$.
We get $\mu(French) = 0$, $\mu(mathematics) = 0.3$ $\mu(\{mathematics, French\}) = 1$.
A student is represented by $x = (x_m, x_f)$ where x_m (resp. x_f) is the evaluation according to mathematics (French). There are two possible cases:
if $x_f \succeq x_m$, then $\oint_\mu(x) = max(min(x_m,1), min(x_f,0)) = x_m$,
or $x_m \succeq x_f$, and $\oint_\mu(x) = max(min(x_f,1), min(x_m,0.3)) = max(x_f, min(x_m,0.3))$.
In both cases the rules are satisfied.

Moreover, Greco *et al.* [10] do not provide an explicit representation of a Sugeno integral in N-based, nor Δ-based possibilistic logic.

6 Concluding Remarks

We have shown that it was possible to associate a Sugeno integral with an equivalent possibilistic logic description. This is clearly of interest for laying bare the meaning of such an integral, and describing the requirements characterizing the objects having a high global evaluation. It would be of interest to internalize the scale itself. To do it, we need to extend the results of Section 3 to non-Boolean alternatives. Moreover, the representation of Sugeno integrals by possibilistic logic bases provides a sound basis for defining an entailment between Sugeno integrals [16], on the basis of the possibilistic logic entailment. Besides, Sugeno integrals have a limited discriminating power, especially when using finite scales. It is possible to make them more discriminant [2] using leximin and leximax in place of min and max. Finding a logical counterpart to such a more discriminant representation is another topic for further research.

References

1. Benferhat, S., Dubois, D., Kaci, S., Prade, H.: Bipolar possibility theory in preference modeling: Representation. Fusion and Optimal Solutions, Information Fusion 7, 135–150 (2006)
2. Dubois, D., Fargier, H.: Making discrete Sugeno integrals more discriminant. Inter. J. of Approximate Reasoning 50, 880–898 (2009)
3. Dubois, D., Le Berre, D., Prade, H., Sabbadin, R.: Using possibilistic logic for modeling qualitative decision: ATMS-based algorithms. Fundamenta Informaticae 37(1-2), 1–30 (1999)
4. Dubois, D., Marichal, J.-L., Prade, H., Roubens, M., Sabbadin, R.: The use of the discrete Sugeno integral in decision-making: A survey. Inter. J. of Uncertainty, Fuzziness and Knowledge-based Systems 9, 539–561 (2001)
5. Dubois, D., Prade, H.: Qualitative possibility functions and integrals. In: Pap, E. (ed.) Handbook of Measure Theory, vol. 2, pp. 1469–1521. Elsevier, Amsterdam (2002)
6. Dubois, D., Prade, H.: Possibilistic logic: a retrospective and prospective view. Fuzzy Sets and Systems 144, 3–23 (2004)
7. Gérard, R., Kaci, S., Prade, H.: Ranking alternatives on the basis of generic constraints and examples - A possibilistic approach. In: Veloso, M.M. (ed.) Proc. 20th Inter. Joint Conf. on Artificial Intelligence (IJCAI 2007), Hyderabad, January 6-12, pp. 393–398 (2007)
8. Grabisch, M.: The Möbius transform on symmetric ordered structures and its application to capacities on finite sets. Discrete Mathematics 287, 17–34 (2004)
9. Grabisch, M., Labreuche, C.: A decade of application of the Choquet and Sugeno integrals in multi-criteria decision aid. Annals of Operations Research 175, 247–286 (2010)
10. Greco, S., Matarazzo, B., Slowinski, R.: Axiomatic characterization of a general utility function and its particular cases in terms of conjoint measurement and rough-set decision rules. Europ. J. of Operational Research 158, 271–292 (2004)
11. Labreuche, C.: A general framework for explaining the results of a multi-attribute preference model. Artificial Intelligence 175, 1410–1448 (2011)
12. Marichal, J.-L.: Aggregation Operations for Multicriteria Decision Aid. Ph.D.Thesis, University of Liège, Belgium (1998)
13. Mesiar, R.: k-order Pan-discrete fuzzy measures. In: Proc. 7th Inter. Fuzzy Systems Assoc. World Congress (IFSA 1997), Prague, June 25-29, vol. 1, pp. 488–490 (1997)
14. Sugeno, M.: Theory of Fuzzy Integrals and its Applications, Ph.D. Thesis, Tokyo Institute of Technology, Tokyo (1974)
15. Sugeno, M.: Fuzzy measures and fuzzy integrals: a survey. In: Gupta, M.M., Saridis, G.N., Gaines, B.R. (eds.) Fuzzy Automata and Decision Processes, pp. 89–102. North-Holland, Amsterdam (1977)
16. Prade, H., Rico, A.: Describing acceptable objects by means of Sugeno integrals. In: Martin, T., et al. (eds.) Proc. 2nd IEEE Inter. Conf. of Soft Computing and Pattern Recognition (SoCPaR 2010), Cergy Pontoise, Paris, December 7-10 (2010)
17. Prade, H., Rico, A., Serrurier, M.: Elicitation of Sugeno integrals: A version space learning perspective. In: Rauch, J., Raś, Z.W., Berka, P., Elomaa, T. (eds.) ISMIS 2009. LNCS, vol. 5722, pp. 392–401. Springer, Heidelberg (2009)
18. Prade, H., Rico, A., Serrurier, M., Raufaste, E.: Elicitating Sugeno integrals: Methodology and a case study. In: Sossai, C., Chemello, G. (eds.) ECSQARU 2009. LNCS, vol. 5590, pp. 712–723. Springer, Heidelberg (2009)

Decomposition of Possibilistic Belief Functions into Simple Support Functions

Matthieu Chemin, Agnès Rico, and Henri Prade

Abstract. In Shafer evidence theory some belief functions, called separable belief functions, can be decomposed in terms of simple support functions. Moreover this decomposition is unique. Recently, a qualitative counterpart to Shafer evidence theory has been proposed. The mass functions in Shafer (addition-based) evidence theory are replaced by basic possibilistic assignments. The sum of weights is no longer 1, but their maximum is equal to 1. In such a context, a maxitive counterpart to belief functions, called possibilistic belief functions can be defined, replacing the addition by the maximum. The possibilistic evidence framework provides a general setting for describing imprecise possibility and necessity measures. This paper investigates a qualitative counterpart of the result about the decomposition of belief functions. Considering the qualitative Möbius transform, conditions for the existence of a decomposition of possibilistic belief functions into simple support functions are presented. Moreover the paper studies the unicity of such a decomposition.

1 Introduction

In a very recent paper [10], the two last authors have developed a qualitative representation setting that parallels the quantitative framework of Shafer evidence theory [13] in many respects. This setting is based on the idea of a basic possibility

Matthieu Chemin
Université Claude Bernard Lyon 1, 43 bld du 11 Novembre, 69100 Villeurbanne, France
e-mail: matthieu.chemin@ens-lyon.fr

Agnès Rico
ERIC, 5 av Pierre Mendès-France, 69676 Bron, France
e-mail: agnes.rico@univ-lyon1.fr

Henri Prade
IRIT, 118 route de Narbonne, 31062 Toulouse cedex 09, France
e-mail: prade@irit.fr

B. De Baets et al. (Eds.): Eurofuse 2011, AISC 107, pp. 31–42, 2011.
springerlink.com

assignment, which was originally suggested in [4, 5], and also partially considered in a few works [15, 3].

As pointed out in [10], this setting can be understood in terms of imprecise possibilities, just as belief and plausibility can be seen as particular lower and upper probabilities. In evidence theory, separable belief functions which are decomposable into a combination of simple support functions, are of particular interest. This paper investigates the counterpart of these notions in the setting of possibilistic evidence theory.

The paper is organized as follows. The next section recalls the definition of the possibilistic counterpart of belief and plausibility functions, and then discusses their interpretation as lower and upper bounds of an imprecisely known possibility measure. By duality, minitive set functions provide upper and lower bounds for an imprecisely known necessity measure, which generalizes the possibilistic logic representation setting. Then the notion of qualitative Möbius transform, and the conjunctive and disjunctive combination rules of the possibilistic evidence setting are recalled. The second half of the paper is devoted to the condition of existence of a decomposition into simple support functions and to a discussion of its unicity.

2 Possibilistic Belief and Plausibility Functions and Imprecise Possibilities

In possibilistic evidence theory, given a universe $\Omega = \{1, \cdots, N\}$ information is represented by means of a mapping σ, called basic possibility assignment, from 2^{Ω} to the unit interval $[0,1]$, which is such that $\bigvee_{A \subseteq \Omega} \sigma(A) = 1$. If the information is consistent then $\sigma(\emptyset) = 0$ is assumed and we say that σ is normal. In the following all basic possibility assignments are normal. We define $\mathscr{F}_{\sigma} = \{A \subseteq \Omega | \sigma(A) > 0\}$. $\sigma(A)$ is the possibility that the epistemic state we want to represent is properly represented by A. Similarly to Shafer evidence theory, a basic possibility assignement σ defines the following possibilistic belief and plausibility functions:

Definition 1. $\forall A \subseteq \Omega, Bel^{pos}(A) = \bigvee_{B \subseteq A} \sigma(B)$, $Bel^{pos}(\emptyset) = 0$, $Bel^{pos}(\Omega) = 1$; $\forall A \subseteq \Omega, Pl^{pos}(A) = \bigvee_{\emptyset \neq A \cap B} \sigma(B), Pl^{pos}(\emptyset) = 0, Pl^{pos}(\Omega) = 1$.

Our aim is to study decomposable possibilistic belief functions. The definition of simple support functions present in Shafer evidence theory can be restated in the possibilistic context.

Definition 2. *A set function $\mu : 2^{\Omega} \to [0,1]$ is a simple support function focusing on the set $S \subseteq \Omega$ if and only if*

$$\mu(A) = \begin{cases} 0 & if\ S \not\subseteq A \\ s & if\ S \subset A\ but\ A \neq \Omega \\ 1 & if\ A = \Omega. \end{cases}$$

In Shafer evidence theory, a basic probability assignment m defines an imprecise probability in the sense that m is compatible with any probability distribution p obtained by sharing the masses $m(A)$ between the elements ω in A (assuming Ω finite):

$$\text{If } \forall A \in 2^{\Omega}, \exists k_A \in [0,1]^A \text{ such that } m(A) = \Sigma_{\omega \in A} k_A(\omega), \text{ and } p(\omega) = \Sigma_{A \ni \omega} k_A(\omega).$$

Clearly $\Sigma_{A \neq \emptyset} m(A) = 1$ ensures that $\Sigma_{\omega \in \Omega} p(\omega) = 1$. Thus, viewing each $m(A)$ as a mass that can be shared and freely reallocated to the elements of A, m is associated to the family of probability distributions p that are compatible with m in the above sense. Then m models a particular imprecise probability system. Letting $P(A) = \Sigma_{\omega \in A} p(\omega)$, it can be checked that

$$Bel(A) = \Sigma_{\emptyset \neq B \subseteq A} m(B) \leq P(A) \leq Pl(A) = \Sigma_{B \cap A \neq \emptyset} m(B).$$

Indeed, $Bel(A) = \Sigma_{\emptyset \neq B \subseteq A} \Sigma_{\omega \in B} k_B(\omega) \leq \Sigma_{\omega \in A} \Sigma_{B \ni \omega} k_B(\omega) = P(A)$.

We may similarly try to define an imprecise possibility structure. Let us consider a basic possibility assignment σ and $\mathscr{U}_{\sigma} = \bigcup_{i|\sigma(A_i)>0} A_i$. The situation is formally similar to the probabilistic case, if we want to interpret the $\sigma(A)$'s as providing an imprecise possibility specification. If it is so, there should exist a possibility distribution π and mappings $\kappa_A \in [0,1]^A$ such that such that $\sigma(A) = \max_{\omega \in A} \kappa_A(\omega)$, and $\pi(\omega) = \max_{B \ni \omega} \kappa_B(\omega)$. Note that $\Pi(A) = \max_{\omega \in A} \pi(\omega) = \max_{\omega \in A} \max_{B \ni \omega} \kappa_B(\omega) \geq \sigma(A)$. Note also that $\forall \omega \in \Omega - \mathscr{U}_{\sigma}, \pi(\omega) = 0$. Clearly $\max_{A \neq \emptyset} \sigma(A) = 1$ ensures that $\max_{\omega \in \Omega} \pi(\omega) = 1$. Thus, $\sigma(A)$ should be understood as a contribution to the specification of the possibility of A. Let us show that

$$Bel^{pos}(A) \leq \Pi(A) \leq Pl^{pos}(A).$$

Indeed, $Bel^{pos}(A) = \bigvee_{B \subseteq A} \sigma(B) \leq \bigvee_{B \subseteq A} \Pi(B) = \Pi(A)$. Let B_0 denote the set such that $A \cap B_0 \neq \emptyset$ and $\sigma(B_0) = \bigvee_{B \cap A \neq \emptyset} \sigma(B)$. For all $\omega \in A$, $\exists B_1$ such that $\omega \in B_1$ and $\pi(\omega) = \kappa_{B_1}(\omega)$. $\kappa_{B_1}(\omega) \leq \sigma(B_1) \leq \sigma(B_0)$ so $\forall \omega \in A, \pi(\omega) \leq \sigma(B_0)$ which entails $\Pi(A) \leq \sigma(B_0) = \bigvee_{B \cap A \neq \emptyset} \sigma(B) = Pl^{pos}(A)$.

Example. $\mathscr{F}_{\sigma} = \{A, \Omega\}$. Then $\sigma(A) = \max_{\omega \in A} \kappa_A(\omega)$, $\sigma(\Omega) = \max_{\omega \in \Omega} \kappa_{\Omega}(\omega)$, and $\pi(\omega) = \max(\kappa_A(\omega), \kappa_{\Omega}(\omega))$. By convention, $\kappa_A(\omega) = 0$ if $\omega \notin A$. Assume $\sigma(A) < \sigma(\Omega)$, then $\sigma(A) \leq \Pi(A) \leq \sigma(\Omega) = \Pi(\Omega) = 1$. If $\sigma(A) \geq \sigma(\Omega)$, then $\Pi(A) = max(\sigma(A), \max_{\omega \in A} \kappa_{\Omega}(\omega)) = \sigma(A)$, and $\Pi(\Omega) = max(\max_{\omega \in \Omega} \kappa_A(\omega), \sigma(\Omega)) = \sigma(A)$.

Observe that since $Bel^{pos}(A) = \bigvee_{B \subseteq A} \sigma(B) \leq \Pi(A) \leq Pl^{pos}(A) = \bigvee_{B \cap A \neq \emptyset} \sigma(B)$ we have by duality, since $N(A) = 1 - \Pi(\overline{A})$ defines the necessity measure associated to Π,

$$\bigwedge_{B \subseteq \overline{A}} 1 - \sigma(B) \geq N(A) \geq Pl^{pos}(A) = \bigwedge_{B \cap \overline{A} \neq \emptyset} 1 - \sigma(B).$$

The possibilistic logic situation [7] is retrieved when the upper bound of $N(A)$ is 1, i.e., when $\bigwedge_{B \subseteq \overline{A}} 1 - \sigma(B) = 1$, which means $\forall B \subseteq \overline{A}, \sigma(B) = 0$. This intuitively expresses that nothing that is inside $\overline{A}$ is possible for sure, which makes possible that $N(A)$ reaches 1, which is intuitively satisfactory. This shows that while the possibilistic logic setting corresponds to the representation of families of possibility measures having a unique maximal element (corresponding to the least specific possibility distribution compatible with constraints of the form $N(A) \geq \alpha$), the possibilistic evidence setting allows us to represent more general families of possibility measures.

3 Qualitative Möbius Transform

In Shafer evidence theory, a belief function *Bel* is associated with only one mass function *m*. The counterpart of this property is not satisfied in the maxitive setting. More precisely, in Shafer's evidence theory, the mass function associated to a belief function is computed with the Möbius transform. In our context, the qualitative Möbius transform [12, 11, 8, 9, 3] associates to a possibilistic belief function Bel^{pos} an interval of basic possibility assignments. Let us recall the definition of the qualitative Möbius transform.

Definition 3. *Let* $v : 2^{\Omega} \to [0,1]$ *be an increasing set function, its qualitative Möbius transform is* $\{\sigma | \forall A \subseteq \Omega \; v(A) = \bigvee_{B \subseteq A} \sigma(B)\} = \{\sigma | \sigma \in [\sigma_*, \sigma^*]\}$
where σ_* *and* σ^* *are the basic possibility assignments defined as follows:*

$\forall A \subseteq \Omega, \sigma^*(A) = v(A)$,
$\forall A \subseteq \Omega, \sigma_*(A) = 0$ *if* $\exists B \subset A$ *s.t.* $v(B) = v(A)$ *and* $\sigma_*(A) = v(A)$ *otherwise.*

Note that the sets A satisfying $\sigma_*(A) \neq 0$ are such that $\forall B \subset A$, $\sigma_*(B)$ is either 0 or $v(B)$ with $v(B) < v(A)$. So for all A, B such that $\sigma_*(A) \neq 0$, $\sigma_*(B) \neq 0$ and $B \subset A$ we have $\sigma_*(B) < \sigma_*(A)$.

In the following $[\sigma_*, \sigma^*]$ will denote the qualitative Möbius transform of a given possibilistic belief function Bel^{pos}.

The sets A such that $\sigma_*(A) \neq 0$ are called the focal elements of Bel^{pos} and the set of the focal elements is denoted $\mathscr{F}_{\sigma_*}$. Note that σ_*, which is unique, plays the same role as the mass function in the additive case; for more details look at [2].

For a given Bel^{pos}, there is no unicity of the basic possibility assignment but one σ can define only one possibilistic belief function. So it seems natural to define the following equivalence.

Definition 4. *Let* σ_1, σ_2 *be two basic possibility assignments.* σ_1 *and* σ_2 *are equivalent if and only if* $Bel^{pos}_{\sigma_1} = Bel^{pos}_{\sigma_2}$. *This equivalence is denoted* $\sigma_1 \sim \sigma_2$.

The qualitative Möbius transform of a simple support function μ focusing on S entails two set functions: σ^* and σ_*. If $\mu(A) = s$ if $S \subseteq A$, then we have $\sigma^* = \mu$ and

$$\forall A \subseteq \Omega, \; \sigma_*(A) = \begin{cases} s & \text{if } S = A \\ 1 & \text{if } A = \Omega \\ 0 & \text{otherwise.} \end{cases}$$

4 The Combination Rules

In Shafer's evidence theory, beside the well-known Dempster rule of combination [13] which corresponds to the conjunction of random sets, other combination laws have been proposed for fusing two mass functions. In particular, the disjunctive rule of combination appears to be another basic rule [6]. These rules have counterparts in the qualitative maxitive setting:

Definition 5. *Let σ_1 and σ_2 be two basic possibility assignments.*

The conjunctive combination rule is:
$(\sigma_1 \otimes \sigma_2)(\emptyset) = 0,$
$\forall A \subseteq \Omega,\ (\sigma_1 \otimes \sigma_2)(A) = \bigvee_{B\cap C=A\neq\emptyset} \sigma_1(B) \wedge \sigma_2(C).$
The disjunctive combination rule is:
$(\sigma_1 \oplus \sigma_2)(\emptyset) = 0,$
$\forall A \subseteq \Omega,\ (\sigma_1 \oplus \sigma_2)(A) = \bigvee_{B\cup C=A\neq\emptyset} \sigma_1(B) \wedge \sigma_2(C).$

Note that the first combination rule was already suggested in [5].

The set function $\sigma_1 \otimes \sigma_2$ is a basic possibility assignment if and only if there exists B and C such that $B \cap C \neq \emptyset$ and $\sigma_1(B) = \sigma_2(C) = 1$.

In our context $\sigma_1 \oplus \sigma_2$ is a basic possibility assignment because the basic possibility assignments are supposed to be normal.

Proposition 1. *The combination rule $\otimes$ is commutative, associative, and has a neutral element.*

Proof. • The definition of $\otimes$ implies the commutativity.

- Let $\sigma_0 : 2^{\Omega} \to [0,1]$ be the basic possibility assignment defined by $\sigma_0(A) = 1$ if $A = \Omega$ and 0 otherwise. Hence for all σ we have $\sigma_0 \otimes \sigma(A) = \bigvee_{B\cap C=A} \sigma_0(B) \wedge \sigma(C)$. If $B \neq \Omega$ then $\sigma_0(B) \wedge \sigma(C) = 0$ and if $B = \Omega, \sigma_0(B) \wedge \sigma(C) = \sigma(C)$ so $\sigma_0 \otimes \sigma(A) = \sigma(A)$ which entails that σ_0 is the neutral element according to $\otimes$.
- To prove the associativity, we consider $\sigma_1, \sigma_2, \sigma_3$, three basic possibilistic assignments,
  $((\sigma_1 \otimes \sigma_2) \otimes \sigma_3)(A) = \bigvee_{B'\cap D=A} (\sigma_1 \otimes \sigma_3)(B') \wedge \sigma_3(D)$
  $= \bigvee_{B'\cap D=A} (\bigvee_{B\cap C=B'} \sigma_1(B) \wedge \sigma_3(C)) \wedge \sigma_3(D)$
  $= \bigvee_{B\cap C\cap D=A} \sigma_1(B) \wedge \sigma_2(C) \wedge \sigma_3(D).$
  Similarly we get $(\sigma_1 \otimes (\sigma_2 \otimes \sigma_3))(A) = \bigvee_{B\cap C\cap D=A} \sigma_1(B) \wedge \sigma_2(C) \wedge \sigma_3(D)$ which entails the associativity of $\otimes$.

Moreover we can prove the following technical result.

Property 1. Let $\sigma_1, \cdots, \sigma_n$ be n basic possibility assignments.
$\sigma_1 \otimes \cdots \otimes \sigma_n(A) = \bigvee_{\cap_{k=1}^{n} X_k = A} \bigwedge_{k=1}^{n} \sigma_k(X_k).$

Proof. According to the definition of $\otimes$, the property is true for $n = 2$.

Let us suppose that the property is true for $n - 1$ and prove that it is true for n.

$$\begin{aligned}
\sigma_1 \otimes \cdots \otimes \sigma_n(A) &= (\sigma_1 \otimes \cdots \otimes \sigma_{n-1}) \otimes \sigma_n(A) \\
&= \bigvee_{B \cap C = A} (\sigma_1 \otimes \cdots \otimes \sigma_{n-1})(B) \wedge \sigma_n(C) \\
&= \bigvee_{B \cap C = A} [\bigvee_{\cap_{k=1}^{n-1} X_k = B} \wedge_{k=1}^{k-1} \sigma_k(X_k)] \wedge \sigma_n(C) \\
&= \bigvee_{B \cap C = A} \bigvee_{\cap_{k=1}^{n-1} X_k = B} [\wedge_{k=1}^{n-1} \sigma_k(X_k) \wedge \sigma_n(C)] \\
&= \bigvee_{\cap_{k=1}^{n} X_k = A} \wedge_{k=1}^{n} \sigma_k(X_k).
\end{aligned}$$

Note that the result above implies that the combination of n basic possibilistic assignments associated to simple support functions, or such that $\sigma(\Omega) = 1$, remains a basic possibility assignment.

Proposition 2. *Let $Bel_1^{pos}, \cdots, Bel_n^{pos}$ be simple support functions focusing respectively on $S_1, \cdots, S_n$.*

If we denote $\sigma_1, \cdots, \sigma_n$ some basic possibility assignments associated respectively to $Bel_1^{pos}, \cdots, Bel_n^{pos}$ then $\sigma_1 \otimes \cdots \otimes \sigma_n \sim \sigma_{1*} \otimes \cdots \otimes \sigma_{n*}$.

Proof. We have to prove that $Bel^{pos}_{\sigma_1 \otimes \cdots \otimes \sigma_n} = Bel^{pos}_{\sigma_{1*} \otimes \cdots \otimes \sigma_{n*}}$.

- Let us prove $Bel^{pos}_{\sigma_1 \otimes \cdots \otimes \sigma_n} \geq Bel^{pos}_{\sigma_{1*} \otimes \cdots \otimes \sigma_{n*}}$.
  $\forall A \subseteq \Omega, Bel^{pos}_{\sigma_1 \otimes \cdots \otimes \sigma_n}(A) = \bigvee_{B \subseteq A} \sigma_1 \otimes \cdots \otimes \sigma_n(B) = \bigvee_{B \subseteq A} \bigvee_{\cap_{i=1}^n X_i = B} (\bigwedge_{i=1}^n \sigma_i(X_i))$
  so $Bel^{pos}_{\sigma_1 \otimes \cdots \otimes \sigma_n}(A) = \bigvee_{\cap_{i=1}^n X_i \subseteq A} (\bigwedge_{i=1}^n \sigma_i(X_i))$.
  Moreover for all $i \in \{1, \cdots, n\}$, $\sigma_i \geq \sigma_{i*}$ which entails
  $Bel^{pos}_{\sigma_1 \otimes \cdots \otimes \sigma_n}(A) \geq \bigvee_{\cap_{i=1}^n X_i \subseteq A} (\bigwedge_{i=1}^n \sigma_{i*}(X_i))$ i.e. $Bel^{pos}_{\sigma_1 \otimes \cdots \otimes \sigma_n} \geq Bel^{pos}_{\sigma_{1*} \otimes \cdots \otimes \sigma_{n*}}$.
- Let us prove $Bel^{pos}_{\sigma_1 \otimes \cdots \otimes \sigma_n} \leq Bel^{pos}_{\sigma_{1*} \otimes \cdots \otimes \sigma_{n*}}$.
  We have $Bel^{pos}_{\sigma_1 \otimes \cdots \otimes \sigma_n}(A) = \bigvee_{\cap_{i=1}^n X_i \subseteq A} (\bigwedge_{i=1}^n \sigma_i(X_i))$ so
  - If $Bel^{pos}_{\sigma_1 \otimes \cdots \otimes \sigma_n}(A) = 0$ then the first inequality proved above entails that $Bel^{pos}_{\sigma_{1*} \otimes \cdots \otimes \sigma_{n*}}(A) = 0$.
  - If $Bel^{pos}_{\sigma_1 \otimes \cdots \otimes \sigma_n}(A) > 0$, then $\bigvee_{\cap_{i=1}^n X_i \subseteq A} (\bigwedge_{i=1}^n \sigma_i(X_i)) > 0$ which implies

$$\exists X_1, \cdots, X_n \in 2^{\Omega}, \text{ such that } \begin{cases} \cap_{i=1}^n X_i \subseteq A \\ \wedge_{i=1}^n \sigma_i(X_i) = Bel^{pos}_{\sigma_1 \otimes \cdots \otimes \sigma_n}(A) > 0 \end{cases}$$

For all i, $\sigma_i(\Omega) = 1$, so we just keep the $X_i \neq \Omega$, so:

$$\exists X_{i_1}, \cdots, X_{i_l} \neq \Omega \text{ such that } \begin{cases} \cap_{k=1}^l X_{i_k} \subseteq A \\ \wedge_{k=1}^l \sigma_{i_k}(X_{i_k}) = Bel^{pos}_{\sigma_1 \otimes \cdots \otimes \sigma_n}(A). \end{cases}$$

Other side for all i, Bel_i^{pos} is a simple support function focusing on S_i. So if $\sigma_i(A) \neq 0$ and $A \neq \Omega$, then $\sigma_i(A) \leq \sigma_{i*}(S_i)$[1].

We have $Bel^{pos}_{\sigma_1 \otimes \cdots \otimes \sigma_n}(A) \neq 0$, which entails for all i_k, $\sigma_{i_k}(X_{i_k}) > 0$ and we have supposed $X_{i_k} \neq \Omega$, so applying the remark above to i_k, we obtain $\sigma_{i_{k*}}(S_{i_k}) \geq \sigma_{i_k}(X_{i_k})$. This inequality is true for each i_k so $\wedge_{k=1}^l \sigma_{i_{k*}}(S_{i_k}) \geq \wedge_{k=1}^l \sigma_{i_k}(X_{i_k})$.

[1] Bel_i^{pos} is a simple support function, whose qualitative Möbius transform is $[\sigma_{i*}, Bel_i^{pos}]$.

Moreover $S_{i_k} \subseteq X_{i_k}$ entails $\cap_{k=1}^{l} S_{i_k} \subseteq \cap_{k=1}^{l} X_{i_k}$ i.e $\wedge_{k=1}^{l} \sigma_{i_k *}(S_{i_k}) \geq Bel^{pos}_{\sigma_1 \otimes \cdots \otimes \sigma_n}(A)$ so we get $Bel^{pos}_{\sigma_{1*} \otimes \cdots \otimes \sigma_{n*}} \geq Bel^{pos}_{\sigma_1 \otimes \cdots \otimes \sigma_n}$.

The previous proposition ensures the consistency of the following definition.

Definition 6. *Let $Bel_1^{pos}, \cdots, Bel_n^{pos}$ be simple support functions focusing respectively on $S_1, \cdots, S_n$ and let $\sigma_1, \cdots, \sigma_n$ be associated basic possibility assignements.*

Hence $Bel_1^{pos} \otimes \cdots \otimes Bel_n^{pos}$ denotes the possibilistic belief function associated to the basic possibility assignment $\sigma_1 \otimes \cdots \otimes \sigma_n$.

$Bel_1^{pos} \otimes \cdots \otimes Bel_n^{pos}$ does not depend on the σ_i's chosen. The operation $\otimes$ on simple support functions has the same properties as the one defined on the basic possibility assignments, i.e., $\otimes$ is associative, commutative, and has a neutral element. Using the previous properties we are going to decompose the possibilistic belief functions into product (in the sense of $\otimes$) of simple support functions.

5 Conditions for the Existence of a Decomposition

This section examines under what conditions a decomposition into simple support functions exists for a possibilistic belief function.

Proposition 3. *Let $Bel^{pos} : 2^{\Omega} \to [0,1]$ be a possibilistic belief function.*

There is an equivalence between the two following properties:

- *Bel^{pos} can be decomposed into combination of simple support functions,*
- *$\forall A, A' \in \mathscr{F}_{\sigma_*}, \quad A \cap A' \neq \emptyset \Rightarrow A \subseteq A'$ or $A' \subseteq A$,*

where $\mathscr{F}_{\sigma_*}$ is the set of the focal elements of Bel^{pos}.

Proof. We consider $Bel^{pos} : 2^{\Omega} \to [0,1]$ a possibilistic belief function.

- Suppose that Bel^{pos} satisfies the property: $\forall A, A' \in \mathscr{F}_{\sigma_*}, \quad A \cap A' \neq \emptyset \Rightarrow A \subseteq A'$ or $A' \subseteq A$, and prove that Bel^{pos} can be decomposed into a combination of simple support functions.

  We denote $n = \text{card}(\mathscr{F}_{\sigma_*})$, $\mathscr{F}_{\sigma_*} = \{A_i\}_{i=1,\cdots,n}$ and for all $i \in \{1, \cdots, n\}$ we define

$$\forall A \subseteq \Omega, \; \sigma_i(A) = \begin{cases} \sigma_*(A_i) & \text{if } A = A_i \\ 1 & \text{if } A = \Omega \\ 0 & \text{otherwise.} \end{cases}$$

  We consider $\sigma = \sigma_1 \otimes \cdots \otimes \sigma_n$. Let us prove that $\sigma \in [\sigma_*, \sigma^*]$. We already have $\sigma(\emptyset) = 0$ and $\sigma(\Omega) = 1$.

  - Let us prove that $\sigma \geq \sigma_*$. We consider $A \in 2^{\Omega} \setminus \{\emptyset, \Omega\}$.
    - If $A \notin \mathscr{F}_{\sigma_*}$, then $\sigma_*(A) = 0$ which entails $\sigma(A) \geq \sigma_*(A)$.
    - If $A \in \mathscr{F}_{\sigma_*}$, then there exists i such that $A = A_i$ so $\sigma(A) = \sigma(A_i) = \bigvee_{\cap_{j=1}^{n} X_j = A_i} \wedge_{j=1}^{n} \sigma_j(X_j)$. So considering $\{X_j\}_{j=1,\cdots,n}$ defined by $j \neq i, X_j = \Omega$ and $X_i = A_i$, we obtain $\sigma(A) \geq \sigma_i(A_i)$ so $\sigma(A) \geq \sigma_*(A_i)$ i.e., $\sigma(A) \geq \sigma_*(A)$.

So we have proved that $\sigma \geq \sigma_*$.

- Le us prove that $\sigma \leq \sigma^*$.
  Reasoning by contradiction, suppose that there exists $A \subseteq \Omega$ such that $\sigma(A) > \sigma^*(A)$. We necessarily have $A \neq \emptyset$, because $\sigma(\emptyset) = 0$. So according to the definition of σ we have $\exists X_1, \cdots, X_n \in 2^\Omega$ such that $\cap_{i=1}^n X_i = A$ and $\wedge_{i=1}^n \sigma_i(X_i) > \sigma^*(A)$ which implies

$$\exists X_1, \cdots, X_n \in 2^\Omega, \text{ such that } \begin{cases} \cap_{i=1}^n X_i = A \\ \forall i \in \{1, \cdots, n\},\ \sigma_i(X_i) > \sigma^*(A) \geq 0. \end{cases}$$

  As we have $\forall i \in \{1, \cdots, n\},\ \sigma_i(X_i) > 0$,

$$\exists X_1, \cdots, X_n \in \mathscr{F}_{\sigma_*} \cup \{\Omega\}, \text{ such that } \begin{cases} \cap_{i=1}^n X_i = A \\ \forall i \in \{1, \cdots, n\},\ \sigma_i(X_i) > \sigma^*(A) \geq 0. \end{cases}$$

  For all i, $\sigma_i(\Omega) = 1$, so we can only keep the sets different from Ω, i.e., the focal elements of σ_*:

$$\exists A_{i_1}, \cdots, A_{i_l} \in \mathscr{F}_{\sigma_*} \text{ such that } \begin{cases} \cap_{k=1}^l A_{i_k} = A \\ \forall i_k,\ \sigma_{i_k}(A_{i_k}) > \sigma^*(A) \geq 0. \end{cases}$$

  We have supposed that if the intersection of two focal elements is not empty, then one is included in the other. So between the A_{i_k} there exists one A_{i_0} such that $A = A_{i_0}$. We have assumed $\sigma_{i_0}(A_{i_0}) = \sigma_*(A_{i_0})$ so $\sigma_*(A) > \sigma^*(A)$, which contradicts $\sigma_* \leq \sigma^*$. To conclude we have $\sigma \leq \sigma^*$.

We have proved that $\sigma \in [\sigma_*, \sigma^*]$ i.e $\forall A \subseteq \Omega$, $Bel^{pos}(A) = \bigvee_{B \subseteq A} \sigma(B)$ which entails the existence of a decomposition.

- Let us prove the second implication. Suppose that there exists $\sigma \in [\sigma_*, \sigma^*]$, such that $\sigma = \sigma_1 \otimes \cdots \otimes \sigma_m$, where the σ_i's are associated to simple support functions.
  The property 2 allows us to choose σ_i as follows:

$$\forall A \subseteq \Omega,\ \sigma_i(A) = \begin{cases} s_i & \text{if } S_i = A \\ 1 & \text{if } A = \Omega \\ 0 & \text{otherwise.} \end{cases}$$

Reasoning by contradiction, suppose that there exists $A_1, A_2 \in \mathscr{F}_{\sigma_*}$, such that $A = A_1 \cap A_2 \neq \emptyset$ with $A \neq A_1$ and $A \neq A_2$.

Hence we can write $\sigma(A_1) = \bigvee_{\cap_{k=1}^m X_k = A_1} \wedge_{k=1}^m \sigma_k(X_k)$. So $\exists X_{k_1}, \cdots, X_{k_l} \subset \Omega$ such that $\cap_{i=1}^l X_{k_i} = A_1$ and $\wedge_{i=1}^l \sigma_{k_i}(X_{k_i}) = \sigma(A_1)$. Moreover $A_1 \in \mathscr{F}_{\sigma_*}$, implies $\sigma(A_1) = \sigma_*(A_1) = \sigma^*(A_1)$.

So $\exists X_{k_1}, \cdots, X_{k_l} \subset \Omega$, such that $\cap_{i=1}^l X_{k_i} = A_1$ and $\wedge_{i=1}^l \sigma_{k_i}(X_{k_i}) = \sigma^*(A_1) > 0$.

Similarly $\exists X_{k'_1}, \cdots, X_{k'_o} \subset \Omega$, such that $\cap_{i=1}^o X_{k'_i} = A_2$ and $\wedge_{i=1}^o \sigma_{k'_i}(X_{k'_i}) = \sigma^*(A_2) > 0$.

Moreover if we have $k_i = k'_j$, then $\sigma_{k_i}(X_{k_i}) > 0$ and $\sigma_{k_i}(X_{k'_j}) > 0$. As we have $X_{k_i} \neq \Omega$ and $X_{k'_j} \neq \Omega$, by hypothesis on the σ_i, we have $X_{k_i} = X_{k'_j}$.

Another way, $\sigma(A) = \bigvee_{\cap_{k=1}^{n} X_k = A} \wedge_{k=1}^{n} \sigma_k(X_k)$ and $A_1 \cap A_2 = A$, where $(\cap_{i=1}^{l} X_{k_i}) \cap (\cap_{j=1}^{o} X_{k'_j}) = A$ which implies that $\sigma(A) \geq (\bigwedge_{i=1}^{l} \sigma_{k_i}(X_{k_i}) \wedge (\bigwedge_{j=1}^{o} \sigma_{k'_j}(X_{k'_j})$ i.e. $\sigma(A) \geq \sigma^*(A_1) \wedge \sigma^*(A_2)$. $\sigma^* \geq \sigma$ entails $\sigma^*(A) \geq \sigma^*(A_1) \wedge \sigma^*(A_2)$. So there exists $B \in \mathscr{F}_{\sigma_*}$, $B = A_1$ or $B = A_2$, such that $\sigma^*(A) \geq \sigma^*(B)$, with $A \subset B$. To conclude we have $Bel^{pos}(A) \geq Bel^{pos}(B)$ and $B \notin \mathscr{F}_{\sigma_*}$ which is absurd. So necessarily either $A = A_1$ or $A = A_2$.

6 Minimal Decompositions

The previous section has presented under what conditions a possibilistic belief function can be decomposed into a combination of simple support functions. In this part we are going to study if such a decomposition is unique.

Proposition 4. *The decomposition of possibilistic belief functions into simple support functions is not unique.*

Proof. Let us present an example. We consider $\Omega = \{1,2,3,4\}$, $S = \{1\}$, $s = 1/2$ and the simple support function:

$$\mu(A) = \begin{cases} 0 & \text{if } S \not\subseteq A \\ s & \text{if } S \subseteq A \text{ but } A \neq \Omega \\ 1 & \text{if } A = \Omega \end{cases}$$

μ can be realised with the following basic possibilistic assignment:

$$\forall A \subseteq \Omega,\ \sigma_*(A) = \begin{cases} s & \text{if } A = S \\ 1 & \text{if } A = \Omega \\ 0 & \text{otherwise.} \end{cases}$$

$\mu = Bel_{\sigma_*}^{pos}$ is a simple support function so it is a first decomposition.

If we consider $\sigma_1, \sigma_2 : 2^{\Omega} \to [0,1]$ the following basic possibility assignments:

$$\forall A \subseteq \Omega,\ \sigma_1(A) = \begin{cases} s & \text{if } A = \{1,2\} \\ 1 & \text{if } A = \Omega \\ 0 & \text{otherwise} \end{cases} \quad \text{and } \forall A \subseteq \Omega,\ \sigma_2(A) = \begin{cases} s & \text{if } A = \{1,3\} \\ 1 & \text{if } A = \Omega \\ 0 & \text{otherwise} \end{cases}$$

then we have $\sigma_1 \otimes \sigma_2(\{1\}) = s$ and $\sigma_1 \otimes \sigma_2(A) = 0$ if $\{1\} \not\subset A$ and $A \neq \Omega$.

So a second decomposition is $Bel_{\sigma}^{pos} = Bel_{\sigma_1 \otimes \sigma_2}^{pos} = Bel_{\sigma_1}^{pos} \otimes Bel_{\sigma_2}^{pos}$.

The above example shows that there exists decompositions with different number of simple support functions. We can study the minimum number of simple support functions present in the decomposition of a given possibilistic belief function Bel^{pos}. We know that this number is lower or equal to the number of focal elements: Indeed, in the proof of the Proposition 3 there is a decomposition based on a simple support function by focal element.

Proposition 5. *Let Bel^{pos} be a possibilistic belief function which has a decomposition into product of simple support functions.*

If we denote $n = \text{card}(\mathscr{F}_{\sigma_*})$ and m the number of the simple support functions present in the decomposition of Bel^{pos} then we have $m \geq n$.

Proof. Let $Bel^{pos} = \mu_{\sigma_1} \otimes \cdots \otimes \mu_{\sigma_m}$ be one decomposition of Bel^{pos}, where the μ_{σ_i} are simple support functions focusing on S_i with σ_i defined as follows

$$\forall A \subseteq \Omega,\ \sigma_i(A) = \begin{cases} s_i & \text{if } S_i = A \\ 1 & \text{if } A = \Omega \\ 0 & \text{otherwise.} \end{cases}$$

Let us prove that $m \geq n$.

For all $A \subseteq \Omega$, $Bel^{pos}(A) = \bigvee_{B \subseteq A} \sigma_1 \otimes \cdots \otimes \sigma_m(B) = \bigvee_{\cap_{i=1}^m X_i \subseteq A} \wedge_{i=1}^m \sigma_i(X_i)$.

Particularly using the definition of σ_*, we obtain for all $F \in \mathscr{F}_{\sigma_*}$

$\sigma_*(F) = \bigvee_{\cap_{i=1}^m X_i = F} \wedge_{i=1}^m \sigma_i(X_i)$:

Indeed, we have $\sigma_*(F) = \bigvee_{\cap_{i=1}^m X_i \subseteq F} \wedge_{i=1}^m \sigma_i(X_i)$ but if the maximum is obtained for a set strictly contained in F, F cannot be a focal element.

So as the maximum is obtained for F, $\exists X_{i_1}, \cdots, X_{i_l}$ such that $\cap_{k=1}^l X_{i_k} = F$ and $\wedge_{k=1}^l \sigma_{i_k}(X_{i_k}) = \sigma_*(F)$ wich entails there exists k_0 such that $\sigma_{k_0}(X_{k_0}) = \sigma_*(F) > 0$. According to the definition of σ_i, we have $X_{k_0} = S_{k_0}$ and so $\sigma_{k_0}(S_{k_0}) = \sigma_*(F)$.

To conclude for any $F \in \mathscr{F}_{\sigma_*}$, there exists k_F, such that $\sigma_{k_F}(S_{k_F}) = \sigma_*(F)$.

- If $F, F' \in \mathscr{F}_{\sigma_*}$ satisfy $F \cap F' \neq \emptyset$, then according to the condition for having a decomposition, either $F \subset F'$, or $F' \subset F$ which entails $\sigma_*(F) \neq \sigma_*(F')$.
- If $F, F' \in \mathscr{F}_{\sigma_*}$ satisfy $F \cap F' = \emptyset$ then if $S_{k_F} = S_{k'_F}$ this set would be in the intersection which is the empty set.

To conclude if $F, F' \in \mathscr{F}_{\sigma_*}$ are different, then the associated S_k are different which implies $m \geq n$.

It can not exist a decomposition with a number of simple support functions lower than $\text{card}(\mathscr{F}_{\sigma_*})$. The proof of the Proposition 3 explains that there is always one decomposition with $\text{card}(\mathscr{F}_{\sigma_*})$ simple support functions which entails the following proposition.

Proposition 6. *If Bel^{pos} can be decomposed into a combination of simple support functions then the decomposition based on the focal elements contains a minimal number of simple support functions.*

Note that there is no unicity of the decomposition containing $\text{card}(\mathscr{F}_{\sigma_*})$ simple support functions as it is presented in the following example.

Example $\Omega = \{1,2,3,4\}$ and Bel^{pos} is such that $\mathscr{F}_{\sigma_*} = \{A_1, A_2, A_3\}$ with

- $A_1 = \{1\}$, $A_2 = \{1,2,3\}$ and $A_3 = \{3\}$,
- $\sigma_*(A_1) = \sigma_*(A_3) = 1/4$, and $\sigma_*(A_2) = 1/2$.

We define $\forall i \in \{1,2,3\}$, $\forall A \subseteq \Omega$, $\sigma_i(A) = \begin{cases} \sigma_*(A_i) & \text{if } A = A_i \\ 1 & \text{if } A = \Omega \\ 0 & \text{otherwise.} \end{cases}$

Hence a decomposition of Bel^{pos} is $Bel^{pos} = \mu_{\sigma_1} \otimes \mu_{\sigma_2} \otimes \mu_{\sigma_3}$.

Now if we consider $S_1 = \{1\}$, $S_2 = \{1,2,3\}, S_3 = \{3,4\}$ and

$\forall i \in \{1,2,3\}$, $\forall A \subseteq \Omega$, $\sigma'_i(A) = \begin{cases} \sigma_*(A_i) & \text{if } A = S_i \\ 1 & \text{if } A = \Omega \\ 0 & \text{otherwise} \end{cases}$

then we have $\sigma'_1 \otimes \sigma'_2 \otimes \sigma'_3(A_1) = \sigma'_1(S_1) = 1/4$, $\sigma'_1 \otimes \sigma'_2 \otimes \sigma'_3(A_2) = \sigma'_2(S_2) = 1/2$ and $\sigma'_1 \otimes \sigma'_2 \otimes \sigma'_3(A_3) = \sigma'_2(S_2) \wedge \sigma'_3(S_3) = 1/4$.

In such a context it is easy to verify the equality $Bel^{pos} = Bel^{pos}_{\sigma'_1 \otimes \sigma'_2 \otimes \sigma'_3}$.

So we have got a second decomposition $Bel^{pos} = \mu_{\sigma'_1} \otimes \mu_{\sigma'_2} \otimes \mu_{\sigma'_3}$.

Moreover $\mu_{\sigma_3}(A_3) \neq \mu_{\sigma'_3}(A_3)$ entails $\mu_{\sigma_3} \neq \mu_{\sigma'_3}$ so the presented decompositions are different and contain card$(\mathscr{F}_{\sigma_*})$ simple support functions.

7 Conclusion

In this paper we have studied the existence of the decomposition of possibilistic belief functions into a qualitative product of simple support functions. A necessary and sufficient condition for the existence of such a decomposition has been provided. There is no unicity of the decomposition, but if the condition is satisfied, we can compute a decomposition with a number of simple support functions equal to the number of the focal elements of the possibilistic belief function. Moreover this number is the minimal number of functions which can be present in a decomposition.

This qualitative counterpart has been obtained by substituting the sum by the maximum. We might also replace the sum by the minimum, which would entail the definition of the functions $Bel^{nec}(A) = \wedge_{B \subseteq A} \sigma(B)$. In such a context σ should be interpreted as an imprecise necessity qualification, as mentioned at the end of section 2. This raises the question of the counterpart of the decomposition result in the "minitive" setting.

Besides, Smets [14] has shown that any belief function can be decomposed into simple support functions if negative weights are allowed in the basic assignment. Does a counterpart to this result exist in the possibilistic evidence setting? What would be the relation with the decomposable signed fuzzy measures studied in [1]? These are questions for further research.

References

1. Mihailović, B., Pap, E.: Decomposable signed fuzzy measures. In: Proc. of EUSFLAT 2007, Ostrava, Czech Rep, pp. 265–269 (2007)
2. Dubois, D.: Fuzzy measures on finite scales as families of possibility meas. In: Proc. European Society For Fuzzy Logic and Technology (EUSFLAT-LFA), Aix-Les-Bains, France (July 2011)
3. Dubois, D., Fargier, H.: Capacity refinements and their application to qualitative decision evaluation. In: Sossai, C., Chemello, G. (eds.) ECSQARU 2009. LNCS, vol. 5590, pp. 311–322. Springer, Heidelberg (2009)
4. Dubois, D., Prade, H.: Upper and lower possibilities induced by a multivalued mapping. In: Sanchez, E. (ed.) Proc. IFAC Symp. on Fuzzy Information, Knowledge Representation and Decision Analysis, Fuzzy Information, Knowledge Representation and Decision Analysis, July 19-21, pp. 152–174. Pergamon Press, Sanchez (1984)
5. Dubois, D., Prade, H.: Evidence measures based on fuzzy information. Automatica 21, 547–562 (1985)
6. Dubois, D., Prade, H.: A set-theoretic view of belief functions. Logical operations and approximations by fuzzy sets. Int. J. General Systems 12, 193–226 (1986)
7. Dubois, D., Prade, H.: Possibilistic logic: a retrospective and prospective view. Fuzzy Sets and Systems 144, 3–23 (2004)
8. Grabisch, M.: The symmetric Sugeno integral. Fuzzy Sets and Systems 139, 473–490 (2003)
9. Grabisch, M.: The Moebius transform on symmetric ordered structures and its application to capacities on finite sets. Discrete Mathematics 287, 17–34 (2004)
10. Prade, H., Rico, A.: Possibilistic evidence. In: Liu, W. (ed.) ECSQARU 2011. LNCS, vol. 6717, pp. 713–724. Springer, Heidelberg (2011)
11. Marichal, J.-L.: Aggregation Operations for Multicriteria Decision Aid. Thesis, University of Liège, Belgium (1998)
12. Mesiar, R.: k-order Pan-discrete fuzzy measures. In: Proc. 7th Inter. Fuzzy Systems Assoc. World Congress (IFSA 1997), Prague, June 25-29, vol. 1, pp. 488–490 (1997)
13. Shafer, G.: A Mathematical Theory of Evidence. Princeton University Press, Princeton (1976)
14. Smets, P.: The canonical decomposition of a weighted belief. In: Proc. of the 14th Inter. Joint Conf. on Artificial Intelligence (IJCAI 1995), Montreal, August 20-25, pp. 1896–1901 (1995)
15. Tsiporkova, E., De Baets, B.: A general framework for upper and lower possibilities and necessities. Inter. J. of Uncert., Fuzz. and Knowledge-Based Syst. 6, 1–34 (1998)

Generalized Attanasov's Operators Defined on Lattice Intervals*

I. Lizasoain and C. Moreno

Abstract. In this paper we give a definition of an OWA operator on any complete lattice that generalizes the notion of an OWA operator in the real case. In addition we introduce a class of functions defined on lattice intervals by weakening the generalized Atanassov's K_α operators. We show that under certain conditions these functions provide a binary OWA operator.

1 Introduction

In [1] Atanassov introduces a class of operators that associate a real value to each real interval. This allows to associate a fuzzy set to any interval-valued fuzzy set. In [2], Bustince and other authors show that, under certain conditions, Atanassov's K_α operators, which act on intervals, provide the same numerical results than the binary OWA operators introduced by Yager in [4]. They generalize Atanassov's operators by introducing the generalized Atanassov's operators, a class of operators that include binary OWA operators.

In this paper we introduce the notion of n-ary OWA operators defined on a complete lattice L and show that they constitute a class of aggregation functions that has the Yager's OWA operator as a particular case.

In addition, the notion of generalized operators of [2] is weakened in order to extend it to the case of any complete lattice.

Section 3 shows that a binary OWA operator restricted to the set of lattice intervals, agrees with certain generalized K_α operator.

Section 4 shows that any generalized K_α operator provides, for each $\alpha \in L$, a binary aggregation function $G\mathbb{K}_\alpha : L^2 \to L$ that satisfies properties similar to those of an OWA binary operator.

I. Lizasoain · C. Moreno
Universidad Pública de Navarra, 31006 Pamplona (Spain)
e-mail: {ilizasoain, cristina.moreno}@unavarra.es

* The authors are partially supported by MTM2010-19938-C03-03.

B. De Baets et al. (Eds.): Eurofuse 2011, AISC 107, pp. 43–51, 2011.
springerlink.com

2 Preliminaries

Throughout this section $\mathscr{L} = (L, \leq_L)$ will be a bounded lattice, i.e. a partially ordered set such that for each pair of elements $a, b \in L$ there exists an infimum $a \wedge b \in L$ and a supremum $a \vee b \in L$. We will denote by $0_{\mathscr{L}}$ the lowest element of L and by $1_{\mathscr{L}}$ the greatest element of L.

Definition 1. *A map $T : L \times L \to L$ is said to be a t-norm in $\mathscr{L}$ if it is commutative, associative, increasing and has neutral element $1_{\mathscr{L}}$.*

Remark 1. 1. If $T : L \times L \to L$ is a t-norm in $\mathscr{L}$, then

$$T(a, b) \leq_L T(a, 1_{\mathscr{L}}) = a \text{ for any } a, b \in L.$$

2. In particular, for any $b \in L$, $T(0_{\mathscr{L}}, b) \leq 0_{\mathscr{L}}$ and so $T(0_{\mathscr{L}}, b) = 0_{\mathscr{L}}$.

Definition 2. *A map $S : L \times L \to L$ is said to be a t-conorm in $\mathscr{L}$ if it is commutative, associative, increasing and has neutral element $1_{\mathscr{L}}$.*

Remark 2. 1. If $S : L \times L \to L$ is a t-conorm in $\mathscr{L}$, then

$$a = S(a, 0_{\mathscr{L}}) \leq S(a, b) \text{ for any } a, b \in \mathscr{L}.$$

2. In particular, for any $b \in \mathscr{L}$, $1_{\mathscr{L}} \leq_L S(1_{\mathscr{L}}, b)$ and so $S(1_{\mathscr{L}}, b) = 1_{\mathscr{L}}$.
3. Because of the associativity of a t-conorm S, we can write $S(a,b,c)$ to denote $S(S(a, b), c)$. If $Y = \{a_1, a_2, \cdots, a_n\}$ is a finite subset of L, we write $S\{a \mid a \in Y\}$ to denote $S(a_1, a_2, \ldots, a_n)$. If $Y = \{a\}$, then we understand that $S\{a \mid a \in Y\} = a$. The same can be said for any t-norm T.

Example 1. In any bounded lattice $\mathscr{L} = (L, \leq_L)$, the meet (greatest lower bound) is a t-norm, denoted by $\wedge$, and the join (least upper bound) is a t-conorm, denoted by $\vee$.

We denote by (D) the following distributive property:

$$T(a, S(b,c)) = S(T(a,b), T(a,c)) \text{ for any } a, b, c \in L. \qquad (D)$$

Notice that the t-norm given by the meet and the t-conorm given by the join in a complete lattice $\mathscr{L} = (L, \leq_L)$ do not always satisfy (D). Indeed, property (D) for the meet and the join is equivalent to the residuation principle in $\mathscr{L}$ (see [3]).

3 n-ary OWA Operators

In this section we introduce the notion of an n-ary OWA operator defined on a complete lattice. In addition, we weaken the notion of GK_α operator given in [2] in order to find examples for any complete lattice. At last, we show that a binary OWA operator agrees with a GK_α operator on any lattice interval.

Throughout this section T and S will denote respectively a t-norm and a t-conorm defined in a complete lattice $\mathscr{L} = (L, \leq_L)$ that satisfy the distributive property (D).

Definition 3. *Let $\mathscr{L} = (L, \leq_L)$ be a lattice. An n-ary aggregation fuction is a function $M : L^n \to L$ such that:*

1. $M(a_1,\ldots,a_n) \leq M(a'_1,\ldots,a'_n)$ whenever $a_i \leq a'_i$ for $1 \leq i \leq n$.
2. $M(0_{\mathscr{L}},\ldots,0_{\mathscr{L}}) = 0_{\mathscr{L}}$ and $M(1_{\mathscr{L}},\ldots,1_{\mathscr{L}}) = 1_{\mathscr{L}}$.

An n-ary aggregation function M is said to be idempotent *if $M(a,\ldots,a) = a$ for every $a \in L$. It is said to be* symmetric *if $M(a_1,\ldots,a_n) = M(a_{\sigma(1)},\ldots,a_{\sigma(n)})$ for every permutation σ of the set $\{1,\ldots,n\}$.*

Our next definition is a generalization for lattices of the n-ary OWA operators defined by Yager in [4].

Definition 4. *A function $F : L^n \to L$ is called an n-*ary OWA operator *if there exists a* weighting vector $(\alpha_1,\ldots,\alpha_n) \in L^n$ *with $S(\alpha_1,\ldots,\alpha_n) = 1_{\mathscr{L}}$ such that*

$$F(a_1,\ldots,a_n) = S(T(\alpha_1,b_1),\cdots,T(\alpha_n,b_n)) \text{ for any } (a_1,\ldots,a_n) \in L^n$$

where, for $1 \leq k \leq n$, $b_k = \vee\{a_{j_1} \wedge \cdots \wedge a_{j_k} \mid \{j_1,\ldots,j_k\} \subseteq \{1,\ldots,n\}\}$.
Notice that

$$a_1 \wedge \cdots \wedge a_n = b_n \leq b_{n-1} \leq \cdots \leq b_1 = a_1 \vee \cdots \vee a_n.$$

The n-ary OWA operator $\hat{F} : L^n \to L$ with weighting vector $(\alpha_n,\ldots,\alpha_1)$ is called the dual operator *of F.*

Remark 3. Notice that, if L is the real interval $[0,1]$ with the usual order $\leq$, $T(a,b) = ab$ for every $a,b \in [0,1]$, $S(a,b) = \min\{a+b,1\}$ for every $a,b \in [0,1]$ and $(\alpha_1,\ldots,\alpha_n)$ is a weighting vector in L^n with $\alpha_1 + \cdots + \alpha_n = 1$, then $F : L^n \to L$ is the OWA operator given by Yager in [4].

Proposition 1. *Let $F : L^n \to L$ be an n-ary OWA operator with weighting vector $(\alpha_1,\ldots,\alpha_n) \in L^n$. Then*

1. F is a symmetric n-ary aggregation function.
2. F is idempotent.
3. $a_1 \wedge \cdots \wedge a_n \leq F(a_1,\ldots,a_n) \leq a_1 \vee \cdots \vee a_n$ for every $(a_1,\ldots,a_n) \in L^n$.

Proof. 1. If $a_i \leq a'_i$ for every $1 \leq i \leq n$ and $1 \leq k \leq n$, then $a_{j_1} \wedge \cdots \wedge a_{j_k} \leq a'_{j_1} \wedge \cdots \wedge a'_{j_k}$ for every $\{j_1,\ldots,j_k\} \subseteq \{1,\ldots,n\}$ and consequently $b_k \leq b'_k$.
Hence, $T(\alpha_k,b_k) \leq T(\alpha_k,b'_k)$ for any $1 \leq k \leq n$ and then

$$S(T(\alpha_1,b_1),\cdots,T(\alpha_n,b_n)) \leq S\left(T(\alpha_1,b'_1),\cdots,T(\alpha_n,b'_n)\right).$$

It is easy to check that $F(0_{\mathscr{L}},\ldots,0_{\mathscr{L}}) = 0_{\mathscr{L}}$, that $F(1_{\mathscr{L}},\ldots,1_{\mathscr{L}}) = 1_{\mathscr{L}}$ and that F is symmetric.

2. If $a \in L$, then $b_k = a$ for every $1 \le k \le n$ and hence

$$\begin{aligned} F(a,\dots,a) &= S(T(\alpha_1,a),\cdots,T(\alpha_n,a)) \\ &= T(S(\alpha_1,\dots,\alpha_n),a) = T(1_{\mathscr{L}},a) = a. \end{aligned}$$

3. Let $(a_1,\dots,a_n) \in L^n$. For any $1 \le k \le n$, we have $b_n \le b_k \le b_1$. Therefore,

$$\begin{aligned} a_1 \wedge \cdots \wedge a_n &= b_n = T(b_n,1_{\mathscr{L}}) = T(b_n,S(\alpha_1,\cdots,\alpha_n)) \\ &= S(T(\alpha_1,b_n),\cdots,T(\alpha_n,b_n)) \le S(T(\alpha_1,b_1),\cdots,T(\alpha_n,b_n)) \\ &\le S(T(\alpha_1,b_1),\cdots,T(\alpha_n,b_1)) = T(b_1,S(\alpha_1,\cdots,\alpha_n)) \\ &= T(b_1,1_{\mathscr{L}}) = b_1 = a_1 \vee \cdots \vee a_n. \end{aligned}$$

Therefore,

$$a_1 \wedge \cdots \wedge a_n \le F(a_1,\dots,a_n) \le a_1 \vee \cdots \vee a_n.$$

The next result shows that the meet and the join are particular cases of an n-ary OWA operator.

Proposition 2. *1. If $F : L^n \to L$ is an n-ary OWA operator with weighting vector $(1_{\mathscr{L}},0_{\mathscr{L}},\dots,0_{\mathscr{L}}) \in L^n$, then $F(a_1,\dots,a_n) = a_1 \vee \cdots \vee a_n$ for any $(a_1,\dots,a_n) \in L^n$.*
2. If $F : L^n \to L$ is an n-ary OWA operator with weighting vector $(0_{\mathscr{L}},\dots,0_{\mathscr{L}},1_{\mathscr{L}}) \in L^n$, then $F(a_1,\dots,a_n) = a_1 \wedge \cdots \wedge a_n$ for any $(a_1,\dots,a_n) \in L^n$.

Proof. 1. Notice that $S(1_{\mathscr{L}},0_{\mathscr{L}},\dots,0_{\mathscr{L}}) = 1_{\mathscr{L}}$. Moreover, for any $(a_1,\dots,a_n) \in L^n$,

$$\begin{aligned} F(a_1,\dots,a_n) &= S(T(1_{\mathscr{L}},b_1),T(0_{\mathscr{L}},b_2),\cdots,T(0_{\mathscr{L}},b_n)) \\ &= S(b_1,0_{\mathscr{L}},\cdots,0_{\mathscr{L}}) = b_1 = a_1 \vee \cdots \vee a_n. \end{aligned}$$

2. Analogously $S(0_{\mathscr{L}},\dots,0_{\mathscr{L}},1_{\mathscr{L}}) = 1_{\mathscr{L}}$. In addition, for any $(a_1,\dots,a_n) \in L^n$,

$$\begin{aligned} F(a_1,\dots,a_n) &= S(T(0_{\mathscr{L}},b_1),\cdots,T(0_{\mathscr{L}},b_{n-1}),\cdots,T(1_{\mathscr{L}},b_n)) \\ &= S(0_{\mathscr{L}},\cdots,0_{\mathscr{L}},b_n) = b_n = a_1 \wedge \cdots \wedge a_n. \end{aligned}$$

Proposition 3. *Let $F : L^2 \to L$ be a binary OWA operator with weighting vector (α,β) such that $S(\alpha,\beta) = 1_{\mathscr{L}}$. Then*

1. $F(0_{\mathscr{L}},1_{\mathscr{L}}) = \alpha$.
2. If $\alpha = 0_{\mathscr{L}}$ and $\beta = 1_{\mathscr{L}}$, then $F(a,b) = a \wedge b$.
3. If $\alpha = 1_{\mathscr{L}}$ and $\beta = 0_{\mathscr{L}}$, then $F(a,b) = a \vee b$.

Proof. It is straightforward.

Let $\mathscr{L} = (L,\le_L)$ be a complete lattice. Denote by $[a_0, a_1]$ any closed interval of $\mathscr{L}$,

$$[a_0, a_1] = \{b \in L \mid a_0 \le b \le a_1\}.$$

Consider the set of all the closed intervals of $\mathscr{L}$,

$$L^I = \{\mathbf{a} = [a_0, a_1] \mid a_0, a_1 \in L, a_0 \leq_L a_1\},$$

with the partial order given by

$$[a_0, a_1] \leq_{L^I} [a'_0, a'_1] \iff a_0 \leq_L a'_0 \text{ and } a_1 \leq_L a'_1.$$

It is known that $\mathscr{L}^I = (L^I, \leq_{L^I})$ is a complete lattice (see [3]). In addition, if $0_{\mathscr{L}}$ and $1_{\mathscr{L}}$ are respectively the lowest and the greatest element of $\mathscr{L}$, then the intervals $[0_{\mathscr{L}}, 0_{\mathscr{L}}]$ and $[1_{\mathscr{L}}, 1_{\mathscr{L}}]$ are the minimum and the maximum of $\mathscr{L}^I$ respectively.

The next concept weakens the notion of generalized K_α operator given in [2].

Definition 5. *Let $\mathscr{L} = (L, \leq_L)$ be a complete lattice. A* generalized K_α *operator or a GK_α operator is a map $GK : L \times L^I \to L$ mapping any $(\alpha, \mathbf{a})$ to $GK_\alpha(\mathbf{a}) \in L$ and satisfying*

1. *If $a_0 = a_1$, then $GK_\alpha([a_0, a_1]) = a_0$ for any $\alpha \in L$.*
2. *$GK_{0_{\mathscr{L}}}([a_0, a_1]) = a_0$ and $GK_{1_{\mathscr{L}}}([a_0, a_1]) = a_1$ for any $[a_0, a_1] \in L^I$.*
3. *If $\mathbf{a} \leq_{L^I} \mathbf{a}'$, then $GK_\alpha(\mathbf{a}) \leq_L GK_\alpha(\mathbf{a}')$ for any $\alpha \in L$.*
4. *$GK_\alpha([0_{\mathscr{L}}, 1_{\mathscr{L}}]) = \alpha$ for any $\alpha \in L$.*

Remark 4. Notice that we do not impose a GK_α operator the condition $GK_\alpha(\mathbf{a}) \leq GK_\beta(\mathbf{a})$ whenever $\alpha \leq \beta$ for every $\mathbf{a} \in L^I$. The main reason is that Example 3 does not satisfy it.

Proposition 4. *Let $\mathscr{L} = (L, \leq_L)$ be a complete lattice and $GK : L \times L^I \to L$ a generalized K_α operator. Then $GK_\alpha([0_{\mathscr{L}}, a]) \leq \alpha$ and $GK_\alpha([a, 1_{\mathscr{L}}]) \geq \alpha$ for every $a \in L$.*

Proof. Let $a \in L$. Since $[0_{\mathscr{L}}, a] \leq [0_{\mathscr{L}}, 1_{\mathscr{L}}]$, then

$$GK_\alpha([0_{\mathscr{L}}, a]) \leq GK_\alpha([0_{\mathscr{L}}, 1_{\mathscr{L}}]) = \alpha.$$

In addition, since $[a, 1_{\mathscr{L}}] \geq [0_{\mathscr{L}}, 1_{\mathscr{L}}]$, then

$$GK_\alpha([a, 1_{\mathscr{L}}]) \geq GK_\alpha([0_{\mathscr{L}}, 1_{\mathscr{L}}]) = \alpha.$$

The following example given in [2] is meaningful for any complete lattice, but it does not always satisfy the first property of Definition 5.

Example 2. Let $\mathscr{L} = (L, \leq_L)$ be a complete lattice. Define

$$GK_\alpha([a_0, a_1]) = \begin{cases} a_1 & \text{if } a_1 \leq \alpha \\ a_0 & \text{if } \alpha \leq a_0 \\ \alpha & \text{otherwise} \end{cases}$$

satisfies all the properties of Definition 5 but the first one.

The following example of GK_α operator is a generalization of the well-known Attanasov's K_α operators.

Example 3. Let $\mathscr{L} = (L, \leq_L)$ be a complete lattice. Suppose that $\beta : L \to L$ is a bijection with $S(\alpha, \beta(\alpha)) = 1_{\mathscr{L}}$ for every $\alpha \in L$, $\beta(0_{\mathscr{L}}) = 1_{\mathscr{L}}$ and $\beta(1_{\mathscr{L}}) = 0_{\mathscr{L}}$. Then the map $GK : L \times L^I \to L$ that maps any $(\alpha, \mathbf{a}) \in L \times L^I$ to

$$GK_\alpha([a_0, a_1]) = S(T(\beta(\alpha), a_0), T(\alpha, a_1))$$

is a generalized K_α operator.

Proposition 5. *The GK_α operator defined in Example 3 satisfies:*

1. $GK_\alpha([\alpha, 1_{\mathscr{L}}]) = \alpha$ *for any* $\alpha \in L$.
2. $GK_\alpha([0_{\mathscr{L}}, \alpha]) = \alpha$ *for any* $\alpha \in L$ *if and only if the t-norm* T *is the meet.*

Proof. 1. Let $\alpha \in L$. Then

$$GK_\alpha([\alpha, 1_{\mathscr{L}}]) = S(T(\beta(\alpha), \alpha), T(\alpha, 1_{\mathscr{L}}))$$
$$= T(S(\beta(\alpha), 1_{\mathscr{L}}), \alpha) = T(1_{\mathscr{L}}, \alpha) = \alpha.$$

2. Let $\alpha \in L$. Then

$$GK_\alpha([0_{\mathscr{L}}, \alpha]) = S(T(\beta(\alpha), 0_{\mathscr{L}}), T(\alpha, \alpha))$$
$$= S(0_{\mathscr{L}}, T(\alpha, \alpha)) = T(\alpha, \alpha).$$

The last value is equal to α for every $\alpha \in L$ if and only if the t-norm T is idempotent. Then, for any $a, b \in L$,

$$T(a,b) \geq T(a \wedge b, b) \geq T(a \wedge b, a \wedge b) = a \wedge b.$$

Since $T(a,b) \leq a$ and $T(a,b) \leq b$, then $T(a,b) \leq a \wedge b$ and so $T(a,b) = a \wedge b$. Conversely, if T is the meet defined on $\mathscr{L}$, then T is idempotent.

Remark 5. 1. If L is the real interval $[0,1]$ with the usual order $\leq$, $T(a,b) = ab$ for every $a, b \in [0,1]$, $S(a,b) = \min\{a+b, 1\}$ for every $a, b \in [0,1]$ and $\beta(\alpha) = 1 - \alpha$ for any $\alpha \in [0,1]$, then the GK_α defined in Example 3 is the Attanasov operator

$$K_\alpha[a_0, a_1] = (1-\alpha)a_0 + \alpha a_1.$$

2. If $F : L^2 \to L$ is the binary OWA operator with weighting vector $(\alpha, \beta(\alpha))$ and GK_α is that of Example 3, then

$$GK_\alpha([a_0, a_1]) = F(a_0, a_1) \text{ for every } [a_0, a_1] \in L^I.$$

However, GK_α is defined only for intervals while F is defined for any pair $(a_0, a_1) \in L^2$.

We show that Example 3 is the unique generalized K_α operator that can be constructed in a certain way.

Proposition 6. *Let $\mathscr{L} = (L, \leq_L)$ be a complete lattice and $GK : L \times L^I \to L$ a generalized K_α operator such that*

$$GK_\alpha([a_0, a_1]) = S(T(\beta(\alpha), a_0), T(\gamma(\alpha), a_1)) \text{ for any } [a_0, a_1] \in L^I.$$

Then $\gamma(\alpha) = \alpha$ for any $\alpha \in L$ and $\beta : L \to L$ is a map such that $S(\alpha, \beta(\alpha)) = 1_{\mathscr{L}}$ for any $\alpha \in L$, $\beta(0_{\mathscr{L}}) = 1_{\mathscr{L}}$ and $\beta(1_{\mathscr{L}}) = 0_{\mathscr{L}}$.

Proof. For any $\alpha \in L$, we have $GK_\alpha([0_{\mathscr{L}}, 1_{\mathscr{L}}]) = \alpha$ and so

$$\alpha = S(T(\beta(\alpha), 0_{\mathscr{L}}), T(\gamma(\alpha), 1_{\mathscr{L}})) = S(0_{\mathscr{L}}, \gamma(\alpha)) = \gamma(\alpha).$$

Moreover, since $GK_\alpha([1_{\mathscr{L}}, 1_{\mathscr{L}}]) = 1_{\mathscr{L}}$, then for any $\alpha \in L$,

$$\begin{aligned} 1_{\mathscr{L}} &= S(T(\beta(\alpha), 1_{\mathscr{L}}), T(\gamma(\alpha), 1_{\mathscr{L}})) = T(S(\beta(\alpha), \gamma(\alpha)), 1_{\mathscr{L}}) \\ &= S(\beta(\alpha), \gamma(\alpha)). \end{aligned}$$

The next result shows a way of building generalized K_α operators starting from one of them.

Theorem 1. *Let $\mathscr{L} = (L, \leq_L)$ be a complete lattice and $GK : L \times L^I \to L$ a generalized K_α operator. If $\tau : L \to L$ is a lattice isomorphism (i.e., a preseving-order bijection), then the map $\overline{GK} : L \times L^I \to L$ given, for any $\alpha \in L$ and any $[a_0, a_1] \in L^I$, by*

$$\overline{GK}_\alpha([a_0, a_1]) = \tau^{-1}\left(GK_{\tau(\alpha)}([\tau(a_0), \tau(a_1)])\right)$$

is a generalized K_α operator.

Proof. It is straightforward.

4 $G\mathbb{K}_\alpha$-Operators

In this section we show that any GK_α operator provides, for each $\alpha \in L$, a binary aggregation function $G\mathbb{K}_\alpha : L^2 \to L$ that satisfies properties similar to an OWA binary operator. Moreover, if GK_α is that defined in Example 3, then $G\mathbb{K}_\alpha$ is exactly the OWA operator with weighting vector $(\alpha, \beta(\alpha))$.

Proposition 7. *Let $\mathscr{L} = (L, \leq_L)$ be a complete lattice and consider a GK_α operator defined on it. If $i : L^2 \to L^I$ is the fuction given by $i(a, b) = [a \wedge b, a \vee b]$ for any $(a, b) \in L^2$, then for any $\alpha \in L$, the function $G\mathbb{K}_\alpha : L^2 \to L$ given by*

$$G\mathbb{K}_\alpha(a, b) = (GK_\alpha \circ i)(a, b) \text{ for any } (a, b) \in L^2$$

is a binary aggregation function that satisfies the following properties:

1. *$G\mathbb{K}_\alpha$ is commutative and idempotent. In particular $G\mathbb{K}_\alpha(0_{\mathscr{L}}, 0_{\mathscr{L}}) = 0_{\mathscr{L}}$ and $G\mathbb{K}_\alpha(1_{\mathscr{L}}, 1_{\mathscr{L}}) = 1_{\mathscr{L}}$*
2. *If $a \leq c$ and $b \leq d$ then $G\mathbb{K}_\alpha(a, b) \leq G\mathbb{K}_\alpha(c, d)$.*

3. $\mathbb{GK}_{0_{\mathscr{L}}}(a,b) = a \wedge b$ *and* $\mathbb{GK}_{1_{\mathscr{L}}}(a,b) = a \vee b$ *for any* $(a,b) \in L^2$.
4. $\mathbb{GK}_\alpha(0_{\mathscr{L}}, 1_{\mathscr{L}}) = \alpha$.
5. $\mathbb{GK}_\alpha(a,b) = \mathbb{GK}_\alpha(a \wedge b, a \vee b)$ *for any* $(a,b) \in L^2$.

Proof. It is an easy checking.

Theorem 2. *Let* $GK : L \times L^I \to L$ *be a generalized* K_α *operator and* $i : L^2 \to L^I$ *the map given by* $i(a,b) = [a \wedge b, a \vee b]$. *Then the map* $\mathbb{GK}_\alpha = GK_\alpha \circ i$ *satisfies the following properties for any* $\alpha \in L$:

1. *If* $\alpha \neq 0_{\mathscr{L}}$ *and* $\mathbb{GK}_\alpha(\alpha, 1_{\mathscr{L}}) = \alpha$, *then* $\mathbb{GK}_\alpha$ *is not strictly increasing.*
2. *If* $\alpha \neq 1_{\mathscr{L}}$ *and* $\mathbb{GK}_\alpha(0_{\mathscr{L}}, \alpha) = \alpha$, *then* $\mathbb{GK}_\alpha$ *is not strictly increasing.*
3. *If* GK_α *is strictly increasing, then* $\mathbb{GK}_\alpha$ *is strictly increasing.*

Proof. 1. We have $0_{\mathscr{L}} < \alpha$. However,

$$\mathbb{GK}_\alpha(0_{\mathscr{L}}, 1_{\mathscr{L}}) = \alpha = \mathbb{GK}_\alpha(\alpha, 1_{\mathscr{L}})$$

by the assumption.

2. $\alpha < 1_{\mathscr{L}}$, but

$$\mathbb{GK}_\alpha(0_{\mathscr{L}}, \alpha) = \alpha = \mathbb{GK}_\alpha(0_{\mathscr{L}}, 1_{\mathscr{L}}).$$

3. It is immediate.

Theorem 3. *Let* $GK : L \times L^I$ *be a generalized* K_α *operator such that* $\mathbb{GK}_\alpha = GK_\alpha \circ i$ *is a bisymmetric operator, i.e.,*

$$\mathbb{GK}_\alpha(\mathbb{GK}_\alpha(a,b), \mathbb{GK}_\alpha(c,d)) = \mathbb{GK}_\alpha(\mathbb{GK}_\alpha(a,c), \mathbb{GK}_\alpha(b,d)) \text{ for every } a,b,c,d \in L.$$

Then $\mathbb{GK}_\alpha(\alpha, 1_{\mathscr{L}}) = \alpha$ *if and only if* $\mathbb{GK}_\alpha(\alpha, 0_{\mathscr{L}}) = \alpha$.

Proof. Assume that $\mathbb{GK}_\alpha(\alpha, 1_{\mathscr{L}}) = \alpha$. Then

$$\begin{aligned}
&\mathbb{GK}_\alpha(\alpha, 0_{\mathscr{L}}) = \mathbb{GK}_\alpha\left((\mathbb{GK}_\alpha(\alpha, 1_{\mathscr{L}}), \mathbb{GK}_\alpha(0_{\mathscr{L}}, 0_{\mathscr{L}})\right) \\
&= \mathbb{GK}_\alpha\left((\mathbb{GK}_\alpha(\alpha, 0_{\mathscr{L}}), \mathbb{GK}_\alpha(1_{\mathscr{L}}, 0_{\mathscr{L}})\right) = \mathbb{GK}_\alpha\left(\mathbb{GK}_\alpha(\alpha, 0_{\mathscr{L}}), \alpha\right) \\
&= \mathbb{GK}_\alpha\left((\mathbb{GK}_\alpha(\alpha, 0_{\mathscr{L}}), \mathbb{GK}_\alpha(\alpha, 1_{\mathscr{L}})\right) \\
&= \mathbb{GK}_\alpha\left((\mathbb{GK}_\alpha(\alpha, \alpha), \mathbb{GK}_\alpha(0_{\mathscr{L}}, 1_{\mathscr{L}})\right) \\
&= \mathbb{GK}_\alpha(\alpha, \alpha) = \alpha.
\end{aligned}$$

If we suppose that $\mathbb{GK}_\alpha(\alpha, 0_{\mathscr{L}}) = \alpha$, it is proven that $\mathbb{GK}_\alpha(\alpha, 1_{\mathscr{L}}) = \alpha$ in a similar way.

Theorem 4. *Let* $GK : L \times L^I$ *be the generalized* K_α *operator defined in Example 3. Then* $\mathbb{GK}_\alpha = GK_\alpha \circ i$ *is the binary OWA operator with weighting vector* $(\alpha, \beta(\alpha))$.

Proof. It is immediate.

References

1. Atanassov, K.: Intuitionistic fuzzy sets. In: VIIth ITKR Session, Deposited in the Central Science and Technology Library of the Bulgarian Academy of Sciences, Sofia, Bulgaria, pp. 1684–1697 (1983)
2. Bustince, H., Calvo, T., de Baets, B., Fodor, J., Mesiar, R., Montero, J., Paternain, D., Pradera, A.: A class of aggregation fuctions encompassing two-dimensional OWA operators. Information Sciences 180, 1977–1989 (2010)
3. Deschrijver, G.: A representation of t-norms in interval-valued L-fuzzy set theory. Fuzzy sets and Systems 159, 1597–1618 (2008)
4. Yager, R.R.: On ordered weighting averaging aggregation operators in multicriteria decision-making. IEEE Transaction on Systems, Man and Cybernetics 18, 183–190 (1988)

Modalities

József Dombi

Abstract. Hedges play an important role in fuzzy theory, although there are relatively few articles on them. Our aim is to provide a theoretical basis not only for hedges, but also for every type of unary operator. One of them is the negation operator, which was presented in an article [14] concerning the DeMorgan class. In our study we will develop unary operators related to other binary operators by demanding that they satisfy certain properties.

1 Introduction

Zadeh introduced modifier functions of fuzzy sets called linguistic hedges. A number of studies [12, 11, 18] have been made which discuss fuzzy logic and fuzzy reasoning with linguistic truth values. However, a systematic view of it has not been presented in the construction of linguistic hedges, which have corresponding reverse effects, such as in the case of "very" and "more or less".

In the early 1970s, Zadeh [32] introduced a class of powering modifiers, which defined the concept of linguistic variables and hedges. He proposed computing with words as an extension of fuzzy sets and logic theory (Zadeh [33, 37, 30]). The linguistic hedges (LHs) change the meaning of primary term values. Many theoretical studies have contributed to the computation with words and to the LH concepts (De Cock and Kerre [11]; Huynh, Ho, and Nakamori [19]; Rubin [27]; Turksen [28]).

As pointed out by Zadeh [34, 35, 36], linguistic variables and terms are closer to human thinking, (which emphasise importance more than certainty) and are used in everyday life. For this reason, words and linguistic terms can be used to model human thinking systems (Liu et al. [25]; Zadeh [31]).

József Dombi
Department of Informatics, University of Szeged, 6720 Szeged, Árpád tér 2., Hungary
e-mail: dombi@inf.u-szeged.hu

B. De Baets et al. (Eds.): Eurofuse 2011, AISC 107, pp. 53–65, 2011.
springerlink.com

Zadeh [32] said that a proposition such as "The sea is very rough" can be interpreted as "It is very true that the sea is rough." Consequently, the sentences "The sea is very rough," "It is very true that the sea is rough," "(The sea is rough) is very true" can be considered equivalent. In fact, truth function modification permits an algorithmic approach to the calculus of deduction in approximate reasoning [2], by strengthening the liaison connection with classical logic. Since in traditional prepositional logic the validity of a reasoning depends on the simple truth proof of logic propositions [5], in a fuzzy logic we have the truth values that determine the fuzzy set associated with the conclusion of a deduction [29].

Basic notions of linguistic variables were formalized in different works by Zadeh in the mid 1970s [34, 35, 36]. These papers sought to provide a mathematical model for linguistic variables.

1.1 Historical Background

Linguistic hedges (LH)

LHs are special linguistic terms by which other linguistic terms are modified. "Very", "more or less", "fairly", and "extremely" are given as examples of LHs (Jang et al. [20]). For example, A^s = "very young" is secondary linguistic that may be produced from the primary linguistic term A = "young" by using LHs (Banks [3]; Jang et al. [20]; Turksen [28]).

Representation of LHs

In the standard (canonical) fuzzy concept an LH or modifier is any operation that changes the meaning of any linguistic term (Banks [3]; Jang et al. [20]). Let A be a continuous linguistic term for an input variable x with MF $\mu_A(x)$. Then A^s is interpreted as a modified version of the original linguistic term, thus

$$A^s := (x, (\mu_A(x))^P)|x \in X, \tag{1}$$

where p denotes the linguistic hedge value of the linguistic term A. Two major modifier operations are commonly used in scientific literature. One of them is the concentration (Jang et al. [20]):

$$CON(A) := A^2. \tag{2}$$

The other is the dilution operations (Jang et al. [20]):

$$DIL(A) := A^{0.5}. \tag{3}$$

Conventionally, CON(A) and DIL(A) are the results of applying hedges "very" and "more or less" to the linguistic term A, respectively. However, there are different and constant LH definitions in the literature, such as "very very" $(p = 4)$, "quite" $(p = 1.25)$, "a little less" $(p = 0.75)$ (Banks [3]; Chatterjee and Siarry [10]; Jang et al. [20]; Turksen [28]).

In the article of Lascio [et al.] [24] they construct a mathematical model for the truth values of the Truth linguistic variable and, on a more general basis, for the values of a Boolean linguistic generic variable, which maintains the natural order relation existing between them.

Cat Ho and Wechler in [8, 9] pointed out the discrepancy between the intuitive use made in the natural language of linguistic truth values and the numerical values obtained using CON and DIL operators.

Shifting modifiers

Another type of fuzzy modifiers, called shifting modifiers, was casually suggested by Lakoff [23] in the 1970's. Hellendoorn [17] and Bouchon [6] then used it in a more formal manner.

2 Introduction to the Pliant Concept

In this section, besides the min/max and the drastic operators, we will be concerned with strict operators and we will look for the general form of $c(x,y)$ and $d(x,y)$. We assume that the following conditions are satisfied:

1. Continuity:
   $c\colon [0,1]\times[0,1]\to[0,1]$ $\qquad d\colon [0,1]\times[0,1]\to[0,1]$
2. Strict monotonous increasing:
   $c(x,y)<c(x,y')$ if $\; y<y' \quad x\neq 0$ $\qquad d(x,y)<d(x,y')$ if $\; y<y' \quad x\neq 0$
3. Compatibility with two-valued logic:
   $c(0,0)=0 \quad c(1,1)=1$ $\qquad d(0,0)=0 \quad d(1,1)=1$
   $c(0,1)=0 \quad c(1,0)=0$ $\qquad d(0,1)=1 \quad d(1,0)=1$
4. Associativity:
   $c(x,c(y,z))=c(c(x,y),z)$ $\; d(x,d(y,z))=d(d(x,y),z)$
5. Archimedean:
   $c(x,x)<x, \quad x\in(0,1)$ $\quad d(x,x)>x, \quad x\in(0,1)$

So

$$c(x,y)=f_c^{-1}(f_c(x)+f_c(y)). \tag{4}$$

Similarly, the strict t-conorm on $(0,1]\times(0,\infty]$ has the form:

$$d(x,y)=f_d^{-1}(f_d(x)+f_d(y)). \tag{5}$$

Here $f_c(x):[0,1]\to[0,\infty]$ and $(f_d(x):[0,1]\to[0,\infty])$ are continuous and strictly increasing (decreasing) monotone functions and they are the generator functions of the strict t-norms and strict t-conorms.

Those familiar with fuzzy logic theory will find that the terminology used here is slightly different from that used in standard texts [22, 7, 1, 4, 26, 16]. This is because I would like to distinguish between fuzzy logic and Pliant logic.

Definition 1. *We say that $\eta(x)$ is a negation if $\eta: [0,1] \to [0,1]$ satisfies the following conditions:*

C1: $\eta: [0,1] \to [0,1]$ is continuous (Continuity)
C2: $\eta(0) = 1$, $\eta(1) = 0$ (Boundary conditions)
C3: $\eta(x) < \eta(y)$ for $x > y$ (Monotonicity)
C4: $\eta(\eta(x)) = x$ (Involution)

From C1, C2 and C3, it follows that there exists a fix point $\nu_* \in [0,1]$ of the negation where

$$\eta(\nu_*) = \nu_* \tag{6}$$

So another possible characterization of negation is when we assign a so-called decision value ν for a given ν_0, i.e. a point (ν, ν_0) can be specified that the curve must intersect. This tells us something about how strong the negation operator is.

$$\eta(\nu) = \nu_0 \tag{7}$$

If $\eta(x)$ has a fix point ν_*, we use the notation $\eta_{\nu_*}(x)$ and if the decision value is ν, then we use the notation $\eta_\nu(x)$. If $\eta(x)$ is used without a suffix then the parameter has no importance in the proofs. Later on we will characterize the negation by the ν_*, ν_0 and ν parameters.

Definition 2. *Generalized operators based on strict t-norms and t-conorms are*

$$c(\mathbf{w},\mathbf{x}) = c(w_1,x_1;w_2,x_2;\ldots;w_n,x_n) = f_c^{-1}\left(\sum_{i=1}^{n} w_i f_c(x_i)\right), \tag{8}$$

$$d(\mathbf{w},\mathbf{x}) = d(w_1,x_1;w_2,x_2;\ldots w_n,x_n) = f_d^{-1}\left(\sum_{i=1}^{n} w_i f_d(x_i)\right), \tag{9}$$

where $w_i \geq 0$.

If $w_i = 1$ we get the t-norm and t-conorm. If $w_i = \frac{1}{n}$, then we get mean operators. If $\sum_{i=1}^{n} w_i = 1$, then we get weighted operators.

Definition 3. *The DeMorgan law holds for the generalized operator based on strict t-norms and strict t-conorms and for negation if and only if the following condition holds,*

$$c(w_1,\eta(x_1);w_2,\eta(x_2);\ldots;w_n,\eta(x_n)) = \eta(d(w_1,x_1;w_2,x_2;\ldots;w_n,x_n)), \tag{10}$$

We call this later on the generalized DeMorgan law.

Theorem 1 (General form of the negation). *We have that $c(\mathbf{w},\mathbf{x})$, $d(\mathbf{w},\mathbf{x})$ and $\eta(x)$ is a DeMorgan triple if and only if*

$$\eta(x) = f^{-1}(k(f(x))), \tag{11}$$

where $f(x) = f_c(x)$ or $f(x) = f_d(x)$ and $k(x)$ is a strictly decreasing continuous function with the property

$$k(x) = k^{-1}(x). \tag{12}$$

where $k\colon [0,\infty] \to [0,\infty]$.

Proof. See [15].

Theorem 2. *$c(x,y)$ and $d(x,y)$ build a DeMorgan system for $\eta_{\nu_*}(x)$ where $\eta_{\nu_*}(\nu_*) = \nu_*$ for all $\nu_* \varepsilon (0,1)$ if and only if*

$$f_c(x) f_d(x) = 1. \tag{13}$$

Proof. See [15].

Definition 4. *If $k(x) = 1/x$, that is*

$$f_c(x) f_d(x) = 1, \tag{14}$$

then we call the generated connectives a multiplicative Pliant system.

Theorem 3. *The general form of the multiplicative Pliant system is*

$$o_\alpha(x,y) = f^{-1}\left(\left(f^\alpha(x) + f^\alpha(y) \right)^{1/\alpha} \right) \tag{15}$$

$$\eta_\nu(x) = f^{-1}\left(f(\nu_0) \frac{f(\nu)}{f(x)} \right) \qquad or \tag{16}$$

$$\eta_{\nu_*}(x) = f^{-1}\left(\frac{f^2(\nu_*)}{f(x)} \right), \tag{17}$$

where $f(x)$ is the generator function of the strict t-norm operator and $f : [0,1] \to [0,\infty]$ is a continuous and strictly decreasing function.

Proof. See [15].

Because the generator function is determined up to a multiplicative constant, we can arrange it such that

$$f(\nu_0) = 1$$

and so

$$\eta_\nu(x) = f^{-1}\left(\frac{f(\nu)}{f(x)} \right) \tag{18}$$

If $f(\nu_0) = 1$ and if $\nu_0 = \nu$, then we get:

$$\eta(x) = f^{-1}\left(\frac{1}{f(x)}\right) \tag{19}$$

Definition 5 (Drastic negation).

$$\eta_1(x) = \begin{cases} 1 \; if \; x \neq 1 \\ 0 \; if \; x = 1 \end{cases} \quad \eta_0(x) = \begin{cases} 1 \; if \; x = 1 \\ 0 \; if \; x \neq 1 \end{cases}$$

$\eta_0(x)$ is the strictest negation, while $\eta_1(x)$ is the least strict negation
ν is the neutral value of the negation and can be interpreted as the strictness of the negation, i.e. if $\nu_1 < \nu_2$ then $\eta_{\nu_1}(x)$ is a stricter negation than $\eta_{\nu_2}(x)$

3 Modalities Induced by Two Negations

The linguistic hedges "very" or "very very" express the modal hedge necessity; and, similarly, the hedge "more or less" expresses the possibility hedge.

From this starting point, the hedges used in fuzzy logic are based on an extension of modal logic to the continuous case. We begin with the negation operator and we use two types of this operator; one that is strict, and one that is less strict. We will show that with these two negation operators we can define the modal hedges.

Modal logic, which is an area of mathematical logic, can be viewed as a logical system obtained by adding logical symbols and inference rules. From a semantics viewpoint, modal logic can also be viewed as a part of a logical system.

This issue is related in part to linguistic hedges to corresponding reverse effects and also to the modal operators with mutually reverse modal concepts. We will construct linguistic modal hedges called necessity and possibility hedges. The construction is based on the fact that modal operators can be realized by combining two kinds of negation operators, i.e. negation in the reverse sense of classical logic and negation in the reverse sense of intuitionistic logic, which is a strict negation operator.

In intuitionistic logic, another kind of negation operator also has to be taken into account because the law of Excluded Middle does not hold. $\sim_1 x$ does not imply "x is not", although "there exists a path such that x is a contradiction." In other words, $\sim_1 x$ is a stronger negation than $\sim_2 x$. Because $\sim_1 x$ in modal logic, it means "x is impossible".

$$\sim_1 x = \Box \sim_2 x \tag{20}$$

One can define the necessity hedge by $\Box x$ and the possibility hedge by $\Diamond x$, which have mutually reverse effects.

We will show that both operators belong to the same class of unary functions, and that because they have a common form in the Pliant system, we will denote both of them by $\tau_\nu(x)$. Depending on the ν value, we get the necessity hedge or the possibility hedge.

As we mentioned above, in modal logic we have two more operators than the classical logic case: namely necessity and possibility; and in modal logic there are two basic identities. These are:

$$\sim_1 x = impossible(x) = necessity(not(x)) = \Box \sim_2 x \tag{21}$$

$$\Diamond x = possible(x) = not(impossible(x)) = \sim_2 (\sim_1 x) \tag{22}$$

In our context, we model $impossible(x)$ with a stricter negation operator than $not(x)$.

In modal logic $\Box x$ means that x is necessarily valid. If we negate x, then "*necessarily not*$_1$ *x*", $\Box \sim_1$ has the meaning "*impossible*" and we suppose "*not*$_2$ *impossible*" is "*possible*", so

$$\Diamond x = \sim_2 \Box \sim_1 x. \tag{23}$$

This serves as a definition of the possibility operator.

If in Eq.(21) we replace x by $\sim_2 x$ and using the fact that $\sim_2 x$ is involutive, we get

$$\Box x = \sim_1 (\sim_2 x), \tag{24}$$

and with Eq.(23), we have

$$\Diamond x = \sim_2 (\sim_1 x). \tag{25}$$

It is also obvious in modal logic that:

$$\Box x \leq x \leq \Diamond x \tag{26}$$

In Pliant logic there are several types of negation operators [14, 13] that can be distinguished by the neutral value. If ν is small we can say that negation operator is strict; otherwise it is not strict. Using this, we can apply Eq.(24) and Eq.(25) to the Pliant concept.

Based on the above considerations, we can formally define the necessity and possibility modifiers.

Definition 6. *On the basis of Eq.(24), we get:*

$$\tau_\Box(x) = \eta_{\nu_1}(\eta_{\nu_0}^{-1}(x)) = \eta_{\nu_1}(\eta_{\nu_0}(x)), \tag{27}$$

where $\nu_1 < \nu_0$, *and* $\tau_\Box(x)$ *the necessity operator.*

We can use Eq.(16) and express $\tau_\Box(x)$ like so

$$\tau_\Box(x) = f^{-1}\left(f(\nu_0)\frac{f(\nu_1)}{f(\nu_0)f(\nu_0)}f(x)\right)$$

then we can rewrite it in the following form:

$$\tau_{\square}(x) = f^{-1}\left(f(\nu_0)\frac{f(x)}{f(\nu_{\square})}\right), \tag{28}$$

where

$$\nu_{\square} = f^{-1}\left(\frac{f(\nu_0)}{f(\nu_1)}\right)$$

Applying similar reasoning, we can get the possibility operator using Eq.(25). Now our results can be summarized by the following theorem:

Theorem 4. *Let $\tau_{\square}(x)$ and $\tau_{\Diamond}(x)$ be the necessity modifier and possibility modifier, respectively. Then on the basis of eqs.(24) and (25), we have*

$$\tau_I(x) = f^{-1}\left(f(\nu_0)\frac{f(x)}{f(\nu_I)}\right)$$

(where I= $\{\square,\Diamond\}$.)

They both have a common form where $\nu_{\Diamond} < \nu_0 < \nu_{\square}$

A more general concept can be stated as a set of definitions.

Definition 7. *A modal hedge means that*

$$\tau_{\nu_1,\nu_2}(x) = \eta_{\nu_1}\left(\eta_{\nu_2}(x)\right), \tag{29}$$

where ν_1 and ν_2 are neutral values. If $\nu_1 < \nu_2$, then $\tau_{\nu_1,\nu_2}(x)$ is a necessity operator and if $\nu_2 < \nu_1$, then $\tau_{\nu_1,\nu_2}(x)$ is a possibility operator.

From the above definition, we get

$$\tau_{\nu_1,\nu_2}(x) = f^{-1}\left(f(\nu_1)\frac{f(x)}{f(\nu_2)}\right), \tag{30}$$

This can be rewritten as

$$\tau_{\nu,\nu_0}(x) = f^{-1}\left(f(\nu_0)\frac{f(x)}{f(\nu)}\right). \tag{31}$$

We call this the general form of the hedges and in this case it is not hard to show that if $\nu_0 < \nu$, $\tau_\nu(x)$ is the necessity hedge and if $\nu < \nu_0$ it is the possibility hedge. If $f(\nu_0) = 1$, then

$$\tau_\nu(x) = f^{-1}\left(\frac{f(x)}{f(\nu)}\right). \tag{32}$$

Definition 8. *We call a necessity (possibility) hedge a dual hedge if* $v_1 = \eta(v_2)$, *i.e.*

$$v_1 = f^{-1}\left(\frac{1}{f(v_2)}\right)$$

From this, the necessity operator is

$$\tau_{\Box}(x) = f^{-1}\left(\frac{f(x)}{f(v)}\right),$$

and the corresponding possibility operator is (if $v < v_0$*)*

$$\tau_{\Diamond} = f^{-1}\left(f(v)f(x)\right).$$

If we use the definition of $\tau(x) = \Box x$ or $\Diamond x$ (i.e. necessity, or possibility x), then we can introduce different necessity and possibility operators.

$$\begin{aligned} \Box^2 x &= \Box(\Box(x)) = \tau_{\Box}(\tau_{\Box}(x)) \\ \Diamond^2 x &= \Diamond(\Diamond(x)) = \tau_{\Diamond}(\tau_{\Diamond}(x)) \end{aligned} \tag{33}$$

We will use the following notation:

- If $v > v_0$ and $f(v_0) = 1$

$$\tau_{\Box}(x) = \Box(x) = f^{-1}\left(\frac{f(x)}{f(v)}\right) = \tau_v(x) \quad \text{a necessity} \tag{34}$$

$$\tau_{\Diamond}(x) = \Diamond(x) = f^{-1}(f(v)f(x)) = \tau_{\eta(v)}(x) \quad \text{a possibility} \tag{35}$$

- If $v < v_0$

$$\tau_{\Diamond}(x) = \Diamond(x) = f^{-1}\left(\frac{f(x)}{f(v)}\right) = \tau_v(x) \quad \text{a necessity}$$

$$\tau_{\Box}(x) = \Box(x) = f^{-1}(f(v)f(x)) = \tau_{\eta(v)}(x) \quad \text{a possibility}$$

Definition 9. *We call graded modalities a k composition of the modalities.*

$$\tau_{\Box}(\tau_{\Box}(\ldots \quad \tau_{\Box}(x))) = \Box(\underbrace{\Box(\ldots \quad \Box}_{K}(x))\ldots) = \Box^K(x) \tag{36}$$

$$\tau_{\Diamond}(\tau_{\Diamond}(\ldots \quad \tau_{\Box}(x))) = \Diamond(\underbrace{\Diamond(\ldots \quad \Diamond}_{K}(x))\ldots) = \Diamond^K(x) \tag{37}$$

Definition 10

drastic necessity
$$\tau_1(x) = \Box_1(x) = \begin{cases} 1 & \text{if } x = 1 \\ 0 & \text{if } x \neq 1 \end{cases} \tag{38}$$

drastic possibility
$$\tau_0(x) = \Diamond_0(x) = \begin{cases} 0 & \text{if } x = 0 \\ 1 & \text{if } x \neq 0 \end{cases} \tag{39}$$

On the basis of eqs.(34) and (35), it is easy to verify the following properties.

4 Basic Properties of Modalities

1. $\tau_\Box(\tau_\Diamond(x)) = \Box(\Diamond(x)) = x$
2. $\tau_\Diamond(\tau_\Box(x)) = \Diamond(\Box(x)) = x$
3. $\tau_1(\tau_0(x)) = \Box_1(\Diamond_0(x)) = \Diamond_1(x)$
4. $\tau_0(\tau_1(x)) = \Diamond_0(\Box_1(x)) = \Box_0(x)$
5. $\tau_\Box^n(\tau_\Box^m(x)) = \Box^n(\Box^m(x)) = \Box^{n+m}(x)$
6. $\tau_\Diamond^n(\tau_\Diamond^m(x)) = \Diamond^n(\Diamond^m(x)) = \Diamond^{n+m}(x)$
7. $\tau_\Diamond^n(\tau_\Box^m(x)) = \Diamond^n(\Box^m(x)) = \begin{cases} \Diamond^{n-m}(x) & \text{if} \quad n-m > 0 \\ x & \text{if} \quad n = m = 0 \\ \Box^{m-n}(x) & \text{if} \quad n-m < 0 \end{cases}$
8. $\tau_\Box^n(\tau_\Diamond^m(x)) = \Box^n(\Diamond^m(x)) = \begin{cases} \Box^{n-m}(x) & \text{if} \quad n-m > 0 \\ x & \text{if} \quad n = m = 0 \\ \Diamond^{m-n}(x) & \text{if} \quad n-m < 0 \end{cases}$
9. $\lim\limits_{K\to\infty} \tau_\Box^K(x) = \tau_1(x)$
10. $\lim\limits_{K\to\infty} \tau_\Diamond^K(x) = \tau_0(x)$

4.1 Limit of Modalities

Here we introduce the drastic modality operator:

$$\tau_1(x) = \begin{cases} 1 & \text{if } x = 1 \\ 0 & \text{if } x < 1 \end{cases}$$

and

$$\tau_0(x) = \begin{cases} 0 & \text{if } x = 0 \\ 1 & \text{if } x > 0 \end{cases}$$

Theorem 5. *Let*

$$\tau_{\nu,\nu_0}(\tau_{\nu,\nu_0}(\tau_{\nu,\nu_0}\dots\tau_{\nu,\nu_0}(x))) = \underbrace{\tau\circ\tau\circ\dots\tau\circ\tau}_{M}(x) = \tau^M(x)$$

If $\nu_0 < \nu$ *then:*

$$\lim_{M\to\infty}\tau^M(x) = \tau_1(x)$$

If $\nu_0 > \nu$ *then:*

$$\lim_{M\to\infty}\tau^M(x) = \tau_0(x)$$

Proof. Because

$$\tau^M(x) = f^{-1}\left(f(\nu_0)\frac{f(x)}{f(\nu)}\right), \tag{40}$$

where

$$\nu_M = f^{-1}\left(f(\nu_0)\left(\frac{f(\nu)}{f(\nu_0)}\right)^M\right) \tag{41}$$

If $\nu < \nu_0$ then $f(\nu_0) > f(\nu)$

$$\lim_{M\to\infty}\nu_M = 0 \tag{42}$$

In a similar way, we can get the result when $\nu_0 < \nu$.

If $\nu = \nu_0$, then $\nu_M = \frac{1}{2}$.

5 Conclusion

In this article we give a general and theoretical basis for modalities. Here we define the necessity and possibility operators and we define them by a generator function of the Pliant operators. The operator system is so called Pliant system because the construction is based on different negation operations and in the Pliant system infinitely many negations are consistent with the DeMorgan identity.

Acknowledgements. This study was partially supported by the TAMOP-4.2.1/B-09/1/KONV- 2010-0005 program of the Hungarian National Development Agency.

References

1. Alsina, C., Schweizer, B., Frank, M.J.: Associative functions: triangular norms and copulas. Word Scientific Publishing, Singapore (2006)
2. Baldwin, J.F.: A new approach to approximate reasoning using a fuzzy logic. Fuzzy Sets and Systems 2, 309–325 (1979)
3. Banks, W.: Mixing crisp and fuzzy logic in applications. In: WESCON 1994 Idea/microelectronics Conference Record, Anaheim, CA, pp. 94–97 (1994)

4. Beliakov, G., Pradera, A., Calvo, T.: Aggregation Functions: A Guide for Practitioners. Studies in Fuzziness and Soft Computing, vol. 221. Springer, Heidelberg (2007)
5. Bergmann, M., Moor, J., Nelson, J.: Logic book. McGraw-Hill, New York (1990)
6. Bouchon-Meunier, B.: La Logique Floue. Que sais-je? vol. 2702, Paris (1993)
7. Calvo, T., Mayor, G., Mesiar, R.: Aggregation Operators. New Trends and Applications Studies in Fuzziness and Soft Computing, vol. 97 (2002)
8. Cat Ho, N., Wechler, W.: Hedge algebras: An algebraic approach to structure of sets of linguistic truth values. Fuzzy Sets and Systems 35, 281–293 (1990)
9. Cat Ho, N., Wechler, W.: Extended hedge algebras and their application to fuzzy logic. Fuzzy Sets and Systems 52, 259–281 (1992)
10. Chatterjee, A., Siarry, P.: A PSO-aided neuro-fuzzy classifier employing linguistic hedge concepts. Expert Systems with Applications 33(4), 1097–1109 (2007)
11. De Cock, M., Kerre, E.E.: Fuzzy modifiers based on fuzzy relations. Information Sciences 160, 173–199 (2004)
12. Cox, E.: The fuzzy systems handbook: a practitioner's guide to building, using, and maintaining fuzzy systems. Academic Press Professional, San Diego (1994)
13. Dombi, J., Gera, Z.: On aggregative operators. Information Science, under Review Process
14. Dombi, J.: DeMorgan systems with infinite number of negation. Information Science (appears in 2011)
15. Dombi, J.: DeMorgan systems with an infinitely many negations in the strict monotone operator case. Information Sciences (accepted 2011)
16. Grabisch, M., Marichal, J.-L., Mesiar, R., Pap, E.: Aggregation Functions. In: Encyclopedia of Mathematics and Its Applications, vol. 127. Cambridge University Press, Cambridge (2009)
17. Hellendoorn, H.: Reasoning with fuzzy logic. Ph.D. Thesis, T.U. Delft (1990)
18. Horikawa, S., Furuhashi, T., Uchikawa, Y.: A new type of fuzzy neural network based on a truth space approach for automatic acquisition of fuzzy rules with linguistic hedges. International Journal of Approximate Reasoning 13, 249–268 (1995)
19. Huynh, V.N., Ho, T.B., Nakamori, Y.: A parametric representation of linguistic hedges in Zadeh's fuzzy logic. International Journal of Approximate Reasoning 30, 203–223 (2002)
20. Jang, J.S.R., Sun, C.T., Mizutani, E.: Neuro-fuzzy and soft computing. Prentice Hall, Upper Saddle River (1997)
21. Kerre, E.E., De Cock, M.: Linguistic modifiers: an overview. In: Chen, G., Ying, M., Cai, K.-Y. (eds.) Fuzzy Logic and Soft Computing, pp. 69–85. Kluwer Academic Publishers, Dordrecht (1999)
22. Klement, E.P., Mesiar, R., Pap, E.: Triangular norms. Kluwer, Dordrecht (2000)
23. Lakoff, G.: Hedges: A study in meaning criteria and the logic of fuzzy concepts. Journal of Philosophical Logic 2, 458–508 (1973)
24. Di Lascio, L., Gisolfi, A., Loia, V.: A new model for linguistic modifiers. International Journal of Approximate Reasoning 15(1), 25–47 (1996)
25. Liu, B.D., Chen, C.Y., Tsao, J.Y.: Design of adaptive fuzzy logic controller based on linguistic-hedge concepts and genetic algorithms. IEEE Transactions on Systems Man and Cybernetics, Part B 31(1), 32–53 (2001)
26. Mesiar, R., Kolesrov, A., Calvo, T., Komornkov, M.: A Review of Aggregation Functions. In: Fuzzy Sets and Their Extension: Representation, Aggregation and Models. Studies in Fuzziness and Soft Computing, vol. 220 (2008)
27. Rubin, S.H.: Computing with words. IEEE Transactions on Systems Man and Cybernetics, Part B 29(4), 518–524 (1999)

28. Trksen, I.B.: A foundation for CWW: Meta-linguistic axioms. In: IEEE Fuzzy Information, Processing NAFIPS 2004, pp. 395–400 (2004)
29. Yager, R.R.: Approximate reasoning as a basis for rule-based expert-systems. IEEE Trans. Systems Man Cybernet, SMC 14, 636–643 (1984)
30. Zadeh, L.A.: The role of fuzzy logic in the management of uncertainty in expert systems. Fuzzy Sets and Systems 8(3), 199–227 (1983)
31. Zadeh, L.A.: Quantitative fuzzy semantics. Information Sciences 3, 159–176 (1971)
32. Zadeh, L.A.: A fuzzy-set - theoretic interpretation of linguistic hedges. Journal of Cybernetics 2(3), 4–34 (1972)
33. Zadeh, L.A.: Fuzzy logic = computing with words. IEEE Trans. Fuzzy Systems 4, 103–111 (1996)
34. Zadeh, L.A.: The concept of a linguistic variable and its application to approximate reasoning, parts 1. Information Sciences 8, 199–249 (1975)
35. Zadeh, L.A.: The concept of a linguistic variable and its application to approximate reasoning, parts 2. Information Sciences 8, 301–357 (1975)
36. Zadeh, L.A.: The concept of a linguistic variable and its application to approximate reasoning, parts 3. Information Sciences 9, 43–80 (1975)
37. Zadeh, L.A.: From computing with numbers to computing with words-From manipulation of measurements to manipulation of perceptions. IEEE Transactions on Circuits and Systems-I: Fundamental Theory and Applications 45(1), 105–119 (1999)

On the Properties of Probabilistic Implications

Przemysław Grzegorzewski

Abstract. A new family of implication operators, called probabilistic implications, are discussed. The suggested implications are based on conditional copulas and make a bridge between probability theory and fuzzy logic. It is shown that probabilistic fuzzy implications have some interesting properties, especially those connected with the dependence structure of the underlying environment. Therefore, it seems that probabilistic implications might be a useful tool in approximate reasoning, knowledge extraction and decision making.

1 Introduction

Multi-valued implications, known in the literature as fuzzy implications, are very interesting not only from the theoretical point on view but also because of many diverse applications. Firstly, they play a key role in approximate reasoning and fuzzy control. For instance, fuzzy rules of inference are modelled by fuzzy implications. As a typical example let us consider a generalization of the classical modus ponens which states that given two true propositions (premises): A and conditional claim $A \to B$, the consequent B is also true. Using IF-THEN rules we may express modus ponens by the following scheme:

$$\begin{array}{ll} \text{Rule:} & \text{IF } X \text{ is } A\text{, THEN } Y \text{ is } B \\ \text{Fact:} & X \text{ is } A \\ \text{Conclusion:} & Y \text{ is } B \end{array} \quad (1)$$

To perform approximate reasoning traditional inference schemas were appropriately generalized. In particular, the classical modus ponens have been extended to

Przemysław Grzegorzewski
Faculty of Mathematics and Information Science, Warsaw University of Technology,
Plac Politechniki 1, 00-661 Warsaw, Poland and Systems Research Institute,
Polish Academy of Sciences, Newelska 6, 01-447 Warsaw, Poland
e-mail: pgrzeg@ibspan.waw.pl

B. De Baets et al. (Eds.): Eurofuse 2011, AISC 107, pp. 67–78, 2011.
springerlink.com

fuzzy logic under the inference pattern called generalized modus ponens which may be expressed in the following way:

$$\begin{array}{ll} \text{Rule:} & \text{IF } X \text{ is } A, \text{ THEN } Y \text{ is } B \\ \text{Fact:} & X \text{ is } A' \\ \text{Conclusion:} & Y \text{ is } B' \end{array} \tag{2}$$

where A, A', B and B' are fuzzy sets. According to (2) scheme the conclusion B' might be done by

$$B'(y) = \sup_{x \in X} T\left(A'(x), I(A(x), B(y))\right),$$

where I is a fuzzy implication and T stands for a t-norm (see, e.g. [11]).

Further on fuzzy implications are applied as fuzzy subsethood measures. For example, one may define a measure of subsethood of A in B as

$$S(A,B) = \inf_{x \in X} I(A(x), B(x)),$$

where I is, as above, a fuzzy implication (see [3]).

One can specify other interesting applications of fuzzy implications like similarity-based reasoning [16], fuzzy mathematical morphology [4], fuzzy relation equations [5], computing with words [17], etc. As a result we have obtained a broad theory of fuzzy implications (see [1]).

Our knowledge of complex systems is often incomplete. It is due not only to general complexity of these systems but often is also entailed by the deficiency of our measuring instruments, including our senses. Therefore, we have to rely rather on perceptions than on precise data and on imperfect knowledge delivered by expert's statements usually formulated not in mathematical terms but in natural language. However, one has to be aware that uncertainty appears in many systems not only because of imprecision but is rather an immanent effect of randomness. Moreover, in most cases we have to cope with imperfect knowledge which abounds with both kinds of uncertainty: imprecision and randomness. Hence our inference based on if-then rules should also comprise with these two sources of uncertainty. Therefore, another type of implication that takes into consideration both imprecision modelled by fuzzy concepts and randomness described by tools originated in probability theory would be desirable. A construction of such concept, called probabilistic implication, was suggested in [9]. Now it is time to explore more deeply the properties of probabilistic implications. And this is the main goal of the present contribution.

The paper is organized as follows. In Sec. 2 we recall some information on fuzzy implications. Some remarks how to combine probability and implication are given in Sec. 3, while basic information on copulas are recalled in Sec. 4. Then, in Sec. 5 we present probabilistic implication and discuss when they are also fuzzy implications. Further on we explore the relation between probabilistic implications and the dependence structure in underlying stochastic environment.

2 Fuzzy Implications

In classical logic the implication can be defined in different ways depending on the mathematical framework, physical nature or philosophical background in which it appears. Anyway, despite this circumstances the so-called truth tables have to be and really are identical. It is not surprising that starting from different classical definitions and using diversity of methods we may obtain a great number of fuzzy implications. Anyway all of them should satisfy some basic requirements that can be perceived as a generalization of the classical truth table. These fundamental properties form the definition of fuzzy implication. Below we recall the definition given in [1] (also equivalent to those proposed in [6] or [10]).

Definition 1. *A function* $I : [0,1]^2 \to [0,1]$ *is called a fuzzy implication if it satisfies the following conditions for all* $x, x_1, x_2, y, y_1, y_2 \in [0,1]$
(I1) if $x_1 \leq x_2$ *then* $I(x_1,y) \geq I(x_2,y)$
(I2) if $y_1 \leq y_2$ *then* $I(x,y_1) \leq I(x,y_2)$
(I3) $I(0,0) = 1$
(I4) $I(1,1) = 1$
(I5) $I(1,0) = 0$.

Note that the property (I1) means that $I(\cdot,y)$ is nonincreasing and captures the left antitonicity of the function I, i.e. a decrease in truth value of the antecedent increases its efficacy to state more about the truth value of its consequent. The axiom (I2) shows that $I(x,\cdot)$ is nondecreasing which reflects the right isotonicity of the overall truth value as a direct function of the consequent. Finally, axioms (I3) – (I5) correspond to basic properties of the classical implication. For more details on fuzzy implications we refer the reader to [1].

3 Implication and Probability

It is worth noting that the imprecision is responsible only for aspect of uncertainty. It seems that the second main source of problems is connected with randomness. These two main grounds of uncertainty may appear together or alone but they are themselves the outputs of quite different mechanisms and hence are described and analyzed using different methodology. Thus now let us have a look on implication from the probabilistic perspective.

Roughly speaking, according to modus ponens inferential rule we may expect that provided A is "surely" true and the implication $A \to B$ is also "surely" true then B is true. However, in practice we often cannot be completely sure that given event surely occur. In many situations instead of determined premises we have only some confidence that they are true. In other words we may neither be completely sure that A nor that the the implication $A \to B$ is 100% true. We can only estimate the probability $P(A)$ that A is true and the probability $P(A \to B)$ that the implication $A \to B$ holds. Anyway, due to appropriate probabilistic inference we can transform these two probabilities into desired conclusion on the probability related to B. However, the particular inference depends on the interpretation of the probability of an implication. It seems that the most natural approach is to interpret the probability of an implication as the conditional probability $P(B|A)$, i.e.

$$P(B|A) = \frac{P(B \cap A)}{P(A)}, \tag{3}$$

where A should not be impossible, i.e. $P(A) > 0$. Hence, if we know the probability $P(A)$ of the premise A and the probability $P(B|A)$ of the implication then, due to that formula, we can find the probability $P(B \cap A)$ that both B and A are true, i.e.

$$P(B \cap A) = P(B|A) \cdot P(A).$$

Applying more sophisticated Bayesian analysis we may develop that inference. However, although conditional probability given by (3) provides a natural and convenient interpretation of the implication it cannot be treated as an implication operator. Therefore we are still looking for the adequate extension of the classical implication operator into probabilistic environment. In the next section we will show how to construct a probabilistic implication operator which is actually a fuzzy implication.

Let us also mention that the conditional probability is not the only way that combines implication with randomness. The other formalization corresponding to "material implication" interprets the probability of implication $A \to B$ as $P(B \text{ or } \neg A)$. For the comparison of these two probabilistic perspectives we refer the reader to [13]. The overview of probability of implication applications in artificial intelligence can be found e.g. in [7, 8].

4 Copulas

One of the reasons that (3) cannot serve directly for defining an implication operator is that its domain is a σ-algebra of the family of all elementary events corresponding to an experiment under study and may vary from one experiment to the other. Thus first of all, when trying to suggest a proper implication operator, we have to get rid of any particular experiment. In probability theory it is usually accessible by the use of a random variables that transform any particular space of elementary events into the real line. Since our goal is to reduce the domain of the desired operator into the unit square we will base our construction on the well-known object called a copula. Let us recall briefly its definition.

Definition 2. *A copula is a function $C : [0,1]^2 \to [0,1]$ which satisfies the following conditions:*

(a) $C(u,0) = C(0,v) = 0$ for every $u, v \in [0,1]$
(b) $C(u,1) = u$ for every $u \in [0,1]$
(c) $C(1,v) = v$ for every $v \in [0,1]$
(d) for every $u_1, u_2, v_1, v_2 \in [0,1]$ such that $u_1 \le u_2$ and $v_1 \le v_2$

$$C(u_2,v_2) - C(u_2,v_1) - C(u_1,v_2) + C(u_1,v_1) \geq 0. \tag{4}$$

Several interesting families of copulas, like Fŕechet's family, Farlie-Gumbel-Morgenstern's family, Marshall-Olkin's family, etc., are considered in the literature. It can be shown that every copula is bounded by the so-called Fŕechet-Hoeffding bounds, i.e. for any copula C and for all $u,v \in [0,1]$

$$W(u,v) \leq C(u,v) \leq M(u,v), \tag{5}$$

where

$$W(u,v) = max\{u+v-1,0\}, \tag{6}$$
$$M(u,v) = min\{u,v\} \tag{7}$$

are also copulas. For more information on copulas we refer the reader to the famous monograph by Nelsen [12].

The importance of the copulas is clarified by the Sklar theorem [15] showing that copulas link joint distribution functions to their one-dimensional margins.

Theorem 1. *Let X and Y be random variables with joint distribution function H and marginal distribution functions F and G, respectively. Then there exists a copula C such that*

$$H(x,y) = C(F(x),G(y)) \tag{8}$$

for all $x,y \in \mathbb{R}$. If F and G are continuous, then C is unique. Otherwise, the copula C is uniquely determined on $Ran(F) \times Ran(G)$. Conversely, if C is a copula and F and G are distribution functions, then the function H defined by (8) is a joint distribution function with margins F and G.

The Sklar theorem shows that we can disregard the particular domains of the random variables under study and transform the whole problem into the unit square which is actually our goal. Actually, C captures all the information about the dependence among the components of (X,Y). Now we have to introduce a conditional structure for copulas. Namely, to describe by the copula method the conditional distribution function of Y given $X \leq x$ such that $P(X \leq x) > 0$ we get

$$P(Y \leq y | X \leq x) = \frac{P(X \leq x, Y \leq y)}{P(X \leq x)} = \frac{H(x,y)}{F(x)} = \frac{C(F(x),G(y))}{F(x)} = \frac{C(u,v)}{u}, \tag{9}$$

where $u = F(x)$ and $v = G(y)$. An this very last concept forms the basis of our proposal for the extension of the logical implication into that probabilistic framework.

Further on let us adopt the convention that $\frac{0}{0} = 1$.

5 Probabilistic Implication Operator

Having in mind our discussion from the previous sections, especially eq. (9), let us consider a function $I_C : [0,1]^2 \to [0,1]$ defined by the following formula

$$I_C(u,v) = \frac{C(u,v)}{u}, \tag{10}$$

where C is any given copula. One may ask whether (10) is correctly defined for $u=0$. However, by the Def. 2 $C(0,v)=0$ for every $v \in [0,1]$ and hence, due to our convention that $\frac{0}{0}=1$, we get $I_C(0,v)=1$. So (10) is correctly defined. Anyway, to avoid any unnecessary discussions further on we will use the following definition (see [9]):

Definition 3. *A function* $I_C : [0,1]^2 \to [0,1]$ *given by*

$$I_C(u,v) = \begin{cases} 1 & \text{if } u=0 \\ \frac{C(u,v)}{u} & \text{if } u>0, \end{cases} \tag{11}$$

where C is a copula, is called a ***probabilistic implication*** *(based on copula C).*

Note, that a function $I_C : [0,1]^2 \to [0,1]$ defined by (11) for any copula C satisfies the following conditions:
(i) $I(0,0)=1$
(ii) $I(1,1)=1$
(iii) $I(1,0)=0$
(iv) if $y_1 \leq y_2$ then $I(x,y_1) \leq I(x,y_2)$
but may not be a fuzzy implication. Actually, one can shown such copula that condition (I1) does not hold. As an example consider the lower Fréchet-Hoeffding bound (6), i.e. $C=W$. Suppose $u_1 = 0.3 < u_2 = 0.5$ and $v=0.9$. Then $I_W(u_1,v) = \frac{2}{3} < I_W(a_2,v) = \frac{4}{5}$ which shows that I_W is not nonincreasing and hence $I_W \notin \mathcal{FI}$.

However, if for every $u_1,u_2,v \in [0,1]$ such that $u_1 \leq u_2$ a copula C satisfies

$$C(u_1,v)u_2 \geq C(u_2,v)u_1 \tag{12}$$

then the probabilistic implication I_C given by (11) is a fuzzy implication (see [9]). Further on probabilistic implications obtained from copulas satisfying condition (12) will be called *probabilistic fuzzy implications*.

Taking into account the whole discussion given above we may conclude that probabilistic fuzzy implications make a bridge between probability and the theory of fuzzy implication operators. Here are some examples of probabilistic fuzzy implications.

It is also worth noting that choosing some specific copulas we may obtain well known fuzzy implications. For example, let us consider the upper Fréchet-Hoeffding bound M given by (7). The probabilistic implication based on $M(u,v)$ is

$$I_M(u,v) = \frac{\min\{u,v\}}{u} = \begin{cases} 1 & \text{if } u \leq v \\ \frac{v}{u} & \text{if } u>v \end{cases} \tag{13}$$

which is nothing else than the Goguen implication, i.e. $I_M(u,v) = I_{GG}(u,v)$.

Let us now consider the product copula $\Pi(u,v)=uv$, which characterizes independent random variables when the distribution functions are continuous. The probabilistic implication based on the product copula is

$$I_{\Pi}(u,v) = \begin{cases} 1 \text{ if } u = 0 \\ v \text{ if } u > 0. \end{cases} \tag{14}$$

One can easily check that the product copula satisfies (12) so the probabilistic implication (14) based on the product copula is a fuzzy implication, i.e. $I_{\Pi} \in \mathscr{F}\mathscr{I}$ (and is known in the literature as the least (S,N)-implication, see [1], p. 57). As it is known (see [1]), if $I \in \mathscr{F}\mathscr{I}$ and $J \in \mathscr{F}\mathscr{I}$ then $I \vee J \in \mathscr{F}\mathscr{I}$. Hence, taking the Rescher implication I_{RS} (see [14])

$$I_{RS}(u,v) = \begin{cases} 1 \text{ if } u \leq v \\ 0 \text{ if } u > v \end{cases} \tag{15}$$

we get $I_{\Pi} \vee I_{RS} \in \mathscr{F}\mathscr{I}$, where

$$I_{\Pi} \vee I_{RS}(u,v) = \begin{cases} 1 \text{ if } u \leq v \\ v \text{ if } u > v. \end{cases} \tag{16}$$

One may easily notice that formula (16) is the famous Gödel implication, i.e. $I_{\Pi} \vee I_{RS}(u,v) = I_{GD}(u,v)$.

As we have mentioned above condition (12) assures that $I_C(\cdot,v)$ is nonincreasing so (I1) is satisfied and I_C is a fuzzy implication. However, sometimes verifying (12) may be tedious. Then the following equivalent criteria may be useful.

Lemma 1. *For any copula C function I_C given by (11) is a fuzzy implication if and only if for almost all u*

$$I_C(u,v) \geq \frac{\partial C(u,v)}{\partial u}. \tag{17}$$

Proof. Each function I_C given by (11) satisfies requirements (I2)-(I5) specified in Def. 1 for any copula C. Hence, it is enough to prove that condition (17) is equivalent to (12), i.e. (17) assures that $I_C(\cdot,v)$ is nonincreasing.

First of all let us notice that the existence of the partial derivative $\frac{\partial C(u,v)}{\partial u}$ is obvious because monotone functions are differentiable almost everywhere. Function $I_C(\cdot,v)$ is nonincreasing if and only if $\frac{\partial I_C(u,v)}{\partial u} \leq 0$. However, simple calculations show that for $u > v$ we get

$$\frac{\partial I_C(u,v)}{\partial u} = \frac{\partial}{\partial u}\left(\frac{C(u,v)}{u}\right) = \frac{1}{u^2}\left(u\frac{\partial C(u,v)}{\partial u} - C(u,v)\right).$$

Hence $\frac{\partial I_C(u,v)}{\partial u} \leq 0$ if and only if $u\frac{\partial C(u,v)}{\partial u} - C(u,v) \leq 0$, i.e. $\frac{\partial C(u,v)}{\partial u} \leq \frac{C(u,v)}{u}$ which is equivalent to (17). □

Example 1. To illustrate possible application of Lemma 1 let us consider the Farlie-Gumbel-Morgenstern family of copulas given by

$$C_{\theta}(u,v) = uv + \theta uv(1-u)(1-v). \tag{18}$$

Here, parameter $\theta \in [-1,1]$ is responsible for the dependence structure. In particular, for $\theta = 0$ we obtain the product copula. Indeed, $C_{\theta}(u,v)|_{\theta=0} = uv = \Pi(u,v)$.

Now let us check whether Farlie-Gumbel-Morgenstern's copulas with any parameter θ also lead to fuzzy implications. Thus let us consider

$$I_{FGM(\theta)}(u,v) = = \begin{cases} 1 & \text{if } u \leq v \\ v + \theta v(1-u)(1-v) & \text{if } u > v \end{cases}$$

For $u > v$ we get

$$\begin{aligned} I_{FGM(\theta)}(u,v) - \frac{\partial C_\theta(u,v)}{\partial u} &= v + \theta v(1-u)(1-v) \\ &- (v + \theta v(1-u)(1-v) - \theta uv(1-v)) = \theta uv(1-v). \end{aligned}$$

It is obvious that $\theta uv(1-v) \geq 0$ for almost all u if and only if $\theta \geq 0$. Therefore function (19) is a probabilistic implication based on the Farlie-Gumbel-Morgenstern copula not for all possible values of parameter θ but only for $\theta \geq 0$. □

6 Probabilistic Implication and the Dependence Structure

Probabilistic implications might be considered and developed in a formal way without any reference to probability theory, just as a particular subfamily of fuzzy implications. However, it seems that such perspective is too restrictive. Taking into account the whole possible probabilistic context we may get a deeper insight into the nature of these implications. Then better understanding would improve our ability for modelling real life processes and solving various practical problems. It is so, especially that the proposed probabilistic implications make a bridge between probability and the theory of fuzzy implication operators.

Identifying and describing dependence is one of the key problems in data analysis, statistics and their applications. Whereas independence is defined simply and unequivocal in probability (relevant join distribution is a product of the corresponding marginals), the dependence appears just as a lack of independence. But actually, we have to do with dependence of different kinds. For example, we say that two rv's X and Y are linearly dependent if and only if there exist such two real numbers β_0 and β_1 that $P(Y = \beta_0 + \beta_1 X) = 1$. The well known Pearson's correlation coefficient is an useful tool for measuring the strength of the linear dependence. On the other hand the notion of positive dependence of two rv's X and Y have been introduced to describe the property that large values of X go together with large values of Y, while small values of X go together with small values of Y, respectively.

Below we recall some basic types of dependence (see [2]).

Definition 4. Let X and Y be random variables. Then

- X and Y are positively quadrant dependent (PQD) if for all $(x,y) \in \mathbb{R}^2$

$$P(X \leq x, Y \leq y) \geq P(X \leq x)P(Y \leq y).$$

- Y is left tail decreasing in X [which is denoted by $LTD(Y|X)$] if $P(Y \leq y|X \leq x)$ is a nonincreasing function of x for all y.

- Y is right tail increasing in X [which is denoted by $RTI(Y|X)$] if $P(Y > y|X > x)$ is a nondecreasing function of x for all y.
- Y is stochastically increasing in X [which is denoted by $SI(Y|X)$] if $P(Y > y|X = x)$ is a nondecreasing function of x for all y.
- X and Y having join density function(or, in the discrete case, join probability mass function) $h(x,y)$ are totally positive of order 2 [which is denoted by TP_2] if for all $x_1, x_2, y_1, y_2 \in \overline{\mathbb{R}}$ such that $x_1 \leq x_2$ and $y_1 \leq y_2$

$$\mathrm{h}(\mathrm{x}_1,\mathrm{y}_1)\mathrm{h}(\mathrm{x}_2,\mathrm{y}_2) \geq \mathrm{h}(\mathrm{x}_2,\mathrm{y}_1)\mathrm{h}(\mathrm{x}_1,\mathrm{y}_2).$$

As it is known, copulas are very useful in modelling and investigating dependence. One of the reasons is that copulas describe several dependence properties of a random variables independently of their marginal behavior. Thus the above dependence concepts may be described (at least when the involved rv's X and Y are continuous) as properties of the corresponding copula C. The proof of the following theorem is straightforward.

Theorem 2. *Let X and Y be continuous random variables with copula C. Then*

- *X and Y are PQD iff $C(u,v) \geq uv$ for every $(u,v) \in [0,1]^2$;*
- *Y is $LTD(Y|X)$ iff $\frac{C(u,v)}{u}$ is nonincreasing in u for any $v \in [0,1]$;*
- *Y is $RTI(Y|X)$ iff $\frac{v-C(u,v)}{1-u}$ is nonincreasing in u for any $v \in [0,1]$;*
- *X and Y are TP_2 iff $C(u_1,v_1)C(u_2,v_2) \geq C(u_2,v_1)C(u_1,v_2)$ for all $u_1,u_2,v_1,v_2 \in [0,1]$ such that $u_1 \leq u_2$ and $v_1 \leq v_2$;*
- *Y is $SI(Y|X)$ iff for any $v \in [0,1]$ and for almost all $u \in [0,1]$ function $\frac{\partial C(u,v)}{\partial u}$ is nonincreasing in u.*

The reader can conclude immediately that the second property is of special importance for our considerations since it coincides with the criteria for the probabilistic implication to be a fuzzy implication.

As it is known, there are relationships among dependence concepts mentioned above. More precisely, we may arrange these notions into the following hierarchy:

$$TP_2 \Longrightarrow SI(Y|X) \Longrightarrow LTD(Y|X) \Longrightarrow PQD \tag{19}$$

and

$$TP_2 \Longrightarrow SI(Y|X) \Longrightarrow RTI(Y|X) \Longrightarrow PQD.$$

For the proof we refer the reader to [2].

Having in mind this short review of the dependence concepts we go back to probabilistic implications. First of all let us adopt the following definition.

Definition 5. *Let X and Y denote two random variables with copula C. Then a probabilistic implication I_C based on the copula C will be called as generated by random variables X and Y.*

Now we can combine some concepts originated in probability theory with those related to theory of implications. In particular, we get immediately quite useful criteria to check whether given probabilistic implication is a fuzzy implication.

Theorem 3. *A probabilistic implication I_C is a fuzzy implication if and only if it is generated by X and Y such that Y is left tail decreasing in X.*

The proof is a straightforward consequence of Th. 2. Thus, by (19), it is easy to prove the following properties:

Theorem 4. *If X and Y are totally positive of order 2 then the probabilistic implication generated by X and Y is a fuzzy implication.*

Theorem 5. *If Y is stochastically increasing in X then the probabilistic implication generated by X and Y is a fuzzy implication.*

To illustrate the possible usefulness of the above theorems let us consider an example.

Example 2 Consider and operator $I : [0,1]^2 \to [0,1]$ defined as follows:

$$\begin{aligned} I(u,v) &= \Big[1+(\theta-1)(u+v) \\ &- \sqrt{[1+(\theta-1)(u+v)]^2-4uv\theta(\theta-1)}\Big] \cdot [2u(\theta-1)]^{-1} \end{aligned} \tag{20}$$

for $u > v$ and $I(u,v) = 1$ for $u \le v$, where $\theta > 0$ and $\theta \ne 1$. One can easily notice that $uI(u,v) = C_\theta(u,v)$, where

$$\begin{aligned} C_\theta(u,v) &= \Big[1+(\theta-1)(u+v) \\ &- \sqrt{[1+(\theta-1)(u+v)]^2-4uv\theta(\theta-1)}\Big] \cdot [2(\theta-1)]^{-1} \end{aligned}$$

is the Plackette family of copulas indexed by the parameter θ. Since

$$\frac{\partial^2 C_\theta(u,v)}{\partial u^2} = \frac{-2\theta(\theta-1)v(1-v)}{\left([1+(\theta-1)(u+v)]^2-4uv\theta(\theta-1)\right)^{3/2}}$$

we may conclude that $\frac{\partial^2 C_\theta(u,v)}{\partial u^2} < 0$ if and only if $\theta > 1$, and hence the copula $C_\theta(u,v)$ is a concave function for $\theta > 1$. Now, let us recall the following lemma (see [12], p.197):

Lemma 2. *Let X and Y be continuous random variables with copula C. Then Y is stochastically increasing in X if and only if $C(u,v)$ is a concave function of u for any $v \in [0,1]$.*

Therefore, by Lemma 2 and Th. 5 the operator (20) is a fuzzy implication. □

The last dependence concept we want to mention here is related to the so-called tail dependence.

Definition 6. *Let X and Y be continuous random variables with cumulative distribution functions F and G, respectively. The lower tail dependence parameter λ_L is given by*

$$\lambda_L = \lim_{t\to 0^+} P\left(Y \le G^{-1}(t) | X \le F^{-1}(t)\right). \tag{21}$$

If $\lambda_L = 0$ then X and Y are called *asymptotically independent in the lower tail.* Thus we get the following theorem.

Theorem 6. *If I_C denotes a probabilistic implication generated by continuous random variables X and Y then X and Y are asymptotically independent in the lower tail.*

Proof. Let $C(u,v) = uI_C(u,v)$ denote a copula corresponding to X and Y. By (21) we get

$$\begin{aligned}\lambda_L &= \lim_{t\to 0^+} P\left(Y \le G^{-1}(t) | X \le F^{-1}(t)\right) = \lim_{t\to 0^+} P\left(G(Y) \le t | F(X) \le t\right) \\ &= \lim_{t\to 0^+} \frac{C(t,t)}{t} = 0,\end{aligned}$$

which means that X and Y are asymptotically independent in the lower tail. □

Another interesting application of the knowledge of the dependence structure background of the probabilistic implications is related to ordering. As we know, in the family of all fuzzy implications we can consider the partial order, i.e. I_2 is said to be stronger than I_1 (which is denoted as $I_2 \succ I_1$) if $I_2(u,v) \ge I_1(u,v)$ for all $u,v \in [0,1]$. It appears that for probabilistic implications relation $I_2 \succ I_1$ is equivalent to the appropriate relation between copulas, i.e. $C_2 \succ_{PDF} C_1$, which means that $C_2(u,v) \ge C_1(u,v)$ for any $(u,v) \in [0,1]^2$. As an example let us consider two Farlie-Gumbel-Morgenstern's implications given by (19) $I_{FGM(\theta_1)}$ and $I_{FGM(\theta_2)}$, where $\theta_1, \theta_2 \ge 0$. By (18) it is obvious that $I_{FGM(\theta_2)}$ is stronger than $I_{FGM(\theta_1)}$, i.e. $I_{FGM(\theta_2)} \succ I_{FGM(\theta_1)}$, if and only if $\theta_2 \ge \theta_1$. The following theorem may be also of interest in this context.

Theorem 7. *Let κ be a measure of concordance. If $C_2 \succ_{PDF} C_1$, where C_2 and C_1 are copulas corresponding to random variables (X_2,Y_2) and (X_1,Y_1), respectively, then $\kappa(X_2,Y_2) \ge \kappa(X_1,Y_1)$.*

The most popular concordance measures are Spearman's rho or Kendall's τ. As an immediate consequence of this theorem and the discussion given above we get

Theorem 8. *Let I_2 and I_1 denote probabilistic fuzzy implications generated by random variables (X_2,Y_2) and (X_1,Y_1), respectively. If I_2 is stronger than I_1 then X_2 and Y_2 are more concordant than X_1 and Y_1.*

7 Conclusions

Probabilistic implications gives a promising link from probability to theory of fuzzy implications that might be useful in approximate reasoning. They have many interesting properties including straightforward relationship with the dependence structure between corresponding random variables. It seems that the awareness of the relationship shown in this paper might be twofold profitable: knowing more about underlying stochastic environment we get a deeper insight into the reasoning schemes modelled with help of corresponding fuzzy implication and, on the other hand, some results obtained for fuzzy implications may also be fruitful for examining and interpreting the behavior of some stochastic events. Moreover, probabilistic implications form valuable tools integrating two sources of uncertainty: imprecision and randomness.

References

1. Baczynski, M., Jayaram, B.: Fuzzy Implications. Springer, Heidelberg (2008)
2. Barlow, R.E., Proschan, F.: Statistical Theory of Reliability and Life Testing. Holt, Rinehart and Winston, Inc. (1975)
3. Bustince, H., Mohedano, V., Barrenechea, E., Pagola, M.: Definition and construction of fuzzy DI-subsethood measures. Information Sciences 176, 3190–3231 (2006)
4. De Beats, B., Kerre, E., Gupta, M.: The fundamentals of fuzzy mathematical morphology, Part 1: Basic concepts. International Journal of General Systems 23, 155–171 (1994)
5. Di Nola, A., Sessa, S., Pedrycz, W., Sanchez, E.: Fuzzy Relation Equations and Their Applications to Knowledge Engineering. Kluwer, Dordrecht (1989)
6. Fodor, J.C., Roubens, M.: Fuzzy Preference Modelling and Multicriteria Decision Support. Kluwer, Dordrecht (1994)
7. Greene, N.: An overview of conditionals and biconditionals in probability. In: Proceedings of the American Conference on Applied Mathematics, MATH 2008, Harvard, USA (2008)
8. Greene, N.: Methods of assessing and ranking probable sources of error. In: Proceedings of the Applied Computing Conference, ACC 2008, Istanbul, Turkey (2008)
9. Grzegorzewski, P.: Probabilistic implications. In: Proceedings of the 7th Conference of the European Society for Fuzzy Logic and Technology (EUSFLAT 2011) and LFA 2011, Aix-Les-Bains, France, pp. 254–258 (2011)
10. Kitainik, L.: Fuzzy Decision Procedures with Binary Relations. Kluwer, Dordrecht (1993)
11. Klir, G.J., Yuan, B.: Fuzzy Sets and Fuzzy Logic. Theory and Applications. Prentice-Hall, Englewood Cliffs (1995)
12. Nelsen, R.B.: An Introduction to Copulas. Springer, New York (1999)
13. Nguyen, H.T., Mukaidono, M., Kreinovich, V.: Probability of Implication, Logical Version of Bayes' Theorem, and Fuzzy Logic Operations. In: Proceedings of the FUZZ-IEEE 2002, Honolulu, Hawaii, pp. 530–535 (2002)
14. Rescher, N.: Many-valued Logic. McGraw-Hill, New York (1969)
15. Schweizer, B., Sklar, A.: Probabilistic Metric Spaces. Elsevier, New York (1983)
16. Yeung, D.S., Tsang, E.C.: A comparative study on similarity-based fuzzy reasoning methods. IEEE Transactions on System, Man and Cybernetics 27, 216–227 (1997)
17. Zadeh, L.: From computing with numbers to computing with words - From manipulation of measurements to manipulation of perceptions. IEEE Transactions on Circuits and Systems - I: Fundamental Theory and Applications 45, 105–119 (1999)

Robustness of N-Dual Fuzzy Connectives

Renata Hax Sander Reiser and Benjamín René Callejas Bedregal

Abstract. The main contribution of this paper is concerned with the robustness of N-dual connectives in fuzzy reasoning. Starting with an evaluation of the sensitivity in n-order function on $[0,1]$, we apply the results in the D-coimplication classes. The paper formally states that the robustness of pairs of mutual dual n-order functions can be compared, preserving properties and the ordered relation of their arguments.

1 Introduction

Fuzzy logic can be conceived as a system of rules of inference fuzzy reasoning which deals with the propagation of variables modeling real systems. So, it plays an important role in the development of applications related to knowledge representations and inferences from imprecise, incomplete, uncertain or even partially true information[9, 15].

Much of human reasoning can be modeled by the fuzzy reasoning which is mainly determined by its internal structures. By using fuzzy logic, fuzzy reasoning takes into account fuzzy connectives linking antecedent fuzzy sets with consequent fuzzy sets in conditional fuzzy-rules. The fuzzy connectives are fuzzy negations, triangular norms (t-norms) and fuzzy implications, including their dual possible constructions (t-conorms, coimplications)[10].

Thus, the notion of robustness or sensitivity in fuzzy logic became important and provides a formal way to model errors and perturbations in a system of rules of inference fuzzy reasoning. Generally, in such approach, rule perturbations should be considered in the modeling of the whole system [12, 21, 27].

Renata Hax Sander Reiser
Name, CDTEC, UFPEL, CP 354, Campus Universitário, 96010-900 Pelotas, Brazil
e-mail: reiser@inf.ufpel.edu.br

Benjamín René Callejas Bedregal
Name, DIMAP, UFRN, Campus Universitário, 59072-970 Natal, Brazil
e-mail: bedregal@dimap.ufrn.br

B. De Baets et al. (Eds.): Eurofuse 2011, AISC 107, pp. 79–90, 2011.
springerlink.com

Since the main results are closely connected to fuzzy connectives, a comprehensive understanding of robustness of fuzzy reasoning is based on some robustness measurements of fuzzy connectives. In [6], the robustness results are formulated in terms of $\delta-$equalities of fuzzy sets and various implicator operators and inference rules are presented. In [25], a discussion of relations between robustness of fuzzy reasoning and fuzzy connectives is based on some logic-oriented equivalence measurements, and the maximum $\delta-$equalities are derived.

In this paper, after reviewing the approach considered in [19], we present an extension of the definition of pointwise sensitivity of n-order function on $U = [0,1]$, i.e. mapping from U^n into U.

Boundary conditions together with 1-place antitonicity and 2-place isotonicity are among the most frequently studied properties proposed in the literature, which usually provide more comprehensive understanding of either implications or coimplications in fuzzy conditional-rules. The main classes with explicit and implicit representations based on aggregation functions and named as S-implications and R-implications, respectively, satisfy all these properties. In addition to the S-implications and R-implications, the QL-implication class, whose elements are not necessarily 1-place monotone functions, are also considered by Li et. al. in [19], describing the pointwise sensitivity of an $n-$order function f on the domain U.

Thus, the extension proposed by the authors considers the $\delta-$sensitivity of fuzzy (co)implications which do not verify the 1-place antitonicity, e.g. Zadeh implication [26], Reichenbach implication [18] and Klir and Yuan implications [15, 16]. We mainly study the robustness of Dishkant (co)implications, which also have explicit representations by aggregation functions and are strictly connected with QL-implications by the contrapositive property and the exchange principal.

In the sequence, following the approach introduced in [19], the paper also investigates the issues of sensitivity of dual constructions of an $n-$order function f on the domain U. We study the robustness of fuzzy logic connectives, mainly focused on classes of fuzzy (co)implications. Coimplications play a special role in intuitionistic fuzzy logic, see for instance [22, 23]. Such algebraic coimplication properties are intensively studied in [11], in order to provide a theoretical background for approximate reasoning applications.

This paper is organized as follows. In Section 2, some preliminaries are presented, mainly reporting the duality relationship and some concepts of fuzzy negations, in order to review N-dual constructions of fuzzy triangular norms and fuzzy implications. Having this in mind, in Section 3 the study of the fuzzy connective sensitivity is considered, based on the refereed work [19]. Then, general results of robustness of $N-$dual fuzzy connectives are introduced, and the sensitivity of mutual $N-$dual pairs of n-order functions is stated. Conclusion and further work are reported in Section 5.

2 Preliminaries

Some basic notions concerning t-(co)norms, fuzzy (co)implications and dual functions are reported based on [13, 14, 9, 8, 17] and [1].

2.1 Fuzzy Negations

Recall that a function $N : U \to U$ is a **fuzzy negation** if

N1 $N(0) = 1$ and $N(1) = 0$;

N1 If $x \geq y$ then $N(x) \leq N(y)$, $\forall x, y \in U$.

Fuzzy negations satisfying the involutive property are called **strong Negations**:

N3 $N(N(x)) = x$, $\forall x \in U$.

And, a continuous fuzzy negation is *strict* when

N3 If $x > y$ then $N(x) < N(y)$, $\forall x, y \in U$.

Strong fuzzy negations are strict, e.g. the standard negation $N_S(x) = 1 - x$.

When $\mathbf{x} = (x_1, x_2, \ldots, x_n) \in U^n$ and N is a fuzzy negation, we use the denotation:

$$N(\mathbf{x}) = N^n(x_1, x_2, \ldots, x_n) = (N(x_1), N(x_2), \ldots, N(x_n)) \tag{1}$$

2.2 Duality Relationship

Let N be a negation on U and $f : U^n \to U$ be a n-order function. The N**-dual function** of $f : U^n \to U$ is given by:

$$f_N(x_1, \ldots, x_n) = N(f(N(x_1), \ldots, N(x_n))). \tag{2}$$

Notice that, when N is involutive, $(f_N)_N = f$, that is the N-dual of f_N is f. In addition, when $f = f_N$, it is clear that f is a self-dual function.

2.3 Triangular (co)norms

A binary function $(S)T : U^2 \to U$ is a **t-(co)norm** (triangular (co)norm) if and only if it satisfies the boundary conditions together with the commutative, associative and monotonic properties, which are given, respectively, for all $x, y, z \in U$ by the following expressions:

T1: $T(x, 1) = x$;
T2: $T(x, y) = T(y, x)$;
T3: $T(x, T(y, z)) = T(T(x, y), z)$;
T4: if $x \leq x'$ and $y \leq y'$, $T(x, y) \leq T(x', y')$.

S1: $S(x, 0) = x$;
S2: $S(x, y) = S(y, x))$;
S3: $S(x, S(y, z)) = S(S(x, y), z))$;
S4: if $x \leq x'$ and $y \leq y'$, $S(x, y) \leq S(x', y')$.

Let N be a fuzzy negation on U. And, the mappings $T_N, S_N : U^2 \rightarrow U$ denoting the N-dual functions of a t-norm T and of a t-conorm S, respectively defined in Eq. (3) and Eq. (4):

$$T_N(x,y) = N(T(N(x),N(y))). \quad (3)$$
$$S_N(x,y) = N(S(N(x),N(y))); \quad (4)$$

2.4 Fuzzy (co)implications

A **(co)implication operator** $(J)I : [0,1]^2 \rightarrow [0,1]$ extend the classical (co)implication function, which means, it satisfies the boundary conditions:
I0: $I(1,1)=I(0,1)=I(0,0)=1, I(1,0)=0$; **J0**: $I(1,1)=I(0,1)=I(0,0)=1, I(1,0)=0$.

According to [4], when $x,y,z \in U^2$, a fuzzy (co)implication is a function satisfying the following properties:

I1: $I(x,y) \geq I(z,y)$ if $x \leq z$;
I2: $I(x,y) \leq I(x,z)$ if $y \leq z$;
I3: $I(0,x) = 1$;
I4: $I(x,1) = 1$;
I5: $I(1,0) = 0$;

J1: $I(x,y) \geq I(z,y)$ if $x \leq z$;
J2: $I(x,y) \leq I(x,z)$ if $y \leq z$;
J3: $J(0,x) = 1$;
J4: $J(0,y) = 0$;
J5: $J(1,0) = 0$.

There exist many classes of (co)implication functions (see, e.g., [8, 9, 1] and [17]). The four most usual ways to define them are the following.

1. **(S,N)**-implications are fuzzy implications denoted as $I_{S,N} : U^2 \rightarrow U$ and defined by
$$I_{S,N}(x,y) = S(N(x),y), \ \forall x,y \in U, \quad (5)$$
when S is a t-conorm and N is a fuzzy negation. They generalize the classical equivalence: $p \rightarrow q \equiv \neg p \vee q$. If N is a strong fuzzy negation, then $I_{S,N}$ is called a strong implication or S-implication. Their $N-$dual functions are **S-coimplications** given by
$$(I_{S,N})_N(x,y) = S_N(N(x),y), \ \forall x,y \in U, \quad (6)$$
where S_N is the $N-$dual function of the t-conorm S.
2. **R-implications** (residuum implications) are implications denoted as $I_T : U^2 \rightarrow U$ and defined by
$$I_T(x,y) = \sup\{z \in U : T(x,z) \leq y\}, \forall x,y \in U, \quad (7)$$
where T is a left-continuous underlying t-norm in the sense of residual lattices. Considering a fuzzy negation N, their $N-$dual functions are **R-coimplications** given by
$$(I_T)_N(x,y) = \inf\{z \in U : T_N(x,z) \geq y\}, \forall x,y \in U, \quad (8)$$
where T_N is the $N-$dual function of T.

3. **QL-implications** (quantum implications) are fuzzy implications denoted as $I_{S,N,T} : U^2 \to U$, given by

$$I_{S,N,T}(x,y) = S(N(x), T(x,y)), \forall x, y \in U, \tag{9}$$

when T is a t-norm, S is a t-conorm and N is a fuzzy negation. They come from Sasaki's operator: $p \to_S q \equiv \neg p \vee (p \wedge q)$. An $N-$dual QL-implication is a coimplication given by

$$(I_{S,T,N})_N(x,y) = S_N(N(x), T_N(x,y)), \forall x, y \in U, \tag{10}$$

4. **D-implications** (Dishkant implications) are fuzzy implications denoted as $I_{S,T,N} : U^2 \to U$, given by:

$$I_{S,T,N}(x,y) = S(T(N(x), N(y)), y), \forall x, y \in U, \tag{11}$$

when T is a t-norm, S is a t-conorm and N is a fuzzy negation. They come from the Dishkant operator: $p \to q \equiv (\neg p \wedge \neg q) \vee q$. When the involved t-conorm is continuous, a QL-implication and a D-implication coincide if and only if they verify contrapositive symmetry [20]. The N-dual D-implication is a (D-)coimplication given by

$$(I_{S,T,N})_N(x,y) = S_N(T_N(N(x), N(y)), y), \forall x, y \in U. \tag{12}$$

3 The Pointwise Sensitivity of Fuzzy Connectives

In the following, the study of a δ sensitivity of $n-$order function f at point $\mathbf{x}$ (or a pointwise sensitivity) on the domain U is considered in order to extend the work introduced in [19] to the class of (co)implications which verifies the $2-$place monotonicity property **I2** (**J2**) but not necessarily verifies the $1-$place monotonicity property **I1** (**J1**).

Definition 1. *[19, Definition 1] Let $f : U^n \to U$ be an $n-$order function, $\delta \in U$ and $\mathbf{x} = (x_1, x_2, \ldots x_n), \mathbf{y} = (y_1, y_2, \ldots y_n) \in U^n$.*

(i) The δ sensitivity of f at point $\mathbf{x}$, denoted by $\Delta_f(\mathbf{x}, \delta)$, is defined by

$$\Delta_f(\mathbf{x}, \delta) = \sup\{|f(\mathbf{x}) - f(\mathbf{y})| : |x_i - y_i|_{i \in \{1,2,\ldots,n\}} \leq \delta\} \tag{13}$$

(ii) The maximum δ sensitivity of f is defined as follows:

$$\Delta_f(\delta) = \bigvee_{\boldsymbol{x} \in U^n} \Delta_f(\boldsymbol{x}, \delta). \tag{14}$$

Robustness can be conceived as a fundamental property of a logical system stating that the conclusions are not essentially changed if the assumption conditions varied within reasonable parameters. And, since fuzzy connectives (mainly implications and coimplications) are important elements in the fuzzy reasoning, the corresponding investigation of the δ sensitivity in such fuzzy connectives, in terms of Definition 1, will be carried out in this section.

Firstly, we compare the robustness of two functions, $f,g : U^n \to U$ based on their δ sensitivity at point $\mathbf{x} \in U^n$. Thus, we analise the δ-sensitivity at point $\mathbf{x} \in U^n$ of binary functions based on the monotonicity property in their both arguments. We conclude analyzing the δ-sensitivity of some special classes of fuzzy implications.

Definition 2. *Let f and g be two $n-$ order connectives. Thus, we can say that:*

(i) f is as robust as g at a point $\mathbf{x} \in U^n$ if for all $\delta > 0$, $\Delta_f(\mathbf{x},\delta) \leq \Delta_g(\mathbf{x},\delta)$;
(ii) f is more robust than g at a point $\mathbf{x} \in U^n$ if there exists $\delta > 0, \Delta_f(\delta) < \Delta_g(\delta)$.

Henceforth, for denotational ease, when $f : U^2 \to U$ and $\mathbf{x} = (x,y) \in U^2$, consider the following denotations:

$$f\lfloor \mathbf{x} \rfloor \equiv f((x-\delta)\vee 0,(y-\delta)\vee 0); \quad f\lfloor \mathbf{x} \rceil \equiv f((x-\delta)\vee 0,(y+\delta)\wedge 1);$$
$$f\lceil \mathbf{x} \rfloor \equiv f((x+\delta)\wedge 1,(y-\delta)\vee 0); \quad f\lceil \mathbf{x} \rceil \equiv f((x+\delta)\wedge 1,(y+\delta)\wedge 1).$$

Proposition 1. *[19, Theorem 2] Let $f : U \to U$ be a reverse order function, such that $x \leq y \Rightarrow f(x) \geq f(y)$, $\delta \in U$ and $x \in U$. The sensitivity of f at point $\mathbf{x}$ is given by*

$$\Delta_f(x,\delta) = [f(x) - f(x+\delta)\wedge 1)] \vee [f((x-\delta)\vee 0) - f(x)]. \tag{15}$$

In particular, Eq. (15) holds for a fuzzy negation function.

Proposition 2. *[19, Theorem 1] Consider $f : U^2 \to U$, $\delta \in U$ and $\mathbf{x} = (x,y) \in U^2$.*

(i) If f is an monotone function, which means, $x \leq x', y \leq y' \Rightarrow f(x,y) \leq f(x',y')$ then the sensitivity of f at point $\mathbf{x}$ is given as

$$\Delta_f(\mathbf{x},\delta) = (f(\mathbf{x}) - f\lfloor \mathbf{x} \rfloor) \vee (f\lceil \mathbf{x} \rceil - f(\mathbf{x})) \tag{16}$$

(ii) If f verifies both properties, $1-$place antimonotonicity and $2-$place monotonicity, which means, $x' \leq x, y \leq y' \Rightarrow f(x,y) \leq f(x',y')$, then the sensitivity of f at point $\mathbf{x}$ is given as

$$\Delta_f(\mathbf{x},\delta) = (f(\mathbf{x}) - f\lceil \mathbf{x} \rfloor) \vee (f\lfloor \mathbf{x} \rceil - f(\mathbf{x})) \tag{17}$$

(iii) If f verifies the $2-$place monotonicity property, which means, $\forall x \in U, y \leq y' \Rightarrow f(x,y) \leq f(x,y')$ then the sensitivity of f at point $\mathbf{x}$ is given as

$$\Delta_f(\mathbf{x},\delta) = \bigvee_{x-\delta \leq x' \leq x+\delta} |f(x',(y-\delta)\vee 0) - f(\mathbf{x})| \vee \bigvee_{x-\delta \leq x' \leq x+\delta} |f(x',(y+\delta)\wedge 1) - f(\mathbf{x})| \tag{18}$$

Proposition 3. *Consider* $f : U^2 \to U$, $\delta \in U$ *and* $\mathbf{x} = (x,y) \in U^2$.

(iv) If f verifies the 1−place antimonotonicity property, which means, $\forall y \in U, x \leq x' \Rightarrow f(x,y) \geq f(x',y)$ then the sensitivity of f at point $\mathbf{x}$ is given as

$$\Delta_f(\mathbf{x},\delta) = \bigvee_{y-\delta \leq y' \leq y+\delta} |f(\mathbf{x}) - f((x-\delta) \vee 0, y')| \vee \bigvee_{y-\delta \leq y' \leq y+\delta} |f(\mathbf{x}) - f((x+\delta) \wedge 1, y')| \quad (19)$$

Proof. Let f be a decreasing function with respect to its first variable. Firstly, consider $|x - x'| \leq \delta$ such that $b = |f(\mathbf{x}) - f(x',y)|$ and

$$c \leq \bigvee_{y-\delta \leq y' \leq y+\delta} |f(\mathbf{x}) - f((x-\delta) \vee 0), y')| \vee \bigvee_{y-\delta \leq y' \leq y+\delta} |f(\mathbf{x}) - f((x+\delta) \wedge 1, y')|.$$

So, if $f(x',y) \geq f(\mathbf{x})$ then $b = f(\mathbf{x}) - f(x',y)$. It holds that

$$b \leq f(\mathbf{x}) - f((x-\delta) \vee 0, y) \leq \bigvee_{y-\delta \leq y' \leq y+\delta} |f(\mathbf{x}) - f((x-\delta) \vee 0, y')| \leq c.$$

And, if $f(x',y) \leq f(\mathbf{x})$, $b = f(x',y) - f(\mathbf{x})$. It follows that

$$b \leq f((x+\delta) \wedge 1, y) - f(\mathbf{x}) \leq \bigvee_{y-\delta \leq y' \leq y+\delta} |f((x+\delta) \wedge 1, y') - f(\mathbf{x})| \leq c$$

Therefore, from both cases, one can conclude that $b \leq c$. In addition, based on Definition 1, it follows that $\Delta_f(\mathbf{x},\delta) = c$. To sum up, if $x' = (x+\delta) \wedge 1$ or $x' = (x+\delta) \wedge 1$, it holds that $b = c$ and $\Delta_f(\mathbf{x},\delta) = c$. □

Based on [19], the next proposition is presented as a consequence of Proposition 2:

Proposition 4. *[19, Corollary 1] Let $I_{S,N}$, I_T and $I_{S,T,N}$ be an (S,N)−implication, an R−implication and a QL−implication, respectively. When $\mathbf{x} \in U^2$ and $\delta \in U$, the statements in the following hold:*

(i) $\Delta_T(\mathbf{x},\delta)$, $\Delta_S(\mathbf{x},\delta)$, the sensitivity of a t-norm T and a t-conorm S at point $\mathbf{x}$, respectively, are both defined by Eq. (16);
(ii) $\Delta_{I_T}(\mathbf{x},\delta)$ and $\Delta_{I_{S,N}}(\mathbf{x},\delta)$, the sensitivity of an R−implication I_T and (S,N)−implication $I_{S,N}$ at point $\mathbf{x}$, respectively, are both defined by Eq. (17);
(iii) $\Delta_{I_{S,N,T}}(\mathbf{x},\delta)$, the sensitivity of a QL−implication $I_{S,N,T}$ at point $\mathbf{x}$, is given by Eq. (18);

Proposition 5. *Let $I_{S,N,T}$ be a D−implication. When $\mathbf{x} \in U^2$ and $\delta \in U$, then:*

(iv) $\Delta_{I_{S,T,N}}(\mathbf{x},\delta)$, the sensitivity of a D−implication $I_{S,T,N}$ at point $\mathbf{x}$, is given by Eq. (19).

Proof. Straightforward from Proposition 1. □

4 The Pointwise Sensitivity of N−Dual Fuzzy Connectives

In the preceding section we described the definitions as foundations to study the pointwise sensitivity of N−dual fuzzy connectives.

Proposition 6. *Let $f : U^2 \to U$ be a second-order function, $\mathbf{x} = (x,y) \in U^n$ and N be the standard fuzzy negation. The following equations hold:*

$(i)\ f_N\lfloor \mathbf{x} \rfloor = N(f\lceil N(\mathbf{x})\rceil)$; $(ii)\ f_N\lfloor \mathbf{x} \rceil = N(f\lceil N(\mathbf{x})\rfloor)$;
$(iii)\ f_N\lceil \mathbf{x} \rfloor = N(f\lfloor N(\mathbf{x})\rceil)$; $(iv)\ f_N\lceil \mathbf{x} \rceil = N(f\lfloor N(\mathbf{x})\rfloor)$.

Proof. The first case proof is presented and other ones can be inferred analogously.

$$\begin{aligned}(i) f_N\lfloor \mathbf{x} \rfloor &= f_N((x-\delta)\vee 0,(y-\delta)\vee 0)] = N(f(N((x-\delta)\vee 0),N((y-\delta)\vee 0)))] \\ &= N(f((1-x+\delta)\wedge 1),(1-y+\delta)\wedge 1)))] \\ &= N(f((N(x)+\delta)\wedge 1,(N(y)+\delta)\wedge 1))] = N(f\lceil N(x),N(y)\rceil)\end{aligned}$$

So, it means that $f_N\lfloor \mathbf{x} \rfloor = N(f\lceil N(\mathbf{x})\rceil)$. □

Taking a strong fuzzy negation N, Theorem 1 states that the sensitivity of a n-order function f at a point $\mathbf{x}$ is equal to the sensitivity of is dual function f_N.

Theorem 1. *Consider $f : U^2 \to U$, $\delta \in U$ and $\mathbf{x} = (x,y) \in U^2$. Let $\Delta_f(\mathbf{x},\delta)$ be the sensitivity of f at point $\mathbf{x}$. If N is the standard fuzzy negation ($N = N_S$ in Eq (1)) and f_N is the N−dual function of f then the sensitivity of f_N at point $\mathbf{x}$ is given by*

$$\Delta_{f_N}(\mathbf{x},\delta) = \Delta_f((N(\mathbf{x})),\delta) \tag{20}$$

Proof. The first case proof is presented since other ones can be inferred in analogous ways. For that, consider $f : U^2 \to U$, $\delta \in U$ and $\mathbf{x} = (x,y) \in U^2$. Based on Proposition 6, if f is an increasing function with respect to its variables then the sensitivity of f_N at point $\mathbf{x}$ is given by

$$\begin{aligned}\Delta_{f_N}(\mathbf{x},\delta) &= [f_N(\mathbf{x}) - f_N\lfloor \mathbf{x} \rfloor] \vee [f_N\lceil \mathbf{x} \rceil - f_N(\mathbf{x})] \\ &= [N(f(N(\mathbf{x})))) - N(f\lceil N(\mathbf{x})\rceil)] \vee [N(f\lfloor N(\mathbf{x})\rfloor)] - N(f(N(\mathbf{x})))] \\ &= [f\lceil N(\mathbf{x})\rceil - f(N(\mathbf{x}))] \vee [f(N(\mathbf{x})) - f\lfloor N(\mathbf{x})\rfloor] = \Delta_f((N(\mathbf{x})),\delta)\end{aligned}$$

So, by Eq. (16), $\Delta_{f_N}(\mathbf{x},\delta) = \Delta_f((N(\mathbf{x}),\delta)$. □

Proposition 7. *Let N be the standard fuzzy negation, f_N be N−dual function related to a function $f : U^2 \to U$, $\delta \in U$ and $\mathbf{x} = (x,y) \in U^2$. The sensitivity of f_N at point $\mathbf{x}$ is defined by the following cases:*

(i) if f is increasing with respect to its variables then:

$$\begin{aligned}\Delta_{f_N}(\mathbf{x},\delta) &= (f(N(\mathbf{x})) - f\lfloor N(\mathbf{x})\rfloor) \vee (f\lceil N(\mathbf{x})\rceil - f(N(\mathbf{x}))) \\ &= (f_N\lceil \mathbf{x} \rceil - f_N(\mathbf{x})) \vee (f_N\lfloor \mathbf{x} \rfloor - f_N(\mathbf{x}))\end{aligned} \tag{21}$$

(ii) *if f is decreasing with respect to its first variable and increasing with its second variable then*

$$\Delta_{f_N}(\mathbf{x},\delta) = (f(N(\mathbf{x})) - f\lceil N(\mathbf{x})\rfloor) \vee (f\lfloor N(\mathbf{x})\rceil - f(N(\mathbf{x}))) \\ = (f_N\lfloor \mathbf{x}\rceil - f_N(\mathbf{x})) \vee (f_N\lceil \mathbf{x}\rfloor - f_N(\mathbf{x})) \quad (22)$$

(iii) *if f is increasing with its second variable then:*

$$\Delta_{f_n}(\mathbf{x},\delta) = \bigvee_{x-\delta\le x'\le x+\delta} |f(N(\mathbf{x})) - f(N(x'),(N(y)+\delta)\wedge 1)| \vee \\ \bigvee_{x-\delta\le x'\le x+\delta} |f(N(\mathbf{x})) - f(N(x'),(N(y)-\delta)\vee 0)| \quad (23)$$

(iv) *if f is decreasing with respect to its first variable then:*

$$\Delta_{f_N}(\mathbf{x},\delta) = \bigvee_{y-\delta\le y'\le y+\delta} |f((N(x)+\delta)\wedge 1, N(y')) - f(N(\mathbf{x}))| \vee \\ \bigvee_{y-\delta\le y'\le y+\delta} |f((N(x)-\delta)\vee 0, N(y')) - f(N(\mathbf{x}))| \quad (24)$$

Proof. Straightforward from Prop. 2 and Theorem 1. □

Proposition 8. *Let $(T)_N$, $(S)_N$, $(I_{S,N})_N$, $(I_T)_N$, $I_{S,N,T})_N$ and $(I_{S,T,N})_N$ be $N-$dual functions related to a t-norm T, a t-conorm S, an $S-$implication $I_{S,N}$, an $R-$implication I_T, a $QL-$implication $I_{S,N,T}$ and a $D-$implication $I_{S,T,N}$, respectively. When $\mathbf{x} = (x,y) \in U^2$ and $\delta \in U$, the statements as follows hold:*

(i) $\Delta_{(T)_N}(\mathbf{x},\delta)$ *and* $\Delta_{(S)_N}(\mathbf{x},\delta)$ *are both defined by Eq. (21);*
(ii) $\Delta_{(I_T)_N}(\mathbf{x},\delta)$ *and* $\Delta_{I_{(S,N)_N}}(\mathbf{x},\delta)$ *are both defined by Eq. (22);*
(iii) $\Delta_{(I_{S,N,T})_N}(\mathbf{x},\delta)$ *is defined by Eq. (23);*
(iv) $\Delta_{(I_{S,T,N})_N}(\mathbf{x},\delta)$ *is defined by Eq. (24).*

Proof. Straightforward from Propositions 7 and 4.

Proposition 9. *Let N be the standard fuzzy negation and $f_N : U^2 \to U$ be the $N-$dual function of a binary connective $f : U^2 \to U$. Then the maximum sensitivities of f and f_N are related by Eq. (25):*

$$\Delta_{f_N}(\delta) = \Delta_f(\delta). \quad (25)$$

Proof. From Theorem 1, we have the following:

$$\Delta_{f_N}(\delta) = \bigvee_{\mathbf{x}\in U^2} \Delta_{f_N}(\mathbf{x},\delta) = \bigvee_{N(\mathbf{x})\in U^2} \Delta_f(N(\mathbf{x}),\delta) = \Delta_f(\delta).$$

Proposition 10. *Let (f,f_N) and (g,g_N) be pairs of mutual $N-$dual functions from $U^n \times U^n$ to U and $\mathbf{x} = (\mathbf{x}_1, \mathbf{x}_2) \in U^n \times U^n$. Thus, we can say that:*

(i) (f_N, f) is a pair as robust as (g_N, g) at $\mathbf{x}$ iff f is as robust as g at the same point;
(ii) (f_N, f) is a pair more robust than (g_N, g) at a point $\mathbf{x}$ iff f is more robust than g at the same point;
(iii) (f_N, f) is a pair as robust as (g_N, g) iff f is as robust as g;
(iv) (f_N, f) is more robust than (g_N, g) iff f is more robust than g.

Proof. Assuming (f,f_N) and (g,g_N) are pairs of mutual $N-$dual functions, we present the proof of the sufficient condition since the converse can be analogously obtained for the first case. So, if f is as robust as g at point $\mathbf{x} = (\mathbf{x}_1, \mathbf{x}_2) \in U^n \times U^n$, it holds that:

$$
\begin{aligned}
(\Rightarrow i)\Delta_{(f,f_N)}((\mathbf{x}_1,\mathbf{x}_2),\delta) &= \sup\{|(f,f_N)(x_{1i},x_{2i}) - (f,f_N)(y_{1i},y_{2i})| : \\
&\qquad |(x_{1i},x_{2i}) - (y_{1i},y_{2i})|_{i\in\{1,2,\dots,n\}} \leq \delta\} \\
&= (\sup\{|f(x_{1i}) - f(y_{1i})| : |x_{1i} - y_{1i}|_{i\in\{1,2,\dots,n\}} \leq \delta\}, \\
&\qquad \sup\{|f(x_{2i}) - f(y_{2i})| : |x_{2i} - y_{2i}|_{i\in\{1,2,\dots,n\}} \leq \delta\}) \\
&= (\Delta_f(\mathbf{x},\delta), \Delta_{f_N}(\mathbf{x}\delta)) \\
&\leq (\Delta_g(\mathbf{x},\delta), \Delta_{g_N}(\mathbf{x}\delta)) \\
&= \sup\{|(g,g_N)(x_{1i},x_{2i}) - (g,g_N)(y_{1i},y_{2i})| : \\
&\qquad |(x_{1i},x_{2i}) - (y_{1i},y_{2i})|_{i\in\{1,2,\dots,n\}} \leq \delta\} \\
&= \Delta_{(g,g_N)}((\mathbf{x}_1,\mathbf{x}_2),\delta).
\end{aligned}
$$

The proof of other cases can be obtained in the same way. □

5 Conclusion

The main contribution of this paper is concerned with the study of robustness on N-dual operators mainly used in fuzzy reasoning. Taking the class of strong fuzzy negation, the paper formally states that the sensitivity of a n-order function f at a point $\mathbf{x} \in [0,1]^n$ coincides with the sensitivity of its dual function f_N at the same point.

Therefore, when this sensitivities are used in the inference process of a fuzzy rule system based on N-dual pairs of connectives, the work of estimating their sensitivity to small changes in fuzzy sets can be reduced in a 50%. Such results also reveal that the robustness of pairs of mutual dual functions can be compared, preserving their properties and the ordered relation of their arguments.

Based on previous work (see, e.g. [2, 3] and [7]) focused on the study of fundamental properties of interval-valued fuzzy (co)implications, our current investigation aims clearly to consider two approaches:

(i) the sensibility of fuzzy inference dependent by considering interval-valued fuzzy rules based on interval-valued fuzzy connectives;

(ii) the extension of the robustness studies in order to consider the main interval-valued classes of (co)implications: S-(co)implications, R-(co)implications, QL-(co)implications and D-(co)implications.

In conclusion, further work will investigate in-depth understanding of robustness property related to representable Atanassovś intuitionistic fuzzy sets, in the sense considered in [5, 22] and [7].

References

1. Baczynski, M., Balasubramaniam, J.: Fuzzy Implications (Studies in Fuzziness and Soft Computing). Springer, Heidelberg (2008)
2. Bedregal, B.C.: On interval fuzzy negations. Fuzzy Sets and Systems 161(17), 2290–2313 (2010)
3. B.B.C.,, D.G.P., Santiago, R.H.N.,, R.R.H.S.: On interval fuzzy S-implications. Information Sciences 180(8), 1373–1389 (2010)
4. Bustince, H., Burillo, P., Soria, F.: Automorphism, negations and implication operators. Fuzzy Sets and Systems 134(2), 209–229 (2003)
5. Bustince, H., Barrenechea, E.,, M.V.: Intuitionistic fuzzy implication opertors – an expression and main properties. International Journal of Uncertainty, Fuzziness and Knowledge-Based Systems (IJUFKS) 12(3), 387–406 (2004)
6. Cai, K.Y.: Robustness of fuzzy reasining and $\sigma-$Equalities of Fuzzy Sets. IEEE Transaction on Fuzzy Systems 9(5), 738–750 (2001)
7. Da Costa, C.G., Bedregal, B.C., Dória Neto, A.D.: Relating De Morgan triples with Atanassovś intuitionistic De Morgan triples via automorphisms. International Journal of Approximate Reasoning 52, 473–487 (2011)
8. De Baets, B.: Coimplications, the forgotten connectives. Tatra Mountains Matematical Publications 12, 229–240 (1997)
9. Dubois, D., Ostasiewisz, W., Prade, H.: Fundamentals of Fuzzy Sets, Fuzzy Sets: History and Basic Notions. In: Dubois, D., Ostasiewisz, W., Prade, H. (eds.) The Handbook of Fuzzy Sets Series, pp. 21–106. Kluwer, Boston (2000)
10. Fodor, J., Roubens, M.: Fuzzy Preference Modelling and Multicriteria Decision Support. Kluwer Academic Publisher, Dordrecht (1994)
11. Gera, Z., Dombi, J.: Type 2 implications on non-interative fuzzy truth values. Fuzzy Sets and Systems 159, 3014–3032 (2008)
12. Jin, J.H., Li, Y., Li, C.: Robustness of fuzzy reasoning via locally equivalence measure. Information Sciences 177(22), 5103–5177 (2007)
13. Klement, E.P., Navara, M.: A survey on different triangular norm-based fuzzy logics. Fuzzy Sets and Systems 101(2), 241–251 (1999)
14. K.E.P.,, M.R., Pap, E.: Triangular Norms. Kluwer Academic Publisher, Dordrechet (2000)
15. Klir, G.J., Folger, T.A.: Fuzzy Sets, Uncertainty and Information. Prentice Hall, Englewood Cliffs (1988)
16. Klir, G.J., Yuan, B.: Fuzzy Sets and Fuzzy logic. Prentice Hall, Englewood Cliffs (1995)
17. Maes, K.C., De Baets, B.: Commutativity and self-duality: Two tales of one equation. International Journal of Approximate Reasoning 50(1), 189–199 (2009)
18. Mas, M., Monserrat, M., Trillas, E.: A survey on fuzzy implication functions. IEEE Transactions on Fuzzy Systems 15(6), 1107–1121 (2007)

19. Li, Y., Li, D., Pedrycz, W., Wu, J.: An approach to measure the robustness of fuzzy reasoning. International Journal of Intelligent Systems 20(4), 393–413 (2005)
20. Mas, M., Monserrat, M., Torrens, J.: $QL-$implications versus D-implications. Kybernetika 42(3), 351–366 (2006)
21. Li, Y.: Approximation and robustness of fuzzy finite automata. International Journal Approximate Reasoning 47(2), 247–257 (2008)
22. Lin, L., Xia, Z.Q.: Intuitionistic fuzzy implication operators: Expressions and properties. Journal of Applied Mathematics and Computing 22(3), 325–338 (2006)
23. Ruiz, D., Torrens, J.: Residual implications and co-implications from idempotent uninorms. Kybernetika 40, 21–38 (2004)
24. Wang, Z., Fang, J.X.: Residual coimplicators of left and right uninorms on a complete lattice. Fuzzy Sets and Systems 160(14), 2086–2096 (2009)
25. Ying, M.S.: Perturbation on fuzzy reasoning. IEEE Transaction on Fuzzy Systems 7, 625–629 (1999)
26. Zadeh, L.A.: Fuzzy sets. Information and Control 15(6), 338–353 (1965)
27. Zhang, L., Cai, K.Y.: Optimal fuzzy reasoning and its robustness analysis. International Journal of Intelligent Systems 19(11), 1033–1049 (2004)

Transitivity and Negative Transitivity in the Fuzzy Setting

Susana Díaz, Bernard De Baets, and Susana Montes

Abstract. A (crisp) binary relation is transitive if and only if its dual relation is negatively transitive. In preference modelling, if a weak preference relation is complete, the associated strict preference relation is its dual relation. It follows from here this well-known result: given a complete weak preference relation, it is transitive if and only if its strict preference relation is negatively transitive.

In the context of fuzzy relations, transitivity is traditionally defined by a t-norm and negative transitivity, by a t-conorm. In this setting, it is also well known that a (valued) binary relation is T-transitive if and only if its dual relation is negatively S-transitive where S stands for the dual t-conorm of the t-norm T. However, in this context there are several proposals to get the strict preference relation from the weak preference relation. Also, there are different definitions of completeness. In this contribution we depart from a reflexive fuzzy relation. We assume that this relation is transitive with respect to a conjunctor (a generalization of t-norms). We consider almost all the possible generators and therefore all the possible strict preference relations obtained from the reflexive relation and we provide a general expression for the negative transitivity that those relations satisfy.

1 Introduction

Preference modeling is usually based on pairwise comparison. Given a set of alternatives, the decision maker is asked to compare them pairwisely. In crisp set theory (see e.g. [1, 14]), there exists a basic relation R on the set of alternatives A, called weak preference relation, such that for any two alternatives a and b, $(a,b) \in R$ expresses that alternative a is considered to be at least as good as alternative b. That

Susana Díaz · Susana Montes
Dept. Statistics and O. R., University of Oviedo
e-mail: {diazsusana,montes}@uniovi.es

Bernard De Baets
Dept. Appl. Math., Biometrics and Process Control, Ghent University
e-mail: bernard.debaets@ugent.be

B. De Baets et al. (Eds.): Eurofuse 2011, AISC 107, pp. 91–100, 2011.
springerlink.com

relation covers all the possible answers of the decision maker over any pair of alternatives. The relation can be decomposed into a strict preference relation P, an indifference relation I and an incomparability relation J. The triplet (P,I,J) is called preference structure. In order to check the coherence of the answers provided by the decision maker, different properties are defined. Maybe the most important one is transitivity. When the decision maker is able to compare any pair of alternatives, this is, when the weak preference relation is complete, another important property is negative transitivity. It is well known that the transitivity of a complete R is characterized by the negative transitivity of P. In fact, it is known that for any relation Q, it is transitive if and only if its dual is negatively transitive. Under completeness, the strict preference relation P becomes the dual relation of R and the result follows.

The main drawback of crisp relations is that they lack of flexibility. They do not express accurately human decisions, since human decisions could sometimes be framed by the mere use of levels of preference, instead of being "clear answers". Fuzzy relations were introduced to solve this problem. They allow the decision maker to express a degree of preference rather than its presence or absence only. A key notion in this context is the notion of a fuzzy preference structure, see [3] for a historical account of its development. In the fuzzy sets context transitivity plays also an important role as a property asso- ciated to rationality. Traditionally, transitivity of a fuzzy relation is defined with respect to a triangular norm and negative transitivity with respect to a triangular conorm. In this setting, it is also known (see e.g. [12]) that a valued binary relation is T-transitive if and only if its dual relation is negatively S-transitive, where S is the dual triangular conorm of T . So, also in this context, if the strict preference relation is obtained from the weak preference relation as its dual, the connection is guaranteed. However, for fuzzy relations, there are multiple ways to obtain the preference structure from the weak preference relation. Also, there are different ways to define completeness of a fuzzy relation. In this contribution we are interested in the connection between the transitivity of the weak preference relation and the negative transitivity of any associated strict preference relation. As a starting point, we have not considered any completeness condition, but an important one: weak completeness. Also, following the general study we have carried out in previous works [6, 7, 8, 10], we have not restricted to t-norms to define transitivity, but we have considered any conjunctor. In this general setting, we study the strongest type of negative transitivity we can assure for the strict preference relation.

The contribution is structured in four sections. In Section 2 we focus on weak preference relations and the associated preference structure. We both recall the association in the classical or crisp case and in the fuzzy context. Section 3 is devoted to conjunctors and other operators that play a key role in this contribution. In Section 4 the result obtained in general for the negative transitivity of the strict preference relation is provided. We also discuss there some particular cases. A brief conclusion is provided in Section 5. We also discuss there some of the most important open points.

2 Preference Structures

Consider a decision maker who is presented a set of alternatives A. Let us suppose that this person compares the alternatives two by two. Given two alternatives, the decision maker can act in one of the following three ways: (i) he/she clearly prefers one to the other; (ii) the two alternatives are indifferent to him/her; (iii) he/she is unable to compare the two alternatives. According to these cases, we can define three (binary) relations on A: the strict preference relation P, the indifference relation I and the incomparability relation J. Thus, for any $(a,b) \in A^2$, we classify:

$$(a,b) \in P \quad \Leftrightarrow \quad \text{he/she prefers } a \text{ to } b;$$
$$(a,b) \in I \quad \Leftrightarrow \quad a \text{ and } b \text{ are indifferent to him/her;}$$
$$(a,b) \in J \quad \Leftrightarrow \quad \text{he/she is unable to compare } a \text{ and } b.$$

We recall that for a relation Q on A, its converse is defined as $Q^t = \{(b,a) \mid (a,b) \in Q\}$, its complement as $Q^c = \{(a,b) \mid (a,b) \notin Q\}$ and its dual as $Q^d = (Q^t)^c$. One easily verifies that P, I, J and P^t establish a particular partition of A^2 [14].

Definition 1. *A preference structure on A is a triplet (P,I,J) of relations on A that satisfies:*

(i) P is irreflexive, I is reflexive and J is irreflexive;
(ii) P is asymmetric, I and J are symmetric;
(iii) $P \cap I = \emptyset$, $P \cap J = \emptyset$ and $I \cap J = \emptyset$;
(iv) $P \cup P^t \cup I \cup J = A^2$.

A preference structure (P,I,J) on A is characterized by the reflexive relation $R = P \cup I$, called large preference relation, in the following way:

$$(P,I,J) = (R \cap R^d, R \cap R^t, R^c \cap R^d).$$

Conversely, for any reflexive relation R on A, the triplet (P,I,J) constructed in this way from R is a preference structure on A such that $R = P \cup I$. As R is the union of the strict preference relation and the indifference relation, $(a,b) \in R$ means that a is at least as good as b.

A relation Q on A is called complete if $(a,b) \in Q \vee (b,a) \in Q$, for all $(a,b) \in A^2$. In the crisp sets context, the completeness of the weak preference relation is characterized by the absence of incomparability in the associated preference structure. THi si, the weak preference relation is complete if and only if its associated incomparability relation J is empty.

A relation Q on A is called transitive if $((a,b) \in Q \wedge (b,c) \in Q) \Rightarrow (a,c) \in Q$, for any $(a,b,c) \in A^3$. A relation Q on A is called negatively transitive if $(a,c) \in Q \Rightarrow ((a,b) \in Q \vee (b,c) \in Q)$, for any $(a,b,c) \in A^3$. This property is frequently found as $((a,b) \notin Q \wedge (b,c) \notin Q) \Rightarrow (a,c) \notin Q)$.

The transitivity of the large preference relation R can be characterized as follows [1].

Theorem 1. *For any reflexive relation R with associated preference structure (P,I,J) it holds that*

$$R \text{ is transitive } \Leftrightarrow P \text{ is negatively transitive.}$$

Finally, we recall an important characterization of preference structures. Let us identify relations with their characteristic mappings, then Definition 1 can be written in the following minimal way [5]: I is reflexive and symmetric, and for any $(a,b) \in A^2$ it holds that

$$P(a,b) + P^t(a,b) + I(a,b) + J(a,b) = 1\,.$$

Classical, also called crisp, preference structures can therefore also be considered as Boolean preference structures, employing 1 and 0 for describing presence or absence of strict preference, indifference and incomparability.

2.1 Additive Fuzzy Preference Structures

A serious drawback of classical preference structures is their inability to express intensities. In contrast, in fuzzy preference modelling, strict preference, indifference and incomparability are a matter of degree. These degrees can take any value in the unit interval $[0,1]$ and fuzzy relations are used for capturing them [12].

The intersection of fuzzy relations is defined pointwisely based on some triangular norm (t-norm for short), *i.e.* an increasing, commutative and associative binary operation on $[0,1]$ with neutral element 1. The three most important t-norms are the minimum operator $T_{\mathbf{M}}(x,y) = \min(x,y)$, the algebraic product $T_{\mathbf{P}}(x,y) = x \cdot y$ and the Łukasiewicz t-norm $T_{\mathbf{L}}(x,y) = \max(x+y-1,0)$. Another important t-norm is the drastic product defined by

$$T_{\mathbf{D}}(x,y) = \begin{cases} \min(x,y) & \text{, if } \max(x,y) = 1, \\ 0 & \text{, otherwise.} \end{cases}$$

According to the usual ordering of functions, the above t-norms can be ordered as follows: $T_{\mathbf{D}} \leq T_{\mathbf{L}} \leq T_{\mathbf{P}} \leq T_{\mathbf{M}}$. In fact, the greatest t-norm is the minimum operator and the smallest t-norm is the drastic product.

Similarly, the union of fuzzy relations is based on a t-conorm, *i.e.* an increasing, commutative and associative binary operation on $[0,1]$ with neutral element 0. T-norms and t-conorms come in dual pairs: to any t-norm T there corresponds a t-conorm S through the relationship $S(x,y) = 1 - T(1-x, 1-y)$. For the above three t-norms, we thus obtain the maximum operator $S_{\mathbf{M}}(x,y) = \max(x,y)$, the probabilistic sum $S_{\mathbf{P}}(x,y) = x + y - xy$ and the Łukasiewicz t-conorm (bounded sum) $S_{\mathbf{L}}(x,y) = \min(x+y,1)$. For more background on t-norms and t-conorms and the notations used in this paper, we refer to [13].

T-conorms are used to define completeness. A fuzzy relation Q on A is S-complete if $S(Q(a,b), Q(b,a)) = 1$ for all $(a,b) \in A^2$ (see for example [2]). The two most important types of completeness are defined by the Łukasiewicz and maximum t-conorm:

- A fuzzy relation Q on A is called weakly complete if it is $S_{\mathbf{L}}$-complete: $Q(a,b)+Q(b,a)=1$ for all $(a,b)\in A^2$.
- A fuzzy relation Q on A is called strongly complete if it is $S_{\mathbf{M}}$-complete: $max(Q(a,b),Q(b,a))=1$.

In this work, we will focus on weak completeness as it shows an important property we recall in Proposition 1.

The definition of a fuzzy preference structure has been a topic of debate during several years (see e.g. [12, 15, 16]). Accepting the *assignment principle* — for any pair of alternatives (a,b) the decision maker is allowed to assign at least one of the degrees $P(a,b)$, $P(b,a)$, $I(a,b)$ and $J(a,b)$ freely in the unit interval — has finally led to a fuzzy version of Definition 1 with intersection based on the Łukasiewicz t-norm and union based on the Łukasiewicz t-conorm. Interestingly, a corresponding minimal definition is identical to the classical one provided we replace crisp relations by fuzzy relations: a triplet (P,I,J) of fuzzy relations on A is a fuzzy preference structure on A if and only if I is reflexive and symmetric, and for any $(a,b)\in A^2$ it holds that

$$P(a,b)+P^t(a,b)+I(a,b)+J(a,b)=1\,,$$

where $P^t(a,b)=P(b,a)$. This identity explains the name *additive fuzzy preference structures*.

Another topic of controversy has been how to construct such a fuzzy preference structure from a reflexive fuzzy relation. The most recent and most successful approach is that of De Baets and Fodor based on (indifference) generators [4].

Definition 2. *A generator i is a commutative binary operation on the unit interval $[0,1]$ that is bounded by the Łukasiewicz t-norm $T_{\mathbf{L}}$ and the minimum operator $T_{\mathbf{M}}$,* i.e. $T_{\mathbf{L}}\le i\le T_{\mathbf{M}}$.

Note that generators are not necessarily t-norms, albeit having neutral element 1. For any reflexive fuzzy relation R on A it holds that the triplet (P,I,J) of fuzzy relations on A defined by:

$$\begin{aligned}
P(a,b) &= R(a,b)-i(R(a,b),R(b,a))\,,\\
I(a,b) &= i(R(a,b),R(b,a))\,,\\
J(a,b) &= i(R(a,b),R(b,a))-(R(a,b)+R(b,a)-1)\,.
\end{aligned}$$

is an additive fuzzy preference structure on A such that $R=P\cup_{S_{\mathbf{L}}} I$, *i.e.* $R(a,b)=P(a,b)+I(a,b)$.

Recall that a binary operation $f:[0,1]^2\to[0,1]$ is 1-Lipschitz continuous if

$$|f(x_1,y_1)-f(x_2,y_2)|\le|x_1-x_2|+|y_1-y_2|\,,$$

for any $(x_1,x_2,y_1,y_2)\in[0,1]^4$. We proved in [8] that the 1-Lipschitz property plays an important role in the study of the propagation of the transitivity from a weak preference relation to its associated strict preference and indifference relation. In this contribution, it plays again an important role. Let us recall that the most important family of generators are the Frank t-norms (see e.g. [12]) and they satisfy the 1-Lipschitz continuity. In particular, the two most employed generators, the Łukasiewicz and the minimum t-norms, are 1-Lipschitz. The Łukasiewicz operator plays a very special role as the following result shows.

Proposition 1. *Consider an additive fuzzy preference structure (P,I,J) generated from a reflexive fuzzy relation R by means of a generator i. Then*

$$R \text{ is weakly complete and } i = T_{\mathbf{L}} \Leftrightarrow J = \emptyset.$$

Observe also that for this particular generator, the additive fuzzy preference structure obtained from a reflexive relation R is

$$(P,I) = (R^d, R + R^t - 1)$$

So the equivalence known between the transitivity of the weak preference relation and the negative transitivity of its associated strict preference relation holds when the weak preference relation is weakly complete and the generator is the Łukasiewicz t-norm.

3 Conjunctors

In this section we recall and introduce some operators that play a key role in the connection between transitivity and negative transitivity. The first important operators are conjunctors and disjunctors that allow to generalize the classical definitions of transitivity and negative transitivity in the fuzzy sets context.

3.1 Generalizing T-Transitivity

The usual way of defining the transitivity of a fuzzy relation is with respect to a t-norm T: a fuzzy relation Q on A is called T-transitive if $T(Q(a,b),Q(b,c)) \leq Q(a,c)$ for any $(a,b,c) \in A^3$. However, the restriction to t-norms is questionable. On the one hand, even when the large preference relation R is T-transitive with respect to a t-norm T, the transitivity of the generated P and I cannot always be expressed with respect to a t-norm [7, 9, 10]. On the other hand, the results presented in the following sections also hold when R is transitive with respect to a more general operation. From the point of view of fuzzy preference modelling, it is not that surprising that the class of t-norms is too restrictive, as a similar conclusion was drawn when identifying suitable generators, as was briefly explained in the previous section. There, continuity, *in casu* the 1-Lipschitz property, was more important than associativity. As discussed in [9, 10], suitable operations for defining the transitivity of fuzzy relations are conjunctors.

Definition 3. *A conjunctor f is an increasing binary operation on $[0,1]$ that coincides on $\{0,1\}^2$ with the Boolean conjunction.*

The smallest conjunctor c_S and greatest conjunctor c_G are given by

$$c_S(x,y) = \begin{cases} 0 & \text{, if } \min(x,y) < 1, \\ 1 & \text{, otherwise,} \end{cases}$$

and

$$c_G(x,y) = \begin{cases} 0 & \text{, if } \min(x,y) = 0, \\ 1 & \text{, otherwise.} \end{cases}$$

Obviously, $c_S \leq T_{\mathbf{D}} \leq T_{\mathbf{M}} \leq c_G$.

Given a conjunctor f, we say that a fuzzy relation Q on A is f-transitive if $f(Q(a,b),Q(b,c)) \leq Q(a,c)$ for any $(a,b,c) \in A^3$. Defining the composition $Q_1 \circ_f Q_2$ with respect to a conjunctor f of two fuzzy relations Q_1 and Q_2 on A by $Q_1 \circ_f Q_2(a,c) = \sup_b f(Q_1(a,b),Q_2(b,c))$, still allows us to use the shorthand $Q \circ_f Q \subseteq Q$ to denote f-transitivity. Clearly, for two conjunctors f and g such that $f \leq g$, it holds that g-transitivity implies f-transitivity. Restricting our attention to reflexive fuzzy relations only, such as large preference relations, not all conjunctors are suitable for defining transitivity. Indeed, for a reflexive fuzzy relation R, we should consider conjunctors upper bounded by $T_{\mathbf{M}}$ only (see [8]).

In the same way as we have generalized classical t-norms, we can generalize t-conorms.

Definition 4. *A disjunctor is an increasing binary operation on $[0,1]$ that coincides on $\{0,1\}^2$ with the Boolean disjunction.*

As t-norms and t-conorms, disjunctors and conjunctors are dual operators. For any conjunctor f, the operator $g(x,y) = 1 - f(1-x, 1-y)$ is a disjunctor and the converse also holds.

Given a disjunctor g we say that a fuzzy relation Q on A is negatively g-transitive if $Q(a,c) = g(Q(a,b),Q(b,c))$ for any $(a,b,c) \in A^3$. This is the definition we will use in the following section. However, observe that, as in the crisp case, this property can also be expressed as follows: $f(Q^c(a,b),Q^c(b,c)) = Q^c(a,c)$ for any $(a,b,c) \in A^3$, where f is the dual conjunctor of the disjunctor g.

3.2 Fuzzy Implications and Related Operations

With a given t-norm T, one usually associates a fuzzy implication (also called R-implication or T-residuum) as a binary operation on $[0,1]$ defined by (see e.g. [12, 13]):

$$\mathscr{I}_T(x,y) = \sup\{z \in [0,1] \mid T(x,z) \leq y\}.$$

When T is left-continuous it holds that $T(x,z) \leq y \Leftrightarrow z \leq \mathscr{I}_T(x,y)$, and $\mathscr{I}_T$ is called the residual implicator of T.

Definition 5. *With a given commutative conjunctor f we associate a binary operation $\mathscr{I}_f$ on the unit interval defined by*

$$\mathscr{I}_f(x,y) = \sup\{z \in [0,1] \mid f(x,z) \leq y\}.$$

The above definition could also be extended to non-commutative operations, but in that case we should distinguish between left and right operations. In this work we will only consider the case of commutative operations (commutative conjunctors or generators). Clearly, $\mathscr{I}_f$ is decreasing in its first argument and increasing in its second argument.

Under a mild condition, the operation $\mathscr{I}_f$ has an interesting logical interpretation.

Definition 6. *An implicator f is a binary operation on $[0,1]$ that is decreasing in its first argument, increasing in its second argument and that coincides on $\{0,1\}^2$ with the Boolean implication.*

Proposition 2. *Consider a commutative conjunctor f, then $\mathscr{I}_f$ is an implicator if and only if $f(1,y) > 0$, for any $y > 0$.*

The condition in the preceding proposition is obviously fulfilled when f has 1 as neutral element.

Other properties of residual implications of left-continuous t-norms can be found, for example, in [13].

In this paper, we associate another binary operations with any generator. This operator will play a key role in the characterization of the negative transitivity of the strict preference relation.

Definition 7. *With a given commutative conjunctor f we associate a binary operation $\mathscr{K}_f$ on the unit interval defined by*

$$\mathscr{K}_f(x,y) = \sup\{z \in [1-y,1] \mid z - f(x,z) = y\}.$$

Despite the previous definition is given for any commutative conjunctor, we will only use it for a particular type of generators. Observe that the set $\{z \in [1-y,1] \mid z - f(x,z) = y\}$ is not always non-empty. For example for $x = 0.5$ and $y = 0$, it becomes $\{z \in [1,1] \mid z - \min(0.5,z) = 0\} = \emptyset$ since for $z = 1$, it does not hold that $1 - \min(0.5,1) = 0$. However, under suitable conditions, the set is not empty.

Lemma 1. *Let i be a 1-Lipschitz increasing generator and let (x,y) satisfy $y \geq 1 - x - i(x,1-x)$, then $\{z \in [1-y,1] | z - f(x,z) = y\}$ is a non-empty set that admits maximum. Moreover, $\mathscr{K}_i(x,y)$ is increasing on its first argument and decreasing on its second argument.*

Observe also that if the operator f is strictly increasing, then the equality $z - f(x,z) = y$ is satisfied by at most one z, so the supremum of such a set is the only value belonging to it. The strict monotonicity of the conjunctor f is not a strong condition. Let us recall that the family of Frank t-norms is a family of strictly increasing operators (except for the case $\lambda = \infty$).

4 Explicit Expression of Negative Transitivity

Once introduced the necessary definitions and notations, we can present the main result of the contribution: the explicit expression of the negative transitivity that can be assured for the strict preference relation P, when the weak preference relation it comes from is transitive with respect to any (commutative) conjunctor.

Theorem 2. *Consider a 1-Lipschitz increasing generator i and a commutative conjunctor h. For any reflexive fuzzy relation R with corresponding strict preference relation P generated by means of i, it holds that*

$$R \text{ is } h\text{-transitive} \quad \Rightarrow \quad P \text{ is negatively } j_h^i\text{-transitive}$$

where

$$j_h^i(x,y) = \sup_{\substack{u \leq 1-x \\ u+i(u,1-u) \geq 1-x \\ v \leq 1-y \\ v+i(v,1-v) \geq 1-y}} f(x,y,u,v) - i(f(x,y,u,v),h(u,v))$$

for

$$f(x,y,u,v) = \min(\mathscr{I}_h(u,\mathscr{K}_i(v,y)),\mathscr{I}_h(v,\mathscr{K}_i(u,x))).$$

Moreover, this is the strongest result possible.

Let us remark that for the particular case of considering as generator the Łukasiewicz operator, the strict preference relation is the dual of the weak preference relation and in that case it is known that the negative transitivity obtained for the strict preference relation is defined by the dual t-conorm of the t-norm defining the transitivity of R. If we replace in the expression obtained in Theorem 2 the generator i by the Łukasiewicz operator, we get a generalization of this well-known result for any conjunctor h.

Corollary 1. *Let $i = T_{\mathbf{L}}$ and consider a commutative conjunctor h. For any reflexive fuzzy relation R with corresponding strict preference relation P generated by means of $i = T_{\mathbf{L}}$, it holds that*

$$R \text{ is } h\text{-transitive} \Rightarrow P \text{ is negatively } h^d\text{-transitive}$$

where

$$h^d(x,y) = 1 - h(1-x,1-y).$$

5 Conclusion

The equivalence between the transitivity of a weak preference relation and the negative transitivity of its associated strict preference relation only holds when they are dual operators. We have studied the negative transitivity satisfied by the strict preference relation in general, when other generators are considered to obtain this relation from a weakly complete weak preference relation. We have provided a general expression that largely generalizes the only result known about the connection between transitivity and negative transitivity. However, the result obtained is just the departing point for a much wider study. The difficult general expression obtained leads to interesting open points. Maybe the first one is to know if that expression admits an easy version for the most popular generators: the Frank t-norms. Also, in this contribution we have provided a result valid for any weakly complete weak preference relation. The study of other completeness conditions, and specially the study of the strong completeness is also an interesting future topic.

Acknowledgements. This work has been partially supported by Project MTM2010-17844.

References

1. Arrow, K.J.: Social Choice and Individual Values. Wiley, Chichester (1951)
2. Bodenhofer, U., Klawonn, F.: A formal study of linearity axioms for fuzzy orderings. Fuzzy Sets and Systems 145, 323–354 (2004)
3. De Baets, B., Fodor, J.: Twenty years of fuzzy preference structures (1978-1997. Belg. J. Oper. Res. Statist. Comput. Sci. 37, 61–82 (1997)
4. De Baets, B., Fodor, J.: Additive fuzzy preference structures: the next generation. In: De Baets, B., Fodor, J. (eds.) Principles of Fuzzy Preference Modelling and Decision Making, pp. 15–25. Academic Press, London (2003)
5. De Baets, B., Van de Walle, B.: Minimal definitions of classical and fuzzy preference structures. In: Proceedings of the Annual Meeting of the North American Fuzzy Information Processing Society, USA, Syracuse, New York, pp. 299–304 (1997)
6. Díaz, S., De Baets, B., Montes, S.: Additive decomposition of fuzzy pre-orders. Fuzzy Sets and Systems 158, 830–842 (2007)
7. Díaz, S., De Baets, B., Montes, S.: On the compositional characterization of complete fuzzy pre-orders. Fuzzy Sets and Systems 159, 2221–2239 (2008)
8. Díaz, S., De Baets, B., Montes, S.: General results on the decomposition of transitive fuzzy relations. Fzzy Optim. Decis. Making 9, 1–29 (2010)
9. Díaz, S., Montes, S., De Baets, B.: Transitive decomposition of fuzzy preference relations: the case of nilpotent minimum. Kybernetika 40, 71–88 (2004)
10. Díaz, S., Montes, S., De Baets, B.: Transitivity bounds in additive fuzzy preference structures. IEEE Trans. on Fuzzy Systems 15, 275–286 (2007)
11. Fishburn, P.C.: Utility Theory for Decision Making. Wiley, New York (1970)
12. Fodor, J., Roubens, M.: Fuzzy Preference Modelling and Multicriteria Decision Support. Kluwer Academic Publishers, Dordrecht (1994)
13. Klement, E.P., Mesiar, R., Pap, E.: Triangular Norms. Kluwer Academic Publishers, Dordrecht (2000)
14. Roubens, M., Vincke, P.: Preference Modelling. Lecture Notes in Economics and Mathematical Systems, vol. 76. Springer, Heidelberg (1998)
15. Van de Walle, B., De Baets, B., Kerre, E.: A plea for the use of Łukasiewicz triplets in the definition of fuzzy preference structures. Part 1: General argumentation. Fuzzy Sets and Systems 97, 349–359 (1998)
16. Van de Walle, B., De Baets, B., Kerre, E.: Characterizable fuzzy preference structures. Annals of Operations Research 80, 105–136 (1998)

Part II

Aggregation Operators

A Characterization Theorem for t-Representable n-Dimensional Triangular Norms

Benjamín Bedregal, Gleb Beliakov, Humberto Bustince, Tomasa Calvo, Javier Fernández, and Radko Mesiar

Abstract. n-dimensional fuzzy sets are an extension of fuzzy sets that includes interval-valued fuzzy sets and interval-valued Atanassov intuitionistic fuzzy sets. The membership values of n-dimensional fuzzy sets are n-tuples of real numbers in the unit interval $[0,1]$, called n-dimensional intervals, ordered in increasing order. The main idea in n-dimensional fuzzy sets is to consider several uncertainty levels in the memberships degrees. Triangular norms have played an important role in fuzzy sets theory, in the narrow as in the broad sense. So it is reasonable to extend this fundamental notion for n-dimensional intervals. In interval-valued fuzzy theory, interval-valued t-norms are related with t-norms via the notion of t-representability. A characterization of t-representable interval-valued t-norms is given in term of inclusion monotonicity. In this paper we generalize the notion of t-representability for n-dimensional t-norms and provide a characterization theorem for that class of n-dimensional t norms.

B. Bedregal
Department of Informatics and Applied Mathematics,
Federal University of Rio Grande do Norte, Natal, Brazil
e-mail: bedregal@dimap.ufrn.br

G. Beliakov
School of Engineering and Information Technology, Deakin University, Burwood, Australia
e-mail: gleb@deakin.edu.au

H. Bustince · J. Fernandez
Department of Automatic and Computation, Public University of Navarra, Pamplona, Spain
e-mail: bustince@unavarra.es, fcojavier.fernandez@unavarra.es

T. Calvo
Department of Automatic and Computation, University of Alcalá, Madrid, Spain
e-mail: tomasa.calvo@uah.es

R. Mesiar
Slovak University of Technology, Bratislava, Slovakia
e-mail: mesiar@math.sk

B. De Baets et al. (Eds.): Eurofuse 2011, AISC 107, pp. 103–112, 2011.
springerlink.com

1 Introduction

Fuzzy sets, from their birth in [27], have been extended in several ways. Some of the such extensions are interval-valued fuzzy sets [28], Atanassov intuitionistic fuzzy sets [1, 2], interval-valued Atanassov intuitionistic fuzzy sets [3], fuzzy multisets or fuzzy bag sets[20, 26] and more recently, n-dimensional fuzzy sets [5, 25]. n-dimensional fuzzy sets consider a n-tuple of ordered values, named n-dimensional intervals, in the unit interval as membership degrees.

On the one hand, n-dimensional fuzzy sets generalize some extensions of fuzzy sets such as interval-valued fuzzy sets and interval-valued Atanassov intuitionistic fuzzy sets. On the other hand, they can be seen as a particular case of fuzzy multisets [5].

Analogously to other extensions of fuzzy set theory, the degrees of n-dimensional fuzzy sets, i.e., n-dimensional intervals, as pointed in [5], admit several interpretations. For example:

1. When n is even, an n-dimensional interval can be seen as a chain of $\frac{n}{2}$ nested intervals representing different uncertainty levels on the membership degree. Thus, the narrowest interval is an approximation of the membership degree which reflects the most optimistic uncertainty[1], whereas the broadest interval reflects the most pessimistic uncertainty. Case n is odd, we can think that we are in a framework where the most optimistic uncertainty is zero, i.e. see the middle value as a degenerate interval. This point of view can bee seen as a generalization of strong and weak uncertainty in bipolar sets [29] in 2004.
2. Memberships degrees given by n different evaluation processes ordered by rigidity.
3. For $n > 1$, an n-dimensional interval can be obtained from a $(n-1)$-dimensional interval considering an uncertainty in one of its components. Thus, for example, given a membership degree a, if we have an uncertainty in the value, we can use a 2-dimensional interval (a_1,a_2) such that $a_1 \leq a \leq a_2$ (see the method in [18]). If now, we have an uncertainty in a_1, then we could use a 3-dimensional interval (a_{1_1},a_{1_2},a_2) such that $a_{1_1} \leq a_1 \leq a_{1_2}$. If we have an uncertainty in a_2, we could use a 4-dimensional interval $(a_{1_1},a_{1_2},a_{2_1},a_{2_2})$ such that $a_{2_1} \leq a_2 \leq a_{2_2}$, and so on.

In [5], some fuzzy operations such as aggregating functions, fuzzy negations, automorphism and t-norms were extended for n-dimensional intervals. In particular t-norms have played an important role in fuzzy set theory in the narrow as in the broad sense because they model the conjunction in fuzzy logic and the intersection of fuzzy sets, and from them other connectives such as t-conorms (which model disjunction and union), residual implications and fuzzy negations (see for example [9]) can be derived. The notion of t-representability of Deschrijver in [12, 13] was generalized for n-dimensional t-norms in [5]. In this paper, we are interested in exploring the notion of t-representability for n-dimensional t-norms. In this sense we

[1] The uncertainty measure of an interval $[a,b]$ is proportional to their length (the radius $\frac{b-a}{2}$ in [21, 22] and the length in [10]).

provide necessary and sufficient conditions, i.e. a characterization of n-dimensional t-representable t-norms and we present some results on the t-representability of the conjugated of a t-representable n-dimensional t-norm.

The paper is organized as follows: In section 2 we provide the definitions and simplest results of t-norms, automorphism and the actions of automorphism to t-norms as well as the basic notion of some extensions of fuzzy sets. In section 3 we review the definition of n-dimensional fuzzy sets, their partial order, projections, and other useful concepts as well as their relation with interval-valued fuzzy sets, Atanassov intuitionistic fuzzy sets and interval-valued intuitionistic fuzzy sets. In section 4 we review the definition of n-dimensional t-norms and the related notion of t-representability and provide a characterization theorem for them. In section 5, based on the definition of n-dimensional automorphisms given in [5] and their characterization in term of automorphism, we provide a characterization of the conjugate of representable n-dimensional t-norms in terms of the conjugates of their representing t-norms. Finally, in section 6 we provide some final remarks and future works.

2 Preliminaries

In the this section we will recall some notions and notations in order to make the text more self-contained.

2.1 t-Norms and Automorphisms

The definitions and results in this section can be found in [19].

Definition 1. *A function $T : [0,1]^2 \to [0,1]$ is a triangular norm, in short t norm if for each $x,y,z \in [0,1]$*

(T1) $T(x,y) = T(y,x)$ *(symmetry);*
(T2) $T(x,T(y,z)) = T(T(x,y),z)$ *(associativity);*
(T3) *if* $x \leq y$ *then* $T(x,z) \leq T(y,z)$ *(monotonicity); and*
(T4) $T(x,1) = x$ *(neutral element).*

From (T1), (T3) and (T4) we can conclude that for each $x \in [0,1]$, $T(x,0) = T(0,x) = 0$ and $T(1,x) = x$. Therefore, all t-norms have the same behaviour on the boundary of the unit square $[0,1]^2$ and generalize the Boolean conjunction when we just consider the values $\{0,1\}$.

Definition 2. *A function $\psi : [0,1] \to [0,1]$ is an automorphism if it is bijective and increasing, i.e. for each $x,y \in [0,1]$ if $x \leq y$ then $\psi(x) \leq \psi(y)$.*

Automorphisms are closed under composition and the inverse of an automorphism is also an automorphism. Summarizing, the set of automorphism with the composition is a group.

Definition 3. *Let $f : [0,1]^n \to [0,1]$ and let ψ be an automorphism. The action of ψ on f is the function $f^\psi : [0,1]^n \to [0,1]$ defined by*

$$f^\psi(x_1,\ldots,x_n) = \psi^{-1}(f(\psi(x_1),\ldots,\psi(x_n))). \tag{1}$$

f^ψ is also called the conjugate of f.

Conjugates, in most of the cases, preserve the main properties of the function. For example, if f is an aggregation function, a t-norm or a (strong) fuzzy negation, then f^ψ is also an aggregation function, a t-norm and a (strong) fuzzy negation, respectively.

2.2 Generalizations of the Notion of Fuzzy Sets

In 1975, in an independent way (see for example [23, 28]), it was proposed to use intervals instead of a single value to describe the membership degree in order to model the possible imprecision when such degree is assigned. This kind of sets are called nowadays interval-valued fuzzy sets (IVFS) and the underlying lattice is $\langle L([0,1]), \leq_{L([0,1])}\rangle$ where $L([0,1]) = \{[a,b] : 0 \leq a \leq b \leq 1\}$ and for each $[a,b],[c,d] \in L([0,1])$, $[a,b] \leq_{L([0,1])} [c,d]$ iff $a \leq c$ and $b \leq d$.

Another important extension of fuzzy sets is nowadays known as Atanassov intuitionistic fuzzy sets (AIFS). AIFS were introduced by Krassimir Atanassov in [1], who used a pair of values (the membership degree and the non-membership degree) in order to model the eventual hesitation when the membership degree is provided. The underlying lattice for AIFS is $\langle L^*, \leq_{L^*}\rangle$ where $L^* = \{(a,b) \in [0,1]^2 : a+b \leq 1\}$ and for each $(a,b),(c,d) \in L^*$, $(a,b) \leq_{L^*} (c,d)$ iff $a \leq c$ and $d \leq b$. Although different from a semantical point of view, both extensions are isomorphic [14, **?**].

In [3], Atanassov and Gargov, proposed an integration of these two extensions, the so called interval-valued intuitionistic fuzzy sets (IVIFS), by considering a pair of intervals to denote the membership and non-memberships degrees. In this way, they model the hesitation in providing the membership degree as well as the imprecision in the membership and non-membership degrees. The underlying lattice is $\langle \mathbb{L}^*, \leq_{\mathbb{L}^*}\rangle$ where $\mathbb{L}^* = \{([a,b],[c,d]) \in L([0,1])^2 : b+d \leq 1\}$ and for any $(X_1,Y_1),(X_2,Y_2) \in \mathbb{L}^*$, $(X_1,Y_1) \leq_{\mathbb{L}^*} (X_2,Y_2)$ iff $X_1 \leq_{L([0,1])} X_2$ and $Y_2 \leq_{L([0,1])} Y_1$.

Several usual notions in fuzzy set theory has been extended for these extensions. In particular, the notion of t-norm was extended in several ways for these extensions (see for example [4, 8, 7, 10, 11, 12]).

3 n-Dimensional Fuzzy Sets

Definition 4 *[25]. Let X be a non empty set and $n \in \mathbb{N}^+ = \mathbb{N} - \{0\}$. An n-dimensional fuzzy set A over X is given by*

$$A = \{(x, \mu_{A_1}(x), \ldots, \mu_{A_n}(x)) \; : \; x \in X\}$$

where the mappings $\mu_{A_i} : X \to [0,1]$, $i = 1,\ldots,n$ *satisfy the condition* $\mu_{A_1} \leq \ldots \leq \mu_{A_n}$. *Each mapping* μ_{A_i} *is called the i-th membership degree of A.*

Analogously to the way Deschrijver and Kerre in [14] have shown that Atanassov intuitionistic fuzzy sets can also be seen as L-fuzzy sets in the sense of Goguen [16], we can also prove that n-dimensional fuzzy sets are a particular class of L-fuzzy sets.

Let $n \geq 1$. We consider the n-dimensional upper simplex:

$$L_n([0,1]) = \{(x_1,\ldots,x_n) \in [0,1]^n : x_1 \leq x_2 \leq \cdots \leq x_n\}. \quad (2)$$

Observe that $L_1([0,1]) = [0,1]$ and $L_2([0,1])$ reduces to the usual lattice $L([0,1])$ of all the closed subintervals of the unit interval $[0,1]$. In general, by considering the natural extension to higher dimensions of the order $\leq$ in $L([0,1])$, we can consider $L_n([0,1])$ as a lattice. Elements of $L_n([0,1])$ are called n-dimensional intervals.

For each $i = 1,\ldots,n$ the i-th projection of $L_n([0,1])$ is the function $\pi_i : L_n([0,1]) \to [0,1]$ defined by $\pi_i(x_1,\ldots,x_n) = x_i$. An element $\mathbf{x} \in L_n([0,1])$ is called degenerate if $\pi_i(\mathbf{x}) = \pi_j(\mathbf{x})$ for each $i,j = 1,\ldots,n$. The set of all degenerate elements of $L_n([0,1])$ will be denoted by $\mathscr{D}_n$. For each $x \in [0,1]$ the degenerate element $(x,\ldots,x)$ will be denoted by $/x/$. An m-ary function $F : L_n([0,1])^m \to L_n([0,1])$ is said to preserve degenerate elements if $F(\mathscr{D}_n^m) \subseteq \mathscr{D}_n$, i.e. if it satisfies the property:

(DP) $F(/x_1/,\ldots,/x_m/) \in \mathscr{D}_N$ for any $x_1,\ldots,x_m \in [0,1]$

Remark 1. As $L_n([0,1])$ generalizes $L([0,1])$, the set $L_n^* = \{(x_1,\ldots,x_n) \in [0,1]^n : \sum_{i=1}^{n} x_i \leq 1\}$ generalizes the set of Atanassov intuitionistic fuzzy values L^*. Moreover the well know isomorphism between L^* and $L([0,1])$ can also be generalized as follows:

$$h(a_1,\ldots,a_n) = (a_1, a_1 + a_2, \ldots, \sum_{i=1}^{n} a_i)$$

Therefore, the results in this paper are also valid for L_n^* and so they could be used to obtain an n-dimensional Atanassov intuitionistic fuzzy sets theory.

Remark 2. Notice also that interval-valued Atanassov intuitionistic fuzzy sets [3] are equivalent to 4-dimensional n-fuzzy sets [25]. In fact, the mapping $h : \mathbb{L}^* \to L_4([0,1])$ defined by

$$h([a,b],[c,d]) = (a, b, 1-d, 1-c) \quad (3)$$

is such an isomorphism.

The product order on $L_n([0,1])$ is defined by:

$$\mathbf{x} \leq \mathbf{y} \text{ iff } \pi_i(\mathbf{x}) \leq \pi_i(\mathbf{y}) \text{ for each } i = 1,\ldots,n \quad (4)$$

Notice that $\langle L_n([0,1]), \leq \rangle$ is a continuous lattice and so is a distributive complete lattice [15] with $[0]$ and $[1]$ being their bottom and top element, respectively.

4 n-Dimensional t-Norms

Thus as the notion of t-norm on $[0,1]$ has been extended for t-norms on $L([0,1])$ (see for example [4, 7, 13, 17]), we can also extend it for $L_n([0,1])$.

Definition 5 *[5]. A function $\mathscr{T} : L_n([0,1])^2 \to L_n([0,1])$ is a n-dimensional t-norm, if for each $\mathbf{x},\mathbf{y},\mathbf{z} \in L_n([0,1])$*

($\mathscr{T}1$) $\mathscr{T}(\mathbf{x},\mathbf{y}) = \mathscr{T}(\mathbf{y},\mathbf{x})$
($\mathscr{T}2$) $\mathscr{T}(\mathbf{x},\mathscr{T}(\mathbf{y},\mathbf{z})) = \mathscr{T}(\mathscr{T}(\mathbf{x},\mathbf{y}),\mathbf{z})$
($\mathscr{T}3$) If $\mathbf{y} \leq \mathbf{z}$ then $\mathscr{T}(\mathbf{x},\mathbf{y}) \leq \mathscr{T}(\mathbf{x},\mathbf{z})$
($\mathscr{T}4$) $\mathscr{T}(\mathbf{x},[1]) = \mathbf{x}$

Theorem 1 *[5]. Let $T_1,\ldots,T_n$ be t-norms such that $T_1 \leq T_2 \leq \ldots \leq T_n$. Then*

$$\widetilde{T_1 \ldots T_n}(\mathbf{x},\mathbf{y}) = (T_1(\pi_1(\mathbf{x}),\pi_1(\mathbf{y})),\ldots,T_n(\pi_n(\mathbf{x}),\pi_n(\mathbf{y}))) \tag{5}$$

is an n-dimensional t-norm.

An n-dimensional t-norm $\mathscr{T}$ is called t-representable if there exist t-norms $T_1 \leq \ldots \leq T_n$ such that $\mathscr{T} = \widetilde{T_1 \ldots, T_n}$. Notice that this notion coincides with the notion of t-representable interval-valued t-norms in [12, 13] when $n = 2$. When $T_1 = T_2 = \ldots = T_n = T$, the n-dimensional t-norm $\widetilde{T_1 \ldots T_n}$ will be denoted by $\widetilde{T}$.

Proposition 1. *A t-representable n-dimensional t-norm $\mathscr{T}$ satisfies (DP) iff $\mathscr{T} = \widetilde{T}$ for some t-norm T.*

Proof. Straightforward. □

Nevertheless, being of the form $\widetilde{T}$ is not a necessary condition for an n-dimensional t-norm to satisfy (DP). In fact, in equation (6), provides an example of non-t-representable n-dimensional t-norm which satisfies (DP).

$$\begin{aligned}\mathscr{T}(\mathbf{x},\mathbf{y}) = (&\min(\pi_1(\mathbf{x}),\pi_1(\mathbf{y})),\ldots,\min(\pi_{n-1}(\mathbf{x}),\pi_{n-1}(\mathbf{y})),\\ &\max(\min(\pi_{n-1}(\mathbf{x}),\pi_n(\mathbf{y})),\min(\pi_n(\mathbf{x}),\pi_{n-1}(\mathbf{y}))))\end{aligned} \tag{6}$$

Let $i = 1,\ldots,n-1$. For any $\mathbf{x},\mathbf{y} \in L_n([0,1])$ we write $\mathbf{x} \subseteq_i \mathbf{y}$ if $\pi_i(\mathbf{y}) \leq \pi_i(\mathbf{x}) \leq \pi_{i+1}(\mathbf{x}) \leq \pi_{i+1}(\mathbf{y})$. We say that an n-dimensional t-norm $\mathscr{T}$ is $\subseteq_i$-monotone if for any $\mathbf{x},\mathbf{y},\mathbf{z} \in L_n([0,1])$, $\mathscr{T}(\mathbf{x},\mathbf{y}) \subseteq_i \mathscr{T}(\mathbf{z},\mathbf{y})$ whenever $\mathbf{x} \subseteq_i \mathbf{z}$. We say that an n-dimensional t-norm $\mathscr{T}$ is $\subseteq$-monotone if it is $\subseteq_i$-monotone for each $i = 1,\ldots,n-1$.

We say that a n-dimensional t-norm $\mathscr{T}$ is motonone by part if for each $i = 1,\ldots,n$ and $\mathbf{x},\mathbf{y},\mathbf{z} \in L_n([0,1])$, $\pi_i(\mathscr{T}(\mathbf{x},\mathbf{y})) \leq \pi_i(\mathscr{T}(\mathbf{z},\mathbf{y}))$ whenever $\pi_i(\mathbf{y}) \leq \pi_i(\mathbf{z})$.

Theorem 2. *Let $\mathscr{T}$ be an n-dimensional t-norm. Then for each $i = 1,\ldots,n$ the function $T_i : [0,1]^2 \to [0,1]$ defined by:*

$$T_i(x,y) = \pi_i(\mathscr{T}(/x/,/y/)) \tag{7}$$

is a t-norm iff $\mathscr{T}$ is monotone by part and $\subseteq$-monotone.

Proof. ($\Rightarrow$) Clearly $\mathscr{T} = \widetilde{T_1 \ldots T_n}$ and $T_i \leq T_{i+1}$ for each $i = 1, \ldots, n-1$. Thus, by the monotonicity of T_i's, it is straightforward that $\mathscr{T}$ is monotone by part and $\subseteq$-monotone.

($\Leftarrow$) Symmetry, monotonicity and 1-identity of T_i are straightforward from the same properties of $\mathscr{T}$. Still, the associativity needs special attention:

Let $x, y, z \in [0,1]$ and $i = 1, \ldots, n-1$. Since $\mathscr{T}$ is monotone by part, then from equation (7), $\pi_i(\mathscr{T}(/x/, /T_i(y,z)/) \leq \pi_i(\mathscr{T}(/x/, \mathscr{T}(/y/, /z/))$ and so

$$T_i(x, T_i(y,z)) \leq \pi_i(\mathscr{T}(/x/, \mathscr{T}(/y/, /z/)).$$

On the other hand, $/T_i(y,z)/ \subseteq_i \mathscr{T}(/y/, /z/)$ then by $\subseteq$-monotonicity, $\mathscr{T}(/x/, /T_i(y,z)/) \subseteq_i \mathscr{T}(/x/, \mathscr{T}(/y/, /z/))$ and so $\pi_i(\mathscr{T}(/x/, /T_i(y,z)/)) \geq \pi_i(\mathscr{T}(/x/, \mathscr{T}(/y/, /z/)))$.

Thus, $T_i(x, T_i(y,z)) \geq \pi_i(\mathscr{T}(/x/, \mathscr{T}(/y/, /z/))$ and therefore

$$T_i(x, T_i(y,z)) = \pi_i(\mathscr{T}(/x/, \mathscr{T}(/y/, /z/)).$$

Following an analogous reasoning, we have that $T_i(T_i(x,y), z) = \pi_i(\mathscr{T}(\mathscr{T}(/x/, /y/), /z/)$. Hence, by associativity of $\mathscr{T}$ we have that

$$T_i(x, T_i(y,z)) = T_i(T_i(x,y), z).$$

It only remains to prove the case of T_n. Since, $/T_n(y,z)/ \geq \mathscr{T}(/y/, /z/)$ then by monotonicity of $\mathscr{T}$ we have that $\pi_n(\mathscr{T}(/x/, /T_n(y,z)/)) \geq \pi_n(\mathscr{T}(/x/, \mathscr{T}(/y/, /z/)))$ and because it is monotone by part, $\pi_n(\mathscr{T}(/x/, /T_n(y,z)/)) \leq \pi_n(\mathscr{T}(/x/, \mathscr{T}(/y/, /z/)))$. So, $\pi_n(\mathscr{T}(/x/, /T_n(y,z)/)) = \pi_n(\mathscr{T}(/x/, \mathscr{T}(/y/, /z/)))$ and therefore $T_n(x, T_n(y,z)) = \pi_n(\mathscr{T}(/x/, \mathscr{T}(/y/, /z/)))$. Following an analogous reasoning, is possible to prove that $T_n(T_n(x,y), z) = \pi_n(\mathscr{T}(\mathscr{T}(/x/, /y/)), /z/)$. Hence, by associativity of $\mathscr{T}$ we have that

$T_n(x, T_n(y,z)) = T_n(T_n(x,y), z).$ □

Corollary 1. *A n-dimensional t-norm is t-representable iff is monotone by part and* $\subseteq$*-monotone*

Proof. Straightforward from Theorem 2. □

5 n-Dimensional Automorphisms Acting on n-Dimensional t-Norms

A function $\varphi : L_n([0,1]) \to L_n([0,1])$ is an n-dimensional automorphism [5] if φ is bijective and

$$\mathbf{x} \leq \mathbf{y} \text{ iff } \varphi(\mathbf{x}) \leq \varphi(\mathbf{y}) \quad (8)$$

Theorem 3 *[5]. $\varphi : L_n([0,1]) \to L_n([0,1])$. φ is an n-dimensional automorphism iff there exists an automorphism $\psi : [0,1] \to [0,1]$ such that for each $\mathbf{x} \in L_n([0,1])$*

$$\varphi(\mathbf{x}) = (\psi(\pi_1(\mathbf{x})), \ldots, \psi(\pi_n(\mathbf{x}))) \tag{9}$$

In this case we denote φ by $\widetilde{\psi}$.

The next proposition extends the Proposition 46 in [6].

Proposition 2 *[5]. Let ψ be an automorphism. Then*

$$\widetilde{\psi^{-1}} = \widetilde{\psi}^{-1} \tag{10}$$

Given a function $F : L_n([0,1])^n \rightarrow L_n([0,1])$ and an n-dimensional automorphism φ, the action of φ to F is the function $F^{\varphi} : L_n([0,1])^n \rightarrow L_n([0,1])$ defined by

$$F^{\varphi}(x_1, \ldots, x_n) = \varphi^{-1}(F(\varphi(x_1), \ldots, \varphi(x_n))) \tag{11}$$

F^{φ} is said the conjugate of F.

Theorem 4 *[5]. Let $\mathscr{T}$ be an n-dimensional t-norm and φ and n-dimensional automorphism. Then $\mathscr{T}^{\varphi}$ is also an n-dimensional t-norm.*

Theorem 5. *Let $T_1, \ldots, T_n$ be t-norms such that $T_i \leq T_{i+1}$ for each $i = 1, \ldots, n-1$ and ψ an automorphism. Then*

$$\widetilde{T_1 \ldots T_n}^{\widetilde{\psi}} = \widetilde{T_1^{\psi} \ldots T_n^{\psi}}$$

Proof. Let $\mathbf{x}, \mathbf{y} \in L_n([0,1])$. Then

$$\begin{aligned}
&\widetilde{T_1 \ldots T_n}^{\widetilde{\psi}}(\mathbf{x}, \mathbf{y}) = \widetilde{\psi}^{-1}(\widetilde{T_1 \ldots T_n}(\widetilde{\psi}(\mathbf{x}), \widetilde{\psi}(\mathbf{y}))) && \text{by eq. (11)}\\
&= \widetilde{\psi^{-1}}(\widetilde{T_1 \ldots T_n}((\psi(\pi_1(\mathbf{x})), \ldots, \psi(\pi_n(\mathbf{x}))), \\
&\quad (\psi(\pi_1(\mathbf{y})), \ldots, \psi(\pi_n(\mathbf{y}))))) && \text{by prop. 2 and eq. (9)}\\
&= \widetilde{\psi^{-1}}(T_1(\psi(\pi_1(\mathbf{x})), \psi(\pi_1(\mathbf{y}))), \ldots, T_n(\psi(\pi_n(\mathbf{x})), \psi(\pi_n(\mathbf{y})))) && \text{by eq. (5)}\\
&= (T_1^{\psi}(\pi_1(\mathbf{x}), \pi_1(\mathbf{y})), \ldots, T_n^{\psi}(\pi_n(\mathbf{x}), \pi_n(\mathbf{y}))) && \text{by eq. (1) and (9)} \quad \square
\end{aligned}$$

Therefore, the conjugate of a t-representable n-dimensional t-norm also is t-representable. Moreover their representants are the conjugates.

6 Final Remarks

n-dimensional fuzzy sets and, in particular, n-dimensional t-norms, can be useful when it is necessary to consider several levels of uncertainty in the membership degrees, for example by using n different methods of (or experts) evaluations ordered by rigidity. Triangular norms have been applied in several aspects of fuzzy sets theory, and clearly each one of these applications could be extended by means of n-dimensional t-norms if we consider n levels of uncertainty in each membership degree.

In order to continue this work, we will attempt to study some usual properties of t-norm, such as continuity, strictness, cancellation law, existence of zero divisors, idempotency, nilpotent elements and classes of n-dimensional t-norm which extend the class of Archimedean and Lipschitzian t-norms.

Acknowledgements. This paper has been partially supported by the Brazilian research council CNPq (projects 308256/2009-3 and 201118/2010-6), by the National Science Foundation of Spain, (reference TIN 2010-15055) and by the Research Services of the Universidad Pública de Navarra.

References

1. Atanassov, K.T.: Intuitionistic Fuzzy Sets. Fuzzy Sets and Systems 20, 87–96 (1986)
2. Atanassov, K.T.: Intuitionistic Fuzzy Sets. Theory and Applications. Physica-Verlag, Heidelberg (1999)
3. Atanassov, K.T., Gargov, G.: Interval Valued Intuitionistic Fuzzy Sets. Fuzzy Sets and Systems 31, 343–349 (1989)
4. Bedregal, B.C., Takahashi, A.: Interval t-norms as interval representations of t-norms. In: Proc. of IEEE International Conference on Fuzzy Systems (Fuzz-IEEE), Reno, Nevada, May 22–25, pp. 909–914 (2005)
5. Bedregal, B., Beliakov, G., Bustince, H., Calvo, T., Mesiar, R., Paternain, D.: A class of fuzzy multisets with a fixed number of memberships. Information Sciences (submitted)
6. Bedregal, B.C., Dimuro, G.P., Santiago, R.H.N., Reiser, R.H.S.: On interval fuzzy S-implications. Information Sciences 180, 1373–1389 (2010)
7. Bustince, H., Barrenechea, E., Pagola, M.: Generation of interval-valued fuzzy and Atanassov's intuitionistic fuzzy connectives from fuzzy connectives and from K_α operators: Laws for conjunctions and disjunctions, amplitude. Int. J. of Intelligent Systems 23, 680–714 (2008)
8. Bustince, H., Burillo, P.: A theorem for constructing interval-valued intuitionistic fuzzy sets from intuitionistic fuzzy sets. Notes on Intuitionistic Fuzzy Sets 1, 5–16 (1995)
9. Bustince, H., Burillo, P., Soria, F.: Automorphism, negations and implication operators. Fuzzy Sets and Systems 134, 209–229 (2003)
10. Cornelis, C., Deschrijver, G., Kerre, E.E.: Implication in intuitionistic fuzzy and interval-valued fuzzy set theory: construction, classification, application. Int. J. of Approximate Reasoning 35, 55–95 (2004)
11. Da Costa, C.G., Bedregal, B.C., Doria Neto, A.D.: Relating De Morgan triples with Atanassov's intuitionistic De Morgan triples via automorphisms. Int. J. Approximate Reasoning 52(4), 473–487 (2011)
12. Deschrijver, G.: A representation of t-norms in interval-valued L-fuzzy set theory. Fuzzy Sets and Systems 159, 1597–1618 (2008)
13. Deschrijver, G., Cornelis, C.: Representability in interval-valued fuzzy set theory. Int. J. 15(3), 345–361 (2007)
14. Deschrijver, G., Kerre, E.E.: On the relation between some extensions of fuzzy set theory. Fuzzy Sets and Systems 133, 227–235 (2003)
15. Gehrke, M., Walker, C., Walker, E.: De Morgan systems on the unit interval. Int. J. on Intelligent Systems 11, 733–750 (1996)
16. Goguen, J.: L-fuzzy sets. Journal of Mathematics Analisys Applied 18, 145–167 (1967)
17. Jenei, S.: A more efficient method for defining fuzzy connectives. Fuzzy Sets and Systems 90, 25–35 (1997)
18. Jurío, A., Pagola, M., Mesiar, R., Beliakov, G., Bustince, H.: Image magnification using interval information. IEEE Trans. on Image Processing; doi:10.1109/TIP.2011.2158227
19. Klement, E.P., Mesiar, R., Pap, E.: Triangular Norms. Kluwer, Dordrecht (2000)
20. Miyamoto, S.: Multisets and fuzzy multisets. In: Liu, Z.-Q., Miyamoto, S. (eds.) Soft Computing and Human-Centered Machines, pp. 9–33. Springer, Berlin (2000)

21. Moore, R.E., Kearfott, R.B., Cloud, M.J.: Introduction to Interval Analysis. SIAM, Philadelphia (2009)
22. Reiser, R.H.S., Dimuro, G.P., Bedregal, B.C., Santiago, R.H.N.: Interval Valued QL-Implications. In: Leivant, D., de Queiroz, R. (eds.) WoLLIC 2007. LNCS, vol. 4576, pp. 307–321. Springer, Heidelberg (2007)
23. Sambuc, R.: Fonctions ϕ-floues. Application á l'aide au diagnostic en pathologie thyroidienne, Ph.D. Thesis, Université de Marseille, France (1975)
24. Scott, D.S.: Continuous Lattices. Lecture Notes in Mathematics 274, 97–136 (1972)
25. Shang, Y., Yuan, X., Lee, E.S.: The n-dimensional fuzzy sets and Zadeh fuzzy sets based on the finite valued fuzzy sets. Computers and Mathematics with Applications 60, 442–463 (2010)
26. Yager, R.R.: On the theory of bags. Int. J. General Systems 13, 23–37 (1986)
27. Zadeh, L.A.: Fuzzy sets. Information and Control 8, 338–353 (1965)
28. Zadeh, L.A.: The concept of a linguistic variable and its application to approximate reasoning - I. Information Sciences 6, 199–249 (1975)
29. Zhang, W.R., Zhang, L.: Yin Yang bipolar logic and bipolar fuzzy logic. Information Sciences 165, 265–287 (2004)

A Construction Method of Aggregations Functions on the Set of Discrete Fuzzy Numbers

J. Vicente Riera and Joan Torrens

Abstract. In this article we propose a method to construct aggregation functions on the set of discrete fuzzy numbers whose support is a set of consecutive natural numbers contained in the finite chain $L = \{0, 1, \cdots, n\}$ from a couple of aggregation functions also defined on L. In addition, if the pair of discrete aggregation functions fulfills several properties such as associativity, commutativity or idempotence, we show that this new operator will satisfy these properties too. The particular case of uninorms is studied showing that some properties and part of the structure of the uninorms is preserved under the presented construction method. Finally, we provide an application of this last operator in a decision-making problem.

1 Introduction

The theory of aggregation functions has been extensively developed in last decades, mainly because they have a great number of applications which include many subjects not only from Mathematics and Computer Science, but also from many applied fields like economics and social sciences [1, 8]. One of the fields, where aggregation functions are specially applied, is the fuzzy set theory and its applications. Not only because many of the usual fuzzy connectives like t-norms, t-conorms, uninorms, copulas, are special kinds of aggregation functions, but also because many times the data to be aggregated are not crisp numbers, but fuzzy numbers and even subjective qualitative information.

J. Vicente Riera
University of the Balearic Islands, Crta. Valldemossa, km. 7.5 Palma de Mallorca(Spain)
e-mail: jvicente.riera@uib.es

Joan Torrens
University of the Balearic Islands, Crta. Valldemossa, km. 7.5 Palma de Mallorca(Spain)
e-mail: dmijts0@uib.es

B. De Baets et al. (Eds.): Eurofuse 2011, AISC 107, pp. 113–124, 2011.
springerlink.com

Qualitative information is often interpreted to take values in a finite scale like Extremely Good, Very Good, Good, Fair, Bad, Very Bad, Extremely Bad. In these cases, the representative finite chain $L = \{0, 1, \ldots, n\}$ is usually considered to model these linguistic hedges and several researchers have developed an extensive study of aggregation functions on L, usually called *discrete aggregation functions* (see [11, 13, 15]).

Recently, another approach deals with the possibility of extending monotonic operations on L to operations on the set of discrete fuzzy numbers whose support is a set of consecutive natural numbers contained in L [4, 16]. More specifically, the concept of discrete fuzzy number was introduced in [17] as a fuzzy subset of $\mathbb{R}$ with discrete support and analogous properties to a fuzzy number. It is well known that arithmetic and lattice operations between discrete fuzzy numbers defined using the Zadeh's extension principle [10] fail and some approaches have been introduced in order to avoid such a drawback [2, 3]. In particular, it is proved in [3] that the set, $\mathscr{A}_1^L$, of discrete fuzzy numbers whose support is a set of consecutive natural numbers contained in L, is a distributive lattice.

Thus, it becomes natural to study aggregation functions defined on $\mathscr{A}_1^L$ equipped with the usual lattice order. In this way, one approach is the one already commented of extending monotonic operations defined on L to monotonic operations defined on the set $\mathscr{A}_1^L$. This was done for different kinds of aggregation functions in [4, 16]. Following with this idea we want to study in this paper the possibility of constructing aggregation functions on $\mathscr{A}_1^L$ from a pair of aggregation functions F, G on L with $F \leq G$. The special case of uninorms is studied proving that some properties and part of the structure of the uninorm is preserved under the presented construction method. At the end, we show an application of these aggregation functions to get a group consensus opinion.

2 Preliminaries

In this section, we recall some definitions and results that we will use along the paper. More details can be found in [5, 13, 15] for operations on discrete settings, and in [2, 3] for discrete fuzzy numbers.

2.1 Operations on Partially Ordered Sets

Let $(P;\leq)$ be a non-trivial bounded partially ordered set (poset) with 0 and 1 as minimum and maximum elements respectively.

Definition 1. *A uninorm on P is a two-place function $U : P \times P \longrightarrow P$ which is associative, increasing in each place, commutative, and such that there exists some element $e \in P$, called* neutral element, *such that $U(e,x) = x$ for all $x \in P$.*

It is clear that the function U becomes a t-norm when $e = 1$ and a t-conorm when $e = 0$.

An important case is when we take as poset a finite chain L with $n+1$ elements. In such a framework only the number of elements is relevant (see [15]) and so it is usually considered the most simple one, that is, $L = \{0,1,\cdots,n\}$. Operations on L are usually called *discrete* operations. With respect to discrete uninorms, the idempotent ones and those in $U_{\min}$ and $U_{\max}$ have been characterized. Note that in both cases the characterization is quite similar to the case of $[0,1]$ (see [6] and [13], respectively).

2.2 Discrete Fuzzy Numbers

By a fuzzy subset of $\mathbb{R}$, we mean a function $A : \mathbb{R} \to [0,1]$. For each fuzzy subset A, let $A^{\alpha} = \{x \in \mathbb{R} : A(x) \geq \alpha\}$ for any $\alpha \in (0,1]$ be its α-level set (or α-cut). By $supp(A)$, we mean the support of A, i.e. the set $\{x \in \mathbb{R} : A(x) > 0\}$. By A^0, we mean the closure of $supp(A)$.

Definition 2. *[17] A fuzzy subset A of $\mathbb{R}$ with membership mapping $A : \mathbb{R} \to [0,1]$ is called* discrete fuzzy number *if its support is finite, i.e., there exist $x_1,...,x_n \in \mathbb{R}$ with $x_1 < x_2 < ... < x_n$ such that $supp(A) = \{x_1,...,x_n\}$, and there are natural numbers s,t with $1 \leq s \leq t \leq n$ such that:*

1. *$A(x_i)=1$ for any natural number i with $s \leq i \leq t$ (*core*)*
2. *$A(x_i) \leq A(x_j)$ for each natural number i,j with $1 \leq i \leq j \leq s$*
3. *$A(x_i) \geq A(x_j)$ for each natural number i,j with $t \leq i \leq j \leq n$*

In [18], Wang et al. characterized each discrete fuzzy number by means of its α-cuts.

From now on, we will denote by $\mathscr{A}_1^L$ the set of all discrete fuzzy numbers whose support is a subset of consecutive natural numbers of L.

Proposition 1. *[2] Given A,B two discrete fuzzy numbers. There exist two unique discrete fuzzy numbers, denoted by $MIN(A,B)$ and $MAX(A,B)$, whose α-cuts are given by*

$$\min(A,B)^{\alpha} = \{z = \min(x,y), x \in supp(A), y \in supp(B) \text{ such that } \min(\min(A^{\alpha}),\min(B^{\alpha})) \leq z \leq \min(\max(A^{\alpha}),\max(B^{\alpha}))\}$$

and

$$\max(A,B)^{\alpha} = \{z = \max(x,y), x \in supp(A), y \in supp(B) \text{ such that } \max(\min(A^{\alpha}),\min(B^{\alpha})) \leq z \leq \max(\max(A^{\alpha}),\max(B^{\alpha}))\},$$

respectively. In addition, if $A,B \in \mathscr{A}_1^L$ then also $MIN(A,B)$ and $MAX(A,B)$ belong to the set $\mathscr{A}_1^L$.

Remark 1. [3] Using these operations, we can define a partial order on $\mathscr{A}_1^L$ in the usual way:
$A \preceq B$ if and only if $MIN(A,B) = A$.

The following result was proved in [4].

Theorem 1. *The triplet* $(\mathscr{A}_1^L, MIN, MAX)$ *is a bounded distributive lattice where* $N \in \mathscr{A}_1^L$ *(the unique discrete fuzzy number whose support is the singleton* $\{n\}$*) and* $O \in \mathscr{A}_1^L$ *(the unique discrete fuzzy number whose support is the singleton* $\{0\}$*) represent the maximum and the minimum, respectively.*

3 Aggregation of Discrete Fuzzy Numbers

In this section we wish to investigate if it is possible to build aggregation functions on the bounded distributive lattice $\mathscr{A}_1^L$ from a couple of aggregation functions on L. Moreover, we will study the well known relevant special case of the uninorms. Let us begin by recalling the following definition.

Definition 3. *An n-ary aggregation function on a bounded partially ordered set P with minimum element* 0 *and maximum element* 1*, is a function* $F : P^n \to P$ *such that it is increasing in each component,* $F(0,\ldots,0) = 0$ *and* $F(1,\ldots,1) = 1$.

Of course, the number of inputs to be aggregated can be different in each case. Thus, aggregation functions are commonly defined not on P^n, but on $\cup_{n\geq 1} P^n$ and then they are usually called *extended aggregation functions*. An easy way to construct extended aggregation functions is from associative binary aggregation functions. For this reason, from now on, we will focus our study on the binary case and we will deal with the special associative cases of the uninorms.

Let us begin with some notation. For any discrete aggregation function $F : L \times L \to L$ and subsets $X, Y \subseteq L$, we denote by $F(X,Y) = \{F(x,y) \mid x \in X, y \in Y\}$.

Note that aggregation functions on $\mathscr{A}_1^L$ were already constructed from aggregation functions on L in [4, 16]. We recall here the main result in this sense.

Theorem 2. *[4] Let F be an aggregation function on L. Then the function*

$$\begin{aligned}\mathscr{F} : \mathscr{A}_1^L \times \mathscr{A}_1^L &\longrightarrow \mathscr{A}_1^L\\ (A,B) &\longmapsto \mathscr{F}(A,B)\end{aligned}$$

is an aggregation function on $\mathscr{A}_1^L$*, where* $\mathscr{F}(A,B)$ *is the discrete fuzzy number whose* α*-cut sets are the sets* $\mathscr{F}(A,B)^\alpha = \{z \in L \mid \min F(A^\alpha, B^\alpha) \leq z \leq \max F(A^\alpha, B^\alpha)\}$ *for each* $\alpha \in [0,1]$.

Now, we want to proceed in a similar way but from a pair of binary aggregation functions F and G on L with $F \leq G$.

Proposition 2. *Let us consider $A,B \in \mathscr{A}_1^L$ and let F and G be a couple of binary aggregation functions on the finite chain L with $F \leq G$. There exists a unique discrete fuzzy number, denoted by $[\mathscr{F},\mathscr{G}](A,B)$, whose α-cuts are exactly the sets*

$$C_{F,G}^{\alpha}(A,B) = \{z \in L \mid \min F(A^{\alpha},B^{\alpha}) \leq z \leq \max G(A^{\alpha},B^{\alpha})\} \tag{1}$$

Moreover, $[\mathscr{F},\mathscr{G}](A,B) \in \mathscr{A}_1^L$.

The previous proposition allows us to define a binary operation $[\mathscr{F},\mathscr{G}]$ on $\mathscr{A}_1^L$ from any couple of binary aggregation function F and G with $F \leq G$ defined on the finite chain L.

Definition 4. *Let us consider a couple of binary aggregation functions F,G on the finite chain L with $F \leq G$. The binary operation on $\mathscr{A}_1^L$ defined as follows*

$$\begin{aligned}[\mathscr{F},\mathscr{G}] : \mathscr{A}_1^L \times \mathscr{A}_1^L &\longrightarrow \mathscr{A}_1^L \\ (A,B) &\longmapsto [\mathscr{F},\mathscr{G}](A,B)\end{aligned}$$

will be called the extension to $\mathscr{A}_1^L$ *of couple of the discrete aggregation functions F and G, being $[\mathscr{F},\mathscr{G}](A,B)$ the discrete fuzzy number whose α-cuts are as in (1) for each $\alpha \in [0,1]$.*

From now on we will denote by $[\mathscr{F},\mathscr{G}](A,B)^{\alpha}$ the sets $C_{F,G}^{\alpha}(A,B)$ considered in relation (1) for each $\alpha \in [0,1]$.

Proposition 3. *Let $[\mathscr{F},\mathscr{G}] : \mathscr{A}_1^L \times \mathscr{A}_1^L \to \mathscr{A}_1^L$ be the extension of the discrete aggregation functions F and G on L to $\mathscr{A}_1^L$. Let 0 and N be the minimum and the maximum of $\mathscr{A}_1^L$, respectively. Then the following properties hold*

1. *$[\mathscr{F},\mathscr{G}]$ is increasing in each place,*
2. *$[\mathscr{F},\mathscr{G}](O,O) = O$*
3. *$[\mathscr{F},\mathscr{G}](N,N) = N$.*

Thus, given any couple of binary aggregation functions F and G with $F \leq G$ on L, its extension $[\mathscr{F},\mathscr{G}]$ to $\mathscr{A}_1^L$ is a binary aggregation function on $\mathscr{A}_1^L$. When $F = G$, it is clear from the definitions that we obtain $[\mathscr{F},\mathscr{F}] = \mathscr{F}$, that is, extensions by a couple of aggregation functions generalize the original extension of aggregation functions given in [16] (see also Theorem 2). Moreover, in the general case, the extension of a couple (F,G) always leads to an aggregation function between $\mathscr{F}$ and $\mathscr{G}$, as it is proved in the following proposition:

Proposition 4. *Let F,G be a couple of aggregation functions on L with $F \leq G$ and let $[\mathscr{F},\mathscr{G}]$ be its extension. Then $\mathscr{F} \preceq [\mathscr{F},\mathscr{G}] \preceq \mathscr{G}$, that is, $\mathscr{F}(A,B) \preceq [\mathscr{F},\mathscr{G}](A,B) \preceq \mathscr{G}(A,B)$ for all $A,B \in \mathscr{A}_1^L$.*

On the other hand, let us introduce the following notation. For any $a \in L$ we will denote by 1_a the unique discrete fuzzy number whose support is given by the singleton $\{a\}$. Note that with this notation the minimum and the maximum elements of $\mathscr{A}_1^L$ can be also denoted by 1_0 and 1_n, respectively.

Proposition 5. *Let F and G be aggregation functions on L with $F \leq G$. When we restrict $[\mathscr{F},\mathscr{G}]$ to crisp numbers of $\mathscr{A}_1^L$ we obtain the discrete fuzzy number whose α-cut sets are the interval of the finite chain L, $[F(a,b),G(a,b)]$ for all $\alpha \in [0,1]$.*

Proposition 6. *Let $[\mathscr{F},\mathscr{G}] : \mathscr{A}_1^L \times \mathscr{A}_1^L \to \mathscr{A}_1^L$ be the extension of the aggregation functions F and G on L to $\mathscr{A}_1^L$. Then the following properties hold.*

1. *$[\mathscr{F},\mathscr{G}]$ is a commutative aggregation function if and only if F and G are commutative as well.*
2. *$[\mathscr{F},\mathscr{G}]$ is an associative aggregation function if and only if F and G are associative as well.*

Theorem 3. *Let $[\mathscr{F},\mathscr{G}] : \mathscr{A}_1^L \times \mathscr{A}_1^L \to \mathscr{A}_1^L$ be the extension of the couple of aggregation functions F and G to $\mathscr{A}_1^L$. Then, the following properties hold*

1. *e is a common neutral element of F and G if and only if 1_e is a neutral element of $[\mathscr{F},\mathscr{G}]$. That is, $[\mathscr{F},\mathscr{G}](A,1_e) = [\mathscr{F},\mathscr{G}](1_e,A) = A$ for all $A \in \mathscr{A}_1^L$.*
2. *k is an common absorbing element of F and G if and only if 1_k is an absorbing element of $[\mathscr{F},\mathscr{G}]$. That is, $[\mathscr{F},\mathscr{G}](A,1_k) = [\mathscr{F},\mathscr{G}](1_k,A) = 1_k$ for all $A \in \mathscr{A}_1^L$.*
3. *$[\mathscr{F},\mathscr{G}]$ is idempotent if and only if F and G are idempotent.*

It is well known that uninorms are a special case of binary associative aggregation functions on L. So, we will use the results above in order to construct uninorms on $\mathscr{A}_1^L$ from a couple of uninorms of L.

Theorem 4. *Let U and U' be a couple of discrete uninorms on L with $U \leq U'$ and $e \in L$ as common neutral element and let $[\mathscr{U},\mathscr{U}']$ be the extension of the couple of U and U' to $\mathscr{A}_1^L$, defined according to Definition 4. Then, $[\mathscr{U},\mathscr{U}']$ is a uninorm on $\mathscr{A}_1^L$ with neutral element 1_e. Moreover, U and U' are idempotent uninorms if and only if so is its extension $[\mathscr{U},\mathscr{U}']$.*

Remark 2. Note that, in particular, a couple of discrete t-norms on L leads to a t-norm on $\mathscr{A}_1^L$ and a couple of discrete t-conorms on L lead to a t-conorm on $\mathscr{A}_1^L$.

Example 1. Consider the discrete idempotent uninorm [6]

$$U(x,y) = \begin{cases} \min(x,y) & \text{if } y \leq 6-x \\ \max(x,y) & \text{if otherwise} \end{cases}$$

and

$$U'(x,y) = \begin{cases} \min(x,y) & \text{if } (x,y) \in [0,3]^2 \\ \max(x,y) & \text{otherwise} \end{cases}$$

defined on the finite chain $L = \{0,1,2,3,4,5,6\}$ (It is obvious that $U \leq U'$ for all $(x,y) \in [0,6]^2$).

Consider, $A = \{0.3/0, 0.3/1, 1/2, 0.3/3\}, B = \{0.3/2, 0.5/3, 1/4, 0.8/5\} \in \mathscr{A}_1^L$. Then, $[\mathscr{U},\mathscr{U}'](A,B) = \{0.3/0, 0.5/1, 1/2, 1/3, 1/4, 0.8/5\}$.

Note that, since U and U' are idempotent so is $[\mathscr{U},\mathscr{U}']$. Thus, for instance, $[\mathscr{U},\mathscr{U}'](A,A) = A$ and $[\mathscr{U},\mathscr{U}'](B,B) = B$.

Proposition 7. *Let U and U' be a couple of uninorms on the finite chain L with the same neutral element $e \in L$ and $U \leq U'$. Then its extension $[\mathscr{U},\mathscr{U}']$ satisfies:*

1. $[\mathscr{U},\mathscr{U}'](A,N) = N$ *for all* $A \succeq 1_e$ *and* $[\mathscr{U},\mathscr{U}'](A,0) = 0$ *for all* $A \preceq 1_e$.
2. $[\mathscr{U},\mathscr{U}'](0,N) \in \{0,N,\mathbb{L}\}$

where N and 0 denote the maximum and the minimum of the bounded distributive lattice $\mathscr{A}_1^L$ respectively and $\mathbb{L}$ is the discrete fuzzy number whose α-cuts are the proper finite chain $L = [0,n]$ for all $\alpha \in [0,1]$.

Proposition 8. *Let us consider the uninorms*

$$U(x,y) = \begin{cases} T(x,y) \text{ if } (x,y) \in [0,e]^2 \\ S(x,y) \text{ if } (x,y) \in [e,n]^2 \\ \min(x,y) \text{ otherwise} \end{cases}$$

$$U'(x,y) = \begin{cases} T'(x,y) \text{ if } (x,y) \in [0,e]^2 \\ S'(x,y) \text{ if } (x,y) \in [e,n]^2 \\ \min(x,y) \text{ otherwise} \end{cases}$$

on L with neutral element $0 < e < n$, being $T \leq T'$ a pair of t-norms on $[0,e]$ and $S \leq S'$ a pair of t-conorms on $[e,n]$. Let $[\mathscr{U},\mathscr{U}']$ be the extension of the couple U and U' to $\mathscr{A}_1^L$. Then

(i) *If $A,B \preceq 1_e$ then $[\mathscr{U},\mathscr{U}'](A,B) = [\mathscr{T},\mathscr{T}'](A,B)$ where $[\mathscr{T},\mathscr{T}']$ is the extension of the couple T and T' to $\mathscr{A}_1^{[0,e]}$.*
(ii) *If $A,B \succeq 1_e$ then $[\mathscr{U},\mathscr{U}'](A,B) = [\mathscr{S},\mathscr{S}'](A,B)$ where $[\mathscr{S},\mathscr{S}']$ is the extension of the couple S and S' to $\mathscr{A}_1^{[e,n]}$.*
(iii) *If $A \preceq 1_e \preceq B$ then $[\mathscr{U},\mathscr{U}'](A,B) = MIN(A,B) = A$.*

Analogously, it is possible to give a similar result when the uninorms U and U' belong to the set $U_{\max}$.

Remark 3. Note that the previous theorems do not give the complete structure of uninorms in $\mathscr{A}_1^L$ that are extensions of uninorms in $U_{\min}$ and $U_{\max}$ on L. Since the order in $\mathscr{A}_1^L$ is not total there are elements $A,B \in \mathscr{A}_1^L$ not comparable with 1_e. Thus if A or B are one of these elements only the general expression of $\mathscr{U}(A,B)$ works.

4 Decision Making Based on a Couple of Uninorms on $\mathscr{A}_1^L$

In recent years, the issue of aggregation operators has been developed mainly from two points of view. On the one hand, the theoretical study of these operators and their properties, for example [1, 5, 8, 10, 13]. And, on the other hand the possible applications of them in several fields of knowledge such as social science(e.g. decision making [9, 19, 20]), applied sciences (e.g. image processing [7, 12]), educational sciences [16]. In particular, it is well known the use of uninorms defined on the unit interval or on a finite chain as a useful tool in problems of decision making

[14, 19, 20]. In the previous section, we have discussed a method to build uninorms on $\mathscr{A}_1^L$ from a couple of uninorms defined on the finite chain L. Thus, we propose to use these operators obtained from a couple of uninorms on L, in order to get the consensus opinion of two group of experts.

Suppose that a hotel group engages the services of two expert groups to evaluate a possible investment in a foreign country. The first expert group is usually hired by the company for such decisions. The second one is specifically hired in this foreign country only to assess the viability of the investment. The proposed method is presented as follows:

Step 1: Establishing the expert groups $NEG = \{O_1, \cdots, O_r\}$ and $FEG = \{(FO)_1, \cdots, (FO)_k\}$ who carry out the evaluation process of the parameters $\mathbf{P} = \{P_1, \cdots, P_s\}$.

Step 2: Choose the linguistic hedges L which are used to make the evaluation process.

Step 3: Each expert $O_j \in NEG$ (with $j = 1 \cdots r$) performs an assessment $O_j^{P_i} \in \mathscr{A}_1^L$ of all parameters $P_i \in \mathbf{P}$ chosen. And, analogously each expert $(FO)_j \in FEG$ (with $j = 1, \cdots, k$) takes into action an assessment $(FO)_j^{P_i} \in \mathscr{A}_1^L$ of all parameters $P_i \in \mathbf{P}$ chosen.

Step 4: For each parameter $P_i \in \mathbf{P}$, the FEG chooses an uninorm $\mathscr{F}\mathscr{U}_i$ to calculate the aggregation of all valuations $\{(FO)_1^{P_i}, \cdots, (FO)_k^{P_i}\}$. According to the previous election, the NEG chooses another uninorm $\mathscr{U}_i$ to calculate the aggregation of all valuations $\{O_1^{P_i}, \cdots, O_r^{P_i}\}$, fulfilling the order relation $\mathscr{U}_i \preceq \mathscr{F}\mathscr{U}_i$ [1]. These aggregations will be denoted by $C(\mathscr{N}\mathscr{O}, P_i) = \mathscr{U}_i(O_1^{P_i}, \cdots, O_r^{P_i})$ and $C(\mathscr{F}\mathscr{O}, P_i) = \mathscr{F}\mathscr{U}_i(O_1^{P_i}, \cdots, O_k^{P_i})$

Step 5: Now, the company (based on their experience) gets for each parameter P_i the aggregation

$$[\mathscr{U}_i, \mathscr{F}\mathscr{U}_i](C(\mathscr{N}\mathscr{O}, P_i), C(\mathscr{F}\mathscr{O}, P_i))$$

which will be denoted by $C(\mathscr{O}, P_i)$. Finally, the company computes the aggregation of all these valuations $C(\mathscr{O}, P_i)$ using another uninorm $[\mathscr{U}, \mathscr{F}\mathscr{U}]$ with $\mathscr{U} \preceq \mathscr{F}\mathscr{U}$, according to the expression $[\mathscr{U}, \mathscr{F}\mathscr{U}](C(\mathscr{O}, P_1), \cdots, C(\mathscr{O}, P_s))$, in order to obtain a final decision to asses the viability of the investment.

Remark 4. Other possible situation is to consider when the foreign expert group expresses some reluctance to foreign investment. In this case, if $\mathscr{U}_i$ and $\mathscr{F}\mathscr{U}_i$ denote the extension of the uninorms used by the experts groups NEG and FEG respectively to assess the parameter P_i, these two uninorms fulfill the order relation $\mathscr{F}\mathscr{U}_i \preceq \mathscr{U}_i$.

Remark 5. Note that from Proposition 4 the discrete fuzzy number

$$[\mathscr{U}_i, \mathscr{F}\mathscr{U}_i](C(\mathscr{N}\mathscr{O}, P_i), C(\mathscr{F}\mathscr{O}, P_i))$$

can be interpreted as a mean of $C(\mathscr{N}\mathscr{O}, P_i)$ and $C(\mathscr{F}\mathscr{O}, P_i)$.

[1] This order is interpreted as a favorable point of view that the foreign expert group shows on the proposal about receiving potential investments of this company.

The following tables 1, 2 and 3 illustrate the procedure explained previously.

Table 1 National Expert Group Sheet

	National Expert Group			
Expert	P_1	P_2	$\cdots$	P_s
O_1	$O_1^{P_1}$	$O_1^{P_2}$	$\cdots$	$O_1^{P_s}$
O_2	$O_2^{P_1}$	$O_2^{P_2}$	$\cdots$	$O_2^{P_s}$
$\vdots$	$\vdots$	$\vdots$	$\vdots$	$\vdots$
O_r	$O_r^{P_1}$	$O_r^{P_2}$	$\cdots$	$O_r^{P_s}$
$\mathscr{N}\mathscr{O}$	$C(\mathscr{N}\mathscr{O},P_1)$	$C(\mathscr{N}\mathscr{O},P_2)$	$\cdots$	$C(\mathscr{N}\mathscr{O},P_s)$

Table 2 Foreign Expert Group Sheet

	Foreign Expert Group			
Expert	P_1	P_2	$\cdots$	P_s
$(FO)_1$	$(FO)_1^{P_1}$	$(FO)_1^{P_2}$	$\cdots$	$(FO)_1^{P_s}$
$(FO)_2$	$(FO)_2^{P_1}$	$(FO)_2^{P_2}$	$\cdots$	$(FO)_2^{P_s}$
$\vdots$	$\vdots$	$\vdots$	$\vdots$	$\vdots$
$(FO)_k$	$(FO)_k^{P_1}$	$(FO)_k^{P_2}$	$\cdots$	$(FO)_k^{P_s}$
$\mathscr{F}\mathscr{O}$	$C(\mathscr{F}\mathscr{O},P_1)$	$C(\mathscr{F}\mathscr{O},P_2)$	$\cdots$	$C(\mathscr{F}\mathscr{O},P_s)$

Table 3 Consensus Opinion Sheet

	Consensus opinion			
Company	P_1	P_2	$\cdots$	P_s
$C(\mathscr{N}\mathscr{O},\mathbf{P})$	$C(\mathscr{N}\mathscr{O},P_1)$	$C(\mathscr{N}\mathscr{O},P_2)$	$\cdots$	$(C(\mathscr{N}\mathscr{O},P_s)$
$C(\mathscr{F}\mathscr{O},\mathbf{P})$	$C(\mathscr{F}\mathscr{O},P_1)$	$C(\mathscr{F}\mathscr{O},P_2)$	$\cdots$	$(C(\mathscr{F}\mathscr{O},P_s)$
$C(\mathscr{O},\mathbf{P})$	$C(\mathscr{O},P_1)$	$C(\mathscr{O},P_2)$	$\cdots$	$C(\mathscr{O},P_s)$
C.Opinion: $[\mathscr{U},\mathscr{F}\mathscr{U}](C(\mathscr{O},P_1),\cdots,C(\mathscr{O},P_s))$				

Example 2. Assume that the National Expert Group and the Foreign Expert Group are made up of three experts ($NEG = \{O_1,O_2,O_3\}$) and two experts ($FEG = \{(FO)_1,(FO)_2\}$) respectively. Let $\mathbf{P} = \{P_1,P_2,P_3\}$ be the set of parameters which will be evaluated, where P_1= *Risk of the investment*, P_2= *Social-Political Impact Analysis* and P_3= *Tax Benefits*.

Consider the nine linguistic hedges $\mathfrak{L} = \{EB,VB,B,MB,F,MG,G,VG,EG\}$ where the letters refer to the linguistic terms Extremely Bad, Very Bad, Bad, More or Less Bad, Fair, More or Less Good, Good, Very Good and Extremely Good and they are listed in an increasing order: $EB \prec VB \prec B \prec MB \prec F \prec MG \prec G \prec VG \prec EG$. It is obvious that we can consider a bijective application between this ordinal scale $\mathfrak{L}$ and the finite chain $L = \{0,1,2,3,4,5,6,7,8\}$ of natural numbers which keep the order. Furthermore, each normal convex fuzzy subset defined on the ordinal scale $\mathfrak{L}$ can be considered like a discrete fuzzy number belonging to $\mathscr{A}_1^L$, and viceversa.

Suppose that

$O_1^{P_1} = \{0.6/2, 1/3, 0.8/4, 0.7/5\}$ $O_2^{P_1} = \{0.8/6, 0.9/7, 1/8\}$
$O_1^{P_2} = \{0.3/3, 0.6/4, 1/5, 0.7/6\}$ $O_2^{P_2} = \{0.6/5, 0.7/6, 1/7, 0.7/8\}$
$O_1^{P_3} = \{0.7/2, 0.8/3, 1/4, 0.5/5\}$ $O_2^{P_3} = \{0.5/4, 0.7/5, 1/6, 0.7/7, 0.4/8\}$

$$O_3^{P_1} = \{0.4/0, 0.6/1, 1/2, 0.4/3\}$$
$$O_3^{P_2} = \{0.5/3, 0.7/4, 1/5\}$$
$$O_3^{P_3} = \{0.6/2, 0.7/3, 1/4, 0.8/5\}$$

represent the assessments of the National Expert Group corresponding to the chosen parameter P_i. Now, suppose that

$(FO)_1^{P_1} = \{0.4/2, 0.7/3, 1/4, 0.7/5\}$ $(FO)_2^{P_1} = \{0.7/6, 1/7, 0.9/8\}$
$(FO)_1^{P_2} = \{0.4/3, 0.8/4, 1/5, 0.8/6\}$ $(FO)_2^{P_2} = \{0.7/5, 0.8/6, 1/7, 0.8/8\}$
$(FO)_1^{P_3} = \{0.6/2, 0.9/3, 1/4, 0.6/5\}$ $(FO)_2^{P_3} = \{0.6/4, 0.8/5, 0.9/6, 1/7, 0.7/8\}$

represent the assessments of the Foreign Expert Group corresponding to the chosen parameter P_i.

Suppose that the national expert group (NEG) and the foreign expert group (FEG) use the extensions $\mathscr{U}, \mathscr{F}\mathscr{U}$ respectively of the uninorms defined on the finite chain $L = \{0, 1, 2, 3, 4, 5, 6, 7, 8\}$:

$$U(x,y) = \begin{cases} \max(0, x+y-4) & \text{if } (x,y) \in [0,4]^2 \\ \max(x,y) & \text{otherwise} \end{cases}$$

$$FU(x,y) = \begin{cases} \min(x,y) & \text{if } (x,y) \in [0,4]^2 \\ \min(8, x+y-4) & \text{if } (x,y) \in [4,8]^2 \\ \max(x,y) & \text{otherwise.} \end{cases}$$

It is easy to show that $\mathscr{U}$ and $\mathscr{F}\mathscr{U}$ fulfill the order relation $\mathscr{U} \preceq \mathscr{F}\mathscr{U}$ because of the uninorms U and FU satisfy the inequalities $U(x,y) \leq FU(x,y)$ for all $(x,y) \in L^2$.

Now, according to step 4 we obtain:

$$\begin{aligned} C(\mathscr{N}\mathscr{O}, P_1) &= \{0.8/6, 0.9/7, 1/8\} \\ C(\mathscr{N}\mathscr{O}, P_2) &= \{0.6/5, 0.7/6, 1/7, 0.7/8\} \\ C(\mathscr{N}\mathscr{O}, P_3) &= \{0.5/0, 0.5/1, 0.5/2, 0.5/3, 0.5/4, \\ & \quad 0.7/5, 1/6, 0.7/7, 0.4/8\} \end{aligned}$$

and

$$\begin{aligned} C(\mathscr{F}\mathscr{O}, P_1) &= \{0.7/6, 1/7, 0.9/8\} \\ C(\mathscr{F}\mathscr{O}, P_2) &= \{0.7/5, 0.8/6, 0.8/7, 1/8\} \\ C(\mathscr{F}\mathscr{O}, P_3) &= \{0.6/2, 0.6/3, 0.6/4, 0.8/5, \\ & \quad 0.9/6, 1/7, 0.7/8\} \end{aligned}$$

Based on the step 5 we can compute the consensus opinion of each parameter P_i:

$$\begin{aligned} C(\mathscr{O},P_1) &= \{0.7/6, 0.9/7, 1/8\} \\ C(\mathscr{O},P_2) &= \{0.6/5, 0.7/6, 0.8/7, 1/8\} \\ C(\mathscr{O},P_3) &= \{0.5/0, 0.5/1, 0.5/2, 0.5/3, 0.5/4, \\ &\quad 0.7/5, 0.9/6, 1/7, 1/8\} \end{aligned}$$

Finally, we obtain the discrete fuzzy number which expresses the consensus opinion of the two groups of experts:

$$[\mathscr{U}, \mathscr{F}\mathscr{U}](C(\mathscr{O},P_1), C(\mathscr{O},P_2), C(\mathscr{O},P_3)) = \{0.7/6, 0.8/7, 1/8\}$$

being again $[\mathscr{U}, \mathscr{F}\mathscr{U}]$ the extension of the pair $U \leq FU$ considered above. Thus, according to the result previously obtained, the company considers suitable to invest in this foreign country.

5 Conclusion

In this work we have suggested a procedure to build aggregation functions on the set of discrete fuzzy numbers whose support is a set of consecutive natural numbers contained in the finite chain $L = \{0, 1, \cdots, n\}$ from a pair of aggregations functions also defined on L. The particular case of uninorms is studied. At the end, we give an application of this last operator in a decision-making problem.

Acknowledgements. This work has been partially supported by the MTM2009-10962 and MTM2009-10320 project grants, both of them with FEDER support.

References

1. Beliakov, G., Pradera, A., Calvo, T.: Aggregation Functions: A Guide for Practicioners. Studies in Fuzziness and Soft Computing, vol. 221, Springer, Heidelberg (2007)
2. Casasnovas, J., Riera, J.V.: Maximum and minimum of discrete fuzzy numbers. Frontiers in Artificial Intelligence and Applications: Artificial Intelligence Research and Development 163, 273–280 (2007)
3. Casasnovas, J., Riera, J.V.: Lattice properties of discrete fuzzy numbers under extended min and max. In: Proceedings IFSA-EUSFLAT 2009, Lisbon, pp. 647–652 (2009)
4. Casasnovas, J., Riera, J.V.: Extension of discrete t-norms and t-conorms to discrete fuzzy numbers. Fuzzy Sets and Systems 167, 65–81 (2011)
5. De Baets, B., Mesiar, R.: Triangular norms on product lattices. Fuzzy Sets and Systems 104, 61–75 (1999)
6. De Baets, B., Fodor, J., Ruiz-Aguilera, D., Torrens, J.: Idempotent Uninorms on Finite Ordinal Scales. International Journal of Uncertainty, Fuzziness and Knowledge-Based Systems 17(1), 1–14 (2009)

7. Fodor, J., Rudas, J., Bede, B.: Uninorms and Absorbing Norms with Applications to Image Processing. In: SISY 2006 4th Serbian-Hungarian Joint Symposium on Intelligent Systems, pp. 59–72 (2006)
8. Grabisch, M., Marichal, J.L., Mesiar, R., Pap, E.: Aggregation functions. Encyclopedia of Mathematics and its Applications, vol. 127. Cambridge University Press, Cambridge (2009)
9. Herrera, F., Herrera-Viedma, E.: Linguistic decision analysis: Steps for solving decision problems under linguistic information. Fuzzy Sets and System 115, 67–82 (2000)
10. Klir, G., Yuan, B.: Fuzzy sets and fuzzy logic (Theory and applications). Prentice Hall, New Jersey (1995)
11. Kolesarova, A., Mayor, G., Mesiar, R.: Weighted ordinal means. Information Sciences 177, 3822–3830 (2007)
12. Lopez-Molina, C., Bustince, H., Fernandez, J., Couto, P., de Baets, B.: A gravitational approach to edge detection based on triangular norms. Pattern Recognition 43, 3730–3741 (2010)
13. Mas, M., Mayor, G., Torrens, J.: t-operators and uninorms on a finite totally ordered set. Int. J. of Intelligent Systems 14, 909–922 (1999)
14. Mata, F., Martínez, L., Martínez, J.C.: Penalizing manipulation strategies in consensus processes. In: Proceedings of ESTYLF 2008, Mieres, Spain, pp. 485–491 (2008)
15. Mayor, G., Torrens, J.: Triangular norms on discrete settings. In: Klement, E.P., Mesiar, R. (eds.) Logical, Algebraic, Analytic, and Probabilistic Aspects of Triangular Norms, pp. 189–230. Elsevier, Netherlands (2005)
16. Riera, J.V., Torrens, J.: Aggregation of subjective evaluations based on discrete fuzzy numbers. Submitted to Fuzzy Sets and Systems
17. Voxman, W.: Canonical representations of discrete fuzzy numbers. Fuzzy Sets and Systems 54, 457–466 (2001)
18. Wang, G., Wu, C., Zhao, C.: Representation and operations of discrete fuzzy numbers. Southeast Asian Bulletin of Mathematics 28, 1003–1010 (2005)
19. Yager, R.: Defending against strategic manipulation in uninorm-based multi-agent decision making. European Journal of Operational Research 141, 217–232 (2002)
20. Yager, R.: Using Importances in Group Preference Aggregation to Block Strategic Manipulation. In: Studies in Fuzziness and Soft Computing, Aggregation operators, pp. 177–191. Physica-Verlag, Heidelberg (2002)

Defining Aggregation Functions from Negations

I. Aguiló, J. Suñer, and J. Torrens

Abstract. In this paper a method of defining aggregation functions from fuzzy negations is introduced. Any aggregation function obtained from a fuzzy negation by this method is proved to be a commutative semicopula and some properties are investigated. In particular, it is proved that by this method some well known examples of copulas and t-norms can be obtained. Moreover, any commutative semicopula constructed by this method can be always obtained from a negation N which is symmetric with respect to the diagonal. Then, those fuzzy negations N for which the corresponding semicopula is a copula are characterized. Also, several examples of negations N are given such that the corresponding semicopula is a t-norm.

1 Introduction

The study of aggregation functions has been extensively developed in last decades and many applications have been pointed out along this time. Applications which include many subjects not only from Mathematics and Computer Science, but also from many applied fields like economics and social sciences. This great quantity of applications is one of the main reasons for the increasing interest in the topic of aggregation functions and this interest is endorsed by the publication of some monographs dedicated entirely to aggregation functions ([4], [5], [14], [22]).

Aggregation functions have been also studied from the theoretical point of view and many authors have devoted their research to this topic. Usually, aggregation functions are divided into four classes: *conjunctive* those that lie under the minimum, *disjunctive* those that lie over the maximum, *means* or *compensatory* those

I. Aguiló · J. Suñer · J. Torrens
University of the Balearic Islands, Dept. of Math. and Comp. Sc.
e-mail: {isabel.aguilo, jaume.sunyer, dmijts0}@uib.es

B. De Baets et al. (Eds.): Eurofuse 2011, AISC 107, pp. 125–135, 2011.
springerlink.com

that lie in between, and *mixed* those that are not in any of the above cases. It can be pointed out that these classes are not disjoint. In particular, uninorms ([12]) are an important example of mixed aggregations, although the idempotent ones ([3], [21]) are in fact compensatory. Among conjunctive aggregation functions, there are some important classes like t-norms, useful in many fields but specially in fuzzy set theory, copulas and quasi-copulas with applications mainly in statistics ([20]), and semicopulas ([10]). From the theoretical point of view, one of the main questions has been the problem of constructing new aggregation functions (see the first chapter in [5] and references therein), specially in the case of copulas (see [7, 8, 9, 11, 13]).

One approach was given in [19] where a new class of binary aggregation functions is introduced from strong negations[1]. Based on this construction, in this paper we want to generalize it constructing new binary aggregation functions from fuzzy negations in general, not necessarily strong nor even continuous. It will be proved that aggregation functions constructed by this method are always commutative semicopulas and it will be characterized when the constructed functions are in fact, copulas or quasi-copulas. Moreover, several examples are given that lead to a t-norm and so it is also investigated when the constructed aggregation function is a t-norm.

2 Preliminaries

In this section we give some basic results that will be used along the paper.

Definition 1. *A function* $N : [0,1] \to [0,1]$ *is said to be a* fuzzy negation *if it is decreasing with* $N(0) = 1$ *and* $N(1) = 0$. *A fuzzy negation* N *is said to be*

- strict *when it is strictly decreasing and continuous.*
- strong *when it is an involution, i.e.,* $N(N(x)) = x$ *for all* $x \in [0,1]$.

It is well known that any strong negation is strict but not vice versa. Among the most used fuzzy negations we find the classical negation $N_{\mathbf{c}}(x) = 1 - x$, and the weakest and the strongest fuzzy negations respectively given by:

$$N_{\mathbf{wt}}(x) = \begin{cases} 0 & \text{if } x > 0 \\ 1 & \text{if } x = 0 \end{cases} \qquad N_{\mathbf{st}}(x) = \begin{cases} 1 & \text{if } x < 1 \\ 0 & \text{if } x = 1. \end{cases}$$

Clearly the first is strong whereas the other two are not continuous.

Definition 2. *An n-ary aggregation function is a function* $F : [0,1]^n \to [0,1]$ *which is increasing in each variable and such that* $F(0,\ldots,0) = 0$ *and* $F(1,\ldots,1) = 1$. *When* $n = 2$ *it is said that* F *is a binary aggregation function.*

From now on, we deal with binary aggregation functions.

[1] A dual approach can be found in [2].

Definition 3. *A binary aggregation function F is said to be a*

- *semicopula when F has neutral element 1, i.e., when $F(x,1) = F(1,x) = x$ for all $x \in [0,1]$,*
- *quasi-copula when F is 1-Lipschitz, i.e., when*

$$F(x',y') - F(x,y) \leq (x'-x) + (y'-y)$$

for all x,x',y,y' with $x \leq x'$ and $y \leq y'$,

- *copula when F is 2-increasing, i.e., when*

$$F(x,y) - F(x,y') - F(x',y) + F(x',y') \geq 0$$

for all x,x',y,y' with $x \leq x'$ and $y \leq y'$.

A special kind of binary aggregation functions defined from strong negations were introduced in [19] (see also [6] for the current notation). Namely,

Proposition 4. *Given a strong negation N the function*

$$F_N(x,y) = \max(0, x \wedge y - N(x \vee y)) \quad \textit{for all} \;\; x,y \in [0,1]$$

is a continuous, commutative and conjunctive aggregation function with neutral element 1.

3 Aggregation Functions from Negations

Let N be any fuzzy negation and let us define the function F_N as in the case of strong negations. That is, consider

$$F_N(x,y) = \max(0, x \wedge y - N(x \vee y)) \quad \text{for all} \;\; x,y \in [0,1]. \tag{1}$$

This definition remains as a well-defined construction method of new aggregation functions as it is stated in the following proposition.

Proposition 5. *Let N be a fuzzy negation and let F_N be given by equation (1). Then*

i) F_N is a commutative and conjunctive aggregation function.
ii) F_N has 0 as annihilator element and 1 as neutral element.
iii) F_N is a semicopula.
iv) If N is continuous then F_N is continuous as well.

Note that some well known commutative semicopulas are examples of this new construction method. Even some copulas and t-norms are as the following example shows.

Example 6. *If we take the negations $N_{\mathbf{c}}, N_{\mathbf{wt}}$, and $N_{\mathbf{st}}$, the corresponding semicopulas F_N are respectively given by the Łukasiewicz t-norm, $F_{N_{\mathbf{c}}} = T_{\mathbf{L}}$, the minimum t-norm, $F_{N_{\mathbf{wt}}} = T_{\mathbf{M}}$, and the drastic t-norm, $F_{N_{\mathbf{st}}} = T_{\mathbf{D}}$.*

Example 7. *Let N be the fuzzy negation defined by:*

$$N(x) = \begin{cases} 1 & \text{if } x = 0 \\ 1/2 & \text{if } 0 < x < 1/2 \\ 0 & \text{if } x \geq 1/2 \end{cases}$$

Then, the corresponding commutative semicopula F_N is given by:

$$F_N(x,y) = \begin{cases} 0 & \text{if } x,y < 1/2 \\ Min(x,y) & \text{otherwise} \end{cases}$$

which is again a t-norm. Specifically the ordinal sum $(\langle 0, 1/2, T_{\mathbf{D}} \rangle)$ (see [16] for the notation).

The main goal of this paper is to study these functions F_N in depth. To begin with this objective, note that the continuity of the negation N is not necessary to obtain a continuous semicopula, and moreover, different fuzzy negations can lead to the same semicopula as the following example shows.

Example 8. *Let N be the fuzzy negation defined by:*

$$N(x) = \begin{cases} 1 & \text{if } x < 1/2 \\ 1 - x & \text{if } x \geq 1/2 \end{cases}$$

Then, N is not continuous but the corresponding commutative semicopula F_N is given again by the Łukasiewicz t-norm.

The previous examples suggest that only a part of the fuzzy negation N is important in order to construct the corresponding semicopula F_N. Next we will prove that this is effectively so. To do it we need to recall some concepts on Id-symmetrical negations. The following definitions are adapted from [17] (see also [18]).

Definition 9. *Let $N : [0,1] \to [0,1]$ be any fuzzy negation and let G be the graph of N, that is*

$$G = \{(x, N(x)) \mid x \in [0,1]\}.$$

For any point of discontinuity s of N, let s^- and s^+ be the corresponding lateral limits, with the convention $s^- = 1$ when $s = 0$ and $s^+ = 0$ when $s = 1$. Then, we define the completed graph *of N, denoted by $G(N)$, as the set obtained from G by adding the vertical segments from s^- to s^+ in any discontinuity point s.*

Definition 10. *A subset S of $[0,1]^2$ is said to be* Id-symmetrical *if for all $(x,y) \in [0,1]^2$ it holds that*

$$(x,y) \in S \iff (y,x) \in S.$$

The above definition expresses that the subset S of $[0,1]^2$ is symmetrical with respect to the diagonal of the unit square. A similar notion of symmetry can be defined for fuzzy negations as follows (see [17]).

Definition 11. *A fuzzy negation* $N : [0,1] \to [0,1]$ *is called* Id-symmetrical *if its completed graph* $G(N)$ *is Id-symmetrical.*

The following theorem gives a mathematical description of Id-symmetrical negations.

Theorem 12. *Let* $N : [0,1] \to [0,1]$ *be a fuzzy negation. The following items are equivalent:*

i) N is Id-symmetrical
ii) N satisfies the following two conditions:

Condition (A) *For all* $x \in [0,1]$ *it is*

$$\inf\{y \in [0,1] \mid N(y) = N(x)\} \leq N(N(x)) \leq \sup\{y \in [0,1] \mid N(y) = N(x)\}$$

Condition (B) *N is constant, say* $N(x) = s$ *in the interval* $]p,q[$ *with* $p < q$, *where*

$$p = \inf\{y \in [0,1] \mid N(y) = s\}$$

and

$$q = \sup\{y \in [0,1] \mid N(y) = s\},$$

if and only if, $s \in]0,1[$ *is a point of discontinuity of N and it is satisfied that*

$$p = s^{+} \quad \text{and} \quad q = s^{-}.$$

Remark 13. *Note that Condition (A) implies in particular that if N is strictly decreasing and continuous on an interval* $]a,b[\subseteq [0,1]$ *and* $N(]a,b[) =]c,d[$, *then N must be strictly decreasing and continuous also in the interval* $]c,d[$ *and* $N^2(x) = x$ *for all* $x \in]a,b[\cup]c,d[$. *That is, N must be involutive in these points.*

The following result is straightforward from the previous theorem.

Corollary 14. *Let N be an Id-symmetrical fuzzy negation. Then the following items are equivalent:*

i) N is continuous
ii) N is strong.

Now, given any fuzzy negation N, we are able to prove that the aggregation function F_N can be constructed in fact from an Id-symmetrical fuzzy negation.

Theorem 15. *Let N be a fuzzy negation. Then there exists an Id-symmetrical fuzzy negation* $\mathbf{N}$ *such that* $F_{\mathbf{N}} = F_N$, *and in this case*

$$F_{\mathbf{N}}(x,y) = \begin{cases} 0 & \text{when } y < \mathbf{N}(x) \text{ or } (y = \mathbf{N}(x) \text{ and } x \leq N^2(x)) \\ x \wedge y - \mathbf{N}(x \vee y) & \text{otherwise.} \end{cases}$$

Proof. Since the proof is constructive, we give a sketch of it. Given any fuzzy negation N consider

$$a_N = \inf\{x \in [0,1] \mid N(x) \leq x\} \tag{2}$$

If $a_N = 0$, we have that $N(x) \leq x$ for all $x > 0$. Let us prove now that $N(x) = 0$ for all $x > 0$. If there exists $a > 0$ such that $N(a) > 0$, then let $x_0 \in (0,a)$ such that $x_0 < N(a)$. Thus since N is decreasing, $N(x_0) \geq N(a) > x_0$ which is a contradiction. Then $N(x) \leq x$ for all $x > 0$, that is, N must be the weakest fuzzy negation $N_{\mathbf{wt}}$ (which is Id-symmetrical) and F_N is the minimum t-norm (see example 6).

If $a_N > 0$, we define $N' : [a_N, 1] \to [0, a_N]$ in the following way:

- If $x > a_N, N'(x) = N(x)$.
- $N'(a_N) = \begin{cases} a_N & \text{if } N(a_N) \geq a_N \\ N(a_N) & \text{if } N(a_N) < a_N \end{cases}$

Now we can extend N' to an Id-symmetrical fuzzy negation $\mathbf{N}$ just by making a copy of the extended graph of N' symmetric with respect to the diagonal. If N' is non-strict, there are infinitely many possibilities for the fuzzy negation $\mathbf{N}$. In any case, one of these possibilities is given by

$$\mathbf{N}(x) = \begin{cases} N'(x) & \text{if } x \geq a_N \\ \sup\{y \in [a_N, 1] \mid N'(y) \geq x\} & \text{otherwise.} \end{cases}$$

A simple calculation shows that $F_{\mathbf{N}} = F_N$. □

Example 16. *Let us see two examples of the construction given in the previous theorem. The first example consists of a fuzzy negation* N_1 *satisfying that* $N(a_{N_1}) < a_{N_1}$ *and the second one gives a fuzzy negation* N_2 *satisfying that* $N(a_{N_2}) \geq a_{N_2}$*. Both the fuzzy negations* N_1 *and* N_2 *and the corresponding Id-symmetrical negations* $\mathbf{N_1}$ *and* $\mathbf{N_2}$*, constructed as in the proof of Theorem 15, are represented in Figure 1.*

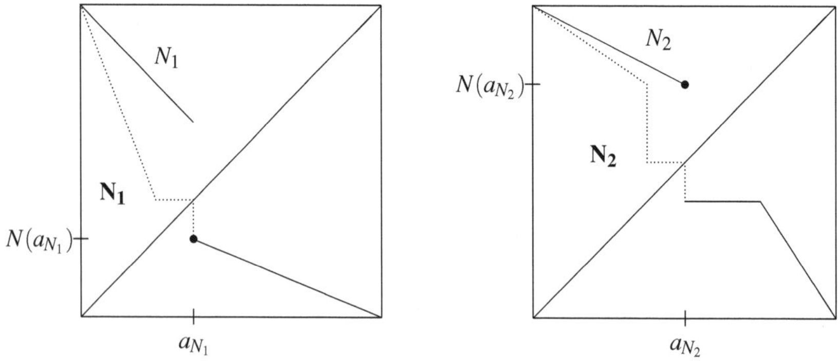

Fig. 1 Two fuzzy negations N_1, N_2 and the corresponding Id-symmetrical negations $\mathbf{N_1}, \mathbf{N_2}$

The previous theorem proves that in order to construct aggregation functions from negations through equation (1), only Id-symmetrical fuzzy negations N must be considered. Thus from now on, we will work only with this kind of fuzzy negations.

Remark 1. Note that from the previous results we have that when N is an Id-symmetrical fuzzy negation then the corresponding semicopula is continuous if and only if N is continuous in $[a_N, 1]$. However, even in this case, when N is non-strict in this interval there are infinitely many Id-symmetrical fuzzy negations deriving the same semicopula F_N.

4 Copulas, Quasi-copulas and t-Norms in the Family of F_N

Recall that for any Id-symmetrical fuzzy negation N, its derived aggregation function F_N is always a semi-copula (see Proposition 5). In the following theorems we investigate when F_N are copulas, quasi-copulas or even t-norms.

Let us begin with the case of quasi-copulas and copulas that become equivalent in our framework.

Theorem 17. *Let N be an Id-symmetrical fuzzy negation and F_N its derived aggregation function. Then the following items are equivalent:*

i) F_N is a copula
ii) F_N is a quasi-copula
iii) N is 1-Lipschitz on $[a_N, 1]$.
iv) $Id + N$ is an increasing function on $[a_N, 1]$.

Example 18. *Let $a \in]0, 1/2]$. Let us consider the following Id-symmetrical fuzzy negation:*

$$N(x) = \begin{cases} \frac{a-1}{a}x + 1 & \text{if } x \in [0, a] \\ \frac{a}{a-1}(x-1) & \text{if } x \in [a, 1] \end{cases}$$

The corresponding aggregation function is given by

$$F_N(x,y) = \max(0, x \wedge y - \frac{a}{1-a}(1 - x \vee y)) \quad \text{for all } x, y \in [0,1]$$

Observe that $a_N = a \leq 1/2$, and thus for all $x \in [a, 1]$,

$$(Id + N)(x) = x + \frac{a}{a-1}(x-1) = x\frac{2a-1}{a-1} - \frac{a}{a-1}$$

is an increasing function. Then, according to the previous theorem, F_N is a copula. Note that in this case F_N coincides with the singular conic copulas introduced by Jwaid et al. (see Example 8 in [15]). Moreover it is clear that the case $a = 1/2$ leads to the Łukasiewicz copula. Of all these copulas, only the Łukasiewicz one is also a t-norm. This can be proved directly but it also follows from Theorem 20 below.[2]

---

[2] Recall that conic t-norms were studied by C. Alsina, see for instance Theorem 3.6.4 in [1].

Example 19

1) Given $a \in [0,1]$, let us consider the following Id-symmetrical fuzzy negation:

$$N(x) = \begin{cases} 1 & \text{if } x = 0 \\ a - x & \text{if } x \in]0,a[\\ 0 & \text{if } x \in [a,1] \end{cases}$$

The corresponding aggregation function is given by

$$F_N(x,y) = \begin{cases} \max(0, x+y-a) & \text{if } (x,y) \in [0,a]^2 \\ \min(x,y) & \text{otherwise} \end{cases}$$

Observe that, for all $x \in [a,1]$, $(Id+N)(x) = x$ and, according to the previous theorem, F_N is a copula. Note that this family of F_N coincides with the family of Mayor-Torrens copulas (see [9]) that are also a well-known family of t-norms (see [16]).

2) Given $a \in [0,1]$, let us consider the following Id-symmetrical fuzzy negation:

$$N(x) = \begin{cases} 1 & \text{if } x = 0 \\ a & \text{if } x \in]0,a[\\ 0 & \text{if } x \in [a,1] \end{cases}$$

The corresponding aggregation function is given by

$$F_N(x,y) = \begin{cases} 0 & \text{if } (x,y) \in [0,a[^2 \\ \min(x,y) & \text{otherwise} \end{cases}$$

Observe that, for all $x \in [a,1]$, $(Id+N)(x) = x$ and, according to the previous theorem, F_N is a copula. Note that in this case F_N is again a t-norm (see Theorem 20 and Example 21).

We can see the graph of the negation of Example 18 in Figure 2 and those of Example 19 in Figure 3.

Examples at the beginning of the paper prove that some t-norms can be derived from fuzzy negations using the construction method presented here. In the nest theorem we present several situations where the obtained aggregation function F_N is in fact a t-norm.

Theorem 20. *Let N be an Id-symmetrical fuzzy negation and F_N its derived aggregation function.*

a) *If N is continuous then F_N is a t-norm if and only if $N = N_{\mathbf{c}}$. Moreover, in this case F_N is the Łukasiewicz t-norm.*

b) *If N is not continuous then in each one of the following cases the corresponding aggregation function F_N is a t-norm:*

b1) *Consider $a = \inf\{x \in]0,1] \mid N(x) = 0\}$ and let N be such that $N(x) \geq a/2$ for all $x < a$ and $N(a) = 0$ or $N(x) \geq a/2$ for all $x \leq a$.*

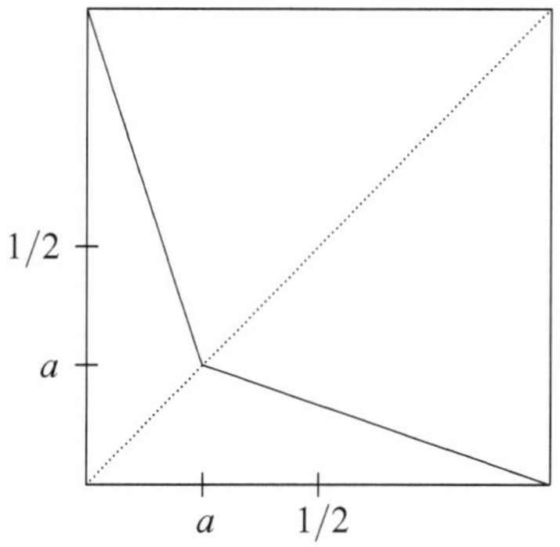

Fig. 2 The Id-symmetrical fuzzy negation of Example 18

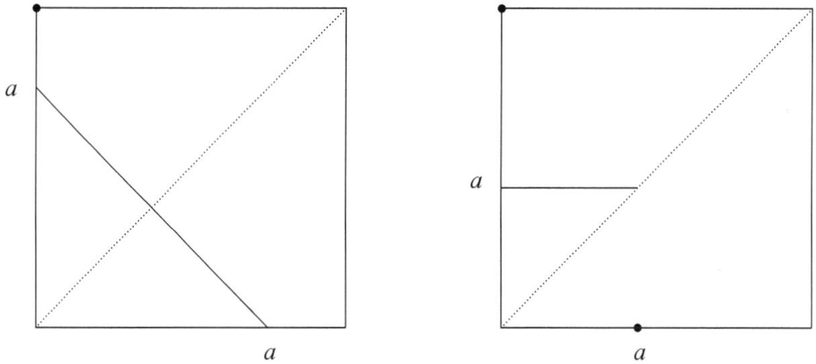

Fig. 3 The Id-symmetrical fuzzy negation of Example 19

b2) *Let* $a \in]0,1[$ *and* N *given by*

$$N(x) = \begin{cases} 1 & \text{if } x = 0 \\ a - x & \text{if } 0 < x < a \\ 0 & \text{if } x \geq a \end{cases}.$$

b3) *Let* $b \in]0,1[$ *and* N *given by*

$$N(x) = \begin{cases} 1 & \text{if } x \leq b \\ b + 1 - x & \text{if } b < x < 1 \\ 0 & \text{if } x = 1 \end{cases},$$

Let us give some new examples of t-norms obtained from Id-symmetrical negations N based on the theorem above.

Example 21. i) *In case* $b1)$ *of the previous theorem we can take for instance*

$$N(x) = \begin{cases} 1 & \text{if } x = 0 \\ a & \text{if } 0 < x < a \\ 0 & \text{if } x \geq a \end{cases} \quad or \quad N(x) = \begin{cases} 1 & \text{if } x = 0 \\ a & \text{if } 0 < x \leq a \\ 0 & \text{if } x > a. \end{cases}$$

In the first case F_N is the ordinal sum t-norm given by $\langle(0,a,T_{\mathbf{D}})\rangle$, which includes the drastic t-norm for the case $a=1$. In the second case F_N is given by

$$F_N(x,y)=\begin{cases}0 & \textit{if}\; x,y\leq a\\ \min(x,y) & \textit{otherwise,}\end{cases}$$

which includes the minimum t-norm for the case $a=0$.

ii) *In case b2) it is easy to see that the corresponding t-norms are the ordinal sums given by $\langle(0,a,T_{\mathbf{L}})\rangle$ where $T_{\mathbf{L}}$ is the Łukasiewicz t-norm. Note that these t-norms are exactly the Mayor-Torrens t-norms (see [16], Section 4.9 and Appendix A.5) and they are given by*

$$F_N(x,y)=\begin{cases}\max(0,x+y-a) & \textit{if}\;(x,y)\in[0,a]^2\\ \min(x,y) & \textit{otherwise.}\end{cases}$$

iii) *In case b3) the corresponding t-norms are given by*

$$F_N(x,y)=\begin{cases}\max(0,x+y-b-1) & \textit{if}\;(x,y)\in[b,1[^2\\ \min(x,y) & \textit{if}\max(x,y)=1\\ 0 & \textit{otherwise.}\end{cases}$$

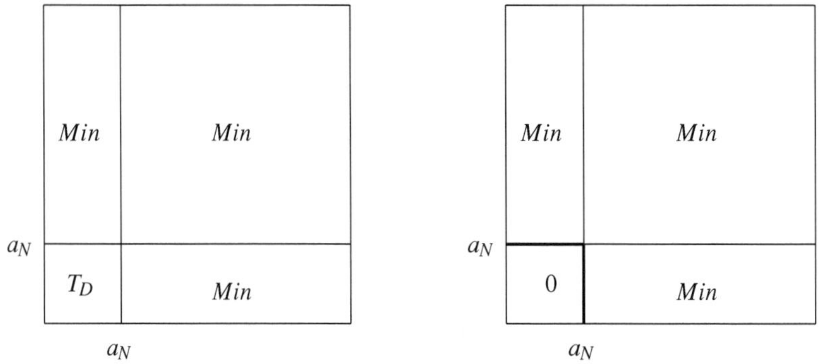

Fig. 4 The t-norms of Example (21-i)

A natural question arise. Are those t-norms the only ones that can be derived from fuzzy negations through the construction method presented here? If not, all t-norms that can be obtained from this method should be characterized. This is part of our future work concerning this topic.

Acknowledgements. The authors are grateful to the referees for their valuable comments, specially for the example of conic copulas (Example 18). This work has been partially supported by the MTM2009-10320 and MTM2009-10962 Spanish project grants, both with FEDER support.

References

1. Alsina, C., Frank, M.J., Schweizer, B.: Associative Functions. Triangular Norms and Copulas. World Scientific, New Jersey (2006)
2. Aguiló, I., Carbonell, M., Suñer, J., Torrens, J.: Dual representable aggregation functions and their derived S-implications. In: Hüllermeier, E., Kruse, R., Hoffmann, F. (eds.) IPMU 2010. LNCS, vol. 6178, pp. 408–417. Springer, Heidelberg (2010)
3. De Baets, B.: Idempotent uninorms. European Journal of Operational Research 118, 631–642 (1999)
4. Beliakov, G., Pradera, A., Calvo, T.: Aggregation Functions: A Guide for Practicioners. Springer, Heidelberg (2007)
5. Calvo, T., Mayor, G., Mesiar, R. (eds.): Aggregation operators. New trends and applications. Studies in Fuzziness and Soft Computing, vol. 97. Physica-Verlag, Heidelberg (2002)
6. Carbonell, M., Torrens, J.: Continuous R-implications generated from representable aggregation functions. Fuzzy Sets and Systems 161, 2276–2289 (2010)
7. Durante, F., Kolesárová, A., Mesiar, R., Sempi, C.: Copulas with given diagonal sections: Novel constructions and applications. International Journal of Uncertainty, Fuzziness and Knowledge-Based Systems 15, 397–410 (2007)
8. Durante, F., Kolesárová, A., Mesiar, R., Sempi, C.: Copulas with given values on a horizontal and a vertical section. Kybernetika 43, 209–220 (2007)
9. Durante, F., Mesiar, R., Sempi, C.: On a family of copulas constructed from the diagonal section. Soft Comput. 10, 490–494 (2006)
10. Durante, F., Sempi, C.: Semicopulae. Kybernetika 41, 315–328 (2005)
11. Erdely, A., González-Barrios, J.M.: On the Construction of Families of Absolutely Continuous Copulas with Given Restrictions. Communications in Statistics - Theory and Methods 35, 649–659 (2006)
12. Fodor, J., Yager, R.R., Rybalov, A.: Structure of uninorms. International Journal of Uncertainty, Fuzziness and Knowledge-Based Systems 5, 411–427 (1997)
13. Fredricks, G.A., Nelsen, R.B.: The Bertino family of copulas. In: Distributions with Given Marginals and Statistical Modeling, pp. 81–92. Kluwer, Dordrecht (2002)
14. Grabisch, M., Marichal, J.L., Mesiar, R., Pap, E.: Aggregation functions. In: Encyclopedia of Mathematics and its Applications, vol. 127. Cambridge University Press, Cambridge (2009)
15. Jwaid, T., De Baets, B., Kalická, J., Mesiar, R.: Conic aggregations functions. Fuzzy Sets and Systems 167, 3–20 (2011)
16. Klement, E.P., Mesiar, R., Pap, E.: Triangular norms. Kluwer Academic Publishers, Dordrecht (2000)
17. Maes, K.C., De Baets, B.: Negation and affirmation: the role of involutive negators. Soft Computing 11, 647–654 (2007)
18. Maes, K.C., De Baets, B.: Orthosymmetrical monotone functions. Bulletin Belgium Mathematical Society 14, 99–116 (2007)
19. Mayor, G., Torrens, J.: On a class of binary operations: Non-strict Archimedean aggregation functions. In: Proceedings of ISMVL 1988, pp. 54–59. Palma de Mallorca, Spain (1988)
20. Nelsen, R.B.: An introduction to copulas. Springer, New York (2006)
21. Ruiz-Aguilera, D., Torrens, J., De Baets, B., Fodor, J.: Some remarks on the characterization of idempotent uninorms. In: Hüllermeier, E., Kruse, R., Hoffmann, F. (eds.) IPMU 2010. LNCS, vol. 6178, pp. 425–434. Springer, Heidelberg (2010)
22. Torra, V., Narukawa, Y.: Modeling decisions. Information fusion and aggregation operators. In: Cognitive Technologies. Springer, Heidelberg (2007)

Discrete Kernel Aggregation Functions

M. Mas, M. Monserrat, and J. Torrens

Abstract. The study of discrete aggregation functions (those defined on a finite chain) with some kind of smoothness has been extensively developed in last years. Smooth t-norms and t-conorms, nullnorms and some kinds of uninorms, copulas and quasi-copulas have been characterized in this context. In this paper discrete aggregation functions with the kernel property (which implies the smoothness property) are investigated. Some properties and characterizations, as well as some construction methods for this kind of discrete aggregation functions are studied. It is also investigated when the marginal functions of a discrete kernel aggregation function fully determine it.

1 Introduction

Aggregation of information is currently a field extensively studied because in any scientific process there is always a step where the aggregation of all collected data must be merged into a representative output. From the mathematical point of view this fusion of information is given by the so-called *aggregation functions*.There are many researchers that have studied aggregation functions in last decades and most of the known results in this topic have been compiled in several monographs entirely devoted to aggregation functions from both the theoretical and the applicational points of view ([2], [3], [8] and [21]).

Aggregation functions are usually divided in four classes (see [2]): *conjunctive* (that lie under the minimum like t-norms, copulas and quasi-copulas), *disjunctive* (that lie over the maximum like t-conorms, dual quasi-copulas, dual copulas), *averaging* or *median* (that lie between minimum and maximum like means, weighted means, OWA's, etc.), and *mixed* (those not included in any of the previous classes

M. Mas · M. Monserrat · J. Torrens
Dept. of Mathematics and Computer Science, University of the Balearic Islands,
07122 Palma de Mallorca, Spain
e-mail: {dmimmg0,dmimma0,dmijts0}@uib.es

B. De Baets et al. (Eds.): Eurofuse 2011, AISC 107, pp. 137–145, 2011.
springerlink.com

like uninorms and nullnorms). However, there are also other important classes of aggregations, like 1-Lipschitz aggregation functions ([9], [10], [11], [13]) or kernel aggregation functions ([13], [14], [15]). Both types of operations are specially interesting, not only because they are always continuous, but also because their close relation with copulas and quasi-copulas.

On the other hand, the study of operations defined on a finite chain (discrete operations) is a field of increasing interest because in many practical situations the rang of calculations and reasonings must be reduced to a finite set of possible values. In this direction, different classes of aggregation functions have been considered in this framework. In most cases this study is carried out on aggregation functions with some kind of *smoothness*, usually considered as the counterpart of continuity (because smoothness is also equivalent to the 1-Lipschitz property) for this kind of operations. Thus, smooth t-norms and t-conorms were characterized in [20], weighted means in [12], uninorms (smooth in some adequate region) and smooth nullnorms in [16], non-commutative versions of the last ones in [6] and [17] respectively, idempotent uninorms in [5], copulas in [19] and quasi-copulas in [1]. Recently, the class of smooth discrete aggregation functions in general has been studied too in [18]. In this last paper smooth aggregation functions on a finite chain are characterized and are used to characterize discrete quasi-copulas through some functional equations.

In this work we want to deal with kernel discrete aggregation functions in a similar way as the study carried out in [14] and [15] for this kind of aggregations but in the fuzzy framework (that is, with values in the unit interval [0,1]). Note that the kernel property implies the smoothness condition and thus, our study will lead to a special kind of smooth discrete aggregation functions with good properties (see Propositions 2, 3, 4 and 5). To do this, we begin by studying some properties and then we give some characterizations. The values of such aggregations on the boundary are analyzed and it is determined when such values fully characterize the whole kernel aggregation function.

2 Preliminaries

In this section we recall some basic definitions and results on discrete aggregation functions. Since in this framework any finite chain with the same number of elements is equivalent (see for instance [20]) we will work with the most simple one with $n+1$ elements:

$$L_n = \{0,1,2,\ldots,n\}.$$

We will denote by $[a,b]$ the sub-chain given by $[a,b] = \{x \in L_n \mid a \leq x \leq b\}$. The maximum of two elements $x,y \in L_n$ will be indistinctly denoted by $\max(x,y)$ or $x \vee y$, and the minimum by $\min(x,y)$ or $x \wedge y$.

Definition 1. *A (binary)* discrete aggregation function *is a function $F : L_n^2 \longrightarrow L_n$ which is increasing in each variable and such that $F(0,0) = 0$ and $F(n,n) = n$.*

Definition 2. *A unary function $f : L_n \to L_n$ is said to be* smooth *when:*

$$|\, f(x) - f(x-1) \,| \leq 1 \quad \text{for all} \quad x \in L_n \quad \text{such that} \quad x \geq 1.$$

Definition 3. *A binary operation $F : L_n^2 \longrightarrow L_n$ is said to be* smooth *when all its vertical and horizontal sections, $F(x,-)$ and $F(-,y)$, are smooth.*

The importance of the smoothness condition lies in the fact that it is used as the counterpart of continuity for discrete operations, since it is equivalent in this framework with the 1-Lipschitz condition:

$$|\, F(x,y) - F(x',y') \,| \leq |\, x - x' \,| + |\, y - y' \,|$$

for all $x, x', y, y' \in L_n$, as it is stated in the following proposition.

Proposition 1. *Let $F : L_n^2 \to L_n$ be any binary aggregation function. The following items are equivalent:*

- *F is smooth.*
- *F is 1-Lipschitz.*

3 Discrete Kernel Aggregation Functions

The following kind of operations was introduced in the framework of $[0,1]$ in [4] and [13], under the name of *kernel aggregation functions*, as a special class of operations which are stable with respect to the standard norm $\| \|_\infty$ in $\mathbb{R}^2$.

Definition 4. *A binary aggregation function $F : L_n^2 \to L_n$ is said to be a* discrete kernel *aggregation function (*DK *for short) when*

$$F(x,y) - F(x',y') \leq \max(x - x', y - y')$$

for all $x, x', y, y' \in L_n$ such that $x' \leq x$ and $y' \leq y$.

Some straightforward properties can be directly deduced from the definition. We list some of them in the following proposition.

Proposition 2. *Let $F : L_n^2 \to L_n$ be a DK aggregation function. Then*

1. *F satisfies the 1-Lipschitz property and so F is smooth.*
2. *F is idempotent, that is, $F(x,x) = x$ for all $x \in L_n$.*
3. *F is averaging, that is, $\min(x,y) \leq F(x,y) \leq \max(x,y)$ for all $x, y \in L_n$.*
4. *The restriction of F to any $[a,b]^2$ is a DK aggregation function on $[a,b]$.*
5. *The minimum is the only commutative (or symmetric) DK aggregation function with $F(n,0) = 0$.*
6. *The maximum is the only commutative (or symmetric) DK aggregation function with $F(n,0) = n$.*

Note that some important classes of aggregation functions are special cases of DK operations. For instance, if we add commutativity and associativity we obtain the following characterization of k-median operations (see [7]) in the discrete case, i.e.,

$$med_k(x,y) = \begin{cases} \max(x,y) & \text{if} \max(\mathrm{x},\mathrm{y}) \leq \mathrm{k} \\ \min(x,y) & \text{if} \min(\mathrm{x},\mathrm{y}) \geq \mathrm{k} \\ k & otherwise. \end{cases}$$

Proposition 3. *Let $F : L_n^2 \to L_n$ be a binary operation on L_n. F is a commutative and associative DK aggregation function if and only if F is a k-median, that is,*

$$F(x,y) = med_k(x,y) \qquad where \quad k = F(0,n) = F(n,0).$$

The following result gives a characterization of DK aggregation functions in general.

Proposition 4. *Let $F : L_n^2 \to L_n$ be a discrete aggregation function on L_n. The following properties are equivalent:*

1. *F is kernel.*
2. *F is* shift-invariant, *i.e., $F(x+a, y+a) \leq a + F(x,y)$ for all $a \in L_n$ such that $x+a, y+a \leq n$.*
3. *$F(x+1, y+1) \leq 1 + F(x,y)$ for all $x, y < n$.*

It is well known that when F, G, H are binary discrete aggregation function on L_n, then the composition function $H(F,G)$ defined by

$$H(F,G)(x,y) = H(F(x,y), G(x,y)) \tag{1}$$

for all $x, y \in L_n$, is a discrete aggregation function too. In this sense, another important property of DK aggregations lies in the fact that they are the only ones that preserve both smoothness and kernel properties through this composition. This is stated in the following proposition.

Proposition 5. *Let $H : L_n^2 \to L_n$ be a discrete aggregation function. Then the function $H(F,G)$ is smooth (or kernel) for all smooth (or kernel) discrete aggregation functions $F, G : L_n^2 \to L_n$ if and only if H is a DK aggregation function.*

In fact construction (1) can be generalized in the following manner. Just note that if F is a DK aggregation and $\alpha \in \mathbb{Z}$ is an integer, then the function $\overline{F} : L_n^2 \to \mathbb{Z}$, given by $\overline{F}(x,y) = F(x,y) + \alpha$ is increasing and satisfies the kernel property. Thus we can give the following result.

Proposition 6. *Let $F, G : L_n^2 \to L_n$ be smooth (or kernel) discrete aggregation functions on L_n and let $H : \mathbb{Z}^2 \to \mathbb{Z}$ be an increasing function satisfying the kernel property. Suppose that there exist $\alpha, \beta \in \mathbb{Z}$ such that $H(\alpha, \beta) = 0$ and $H(n+\alpha, n+\beta) = n$. Consider $\overline{F}, \overline{G} : L_n^2 \to \mathbb{Z}$ given by*

$$\overline{F}(x,y) = F(x,y) + \alpha \quad and \quad \overline{G}(x,y) = G(x,y) + \beta$$

for all $x,y \in L_n$. *Then* $H(\overline{F},\overline{G})$ *is a smooth (or kernel) discrete aggregation function.*

We end this section with a characterization of DK aggregation functions through symmetric DK aggregation functions.

Proposition 7. *Let* $F : L_n^2 \to L_n$ *be a binary operation.* F *es una DK aggregation function if and only if there exist two symmetric DK aggregation functions,* F_1 *y* F_2*, such that*

$$F(x,y) = \begin{cases} F_1(x,y) & if\, x \leq y \\ F_2(x,y) & if\, x > y. \end{cases} \tag{2}$$

From this last proposition the study of DK aggregation functions can be reduced to the symmetric case and we will do it from now on.

4 Construction of DK Aggregation Functions from Their Boundary Values

We have already seen that there is a unique symmetric DK aggregation function, F, such that $F(n,0) = 0$ (the minimum), and only one such that $F(n,0) = n$ (the maximum). Next we want to see that these are the only cases for which the value of F at point $(n,0)$ fully determines the symmetric DK aggregation function F. Let us begin with the following result, that gives for a fixed $a \in L_n$, the maximum and the minimum symmetric DK aggregation functions, F, such that $F(n,0) = a$.

Proposition 8. *Fix an* $a \in L_n$. *The maximum and the minimum symmetric DK aggregation functions satisfying* $F(n,0) = a$ *are respectively given by*

$$F_a(x,y) = \min(x \vee y, x \wedge y + a)$$

and

$$F_a^*(x,y) = \max(x \wedge y, x \vee y - n + a).$$

Proposition 9. *Fix an* $a \in L_n$. *There exists a unique symmetric DK aggregation function* F *satisfying* $F(n,0) = a$ *if and only if* $a = 0$ *or* $a = n$. *In these cases,* F *is the minimum and the maximum respectively.*

Remark 1. In the case of non symmetric DK aggregation functions, using Proposition 7 and the previous one we can deduce that there is a unique DK aggregation function F satisfying $F(n,0) = a$ and $F(0,n) = b$ if and only if $a,b \in \{0,n\}$. Thus, we obtain four different cases: when $a = b = 0$ the only DK is the minimum, when $a = b = n$ the only DK is the maximum, when $a = 0, b = n$, the only DK is the first projection, that is $F(x,y) = x$ for all $x,y \in L_n$, and when $a = n, b = 0$ the only one is the second projection, that is, $F(x,y) = y$ for all $x,y \in L_n$.

Another case for which boundary values determine the whole DK aggregation function is when we deal with the so-called *marginal functions* of an aggregation function F. That is, functions given by

$$f(x) = F(0,x), \qquad \varphi(x) = F(x,0), \qquad g(x) = F(x,n), \quad \text{and} \quad \psi(x) = F(n,x).$$

Obviously, in the case of symmetric DK aggregation functions it is $f = \varphi$ and $g = \psi$. Let us begin by investigating how can be these marginal functions for a DK aggregation function in general (not necessarily symmetric). We will deal with functions f, g, but clearly we can deduce identical results for functions φ and ψ.

Proposition 10. *Let $F : L_n^2 \to L_n$ be a DK aggregation function and let $f(x) = F(0,x)$, $g(x) = F(x,n)$ be its marginal functions. Then*

1. *f, g are increasing and smooth.*
2. *$f(0) = 0$, $f(n) = g(0)$, $g(n) = n$.*
3. *$f(x) \leq x \leq g(x)$ and $g(x) - f(n-x) \leq x$ for all $x \in L_n$.*

Now, given any function f (or g) with these properties we want to find those maximum and minimum symmetric DK aggregation functions with marginal function f (or g), and using this result deduce those cases for which this marginal function fully determines the symmetric DK aggregation function.

Proposition 11. *Let $f : L_n \to L_n$ be a smooth and increasing function with $f(x) \leq x$ for all $x \in L_n$. The maximum and the minimum symmetric DK aggregation functions satisfying $F(0,x) = f(x)$ are respectively given by*

$$F_f(x,y) = x \wedge y + f(x \vee y - x \wedge y)$$

and

$$F_f^*(x,y) = \max(x \wedge y, f(x \vee y)).$$

Proposition 12. *Let $f : L_n \to L_n$ be a smooth and increasing function with $f(x) \leq x$ for all $x \in L_n$. The following items are equivalent:*

- *There exists a unique symmetric DK aggregation function F satisfying $F(0,x) = f(x)$.*
- *f is strictly increasing in the interval $[n - f(n), n]$.*
- *f is given by $f(x) = \max(0, x - n + f(n))$ for all $x \in L_n$.*

In this case, the unique symmetric DK aggregation function with marginal function f is given by

$$F(x,y) = \max(x \wedge y, (x \vee y) - n + f(n)) \quad \text{for all} \quad x, y \in L_n.$$

Similarly, we can deal with the marginal function $g(x) = F(x,n)$ and we obtain completely analogous results. They can be also obtained by duality. Just note that if F is a symmetric DK aggregation function with marginal functions f and g, its dual function given by

$$F^d(x,y) = n - F(n-x, n-y) \qquad \text{for all} \quad x,y \in L_n,$$

is again a symmetric DK aggregation function with marginal functions f^d, g^d given by

$$f^d(x) = F^d(0,x) = n - F(n, n-x) = n - g(n-x)$$

and

$$g^d(x) = F^d(x,n) = n - F(n-x, 0) = n - f(n-x)$$

for all $x \in L_n$. Thus, the results with respect to the marginal function g are the following.

Proposition 13. *Let $g : L_n \to L_n$ be a smooth and increasing function with $x \leq g(x)$ for all $x \in L_n$. The maximum and the minimum symmetric DK aggregation functions satisfying $F(x,n) = g(x)$ are respectively given by*

$$F_g^*(x,y) = \min(x \vee y, g(x \wedge y))$$

and

$$F_g(x,y) = x \vee y - n + g(n - x \vee y + x \wedge y).$$

Proposition 14. *Let $g : L_n \to L_n$ be a smooth and increasing function with $x \leq g(x)$ for all $x \in L_n$. The following items are equivalent:*

- *There exists a unique symmetric DK aggregation function F satisfying $F(x,n) = g(x)$.*
- *g is strictly increasing in the interval $[0, n - g(0)]$.*
- *g is given by $g(x) = \min(n, x + g(0))$ for all $x \in L_n$.*

In this case, the unique symmetric DK aggregation function with marginal function g is given by

$$F(x,y) = \min(x \vee y, x \wedge y + g(0)) \quad \textit{for all} \quad x,y \in L_n.$$

5 Conclusions and Future Work

In this paper kernel aggregation functions defined on the finite chain L_n (DK aggregation functions) were introduced and studied in a similar way as it was done in the framework of $[0,1]$ in ([13], [14] and [15]). Some characterizations have been given for DK aggregation functions in general and, in particular, k-medians have been characterized as those DK aggregation functions that are symmetric and associative. It was proved also that any DK aggregation function is a special combination of two symmetric DK aggregation functions.

Moreover, expressions of the maximum and the minimum symmetric DK aggregation functions satisfying $F(n,0) = a$ for a fixed $a \in L_n$ have been stated, and expressions of the maximum and the minimum symmetric DK aggregation functions with a fixed marginal function $f(x) = F(0,x)$ (or $g(x) = F(x,n)$) have been stated as well. In both cases, such expressions have been used to characterize the cases for which a fixed marginal function f (or g) fully determine the symmetric DK aggregation function.

It is worth to study as a future work also the combined case. That is, given fixed marginal functions f and g, find the maximum and the minimum symmetric DK aggregation functions with marginal functions f and g and study when there is one and only one symmetric DK aggregation function with both marginal functions.

Acknowledgements. This work has been partially supported by the Spanish Grant MTM2009-10320 with FEDER support.

References

1. Aguiló, I., Suñer, J., Torrens, J.: Matrix representation of discrete quasi-copulas. Fuzzy Sets and Systems 159, 1658–1672 (2008)
2. Beliakov, G., Pradera, A., Calvo, T.: Aggregation Functions: A Guide for Practicioners. Springer, Heidelberg (2007)
3. Calvo, T., Mayor, G., Mesiar, R. (eds.): Aggregation operators. New trends and applications, Studies in Fuzziness and Soft Computing, vol. 97. Physica-Verlag, Heidelberg (2002)
4. Calvo, T., Mesiar, R.: Stability of aggregation operators. In: Proceedings of EUSFLAT 2001, Leicester, pp. 475–478 (2001)
5. De Baets, B., Fodor, J., Ruiz-Aguilera, D., Torrens, J.: Idempotent uninorms on finite ordinal scales. International Journal of Uncertainty, Fuzziness and Knowledge-Based Systems 17, 1–14 (2009)
6. Fodor, J.C.: Smooth associative operations on finite ordinal scales. IEEE Trans. on Fuzzy Systems 8, 791–795 (2000)
7. Fung, L., Fu, K.: An axiomatic approach to rational decision-making in fuzzy environment. In: Tanaka, K., Zadeh, L., Fu, K., Shimura, M. (eds.) Fuzzy Sets and Their Applications to Cognitive and Decision Processes, pp. 227–256. Academic Press, New York (1975)
8. Grabisch, M., Marichal, J.L., Mesiar, R., Pap, E.: Aggregation functions. In: The series: Encyclopedia of Mathematics and its Applications, vol. 127, Cambridge University Press, Cambridge (2009)
9. Kalicka, J.: On some construction methods for 1-Lipschitz aggregation functions. Fuzzy Sets and Systems 160, 726–732 (2009)
10. Klement, E.P., Kolesárová, A.: Extension to copulas and quasi-copulas as especial 1-Lipschitz aggregation operators. Kybernetika 41, 329–348 (2005)
11. Kolesárová, A.: 1-Lipschitz aggregation operators and quasi-copulas. Kybernetika 39, 615–629 (2003)
12. Kolesárová, A., Mayor, G., Mesiar, R.: Weighted ordinal means. Information Sciences 177, 3822–3830 (2007)
13. Kolesárová, A., Mordelová, J.: 1-Lipschitz and kernel aggregation operators. In: Proceedings of AGOP 2001, Oviedo, Spain, pp. 77–80 (2001)
14. Kolesárová, A., Mordelová, J., Muel, E.: Kernel aggregation operators and their marginals. Fuzzy sets and Systems 142, 35–50 (2004)
15. Kolesárová, A., Muel, E., Mordelová, J.: Construction of kernel aggregation operators from marginal values. International Journal of Uncertainty, Fuzziness and Knowledge-Based Systems 10, 37–49 (2002)

16. Mas, M., Mayor, G., Torrens, J.: $t-$Operators and uninorms on a finite totally ordered set. International Journal of Intelligent Systems 14, 909–922 (1999)
17. Mas, M., Monserrat, M., Torrens, J.: On left and right uninorms on a finite chain. Fuzzy Sets and Systems 146, 3–17 (2004)
18. Mas, M., Monserrat, M., Torrens, J.: Smooth aggregation functions on finite scales. In: Hüllermeier, E., Kruse, R., Hoffmann, F. (eds.) IPMU 2010. LNCS (LNAI), vol. 6178, pp. 398–407. Springer, Heidelberg (2010)
19. Mayor, G., Suñer, J., Torrens, J.: Copula-like operations on finite settings. IEEE Transactions on Fuzzy Systems 13, 468–477 (2005)
20. Mayor, G., Torrens, J.: Triangular norms in discrete settings. In: Klement, E.P., Mesiar, R. (eds.) Logical, Algebraic, Analytic, and Probabilistic Aspects of Triangular Norms, pp. 189–230. Elsevier, Amsterdam (2005)
21. Torra, V., Narukawa, Y.: Modeling decisions. Information fusion and aggregation operators. In: The series: Cognitive Technologies. Springer, Heidelberg (2007)

On a Generalization of the Notion of a Survival Copula

B. De Baets, H. De Meyer, and R. Mesiar

Abstract. We introduce a transformation that acts on binary aggregation functions and that generalizes the transformation that maps copulas, a well-studied class of binary aggregation functions with a profound probabilistic interpretation, to their associated survival copulas. The new transformation, called double flipping, is the composition of two elementary flipping transformations introduced earlier, each operating on one of the arguments of the aggregation function. We lay bare the relationships between these elementary flipping operations and double flipping. We characterize different subclasses of flippable aggregation functions, in particular aggregation functions that have an absorbing element or that have a neutral element. In this investigation, the key role played by quasi-copulas and their dual operations is highlighted. These findings support the introduction of the term survival aggregation function.

1 Introduction

Let us recall [3] that for a given (binary) aggregation function A, i.e. an increasing function $A : [0,1]^2 \to [0,1]$ that satisfies $A(0,0) = 0$ and $A(1,1) = 1$, flipping A (on

B. De Baets
Dept of Applied Mathematics, Biometrics and Process Control,
Ghent University, Coupure links 653, B-9000 Gent, Belgium
e-mail: bernard.debaets@ugent.be

H. De Meyer
Dept of Applied Mathematics and Computer Science,
Ghent University, Krijgslaan 281 S9, B-9000 Gent, Belgium
e-mail: hans.demeyer@ugent.be

R. Mesiar
Dept of Mathematics and Descriptive Geometry,
Slovak University of Technology, Radlinského 11, 813 68 Bratislava, Slovakia
e-mail: mesiar@math.sk

B. De Baets et al. (Eds.): Eurofuse 2011, AISC 107, pp. 147–156, 2011.
springerlink.com

[0,1]) in the x-argument (x-flipping for short) yields the function $A^{x\text{-flip}} : [0,1]^2 \to [0,1]$, called x-flip of A, defined by

$$A^{x\text{-flip}}(x,y) = A(0,y) + A(1,y) - A(1-x,y)\,.$$

Similarly, flipping A (on [0,1]) in the y-argument (y-flipping for short) yields the function $A^{y\text{-flip}} : [0,1]^2 \to [0,1]$, called y-flip of A, defined by

$$A^{y\text{-flip}}(x,y) = A(x,0) + A(x,1) - A(x,1-y)\,.$$

If $A^{x\text{-flip}}$ (resp. $A^{y\text{-flip}}$) is also an aggregation function, we say that A is x-flippable (resp. y-flippable). If A is both x-flippable and y-flippable, we simply call it *flippable*. From Proposition 6.1 in [3], it follows that an aggregation function A with neutral element 1 is flippable if and only if it is 1-Lipschitz continuous; in particular, quasi-copulas (1-Lipschitz continuous aggregation functions with neutral element 1) and copulas (2-increasing aggregation functions with neutral element 1 and absorbing element 0) are always flippable. Further in this paper, in Theorem 3, we will show that this equivalence extends to aggregation functions with an arbitrary neutral element.

Whether it is meaningful to perform x-flipping and y-flipping in a consecutive manner, depends on the answer to the following questions raised in [3]:

(i) If A is flippable, does it hold that $A^{x\text{-flip}}$ is y-flippable, and that $A^{y\text{-flip}}$ is x-flippable?
(ii) If so, do x-flipping and y-flipping commute?

In the next section, we will provide affirmative answers to these questions. This motivates us to carry out an in-depth study of the composition of x-flipping and y-flipping, called double flipping. In Section 2, we will show that double flipping results in an aggregation function which elegantly generalizes the concept of a survival copula. The most important results of this paper are contained in Section 3, including a representation theorem for flippable aggregation functions with an absorbing element, and one for flippable aggregation functions with a neutral element. Both theorems underline the key role played by quasi-copulas, an observation that is further unravelled in Section 4. We conclude with an outlook in Section 5.

2 Double Flipping

Note that x-flipping and y-flipping are clearly involutive operations. Hence, if A is x-flippable, then obviously $A^{x\text{-flip}}$ also is. The following proposition confirms that the term flippability (standing for simultaneous x-flippability and y-flippability) is appropriate.

Proposition 1. *Consider an aggregation function A. If A is flippable, then $A^{x\text{-flip}}$ is y-flippable and $A^{y\text{-flip}}$ is x-flippable.*

One easily verifies that for a flippable aggregation function A it holds that

$$(A^{x\text{-flip}})^{y\text{-flip}} = (A^{y\text{-flip}})^{x\text{-flip}},$$

in other words, x-flipping and y-flipping are commuting operations when applied to a flippable aggregation function. It is therefore appropriate to reserve a special name for their composition, irrespective of the order of application. Proposition 1 guarantees that the result is an aggregation function as well.

Definition 1. *Consider a flippable aggregation function A. Double flipping of A yields the aggregation function $A^{\#} := (A^{x\text{-flip}})^{y\text{-flip}} = (A^{y\text{-flip}})^{x\text{-flip}}$, called double flip of A, or explicitly, the aggregation function $A^{\#} : [0,1]^2 \to [0,1]$ defined by*

$$\begin{aligned} A^{\#}(x,y) = {} & 1 + A(1,0) + A(0,1) - A(1-x,0) - A(1-x,1) \\ & -A(0,1-y) - A(1,1-y) + A(1-x,1-y)\,. \end{aligned}$$

Just as x-flipping and y-flipping, double flipping is also an involutive operation.

Example 1. Consider a flippable aggregation function A with neutral element 1 (and hence also absorbing element 0), then $A^{\#}$ is given by

$$A^{\#}(x,y) = x + y - 1 + A(1-x,1-y)\,.$$

For a copula A, the function $A^{\#}$ is known as the survival copula of A, which is, as the name indicates, again a copula. It has a specific probabilistic meaning: given two random variables X and Y uniformly distributed on the unit interval $[0,1]$ whose dependence is captured by a copula C, then the dependence of $1-X$ and $1-Y$ is captured by the survival copula of C. Clearly, double flipping extends the 'survival' concept from copulas to flippable aggregation functions, which justifies the title of the present paper.

Above, we have introduced the double flipping operation for flippable aggregation functions. It is therefore important to know when an aggregation function is flippable. To that end, we draw upon some of our results in [3], which make use of the notion of volumes of an aggregation function. Clearly, for an aggregration function to be flippable, it should satisfy the conditions of both Propositions 2 and 3.

Definition 2. *Consider an aggregation function A. The volume V_A of A on the rectangle $[x,x'] \times [y,y'] \subseteq [0,1]^2$ is defined as:*

$$V_A([x,x'] \times [y,y']) = A(x',y') - A(x',y) - A(x,y') + A(x,y)\,.$$

Proposition 2. *An aggregation function A is x-flippable if and only if for any $x \in [0,1]$ and any $[y,y'] \subseteq [0,1]$ it holds that*

$$V_A([0,x] \times [y,y']) \leq A(1,y') - A(1,y)\,, \tag{1}$$

or, equivalently, if and only if

$$V_A([x,1] \times [y,y']) \geq A(0,y) - A(0,y'), \tag{2}$$

or, equivalently, if and only if

$$A(x,y') - A(x,y) \leq A(0,y') - A(0,y) + A(1,y') - A(1,y). \tag{3}$$

Moreover, if A has first order partial derivatives w.r.t. y everywhere in $[0,1]^2$ *then it is x-flippable if and only if for all* $(x,y) \in [0,1]^2$ *it holds that*

$$\frac{\partial A(x,y)}{\partial y} \leq \frac{\partial A(0,y)}{\partial y} + \frac{\partial A(1,y)}{\partial y}. \tag{4}$$

Similarly, we can characterize the aggregation functions that are y-flippable.

Proposition 3. *An aggregation function A is y-flippable if and only if for any* $[x,x'] \subseteq [0,1]$ *and any* $y \in [0,1]$ *it holds that*

$$V_A([x,x'] \times [0,y]) \leq A(x',1) - A(x,1), \tag{5}$$

or, equivalently, if and only if

$$V_A([x,x'] \times [y,1]) \geq A(x,0) - A(x',0), \tag{6}$$

or, equivalently, if and only if

$$A(x',y) - A(x,y) \leq A(x',0) - A(x,0) + A(x',1) - A(x,1), \tag{7}$$

Moreover, if A has first order partial derivatives w.r.t. x everywhere in $[0,1]^2$ *then it is y-flippable if and only if for all* $(x,y) \in [0,1]^2$ *it holds that*

$$\frac{\partial A(x,y)}{\partial x} \leq \frac{\partial A(x,0)}{\partial x} + \frac{\partial A(x,1)}{\partial x}. \tag{8}$$

Note that condition (4) (resp. (8)) expresses that in any point of the unit square the derivative of A in the y-direction (resp. x-direction) should not exceed the sum of the derivatives of A in the y-direction (resp. x-direction) in the points with the same ordinate (resp. abscis) lying on the left and right boundary (resp. upper and lower boundary) of the unit square. Conditions (3) and (7) are nothing else but the discretized versions of (4) and (8), which should be applied when the partial derivatives do not exist everywhere.

Example 2. The smallest aggregation function A_* is given by $A_*(1,1) = 1$ and $A_*(x,y) = 0$ elsewhere, while the greatest one A^* is given by $A^*(0,0) = 0$ and $A^*(x,y) = 1$ elsewhere. Conditions (3) and (7) readily show that both are flippable. Moreover, $(A^*)^\# = A_*$ and $(A_*)^\# = A^*$

3 Flippability for Certain Classes

In this section, we investigate the flippability of aggregation functions that have special properties, such as having an absorbing element and/or a neutral element, or being 1-Lipschitz continuous. These properties are closely related to the concepts of semi-copulas, quasi-copulas and copulas, classes of aggregation functions which already played a privileged role in the case of x-flipping and y-flipping alone.

3.1 Absorbing Element

We start with the most general situation of an aggregation function A that has an absorbing element $a \in [0,1]$, i.e. $A(x,a) = A(a,y) = a$ for all $(x,y) \in [0,1]^2$. In particular, $A(0,a) = A(a,0) = A(1,a) = A(a,1) = A(a,a) = a$ and A being increasing, it follows that $A(x,y) = a$ for all $(x,y) \in [0,a] \times [a,1]$ and for all $(x,y) \in [a,1] \times [0,a]$. Hence, if we divide the unit square into four blocks by means of the lines $x = a$ and $y = a$, then A takes a constant value a on the two non-diagonal blocks. Since $A(1,1) = 1$, the restriction of A to the upper diagonal block is a rescaled aggregation function with absorbing element 0, also called a conjunctor [4].

Similarly, since $A(0,0) = 0$, the restriction of A to the lower diagonal block is a rescaled aggregation function with absorbing element 1, also called a disjunctor (the dual operation of a conjunctor). Summarizing, the structure of an aggregation function with absorbing element $a \in [0,1]$ is as follows:

$$A(x,y) = \begin{cases} a - aU\left(\dfrac{a-x}{a}, \dfrac{a-y}{a}\right) & \text{, if } (x,y) \in [0,a]^2\,, \\ a + (1-a)V\left(\dfrac{x-a}{1-a}, \dfrac{y-a}{1-a}\right) & \text{, if } (x,y) \in [a,1]^2\,, \\ a & \text{, elsewhere}\,, \end{cases}$$

where U and V are conjunctors, called the underlying conjunctors of A. Interesting examples of such aggregation functions are nullnorms [2].

From this general structure, one easily derives a first characterization of a flippable aggregation function with absorbing element a in terms of the flippability of the constituing conjunctors.

Proposition 4. *An aggregation function A with absorbing element a and underlying conjunctors U and V is x-flippable (resp. y-flippable, flippable) if and only if U and V are x-flippable (resp. y-flippable, flippable).*

The following proposition states that 1-Lipschitz continuity is sufficient for ensuring the flippability of an aggregation function with absorbing element a.

Proposition 5. *If an aggregation function A with absorbing element a is 1-Lipschitz continuous, then it is flippable. Moreover, its double flip $A^{\#}$ has neutral element $1-a$.*

The 1-Lipschitz continuity condition in Theorem 5 is sufficient, but not necessary, as is illustrated by the following example.

Example 3. Consider the geometric mean A defined by $A(x,y) = \sqrt{xy}$. Clearly, A is an aggregation function with absorbing element 0, but A is not 1-Lipschitz continuous. Nonetheless, A is flippable since

$$A^{\#}(x,y) = (1-\sqrt{1-x})(1-\sqrt{1-y})$$

is an aggregation function.

Quasi-copulas can be characterized as aggregation functions A with absorbing element 0 and neutral element 1 for which the volume V_A on any rectangle having at least two corners on the boundary of the unit square, is positive [5].

Definition 3. *An aggregation function A is called restricted 2-increasing if $V_A([x,x'] \times [y,y']) \geq 0$ whenever at least one of x,x',y,y' is either equal to 0 or to 1, i.e. for any $0 \leq x \leq x' \leq 1$ and $0 \leq y \leq y' \leq 1$, it holds that*

$$\begin{aligned} V_A([0,x'] \times [y,y']) &\geq 0 \\ V_A([x,x'] \times [0,y']) &\geq 0 \\ V_A([x,1] \times [y,y']) &\geq 0 \\ V_A([x,x'] \times [y,1]) &\geq 0 . \end{aligned}$$

An aggregation function A is called restricted 2-decreasing if $V_A([x,x'] \times [y,y']) \leq 0$ whenever at least one of x,x',y,y' is either equal to 0 or to 1.

The property of restricted 2-increasingness is sufficient to guarantee flippability.

Proposition 6. *Any aggregation function A that is restricted 2-increasing is flippable.*

The condition of restricted 2-increasingness is not necessary to ensure flippability. This is, for instance, the case for the greatest aggregation function A^*, which is restricted 2-decreasing, but not restricted 2-increasing, and flippable. The property of restricted 2-increasingness, however, becomes a necessary condition if the aggregation function has absorbing element 0.

Proposition 7. *An aggregation function A with absorbing element 0 is flippable if and only if it is restricted 2-increasing.*

Next, we provide a representation theorem for flippable aggregation functions with absorbing element 0.

Theorem 1. *An aggregation function A with absorbing element 0 is flippable if and only if there exists a quasi-copula Q and two increasing functions f and g with fixed points 0 and 1 such that A is given by $A(x,y) = Q(f(x),g(y))$.*

From Theorem 1, we can easily derive a representation of flippable aggregation functions with absorbing element 1. To that end, we recall the notion of a dual operation. The most common definition is

$$A^c(u,v) = 1 - A(1-u, 1-v).$$

For a t-norm A, A^c is known as its dual t-conorm, while for a copula A, A^c is called its co-copula.

Corollary 1. *An aggregation function A with absorbing element 1 is flippable if and only if there exists a quasi-copula Q and two increasing functions f and g with fixed points 0 and 1 such that A is given by* $A(x,y) = Q^c(f(x), g(y))$.

We now generalize Theorem 1 and Corollary , dealing with absorbing elements 0 or 1, to the case of an arbitrary absorbing element $a \in [0,1]$, and obtain the following representation theorem.

Theorem 2. *Consider an aggregation function A with absorbing element* $a \in [0,1]$. *The following statements are equivalent:*

(i) A is flippable;
(ii) There exist two quasi-copulas Q_1 *and* Q_2 *and two increasing functions f and g with fixed points 0, a and 1, such that A is given by*

$$A(x,y) = \mathrm{med}(a, Q_1^c(f(x), g(y)), Q_2(f(x), g(y))),$$

where 'med' stands for the median;
(iii) There exists a 1-Lipschitz continuous aggregation function B and two increasing functions f and g with fixed points 0, a and 1, such that A is given by

$$A(x,y) = B(f(x), g(y)).$$

3.2 Neutral Element

The implication $(iii) \Rightarrow (i)$ in Theorem 2 reveals that 1-Lipschitz continuity is sufficient but not necessary for an aggregation function with absorbing element a to be flippable. This condition turns out to be also a necessary condition if instead of having an absorbing element a, the aggregation function has a neutral element $e \in [0,1]$. We first show that it is indeed a necessary condition.

Proposition 8. *Any aggregation function A with neutral element e and is flippable, is 1-Lipschitz continuous. Moreover, its double flip* $A^\#$ *is given by*

$$A^\#(x,y) = x + y - 1 + A(1-x, 1-y)$$

and has absorbing element $1-e$.

Note that this proposition extends the validity of the expression in Example 1 to the case of an arbitrary neutral element e, providing further support for the name 'survival aggregation function' given to the double flip of a flippable aggregation function.

Relying on [6], we can show that 1-Lipschitz continuity is also a sufficient condition. For a 1-Lipschitz continuous aggregation function A, a second dual operation can be considered, namely

$$A^d(u,v) = u + v - A(u,v)\,.$$

For a copula A, A^d is called the dual of the copula. Moreover, the corresponding survival copula can be expressed as $A^\# = (A^d)^c = (A^c)^d$.

Theorem 3. *Consider an aggregation function A with neutral element $e \in [0,1]$. The following statements are equivalent:*

(i) A is flippable;
(ii) There exist two quasi-copulas Q_1 and Q_2 such that A can be expressed as

$$A(x,y) = x + y - \operatorname{med}(e, Q_1(x,y), Q_2^d(x,y)) \tag{9}$$

where 'med' stands for the median;
(iii) A is 1-Lipschitz continuous.

Corollary 2. *Consider a flippable aggregation function A with neutral element $e \in [0,1]$. Then $A^\#$ is given by*

$$A^\#(x,y) = \operatorname{med}(1-e, Q_1^{\#d}(x,y), Q_2^\#(x,y)))\,,$$

where Q_1 and Q_2 are the same quasi-copulas as in (9).

4 A Probabilistic View on Flippability

The probabilistic flavour of flippability of aggregation functions can be grasped from the case of flippable aggregation functions with absorbing element 0. Recall the famous theorem of Sklar [7, 9], which states that a function A is the restriction to $[0,1]^2$ of the cumulative distribution function of a random vector (X,Y) on $[0,1]^2$, if and only if $A(x,y) = C(F(x),G(y))$ for some copula C, where F and G are the cumulative distribution functions of X and Y, respectively. Moreover, for fixed X and Y, the dependence structure of the random vector (X,Y) may vary, which is expressed by the corresponding copula C. For any family $(C_i)_{i\in\mathscr{I}}$ of copulas, $Q = \sup_{i\in\mathscr{I}} C_i$ (and also $\inf_{i\in\mathscr{I}} C_i$) is a quasi-copula (see [7]), and any quasi-copula can be obtained in this manner. However, then for $A_i(x,y) = C_i(F(x),G(y))$, $i \in \mathscr{I}$, we have

$$B(x,y) := \sup_{i\in\mathscr{I}} A_i(x,y) = Q(F(x),G(y))\,.$$

Taking into account that F and G are both left-continuous or both right-continuous (depending on the definition of a cumulative distribution function used), and confronting the preceding expression with Theorem 1, we see that each left- or right-continuous flippable aggregation function B with absorbing element 0 is just a supremum (or, equivalently, an infimum) of a family of cumulative distribution functions of a random vector (X,Y) on $[0,1]^2$ with fixed marginals, or in other words, a family of copulas. Moreover, the double flip/survival aggregation function $B^{\#}$ is the supremum (or, equivalently, an infimum) of a family of corresponding survival copulas.

The paper of Alsina et al. [1], where quasi-copulas were introduced in their original form, gives a hint of another description of flippable aggregation functions with absorbing element 0. Indeed, under the assumption of either left- or right-continuity, they are exactly those aggregation functions B such that for any two continuous distribution functions $U,V:\mathbb{R}\to\mathbb{R}$ with $U(0)=V(0)=0$ and $U(1)=V(1)=1$, there exists a joint distribution function (restricted to $[0,1]^2$) $A:[0,1]^2\to[0,1]$ with marginals $A(x,1)=B(x,1)$ and $A(1,y)=B(1,y)$, so that

$$B(U(t),V(t)) = A(U(t),V(t)),$$

for any $t\in[0,1]$. The curve $\{(U(t),V(t)) \mid t\in[0,1]\}$ is called a track. So B coincides on any track with some joint distribution function A that has the same marginals as B.

5 Conclusion

We have shown that the double flipping, which can be considered as a generalization of the survival copula construction for copulas, is an operation on aggregation functions that has remarkable properties. Double flipping can be considered as more fundamental than the x-flipping and y-flipping operations of which it is composed, as it puts the two arguments of the aggregation function at the same level of importance. We have provided interesting characterizations of the class of flippable aggregation functions with an absorbing and/or neutral element. The fact that quasi-copulas play a prominent role in these characterizations is a key towards probabilistic interpretations of double flipping. At the same time, these findings indicate that generalizing the different flipping operations to n-ary aggregation functions ($n>2$), might be far from straightforward. Indeed, we should keep in mind that different equivalent characterizations of quasi-copulas in 2 dimensions lead to non-equivalent notions in more dimensions [8]. In this sense, the study of multi-flipping operations of n-ary aggregation functions, which we intend to relate to the construction of the survival copula associated with n-copulas [10], might well contribute to a deeper understanding of n-quasi-copulas.

Acknowledgements. This research was sponsored by the Bilateral Scientific Cooperation Flanders–Czech Republic, sponsored by the Special Research Fund of Ghent University (project nr. 011S01106). R. Mesiar was supported by the grants MSM VZ 6198898701, VEGA 1/4209/07 and APVV-0012-07.

References

1. Alsina, C., Nelsen, R., Schweizer, B.: On the characterization of a class of binary operations on distribution functions. Statistics & Probability Letters 17, 85–89 (1993)
2. Calvo, T., De Baets, B., Fodor, J.: The functional equations of Frank and Alsina for uninorms and nullnorms. Fuzzy Sets and Systems 120, 385–394 (2001)
3. De Baets, B., De Meyer, H., Kalická, J., Mesiar, R.: Flipping and cyclic shifting of binary aggregation functions. Fuzzy Sets and Systems 160, 752–765 (2009)
4. Díaz, S., Montes, S., De Baets, B.: Transitivity bounds in additive fuzzy preference structures. IEEE Trans. Fuzzy Systems 15, 275–286 (2007)
5. Genest, C., Quesada-Molina, J.J., Rodríguez-Lallena, J.A., Sempi, C.: A characterization of quasi-copulas. Journal of Multivariate Analysis 69, 193–205 (1999)
6. Kolesárová, A.: 1-Lipschitz aggregation operators and quasi-copulas. Kybernetika 39, 615–629 (2003)
7. Nelsen, R.B.: An Introduction to Copulas, 2nd edn. Springer, New York (2006)
8. Rodríguez-Lallena, J.A., Úbeda-Flores, M.: Some new characterizations and properties of quasi-copulas. Fuzzy Sets and Systems 160, 717–725 (2009)
9. Sklar, A.: Fonctions de répartition à *n* dimensions et leurs marges. Publications de l'Institut de Statistique de l'Université de Paris 8, 229–231 (1959)
10. Wolff, E.F.: N-dimensional measures of dependence. Stochastica 4, 175–188 (1980)

On e-Vertical Generated Implications

S. Massanet and J. Torrens

Abstract. Recently, a new construction method of a fuzzy implication from two given ones, called e-generation method, has been introduced. This method allows to control, up to a certain level, the increasingness on the second variable of the fuzzy implication through an adequate scaling on that variable of the two given implications. In this paper, the main goal is to reproduce the same idea but now on the first variable of the fuzzy implication. The new implications, called e-vertical generated implications, are studied in detail focusing on the preservation of the most common properties of fuzzy implications from the initial ones to the constructed implication.

1 Introduction

Fuzzy implications have probably become the most important operations in fuzzy logic, approximate reasoning and fuzzy control. These operators are not limited to model fuzzy conditionals, but also to make inferences in any fuzzy rule based system ([6, 8]). In addition, they are useful in fuzzy rough sets ([14] and references therein), fuzzy relational equations and fuzzy mathematical morphology ([10]), fuzzy DI-subsethood measures and image processing ([3, 4]), data mining ([16]), computing with words ([10]), fuzzy equivalence relations and fuzzy partitions ([7]).

All these properties emphasize the necessity of having a large repertoire of fuzzy implications at one's disposal. There are many different models of fuzzy implications and more than forty implications have been used just in fuzzy control. In [15], the meaning and the behaviour of a model of fuzzy implications is pointed out as the crucial point to its election. In such a paper, the authors argue that even new models for fuzzy implications could be convenient. In this direction, another

S. Massanet · J. Torrens
Dept. of Math. and Comp. Science, University of the Balearic Islands,
07122 Palma de Mallorca, Spain
e-mail: s.massanet,dmijts0@uib.es

B. De Baets et al. (Eds.): Eurofuse 2011, AISC 107, pp. 157–168, 2011.
springerlink.com

possibility comes from generating a new fuzzy implication from given ones. In this way, e-generated implications has been introduced in [11] and the preservation of the most common properties of fuzzy implications from the initial ones to the constructed implication has been studied in [12]. These implications are based on an adequate scaling on the second variable of the initial implications. Now, an analogous generation method, called e-vertical generation method, with a scaling on the first variable is performed. Some basic properties are studied and their preservation under the new construction method is investigated.

The communication is organized as follows. In the next section we recall the basic definitions and properties of implications needed in the subsequent sections. In Section 3, we present the e-vertical generation method of a fuzzy implication from two given ones and we discuss the basic properties preserved by this new method. Then, the exchange principle and the contrapositive symmetry are deeply studied in Section 4 and Section 5. The paper ends with some conclusions and future work.

2 Preliminaries

We will suppose the reader to be familiar with the theory of t-norms, t-conorms and fuzzy implications (all necessary results and notations can be found in [9] for t-norms and t-conorms, and in [2] for fuzzy implications). To make this work self-contained, we recall here some of the concepts and results employed in the rest of the paper.

2.1 Fuzzy Negations

Definition 1. *(see Definition 1.1 in [5] and Definition 11.3 in [9]) A decreasing function* $N : [0,1] \to [0,1]$ *is called a* fuzzy negation, *if* $N(0) = 1, N(1) = 0$. *A fuzzy negation* N *is called*

(i). strict, if it is strictly decreasing and continuous.
(ii). strong, if it is an involution, i.e., $N(N(x)) = x$ *for all* $x \in [0,1]$.

2.2 Fuzzy Implications

The most accepted definition of fuzzy implication is the following one.

Definition 2. *(see Definition 1.15 in [5], Definition 1.1.1 in [2]) A binary operator* $I : [0,1]^2 \to [0,1]$ *is said to be an* implication function, *or an* implication, *if it satisfies:*

(I1) $I(x,z) \geq I(y,z)$ *when* $x \leq y$, *for all* $z \in [0,1]$.
(I2) $I(x,y) \leq I(x,z)$ *when* $y \leq z$, *for all* $x \in [0,1]$.
(I3) $I(0,0) = I(1,1) = 1$ *and* $I(1,0) = 0$.

Note that, from the definition, it follows that $I(0,x) = 1$ and $I(x,1) = 1$ for all $x \in [0,1]$ whereas the symmetrical values $I(x,0)$ and $I(1,x)$ are not derived from it.

Example 1. In addition of the basic fuzzy implications, that are collected in Table 1.3. in [2], special implications that will be used along this paper are the *least fuzzy implication*

$$I_{Lt}(x,y) = \begin{cases} 1 \text{ if } x = 0 \text{ or } y = 1, \\ 0 \text{ if } x > 0 \text{ and } y < 1, \end{cases}$$

and the *greatest fuzzy implication*

$$I_{Gt}(x,y) = \begin{cases} 1 \text{ if } x < 1 \text{ or } y > 0, \\ 0 \text{ if } x = 1 \text{ and } y = 0. \end{cases}$$

Some specially interesting properties of implications are:

- The *exchange principle*,

$$I(x,I(y,z)) = I(y,I(x,z)), \quad x,y,z \in [0,1]. \tag{\textbf{EP}}$$

- The *left neutrality principle*,

$$I(1,y) = y, \quad y \in [0,1]. \tag{\textbf{NP}}$$

- The *ordering property*,

$$x \leq y \Longleftrightarrow I(x,y) = 1, \quad x,y \in [0,1]. \tag{\textbf{OP}}$$

- The *identity principle*,

$$I(x,x) = 1, \quad x \in [0,1]. \tag{\textbf{IP}}$$

- The *contrapositive symmetry* with respect to a fuzzy negation N,

$$I(x,y) - I(N(y),N(x)), \quad x,y \subset [0,1]. \tag{\textbf{CP}(\textbf{N})}$$

Definition 3. *(see Definition 1.14.15 in [2]) Let I be a fuzzy implication. The function N_I defined by $N_I(x) = I(x,0)$ for all $x \in [0,1]$, is called the* natural negation *of I.*

Finally, let us recall the e-generation method of a fuzzy implication.

Theorem 1. *([11]) Let I_1, I_2 be two implications and $e \in (0,1)$. Then the binary function $I_{I_1-I_2} : [0,1]^2 \to [0,1]$, called the e-generated implication from I_1 and I_2, defined as*

$$I_{I_1-I_2}(x,y) = \begin{cases} 1 & \text{if } x = 0, \\ e \cdot I_1\left(x, \dfrac{y}{e}\right) & \text{if } x > 0, y \leq e, \\ e + (1-e) \cdot I_2\left(x, \dfrac{y-e}{1-e}\right) & \text{if } x > 0, y > e, \end{cases}$$

is a fuzzy implication.

3 *e*-Vertical Generation Method of a Fuzzy Implication

The *e*-vertical generation method of a fuzzy implication from two given ones is based on an adequate scaling on the first variable of the initial implications. Note that the scaled first implication I_1 generates the new implication for $x < e$ and the second one, I_2 for $x \geq e$.

Theorem 2. *Let I_1, I_2 be two implications and $e \in (0,1)$. Then the binary function $I_{I_1|I_2} : [0,1]^2 \to [0,1]$, called the e-vertical generated implication from I_1 and I_2, defined as*

$$I_{I_1|I_2}(x,y) = \begin{cases} e + (1-e) \cdot I_1\left(\frac{x}{e}, y\right) & \text{if } x < e, y < 1, \\ e \cdot I_2\left(\frac{x-e}{1-e}, y\right) & \text{if } x \geq e, y < 1, \\ 1 & \text{if } y = 1, \end{cases}$$

is a fuzzy implication.

Example 2. Let us see some of the *e*-vertical generated implications from some well-known fuzzy implications (see Table 1.3 in [2] for their expressions). The plot of these implications can be viewed in Figure 1.

(i) Taking the Kleene-Dienes implication, the Reichenbach implication and $e = \frac{1}{2}$, we obtain the following *e*-vertical generated implication

$$I_{I_{KD}|I_{RC}}(x,y) = \begin{cases} \max\{1-x, \frac{1+y}{2}\} & \text{if } x < \frac{1}{2}, y < 1, \\ 1 - x - \frac{y}{2} + xy & \text{if } x \geq \frac{1}{2}, y < 1, \\ 1 & \text{if } y = 1. \end{cases}$$

(ii) If we consider the Rescher implication, the Goguen implication and $e = \frac{1}{3}$, we obtain the following *e*-vertical generated implication

$$I_{I_{RS}|I_{GG}}(x,y) = \begin{cases} 1 & \text{if } (x < \frac{1}{3}, 3x \leq y < 1) \text{ or } y = 1, \\ \frac{1}{3} & \text{if } (x < \frac{1}{3}, y < 3x) \text{ or } (x \geq \frac{1}{3}, \frac{3x-1}{2} \leq y < 1), \\ \frac{2y}{9x-3} & \text{if } x \geq \frac{1}{3}, y < \frac{3x-1}{2}. \end{cases}$$

(iii) Taking the Weber implication, the Yager implication and $e \in (0,1)$, we obtain the following family of *e*-vertical generated implications

$$I_{I_{WB}|I_{YG}}(x,y) = \begin{cases} 1 & \text{if } (x < e, y < 1), \text{ or } (x = e, y = 0), \text{ or } y = 1, \\ e \cdot y^{\frac{x-e}{1-e}} & \text{otherwise} \end{cases}$$

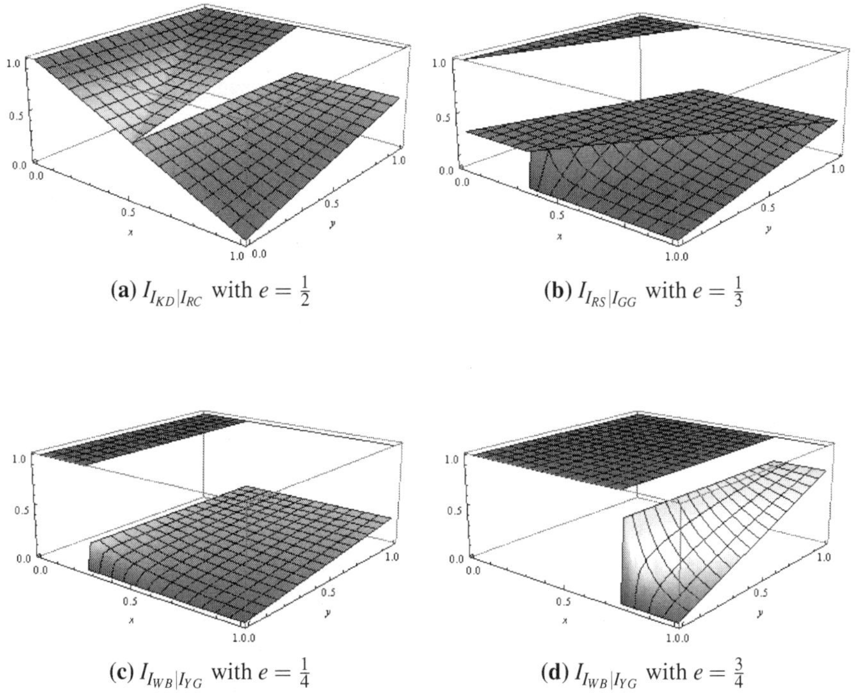

(a) $I_{I_{KD}|I_{RC}}$ with $e = \frac{1}{2}$ **(b)** $I_{I_{RS}|I_{GG}}$ with $e = \frac{1}{3}$

(c) $I_{I_{WB}|I_{YG}}$ with $e = \frac{1}{4}$ **(d)** $I_{I_{WB}|I_{YG}}$ with $e = \frac{3}{4}$

Fig. 1 Plot of some e-vertical generated implications

3.1 Basic Properties

In this section, we study the preservation of the most common properties of fuzzy implications from the two given ones to their corresponding e-vertical generated implication. When a property is not straightforwardly preserved, we study which conditions must satisfy the initial implications in order to be preserved. First of all, these implications allow us to have a degree of control of the decreasingness with respect to the first variable of the fuzzy implication.

Proposition 1. *Let I_1 and I_2 be two implications. Then*

(i) $I_{I_1|I_2}(x,y) \leq e$ *if* $x > e$ *and* $y < 1$,
$I_{I_1|I_2}(e,y) = e$ *if* $y < 1$,
$I_{I_1|I_2}(x,y) \geq e$ *if* $x < e$ *and* $y < 1$.
(ii) When $x < e$ *and* $y < 1$ *then* $I_{I_1|I_2}(x,y) > e \Leftrightarrow I_1(a,b) > 0$ *for* $a,b < 1$.
(iii) When $x > e$ *and* $y < 1$ *then* $I_{I_1|I_2}(x,y) < e \Leftrightarrow I_2(a,b) < 1$ *for* $a > 0, b < 1$.

The next results fully characterize these implications and the e-generated implications.

Theorem 3. *Let I be a fuzzy implication. Then I is an e-vertical generated implication $I_{I_1|I_2}$, if and only if, $I(e,y) = e$ for all $y < 1$. In this case, the initial implications I_1 and I_2 are respectively given by*

$$I_1(x,y) = \frac{I(ex,y) - e}{1-e},$$

$$I_2(x,y) = \begin{cases} \dfrac{I(e+(1-e)\cdot x,y)}{e} & \text{if } y < 1, \\ 1 & \text{if } y = 1. \end{cases}$$

Theorem 4. *Let I be a fuzzy implication. Then I is an e-generated implication $I_{I_1-I_2}$, if and only if, $I(x,e) = e$ for all $x > 0$. In this case, the initial implications I_1 and I_2 are respectively given by*

$$I_1(x,y) = \begin{cases} \dfrac{I(x,ey)}{e} & \text{if } x > 0, \\ 1 & \text{if } x = 0, \end{cases}$$

$$I_2(x,y) = \frac{I(x,e+(1-e)\cdot y) - e}{1-e}.$$

The natural negation of the e-vertical generated implications is generated by the natural negations of the initial implications and it can be continuous in some cases unlike what happened with the natural negation of the e-generated implications.

Proposition 2. *Let I_1 and I_2 be two implications. Then the natural negation of $I_{I_1|I_2}$ is*

$$N_{I_{I_1|I_2}}(x) = \begin{cases} e + (1-e)N_{I_1}\left(\dfrac{x}{e}\right) & \text{if } x < e, \\ e \cdot N_{I_2}\left(\dfrac{x-e}{1-e}\right) & \text{if } x \geq e. \end{cases}$$

Proposition 3. *Let I_1 and I_2 be two implications. Then $N_{I_{I_1|I_2}}$ is continuous, if and only if, N_{I_1} and N_{I_2} are continous.*

The following proposition deals with the satisfaction of the remaining basic properties of fuzzy implications by the e-vertical generated ones. Note that the counterpart of the properties for implications derived from uninorms (see [1, 13]) are also studied. Thus, for a fixed $e \in (0,1)$, we study (NP$_e$), that is $I(e,y) = y$ for all $y \in [0,1]$; (IP$_e$), that is $I(x,x) = e$ for all $x \in (0,1)$ and (OP$_e$), that is $I(x,y) \geq e \Leftrightarrow x \leq y$.

Proposition 4. *Let I_1, I_2 be two implications and $e \in (0,1)$. Then*

(i) $I_{I_1|I_2}$ does not satisfy either (NP) or (NP$_e$).
(ii) $I_{I_1|I_2}(x,y) = 1 \Leftrightarrow y = 1$ or ($x < e$ and $y < 1$ with $I_1\left(\frac{x}{e},y\right) = 1$), i.e., $I_{I_1|I_2}$ does not satisfy (OP).

(iii) $I_{I_1|I_2}(x,y) \geq e \Leftrightarrow y = 1$ *or* $(x < e$ *and* $y < 1)$ *or* $(x \geq e$ *and* $y < 1$ *with* $I_2\left(\frac{x-e}{1-e}, y\right) = 1)$, *i.e.,* $I_{I_1|I_2}$ *does not satisfy (*OP_e*).*

(iv) $I_{I_1|I_2}(x,x) = 1 \Leftrightarrow x = 1$ *or* $(x < e$ *with* $I_1\left(\frac{x}{e}, x\right) = 1)$, *i.e.,* $I_{I_1|I_2}$ *does not satisfy (IP).*

(v) $I_{I_1|I_2}$ *satisfies (*IP_e*) if and only if* $I_1\left(\frac{x}{e}, x\right) = 0$ *when* $0 < x < e$ *and* $I_2\left(\frac{x-e}{1-e}, x\right) = 1$ *when* $e \leq x < 1$.

Finally, the next results show that the e-vertical generated implication is never continuous but under some assumptions we can delimit the points of discontinuity.

Proposition 5. *Let* I_1, I_2 *be two implications. Then* $I_{I_1|I_2}$ *is not continuous at the points* $(x,1)$ *with* $x \geq e$.

Proposition 6. *Let* I_1, I_2 *be two implications. Then* $I_{I_1|I_2}$ *is continuous everywhere except at the points* $(x,1)$ *with* $x \geq e$ *if and only if* I_1 *is continuous everywhere except maybe at* $(1,1)$, I_2 *is continuous everywhere except maybe at the points* $(x,1)$ *with* $x \in [0,1]$ *and* $I_1(1,y) = 0$ *for all* $y < 1$.

4 Exchange Principle

The exchange principle is one of the most studied properties that can be satisfied by an implication. This property is not straightforwardly preserved by the e-vertical generation method (see Example 3) similarly to other generation methods of a fuzzy implication from two given ones, except the e-generation method (see [12]).

Example 3. The e-vertical generated implication

$$I_{I_{KD}|I_{KD}}(x,y) = \begin{cases} e + (1-e)\max\{\frac{e-x}{e}, y\} & \text{if } x < e, y < 1, \\ e\max\{\frac{1-x}{1-e}, y\} & \text{if } x \geq e, y < 1, \\ 1 & \text{if } y = 1, \end{cases}$$

where I_{KD} is the Kleene-Dienes implication, does not satisfy (EP) (just take $0 < x < e$, $y = e$ and $z < 1$), although I_{KD} satisfies it.

The first result gives us necessary and sufficient conditions in order to (EP) holds for these implications.

Proposition 7. *Let* I_1 *and* I_2 *be two implications such that* $I_1(x,y) < 1$ *when* $x > 0$ *and* $y < 1$. *If* $I_1(x,y) = 0$ *for all* $(x,y) \in [0,1]^2$ *with* $0 < x < 1$, $y \leq e$ *and*

$$I_1\left(\frac{x}{e}, e + (1-e)I_1\left(\frac{y}{e}, z\right)\right) = I_1\left(\frac{y}{e}, e + (1-e)I_1\left(\frac{x}{e}, z\right)\right) \tag{1}$$

for all $x, y < e$, $z < 1$ *and* $I_2(x,y) = 1$ *for all* $(x,y) \in [0,1]^2$ *with* $y \geq e$ *and*

$$I_2\left(\frac{x-e}{1-e}, eI_2\left(\frac{y-e}{1-e}, z\right)\right) = I_2\left(\frac{y-e}{1-e}, eI_2\left(\frac{x-e}{1-e}, z\right)\right) \tag{2}$$

for all $x, y \geq e$, $z < 1$, then $I_{I_1|I_2}$ satisfies (EP). Moreover, if $I_2(1,\cdot)$ is continuous, the converse holds too.

In order to characterize the implications that satisfy Equations (1) and (2), we will need two new construction methods of a fuzzy implication from an initial one.

Lemma 1. *Let I be a fuzzy implication. Then*

$$I_I^0(x,y) = \begin{cases} 1 & \text{if } x = 0, \\ 0 & \text{if } x > 0 \text{ and } y \leq e, \\ I\left(ex, \frac{y-e}{1-e}\right) & \text{if } x > 0 \text{ and } y > e, \end{cases}$$

and

$$I_I^1(x,y) = \begin{cases} 1 & \text{if } x = 0, \\ I\left(e+(1-e)x, \frac{y}{e}\right) & \text{if } x > 0 \text{ and } y \leq e, \\ 1 & \text{if } x > 0 \text{ and } y > e, \end{cases}$$

are fuzzy implications.

The implications I_I^0 and I_I^1 satisfy with some assumptions the required equations of Proposition 7 as the following results show.

Proposition 8. *Let I be a fuzzy implication such that $I(x,y) > 0$ when $y > 0$. Then I satisfies (EP) for all $x, y < e$ and $z \in (0,1]$, if and only if, I_I^0 satisfies Equation (1) for all $x, y < e$ and $z \in [0,1]$.*

Proposition 9. *Let I be a fuzzy implication. Then I satisfies (EP) for all $x, y > e$ and $z \in [0,1]$, if and only if, I_I^1 satisfies Equation (2) for all $x, y \geq e$ and $z \in [0,1]$.*

Therefore, from the previous results we can conclude the main result on the satisfaction of (EP) by this kind of implications.

Proposition 10. *Let I_1 and I_2 be two implications such that $I_1(x,y) > 0$ for all $y > 0$ and $I_1(x,y) < 1$ for all $x \leq e$ and $y < 1$. If I_1 satisfies (EP) for all $x, y < e$ and $z \in (0,1]$ and I_2 satisfies (EP) for all $x, y \geq e$ and $z \in [0,1]$ then $I_{I_{I_1}^0|I_{I_2}^1}$ satisfies (EP). Moreover, if $I_2(1,\cdot)$ is continuous, the converse also holds.*

Note that the following corollary, that gives us the preservation of (EP) through the e-vertical generation method, is immediate.

Corollary 1. *Let I_1 and I_2 be two implications such that $I_1(x,y) > 0$ for all $y > 0$ and $I_1(x,y) < 1$ for all $x \le e$ and $y < 1$. If I_1 and I_2 satisfy (EP), then $I_{I^0_{I_1}|I^1_{I_2}}$ satisfies (EP).*

5 Contrapositive Symmetry

Another important property for fuzzy implications is the contrapositive symmetry. We give in this section a similar study for this property as the one given in the previous section for the exchange principle. The e-vertical generated implications satisfying (CP(N)) are not fully characterized, but some interesting results can be obtained. First of all, the following examples show that there exist e-vertical generated implications that do not satisfy (CP(N)) for any negation N and others that satisfy (CP(N)) for an infinity of negations N.

Example 4. (i) The e-vertical generated implication

$$I_{I_{FD}|I_{FD}}(x,y) = \begin{cases} 1 & \text{if } y = 1 \text{ or } (x < e, \frac{x}{e} \le y < 1), \\ \max\{\frac{e-x}{e}, y\} & \text{if } x < e, y < 1, \frac{x}{e} > y, \\ e & \text{if } x \ge e, y < 1, \frac{x-e}{1-e} \le y, \\ e\max\{\frac{1-x}{1-e}, y\} & \text{otherwise.} \end{cases}$$

where I_{FD} is the Fodor implication, does not satisfy (CP(N)) for any negation N, although I_{FD} satisfies (CP(N_C)).

(ii) If we consider the least and the greatest fuzzy implications, then

$$I_{I_{Lt}|I_{Gt}}(x,y) = \begin{cases} 1 \text{ if } x = 0 \text{ or } y = 1, \\ 0 \text{ if } x = 1 \text{ and } y = 0, \\ e \text{ otherwise,} \end{cases}$$

is an e-vertical generated implication that satisfies (CP(N)) for any negation N such that $N(x) \neq 0, 1$ for all $x \in (0,1)$.

Now, if (CP(N)) holds for an e-vertical generated implication, then some necessary conditions must be satisfied as it is stated in the following proposition.

Proposition 11. *Let I_1, I_2 be two implications and N a fuzzy negation. If $I_{I_1|I_2}$ satisfies (CP(N)), then*

(i) $0 < N(x)$ for all $x < 1$.
(ii) $N(x) < 1$ for all $x \in [e,1)$. Furthermore, if N is continuous then $N(x) < 1$ for all $0 < x < 1$.
(iii) $I_2(x,y) = 1$ for all $y \ge N(e)$ and $x \in [0,1]$.

Note that for some specific forms of negations N we can fully determine the initial implications needed to obtain an e-vertical generated implication satisfying (CP(N)).

Proposition 12. *Let N be the following negation*

$$N(x) = \begin{cases} 1 \text{ if } x = 0, \\ e \text{ if } 0 < x < 1, \\ 0 \text{ if } x = 1, \end{cases}$$

and I_1, I_2 two implications. Then

$$I_{I_1|I_2} \text{ satisfies } (CP(N)) \Leftrightarrow I_1 = I_{Lt} \text{ and } I_2 = I_{Gt}.$$

Now, we want to study when the contrapositive symmetry of the initial given implications is preserved by the e-vertical generation method. Assuming that the negation N satisfies some initial requirements, the first result gives us necessary and sufficient conditions for which the e-vertical generated implications satisfy (CP(N)).

Proposition 13. *Let I_1 and I_2 be two implications and N a fuzzy negation such that $N(x) = e \Leftrightarrow x = e$ and $0 < N(x) < 1$ if $x \in (0,1)$. Then $I_{I_1|I_2}$ satisfies (CP(N)) if and only if $I_1(x,y) = 0$ for all $(x,y) \in [0,1]^2$ with $x > 0$, $y \leq e$ and*

$$I_1\left(\frac{x}{e}, y\right) = I_1\left(\frac{N(y)}{e}, N(x)\right) \tag{3}$$

for all $(x,y) \in (0,e) \times (e,1)$, and $I_2(x,y) = 1$ for all $(x,y) \in [0,1]^2$ with $y \geq e$ and

$$I_2\left(\frac{x-e}{1-e}, y\right) = I_2\left(\frac{N(y)-e}{1-e}, N(x)\right) \tag{4}$$

for all $(x,y) \in [e,1] \times [0,e]$.

In order to characterize the implications that satisfy Equations (3) and (4), let us start with a new construction method of a fuzzy implication from an initial one and a negation N.

Lemma 2. *Let I be a fuzzy implication and N a fuzzy negation.*

(i) If $N(x) \in [0,e]$ for all $e < x$ then

$$I^0_{I,N}(x,y) = \begin{cases} 1 & \text{if } x = 0, \\ 0 & \text{if } x > 0, y \leq e, \\ I\left(x, N\left(\frac{N(y)}{e}\right)\right) & \text{if } x > 0, y > e, \end{cases}$$

is a fuzzy implication.

(ii) If $N(x) \in [e,1]$ for all $x < e$ then

$$I^1_{I,N}(x,y) = \begin{cases} 1 & \text{if } y \geq e, \\ I\left(x, N\left(\frac{N(y)-e}{1-e}\right)\right) & \text{if } y < e, \end{cases}$$

is a fuzzy implication.

Lemma 3. *Let I be a fuzzy implication and N a strong negation with $N(e) = e$. Then*
(i) If I satisfies (CP(N)) then

$$I^0_{I,N}\left(\frac{x}{e}, y\right) = I^0_{I,N}\left(\frac{N(y)}{e}, N(x)\right)$$

for all $0 < x < e$ and $e < y < 1$. Moreover, if $I(x,0) = I(1,N(x))$ for all $x \in [0,1]$, the reciprocal holds too.
(ii) I satisfies (CP(N)) if and only if

$$I^1_{I,N}\left(\frac{x-e}{1-e}, y\right) = I^1_{I,N}\left(\frac{N(y)-e}{1-e}, N(x)\right)$$

for all $x \geq e$ and $y \leq e$.

These lemmas are essential in order to give some partial answer to the preservation of the contrapositive symmetry.

Proposition 14. *Let I_1 and I_2 be two fuzzy implications and N a strong fuzzy negation with fixed point e. Then if I_1 and I_2 satisfy (CP(N)), then the e-vertical generated implication $I_{I^0_{I_1,N}|I^1_{I_2,N}}$ satisfies CP(N). Moreover, if $I_1(x,0) = I_1(1,N(x))$, the reciprocal holds too.*

6 Conclusions and Future Work

In this paper, a new generation method of a fuzzy implication from two previous ones, called e-vertical generation method, is introduced and the propagation of the basic properties from the initial implications to the e-vertical generated implication is studied. This method allows a certain control of the decreasingness of the implication with respect to the first variable. In addition, the characterizations of these implications and the recently presented e-generated implications are given in Theorems 3 and 4.

As a future work, it would be worth to study the preservation of the law of importation and the distributivity properties from the initial implications to the e-vertical generated implications and some others. Finally, note that with this method and the e-generation method we can define some kind of ordinal sum of implication functions combining them adequately, that is worth to study.

Acknowledgements. This paper has been partially supported by the Spanish Grant MTM2009-10320 with FEDER support.

References

1. Aguiló, I., Suñer, J., Torrens, J.: A characterization of residual implications derived from left-continuous uninorms. Information Sciences 180(20), 3992–4005 (2010)
2. Baczyński, M., Jayaram, B.: Fuzzy Implications. Springer, Heidelberg (2008)
3. Bustince, H., Mohedano, V., Barrenechea, E., Pagola, M.: Definition and construction of fuzzy DI-subsethood measures. Information Sciences 176, 3190–3231 (2006)
4. Bustince, H., Pagola, M., Barrenechea, E.: Construction of fuzzy indices from fuzzy DIsubsethoodmeasures: application to the global comparison of images. Information Sciences 177, 906–929 (2007)
5. Fodor, J., Roubens, M.: Fuzzy Preference Modelling and Multicriteria Decision Support. Kluwer Academic Publishers, Dordrecht (1994)
6. Gottwald, S.: A Treatise on Many-Valued Logic. Research Studies Press, Baldock (2001)
7. Jayaram, B., Mesiar, R.: I-fuzzy equivalence relations and i-fuzzy partitions. Information Sciences 179(9), 1278–1297 (2009)
8. Kerre, E., Huang, C., Ruan, D.: Fuzzy Set Theory and Approximate Reasoning. Wu Han University Press, Wu Chang (2004)
9. Klement, E., Mesiar, R., Pap, E.: Triangular norms. Kluwer Academic Publishers, Dordrecht (2000)
10. Mas, M., Monserrat, M., Torrens, J., Trillas, E.: A survey on fuzzy implication functions. IEEE Transactions on Fuzzy Systems 15(6), 1107–1121 (2007)
11. Massanet, S., Torrens, J.: *e*-generation method of construction of a new implication from two given ones. Submitted to Fuzzy Sets and Systems (2011)
12. Massanet, S., Torrens, J.: On some properties of e-generated implications. Submitted to Fuzzy Sets and Systems (2011)
13. Massanet, S., Torrens, J.: On some properties of e-generated implications. Submitted to Fuzzy Sets and Systems (2011)
14. Radzikowska, A.M., Kerre, E.E.: A comparative study of fuzzy rough sets. Fuzzy Sets and Systems 126(2), 137–155 (2002)
15. Trillas, E., Mas, M., Monserrat, M., Torrens, J.: On the representation of fuzzy rules. Int. J. Approx. Reasoning 48(2), 583–597 (2008)
16. Yan, P., Chen, G.: Discovering a cover set of ARsi with hierarchy from quantitative databases. Information Sciences 173, 319–336 (2005)

Some Properties of Consistency in the Families of Aggregation Operators

Karina Rojas, Daniel Gómez, J. Tinguaro Rodríguez, and Javier Montero

Abstract. Properties related with aggregation operators functions have been widely studied in literature. Nevertheless, few efforts have been dedicated to analyze those properties related with the family of operators in a global way. What should be the relationship among the members of a family of aggregation operators? Is it possible to build the aggregation of n data with aggregation operators of lower dimension? Should it exist some consistency in the family of aggregation operators? In this work, we analyze two properties of consistency in a family of aggregation operators: *Stability* and *Structural Relevance*. The stability property for a family of aggregation operators tries to force a family to have a stable/continuous definition in the sense that the aggregation of n items should be similar to the aggregation of $n+1$ items if the last item is the aggregation of the previous n items. Following this idea some definitions and results are given. The second concept presented in this work is related with the construction of the aggregation operator when the data that have to be aggregated has an inherent structure. The Structural Relevance property tries to give some ideas about the construction of the aggregation operator when the items are related by means of a graph.

1 Introduction

Aggregation of information appears in a natural way in all kinds of knowledge based systems (see, e.g. [2, 4, 6, 11]). Usually, the aggregation-fusion process produces a reduction in the dimension of the original data since the main aim of this aggregation

Karina Rojas · Daniel Gómez · J. Tinguaro Rodríguez · Javier Montero
Universidad Complutense de Madrid, Plaza de las Ciencias 3, Madrid
e-mail: krpatuelli@yahoo.com, dagomez@estad.ucm.es, {jtrodrig,monty}@mat.ucm.es

B. De Baets et al. (Eds.): Eurofuse 2011, AISC 107, pp. 169–176, 2011.
springerlink.com

is to simplify the information in such a way that the decision maker passes from a complex problem to a simpler one.

An aggregation operator is a real function of n values, which is used to reduce the dimension of the original data in order to simplify the information. The standard definition of a family of aggregation operators in the fuzzy context is $\{A_n : [0,1]^n \rightarrow [0,1], n\varepsilon\mathbb{N}\}$, for every cardinal n of the data.

Many properties have been studied in relation with the aggregation operator functions such as continuity, commutativity, monotonicity, associativity (and a large etcetera). But in contrast, few efforts have been dedicated to research the relations among the members of a family of aggregation operators. As has been pointed in the past for some authors (see for example [1, 5, 7, 8, 9, 12], among others), these common properties (as for example continuity) show us some desirable characteristics related with each aggregation function A_n, but do not give us any information about the consistency of the family of aggregation operators in the sense of the relations that should exist among its members. In practice, it is frequent that some information can get lost, be deleted or added, and each time a cardinality change occurs a new aggregation operator A_m has to be used to aggregate the new collection of m elements, and therefore a relation between $\{A_n\}$ and $\{A_m\}$ [4] do not necessarily exist in a family of aggregation operators. In this context, it seems natural to incorporate some properties to maintain the logical consistency between these families when changes on the cardinality of the data occur, for which we need to be able to build up a definition of family of aggregation operators in terms of its logical consistency, and solve each problem of aggregation without knowing *apriori* the cardinality of the data.

Therefore, it is clear that taking into account that in the framework of aggregation process we have the sequence $A(2)$, $A(3)$, $A(4), \ldots$, it seems logical to study properties that give sense to this sequence in such a way that the presence of an unifying concept behind such an aggregation function is suggested. Otherwise we may have only a bunch of disconnected operators.

For example, it would seem quite strange to propose an aggregation using the minimum for two numbers, the arithmetic mean for three numbers, the geometric mean for four and the median for five. And though it could seem that a formal approach should solve this problem by demanding a conceptual unity through a mathematical formula, it should be noted that the last example allows a trivial compact mathematical formulation (but notice that an aggregation function should offer a solution far beyond the first five cases).

Thus, as it is showed in [12], a debate about those properties that may reflect the importance of considering the aggregation operators functions as a whole family following some idea of consistency is a necessary task. Some references in this line can be found. For example, recursive rules (see [4]) seems to gather the idea that it is possible to build an operator aggregation A_n in a sequential way. The key idea of recursiveness is that, in order to be consistent, an aggregation rule should be based upon an iterative application of binary operators taking advantage of previous aggregations. This idea is also studied in [1, 12] in which the recursive rules are generalized in a more flexible way.

But recursiveness is not the only property related with the consistency of the whole family of operators. In [17, 18], the concept of operativity is analyzed from a computational point of view, trying to capture the consistency in the implementation of the aggregation function.

In this work, we will study two different properties related with the consistency of a family of aggregation operators: *Stability* and *Structural Relevance*. The stability property for a family of aggregation operators tries to force such a family to have a stable/continuous definition in the sense that the aggregation of n items should be similar to the aggregation of $n+1$ items whenever the last item is the aggregation of the previous n items. Following this idea some definitions and results are given.

The second concept presented in this work is related with the construction of an aggregation operator when the data that have to be aggregated possess an inherent structure. In our opinion, the idea and definition of consistency of a aggregation family should also take into account the structure of the data. Data structure refers to the relative position of information units. Some preliminary ideas can be founded in [21], in which the structure of a set of fuzzy classes is included in order to distinguish between some apparently similar situations which however present different underlying structures (as the interval-valued fuzzy sets and the intuitionistic fuzzy sets). In other areas as probability theory, it is clear that the estimation process is dependent of the structure of the data that has to be aggregated. For example, in order to forecast tomorrow's weather, the information about today's weather is more relevant than the information about yesterday's weather. These situations are usually represented as a lineal structure (as in a time series) or by means of graph (as in Markov chains).

2 Stability of a Family of Aggregation Operators

As it is pointed in [4], *stability* of any mathematical model for engineering problems means, roughly speaking, that the "small inputs errors" do not result into "big output error". Stability of an aggregation function is defined in a similar way to continuity, in the sense that small changes in the vector x should not produce big changes in the $A(x)$. Formally, stability is defined in the following way: if $\max|x_i - y_i|$ small then $|A(x) - A(y)|$ also is small. As it has been pointed in the introduction, stability has been defined only for aggregation functions, particularly we could not find in the literature a definition following the general idea of stability for the *whole* family of aggregation operators.

Similarly, as it is concerned with the robustness of the aggregation process, the notion of *stability* proposed here for a family of aggregation operators is also inspired in that of continuity, though it now focuses in the cardinality of the data rather than in the data itself. Therefore, we have considered a property close to the continuity of functions in order to assure robustness in the result of the aggregation. We mean for *Stability* of a family of aggregation operators, for changes minimal in the value aggregated when a change in the cardinality of the data is produced.

DEFINITION 1: Let $\{A_n : [0,1]^n \to [0,1], n \in \mathbb{N}\}$ be a family of aggregation operators. Then, it is said that:

1. It fulfils the property of *Strict Stability from the Right* if for all $n > n_o$ and for all $x_1, \ldots x_{n-1}$ the following holds:

$$A_n(x_1,x_2,\ldots x_{n-1},A_{n-1}(x_1,x_2\ldots,x_{n-1})) = A_{n-1}(x_1,x_2,\ldots,x_{n-1}), x_i \in [0,1]\ \forall i$$

2. It fulfils the property of *Strict Stability from the Left* if for all $n > n_o$ and for all $x_1, \ldots x_{n-1}$ the following holds:

$$A_n(A_{n-1}(x_1,\ldots,x_{n-1}),x_1,\ldots x_{n-1}) = A_{n-1}(x_1,\ldots,x_{n-1}), x_i \in [0,1]\ \forall i$$

3. It fulfils the property of *Strict Stability* if A_n satisfies the two points above.

Let us observe that the Definition 1.1 was defined in [24] for a specifical aggregation function called as a self-identity aggregation function. With the Definition 1.1. we also want to make emphasize in the fact that this property should be defined for the whole family and not for the aggregation function. In this sense, our Definition 1.1 extends the one given in [24]. In addition with this, Definition 1.2 and Definition 1.3 establish some differences among the family of aggregation operators in terms of the structure of the aggregation process.

Previous definitions try to assure that any change in the cardinality of the data will not produce a significant impact in the result of the aggregation.

The operators Minimum (Min) and Maximum (Max) trivially satisfy this stability property from both the right and the left, which are also fulfilled by the arithmetic mean $M_n(x_1,\ldots,x_n) = \sum_{i=1}^n \frac{x_i}{n}$, the geometric mean $G_n(x_1,\ldots,x_n) = (\prod_{i=1}^n x_i)^{1/n}$, the harmonic mean $H_n(x_1,\ldots,x_n) = \frac{n}{\sum_{i=1}^n 1/x_i}$ and the median Md. In addition, any binary idempotent operator with inductive extension forward (A_*) satisfies the property of *Strict Stability by the Right*, and any binary idempotent operator with inductive extension backward (A^*) satisfy the property of *Strict Stability by the Left*. However, some aggregation operators fulfil the property of stability in a way little more weakly, for example the weighted mean $W_n(x_1,\ldots,x_n) = \frac{\sum_{i=1}^n x_i * w_i}{\sum_{i=1}^n w_i}$. In this situation, it holds that A_n converges in law to A_{n-1}, and therefore the stability property is fulfilled in the limit.

DEFINITION 2: Let $\{A_n : [0,1]^n \to [0,1], n \in \mathbb{N}\}$ be a family of aggregation operators. Then, it is said that:

1. It fulfils the property of *Stability by the Right* if the following holds:

$$\lim_{n\to+\infty} |A_n(x_1,x_2,\ldots x_{n-1},A_{n-1}(x_1,x_2,\ldots,x_{n-1})) - A_{n-1}(x_1,x_2,\ldots,x_{n-1})| = 0$$

2. It fulfils the property of *Stability by the Left* if the following holds:

$$\lim_{n\to+\infty} |A_n(A_{n-1}(x_1,\ldots,x_{n-1}),x_1,\ldots x_{n-1}) - A_{n-1}(x_1,\ldots,x_{n-1})| = 0$$

3. It fulfils the property of *Stability* if A_n satisfies the above two points.

Finally, other aggregation operators fulfil the property of stability in a weak sense: for example, the product $A_n(x_1,\ldots,x_n) = \prod_{i=1}^{n} x_i$ and $Q_n(x_1,\ldots,x_n) = \prod_{i=1}^{n} x_i^i$. In this situation, it can be seen that A_n converges almost surely to A_{n-1}, and therefore the stability property is fulfilled by asymptotic convergence.

DEFINITION 3: Let $\{A_n : [0,1]^n \rightarrow [0,1], n \in \mathbb{N}\}$ be a family of aggregation operators. Then, we will say that:

1. It fulfils the property of *Weak Stability from the Right*, if the following holds:

$$\mathbb{P}[\lim_{n\rightarrow+\infty} |A_n(x_1,x_2,\ldots x_{n-1},A_{n-1}(x_1,x_2,\ldots,x_{n-1}))- \\ A_{n-1}(x_1,x_2,\ldots,x_{n-1})| = 0\,] = 1, \forall x_i \sim U(0,1)$$

2. It fulfils the property of *Weak Stability from the Left*, if the following holds:

$$\mathbb{P}[\lim_{n\rightarrow+\infty} [\ |A_n(A_{n-1}(x_1,x_2,\ldots,x_{n-1}),x_1,x_2,\ldots x_{n-1})- \\ A_{n-1}(x_1,x_2\ldots,x_{n-1})| = 0\,] = 1, \forall x_i \sim U(0,1)$$

3. It fulfils the property of *Weak Stability* if A_n satisfies the above two points.

PROPOSITION 1: Let $\{A_n : [0,1]^n \rightarrow [0,1], n \in \mathbb{N}\}$ be a family of aggregation operators. Then the following holds:

1. If the family $\{A_n\}_n$ satisfies the property of $strict stability$ then it also satisfies the property of $stability$.
2. If the family $\{A_n\}_n$ satisfies the property of $stability$ then it satisfies the property of weak stability.

Let us put an example of a family not satisfying any stability property. Let $A_n(x_1,\ldots,x_n)$ be a family defined as the maximum of the data if n is even and as the minimum if n is odd. It is easy to see that this family does not satisfy any stability property.

3 Structural Relevance in the Construction of a Family of Aggregation Operators

The stability concept has been presented in the last section for any family situation. In this section, we show another concept that appears in a natural way when the data that have to be aggregated is related by means of a graph. The structure of the data is concerned with the relative position of their elements. Therefore, if the data elements are interchangeable, the data is unstructured. Any structured data can be represented by a graph, this graph being composed by data elements $x_1,x_2,...,x_n$ and a family of aggregation operators $\{A_n\}$, which can be part of a bigger structure, with other collections of elements and families of operators. For example, if we want to graphically represent a data set with a lineal structure from left to right, where the result of aggregating the two first elements is used in the aggregation with the third element, the resulting graph will have two substructures (Figure 1). On the other hand, if we don't have a lineal structure but rather a nested structure, the resulting graphical representation could look as the one showed in Figure 2.

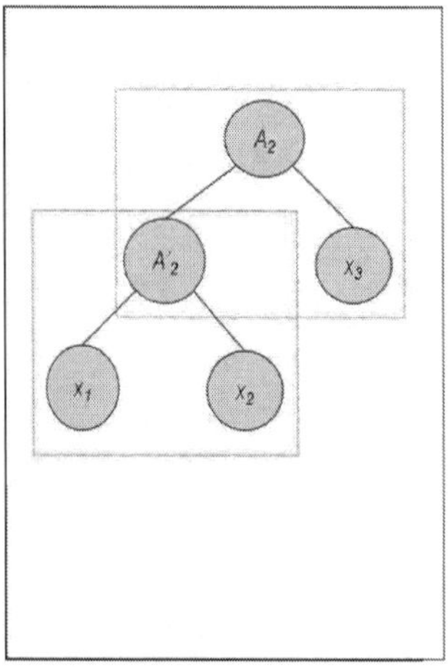

Fig. 1 Representation of a dataset with linear srtucture of the right

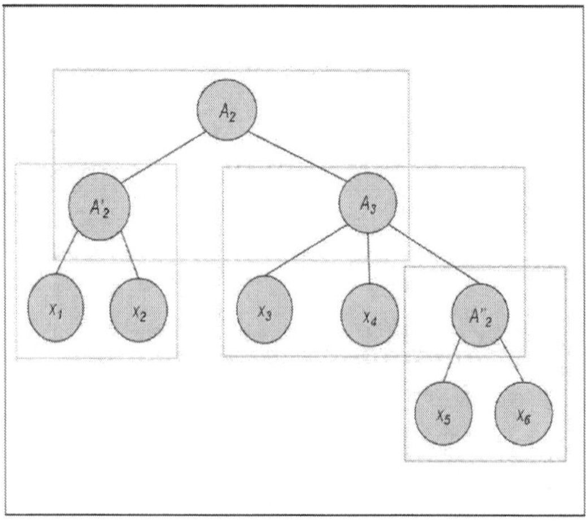

Fig. 2 Representation of a dataset with a nested structure

In these graphs we can see that each substructure is contained only by one family of aggregation operators, and the result of this aggregation is part of the ascendant substructure, where is considered an element more of conjoint of elements. Of course, this only is a way of writing it, also we can write the root family with cardinality three in the first graph or six in the second, but the fulfilling of the stability property is complicated in this way. Therefore, to continue with the construction of a stable family of aggregation operators behind a structured data, we need to consider the representation of the data as pointed above, and for this way, for each substructure we can see that the stability property is fulfilled. This way, the more stable is the family of aggregation operators of each substructure of the data, the more robust will be the aggregated value of a collection of n elements.

In order to give a general idea of what we understand by *structuralrelevance* of a family of aggregation operators, let us say that a family $\{A_n\}_n$ of operators satisfies the structural relevance if it is possible to build a digraph for each n (in the sense of the figures 1 and 2) with two class of nodes (data and aggregation functions of lower dimensions) in such a way that $A_n(x_1,\ldots,x_n)$ is build as a function of operators with lower dimension following the nodes that are directly connected with the operator. For example, following Figure 2 we could establish that

$$A_6(x_1,\ldots,x_6) = A_2\left(A'_2(x_1,x_2),A_3(x_3,x_4,A_2(x_5,x_6))\right).$$

The figure 1 represents a widely known recursive aggregation operator since $A_3(x_1,x_2,x_3) = A_2(A'_2(x_1,x_2),x_3)$. If we impose that $A_2 = A'_2$ we have the classical definition of recursive rule from left to right (see for example [4]). In a more general situation in which we permit that the operators depend on the level of the hierarchical digraph then we have the formal definition of recursive rule given in [1, 12].

4 Conclusions

In our opinion an aggregation family should never be understood just as a family of n-ary operators. Rather, all these aggregation operators must be deeply related following some building procedure throughout the aggregation process. To this aim, we have presented here two properties that follows such an objective. It is clear that we should not define a family of aggregation operators $\{A_n\}$ in which A_2 is the mean, A_3 geometric mean, A_4 is the minimum. Thus, in our opinion the aggregation process demands a conceptual unit idea rather than a mathematical formula.

The stability notion proposed in this paper makes emphasize in the idea of robustness-stability-continuity of the family in the sense that the operator defined for n data items should not differ too much of the operator defined for $n-1$ elements.

Another aspect considered in this paper is related with the structure of the data. The notion of consistency in the relation among the aggregation functions is not trivial and could depend on the structure of the data. In [1] a possible definition of consistency in the framework of recursive rules is done. For more general situations, here we present by means of a graph a mechanism that permits us to build the aggregation function taking into account the structure of the data that has to be aggregated. Nevertheless, the definition proposed here is just a seminal effort and possible modifications coming from a further analysis (we think) merit to be carried out.

Acknowledgment. This research has been partially supported by the Government of Spain, grant TIN2009-07901.

References

1. Amo, A., Montero, J., Molina, E.: Representation of consistent recursive rules. European Journal of Operational Research 130, 29–53 (2001)
2. Beliakov, G., Pradera, A., Calvo, T.: Aggregation Functions, a Guide to Practitioners. Springer, Berlin (2007)
3. Bustince, H., Barrenechea, E., Fernández, J., Pagola, M., Montero, J., Guerra, C.: Aggregation of neighbourhood information by means of interval type 2 fuzzy relations

4. Calvo, T., Kolesarova, A., Komornikova, M., Mesiar, R.: Aggregation operators, properties, classes and construction methods. In: Calvo, T., et al. (eds.) Aggregation Operators New trends ans Aplications, pp. 3–104. Physica-Verlag, Heidelberg (2002)
5. Calvo, T., Mayor, G., Torrens, J., Suñer, J., Mas, M., Carbonell, M.: Generation of weighting triangles associated with aggregation fuctions. International Journal of Uncertainty, Fuzziness and Knowledge-Based Systems 8(4), 417–451 (2000)
6. Carlsson, C.H., Fuller, R.: Fuzzy reaasoning in decision making and optimization. Heidelberg, Springfield-Verlag (2002)
7. Cutello, V., Montero, J.: Hierarchical aggregation of OWA operators: basic measures and related computational problems. Uncertainty, Fuzzinesss and Knowledge-Based Systems 3, 17–26 (1995)
8. Cutello, V., Montero, J.: Recursive families of OWA operators. In: Proceedings FUZZ-IEEE Conference, pp. 1137–1141. IEEE Press, Piscataway (1994)
9. Cutello, V., Montero, J.: Recursive connective rules. International Journal of Intelligent Systems 14, 3–20 (1999)
10. Dubois, D., Gottwald, S., Hajek, P., Kacprzyk, J., Prade, H.: Terminological difficulties in fuzzy set theory - the case of intuitionistic fuzzy sets. Fuzzy Sets and Systems 156, 485–491 (2005)
11. Fung, L.W., Fu, K.S.: An axiomatic approach to rational decision making in a fuzzy environment. In: Zadeh, L.A. (ed.) Fuzzy Sets and their Applications to Cognitive and Decision Processes, pp. 227–256. Academic Press, London (1975)
12. Gómez, D., Montero, J.: A discussion of aggregation functions. Kybernetika 40, 107–120 (2004)
13. Gómez, D., Montero, J., Yáñez, J., Poidomani, C.: A graph coloring algorithm approach for image segmentation. Omega 35, 173–183 (2007)
14. Gómez, D., Montero, J., Yánez, J.: A coloring algorithm for image classification. Information Sciences 176, 3645–3657 (2006)
15. Grabisch, M., Marichal, J., Mesiar, R., Pap, E.: Aggregation Functions, Encyclopedia of Mathematics and its Applications (2009)
16. Kolesárová, A.: Sequential aggregation. In: González, M., et al. (eds.) Proceedings of the Fifth International Summer School on Aggregation Operators, AGOP, Universitat de les Illes Balears, Palma de Mallorca, pp. 183–187 (2009)
17. López, V., Garmendia, L., Montero, J., Resconi, G.: Specification and computing states in fuzzy algorithms. Uncertainty, Fuzziness and Knowledge-Based Systems 16, 301–336 (2008)
18. López, V., Montero, J.: Software engineering specification under fuzziness. Multiple-Valued Logic and Soft Computing 15, 209–228 (2009)
19. López, V., Montero, J., Tinguaro Rodriguez, J.: Formal Specification and implementation of computational aggregation Functions. In: Computational Intelligence, Foundations and Applications Proceedings of the 9th International FLINS Conference 2010, pp. 523–528 (2010)
20. Montero, J.: A note on Fung-Fu's theorem. Fuzzy Sets and Systems 13, 259–269 (1985)
21. Montero, J., Gómez, D., Bustince, H.: On the relevance of some families of fuzzy sets. Fuzzy Sets and Systems 158, 2429–2442 (2007)
22. Montero, J., López, V., Gómez, D.: The role of fuzziness in decision making. Studies in Fuzziness and Soft Computing 215, 337–349 (2007)
23. Yager, R.R.: On ordered weighted averaging aggregation operators in multi-criteria decision making. IEEE Transactions on Systems, Man and Cybernetics 18, 183–190 (1988)
24. Yager, R.R., Rybalov, A.: Nonconmutative self-identity aggregation. Fuzzy Sets and Systems 85, 73–82 (1997)

Part III

Knowledge Extraction

Fuzzy Ontologies to Represent Background Knowledge: Some Experimental Results on Modelling Climate Change Knowledge

Emilio Fdez-Viñas, Mateus Ferreira-Satler, Francisco P. Romero,
Jesus Serrano-Guerrero, Jose A. Olivas, and Natalia Saavedra

Abstract. Ontologies represent a method of sharing and reusing knowledge on the semantic web. Moreover, fuzzy ontologies, i.e., the combination of fuzzy logic and ontologies, may be an interesting tool for representing domain knowledge with the aim of solving problems where uncertainty is present. This paper presents three fuzzy-based ontology models for knowledge representation. These ontologies have been obtained after the automatic analysis of a collection of relevant documents that are related to a specific subject. Some experiments have been carried out to illustrate the feasibility of these approaches.

1 Introduction

The World Wide Web presents new challenges to information retrieval [1]. The rapid growth of digital libraries, such as Internet, makes it difficult to human beings to access useful information conveniently and effectively. This is due to the fact that most of information is embedded in a non-structured or semi-structured way, which makes the search of a particular content a daunting and time consuming task. Ontologies have proven to be successful in handling a machine processable representation of information and is becoming very useful in various fields such as intelligent information extraction and retrieval, cooperative information systems, electronic commerce, and knowledge management [2]. They allow domain knowledge to be captured in an explicit and formal way such that it can be shared among human and computer systems [3]. Ontology can take the simple form of a

Emilio Fdez-Viñas · Mateus Ferreira-Satler · Francisco P. Romero · Jesus Serrano-Guerrero · Jose A. Olivas · Natalia Saavedra
Dept. of Information Systems and Technologies, University of Castilla La Mancha
Ciudad Real, Spain
e-mail: emilio.fdez@alu.uclm.es, msatler@gmail.com,
{FranciscoP.Romero,JoseA.Olivas,Jesus.Serrano}@uclm.es,
nattaliasm18@gmail.com

B. De Baets et al. (Eds.): Eurofuse 2011, AISC 107, pp. 179–191, 2011.
springerlink.com

taxonomy or as a vocabulary with standardized machine interpretable terminology supplemented with natural language definitions. Ontology is often specified in a declarative form by using semantic markup languages such as RDF and OWL [4].

Although ontologies are used in a wide range of applications and have been instrumental in many interoperability projects, they have so far had only limited success [5]. An important bottleneck in developing ontology-based systems stems from the fact that the conceptual formalism that supports by typical ontologies may not be sufficient to represent uncertainty that is commonly found in many application domains [6].

Fuzzy logic concepts are incorporated into the classic ontologies, resulting in fuzzy ontologies that can capture richer semantics than classic crisp ontologies. These fuzzy ontologies include relatedness degree values between concepts. The advantages of fuzzy ontologies include the potential to represent the domain knowledge, this potential has been exploited in this work. Fuzzy and crisp ontologies can be created or extended by domain experts or be reused from existing ones. Alternatively, it would be possible to obtain fuzzy ontologies in a semi-automatic way, as proposed by some works as it can be seen in Section 2. In this paper, three fuzzy-based ontology models for knowledge representation are presented. These ontologies are obtained through analysing a collection of relevant documents about a certain knowledge domain. The first model is based on the KeyWord connection matrix [7], the second model is defined using multiple relationships between concepts and terms into a set of documents and the last method consists of extracting a valued network of relations between concepts in a document collection.

These ontologies has been applied into the PLINIO project [8]. This project intends to solve the problem of managing the huge volume of information on Internet about climate change impacts using Semantic Web methods and technologies as ontologies. Therefore, the purpose of this project is to provide organized and meaningful information about the subject that has appeared on the Web, which can become an important strategic advantage for decission making. Some experiments have been carried out in order to illustrate the feasibility of each ontology in order to represent the knowledge about climate change. *Ecoresearch.net* [1] and "The Environmental Web-portal" (EnviWeb[2]) [9] are both projects similar to PLINIO. All these web applications are based on ontologies and in the case of EnviWeb the method to obtain this ontology is explained in [10]. Personalization of contents, the application of fuzzy ontologies and the use of meta-search engines are distinguish features on PLINIO.

The rest of the paper is organized as follows. A brief overview of some related works on fuzzy ontology construction is presented in section 2. Section 3 provides a detailed description of all stages performed in the proposed strategies to automatically construct fuzzy ontologies. The results of the experiments performed are presented in section 4 and finally, some conclusions and future works are pointed out in section 5.

[1] `http://www.ecoresearch.net/climate`

[2] `http://www.enviweb.cz/`

2 Related Work

Ontologies have been applied as background knowledge in several research fields such as information retrieval [11] and semantic web [12]. Ontologies can be mainly applied for understanding the problem for analysing or representing interestingness on user or patterns. The automation of ontology construction process (also known as Ontology Learning), is concerned with knowledge acquisition and comprises identify concepts and define the relationships among these concepts. Fuzzy ontology acquisition is not a simple task and in some cases consists of extracting information by applying natural language techniques to text. In recent years several methodologies and techniques have been proposed in literature to capture, store and process automatically, in an appropriate and reasonable time, relevant information to build an ontology.

Nikravesh [13] presents a framework for reasoning and deduction in the web. The model focuses on intelligent information and knowledge retrieval through conceptual matching of text and provides a Concept–Based Web–Based Databases for Intelligent Decision Analysis, which aims to address the problem of finding more precise and more relevant information in the web. The model presented can also be used for constructing ontology, i.e., the ontology is automatically constructed from a text document collection by using a term similarity function based on conceptual–latent semantic indexing technique (CLSI). A user profile is also constructed from a text document collection. These components are used for query refinement and for ranking the results of a query.

Lau et al [14] present a text mining methodology for the automatic discovery of fuzzy domain ontology from a collection of online messages posted to blogs, emails, chat rooms, web pages, and so on. The collection of messages is treated as a textual corpus. The method consists of a document parsing (stop word removal, part-of-speech tagging, and entity tagging and stemming), concept extraction (pat tern filtering, text windowing, and mutual information computation), dimensionality reduction (concept pruning and term space reduction), fuzzy relation extraction (computing fuzzy relation membership) and fuzzy taxonomy extraction (taxonomy generation and taxonomy pruning).

An example of application of fuzzy ontologies can be seen in [15], where the authors show how a fuzzy ontology-based approach can improve semantic documents retrieval. The proposal is illustrated using an information retrieval algorithm based on an object-fuzzy concept network. The proposed fuzzy ontology, based on the semantic correlation between concepts, is capable to represent a dynamic knowledge of a domain adapting itself to the context. The correlations are first assigned by experts and then updated after querying or when a document has been inserted into a database.

3 Fuzzy Ontologies Building Methods

In this paper, three ontology models for knowledge representation are presented. The first model is based on the KeyWord connection matrix[7], where the results are represented by a set of relationships between terms. The second model is based on the Multirrelation among concepts and terms in a document collection. This model establishes that two concepts are linked by means of four fuzzy relations. The third approach aims to construct a set of directed graphs where each node represents a term and the edges denote that a term "is related with" another term. All the ontologies have been constructed from the same text corpus as source. In the following sections, the proposed models are explained.

3.1 Keyword Connection Matrix

In the KeyWord connection matrix[7], relevant terms keywords are encoded as a single relation that expresses the similarity degree between terms from a set. This similarity is measured basing on co-occurrence of terms in documents present in the collection. This approach is based on FOG(Framework automatic for Ontology Generator[16]). This framework uses a four-stages strategy to generate fuzzy relations:

a. The first stage is the preprocessing of data contained in text collection. The text collection modeling in only possible for words that are classified as *noun* such as:NN (noun singular o mass), NNP (proper noun singular), NNPS (proper noun plural), NNS (noun plural), NP (proper noun singlar), NPS (proper noun plural).
b. The second stage consists of indexing the word set obtained in previous stage. By indexing the word set, we guarantee independence between the framework and the text collection.
c. The third stage involves information modeling. Each document is represented by a vector of terms. This representation allows to determine if a term is present in a document or to find out which documents contains a particular term. Giving this representation for all documents from the collection, it is possible to build a knowledge representation of the collection. This knowledge can be represented by a symmetrical matrix, where the element W_{ij} determines the relationship between two terms $t_i, t_j \in T$. This relation W_{ij} are restricted to the range [0,1], where 0 indicates no relationship between two keyword and 1 indicates the strongest possible relationship.

$$W_{ij} = \frac{N_{ij}}{N_i + N_j - N_{ij}} \tag{1}$$

In equation 1, N_{ij} represents the number of documents that contains the terms t_i and t_j. N_i is the number of documents that contains the term t_i, and N_j is the number of documents that contains the term t_j. In this case, $W_{ij} = 1$ when $i = j$.

d. The last stage consists of the ontological model-based representation of information obtained in previous stages. In this ontological model are represented all terms and relations identified by the Keyword connection matrix strategy.

3.2 Fuzzy Multi-relationship

This strategy is an adaptation of the Multi-relationship concept networks model proposed by Horng [17]. In this approach, the Multi-Relationship fuzzy are represented by tree kind of relation: Fuzzy Positive association (P), Fuzzy Generalization association (G)and Fuzzy Specialization association (S). The semantic descriptions of this fuzzy relations are shown bellow:

- Fuzzy positive association. Relates concepts which are similar to each other.
- Fuzzy generalization association. Relates a concept to another concept if the former is a part of the latter or the former is a kind of the latter.
- Fuzzy specialization association. This association is the inverse of the fuzzy generalization relationship.

Multi-Relationship modeling is performed similarly to the modeling of Keyword connection matrix using FOG. Both the preprocessing and the indexing of text collection produce the same result. The only change is in the modeling strategy that, in this case, we used the Multi-Relationship technique. The text collection modeling using Multi-Relationship strategy is described as follows. Each word from collection has a relevance weight that is defined by the equation 2:

$$W_{word\_document}(t,d_i) = \frac{\left(0.5+0.5\frac{tf_{it}}{Max\, tf_{ik\,k=1,2,\dots,L}}\right)\log\frac{N}{df_t}}{Max_{j=1,2,\dots,L}\left\{\left(0.5+0.5\frac{tf_{it}}{Max\, tf_{ik\,k=1,2,\dots,L}}\right)\log\frac{N}{df_j}\right\}} \quad (2)$$

where tf_{it} is the frequency of word t in document d_i, df_t is the number of documents that contains the word t, L is the number of words contained in document d_i, and N is the number of documents in the collection. The larger the value of $W_{word\ document}(t,d_i)$ the more important the word t in document d_i.

After calculating the weight of the words in documents, the weight of word t in concept c can be obtained by the equation 3:

$$W_{word\ concept}(t,c) = \frac{\Sigma_{i=1}^{m} W_{word\ document}(t,d_i)}{m} \quad (3)$$

where m is the number of documents that contain the word t and belong to concept c.The fuzzy Multi-relationship model defines a concept as a word which contains other words. If a word of the collection do not contains any other words, this word will not be considered a concept. Since each concept contains particular words from the word set, we can use a mapping function M to represent each concept by showing its corresponding fuzzy subset in the word set W. The mapping function M is defined by:

$$M(c_i) = w_{i1} + w_{i2} + \ldots + w_{ih} \tag{4}$$

where $M : C \longrightarrow [0,1]$,$w_{ij}$ is the weight of word t_j in concept c_i, and h is the number of words in the word set W. If $w_{ij} > 0$, then the word t_j is contained in the concept c_i.

The next step is to identify the direction of relations between concepts. For this, we used the generalization equation:

$$G(c_i, c_j) = \begin{cases} \left(\frac{\sum_{k=1}^{h} min\left(w_{ki}, w_{kj}\right)}{\sum_{k=1}^{h} w_{ki}} \right)^{\frac{WC(c_i)}{max\left(WC(c_i), WC(c_j)\right)}} & \text{if } M_{cj} \neq \phi \\ 1 & \text{if } M_{cj} = \phi \end{cases} \tag{5}$$

where w_{ki} is the weight of word t_k in concept c_i, w_{kj} is the weight of word t_k in concept c_j, $WC(c_i)$ is the number of words contained in concept c_i, $WC(c_j)$ is the number of word contained in concept c_j. The advantage of this approach is that it takes into account the number of words contained in each concept. Therefore, if concept c_j contains more words than concept c_i, the value of $G(c_i, c_j)$ will be increased.

The degree of the concept c_j contained in the concept c_i is determined by $S(c_i, c_j)$. Once the fuzzy specialization relationship is the inverse of the fuzzy generalization relationship, then:

$$S(c_i, c_j) = G(c_j, c_i) \tag{6}$$

Moreover, the degree of fuzzy positive association relationship between concept c_i and concept c_j is represented by $P(c_i, c_j)$, where:

$$P(c_i, c_j) = min(G(c_i, c_j), S(c_i, c_j)) \tag{7}$$

It is posible to see that if both values of $G(c_i, c_j)$ and $S(c_i, c_j)$ are high, then the value of $P(c_i, c_j)$ will also be high. This means that if concept c_i and concept c_j contains almost the same words with similar weight, the degree of fuzzy positive association relationship between concept c_i and concept c_j is high.

At this point, it is posible to define the direction of relations between concepts basing on a threshold α. The direction of the relations can be defined by:

- *Case 1*. If the degree of fuzzy generalization relationship between concept c_i and concept c_j is $(\mu_G) \geq \alpha$, and the degree of fuzzy specification relationship between concept c_i and concept c_j is $(\mu_S) \geq \alpha$, then they are synonymous concepts and should be classified in the same concept class denoted by:

$$c_i \overset{\mu_G, \mu_S, \mu_P}{\leftrightarrow} c_j.$$

- *Case 2*. If the degree of fuzzy generalization relationship between concept c_i and concept c_j is $(\mu_G) \geq \alpha$, and the degree of fuzzy specification relationship between concept c_i and concept c_j is $(\mu_S) < \alpha$, then the concept c_j is more general than concept c_i, which is denoted by:

$$c_i \stackrel{\mu_G,\mu_S,\mu_P}{\longrightarrow} c_j.$$

- *Case 3*. If the degree of fuzzy generalization relationship between concept c_i and concept c_j is $(\mu_G) < \alpha$, and the degree of fuzzy specification relationship between concept c_i and concept c_j is $(\mu_S) \geq \alpha$, then the concept c_i is more general than concept c_j, which is denoted by:

$$c_i \stackrel{\mu_G,\mu_S,\mu_P}{\longleftarrow} c_j.$$

Taking into account the fuzzy positive relationship, the fuzzy generalization relationship and the fuzzy specification relationshipis, it is posible to build a hierarchy with the identified concept.

3.3 Relatednes Degrees

In this proposed strategy, the fuzzy ontology FO is represented by a valued network of relations between terms extracted from a collection of text documents (D). In this context, the FO component may be considered as a set of directed graphs where each node represents a term (t) and the edges denote that a term *"is related with"* another term. A ***Relatedness Degree*** (*RD*) is associated with each edge to represent the strength of the *"is related with"* association. Formally, the fuzzy ontology $FO(\mathfrak{R})$ is defined by the set $\mathfrak{R}$ (Eq. 8) as a collection of relations $R(t_i,t_j,RD)$ where:

$$\mathfrak{R} = \{R_1,R_2,R_3,\ldots,R_n\} \tag{8}$$

To calculate the relatedness degree (*RD*), we used the idea of occurrence frequency. Given a term t_i, the occurrence of t_i in a document d is represented by $ocurr(t_i,d)$ and its membership value is defined by (Eq. 9):

$$ocurr(t_i,d) = \frac{f(t_i,d)}{f_{max}(d)} \tag{9}$$

where $f(t_i,d)$ is the frequency of term t_i in document d and $f_{max}(d)$ is the occurrences number of the most frequent term in document d. In this way, the relatedness degree (*RD*) is defined as (Eq. 10):

$$RD(t_i,t_j) = \frac{\sum_{d\in D} occur(t_i,d) \otimes occur(t_j,d)}{\sum_{d\in D} occur(t_i,d)} \tag{10}$$

where $\otimes$ denotes a fuzzy conjunction operator, defined by *Einstein Product* and *Drastic Product* t-norms.

3.3.1 Building and Updating Process

The automatic construction of fuzzy ontologies using the Relatedness Degrees strategy consists of several stages of data processing, discovering and representation of textual information. The process is divided in the following stages: linguistic pre-processing, term indexing (called pre–ontology), fuzzy ontology construction and fuzzy ontology updating process.

In the *Linguistic Preprocessing* stage, the document collection (D) is transformed into a representation suitable for further analysis. The goal is to extract the textual information in form of individual terms (t). For this, a document transformation is done to deal with the different format types (.doc, .txt, .pdf, xml, etc.). Then, all non–textual information like digits, dates and punctuation marks, is removed (lexical analysis). Finally, two techniques are used to reduce the vocabulary and make the representation of texts more meaningful: Stop lists [18] and Stemming [19]. Language detection and spelling correction processes are also included in this stage.

The *Term Indexing* stage aims to provide an index structure, called pre–ontology, which contains information about all terms generated in the previous stage. Therefore, the proposed pre–ontology (PO) is defined by (Eq. 11):

$$PO = \{(t_1, \zeta_1), (t_2, \zeta_2), (t_3, \zeta_3), \ldots, (t_n, \zeta_n)\} \tag{11}$$

where ζ_k is a *List of Documents Features* where the term (t_k) appears. The stored features are the following: Document ID, Number of occurrences of the most frequent term in the document and a list of the term's positions in the document. Each term position is a tuple (p, s) where s represents the section or paragraph where the term is located and p the index of the term into the corresponding section.

In the next stage, the fuzzy ontology is constructed by first calculating the relatedness degree RD and then generating the relation R for each pair of two distinct terms from the pre–ontology (i.e. $R(t_i,t_j,RD(t_i,t_j))$ and $R(t_j,t_i,RD(t_j,t_i))$). Generated relations are then stored in the fuzzy ontology set $\Re$. Finally, redundant concepts associations and less meaningful information are eliminated by removing from $\Re$ relations where RD value is lower than an $\alpha - cut$. This filtering strategy allows preserving positive, stronger and more significant concept associations. In this research work, the value of $\alpha - cut$ (0.6) was defined based on empirical experiments carried out previously in [20] and [21]. Unrelated terms ($RD = 0$) are automatically excluded in this stage. Ontology is often specified in a declarative form by using semantic markup languages like OWL [4].

In the last stage, the *Fuzzy Ontology Updating Process* allows the inclusion of new information and knowledge when new documents are added in the documents collection. The first step of this process consists of the *linguistic pre–processing* and the *term indexing* for new documents. Then, the relatedness degree RD is calculated and the relation R is generated for each new concept created. Finally, the fuzzy ontology is updated and filtered according to the same statements of *Fuzzy Ontology*

Construction stage. During this process, the relatedness degrees RD_i of previous concepts are recalculated and the relations R_i are updated.

4 Experiments

In this section we present the evaluation of the proposed methodologies. The main objective in this evaluation process is to determine if it each ontology fulfills the proposed objectives, that is, the knowledge extracted from source documents and represented as an ontology is useful and interesting for an expert on Climate Changes.

Two sets of source documents has been used to build ontologies using the previously explained methodologies. The first set of source documents has been the technical papers I, II, III y IV and the special report "The regional impacts of Climate Change: an assessment of vulnerability" written by the IPCC (Intergovernmental Panel of Climate Change)[3] between 1996 and 1997. The second set of source documents is "IPCC Fourth Assessment Report: Climate Change" published in 2007 and three special reports published between 2005 and 2008.

An expert on climate change has evaluated the three methods. The evaluation method has been the following:

1. 15 source relevant concepts on this knowledge domain have been selected (Table 1). In this set of concepts, there are concepts that define the climate system (scale, global), climate system components (glacier, ecosystem, permafrost, land), technical concepts (mitigation, adaptation) and consequences of climate change (flood, vulnerability, impact).

Table 1 List of Analized Concepts

adaptation	biodiversity	ecosystem	flood
glacier	global	greenhouse	impact
land	mitigation	permafrost	resource
scale	vulnerability	weather	

2. For each source relevant concept, 15 related concepts have been extracted from each obtained ontology. These 15 related concepts have been analysed in to obtain a correlation degree with relevant concept. The related concepts are classified as both corrects or doubtful and included or not included in the top 15 related terms.
3. Quality Measures (Eq. 12): **Precision** (P) is defined as the ratio of the correct included concepts to the included concepts , **Recall** (R) is the ratio of the correct included concepts to the correct concepts and **F - measure** (F1) is a combination metric that gives equal weight to both precision and recall:

[3] `http://www.ipcc.ch/`

$$P = \frac{N_{ci}}{N_i} \qquad R = \frac{N_{ci}}{N_c} \qquad F1 = \frac{(2*R*P)}{(R+P)} \tag{12}$$

where N_{ci} are the included correct concepts, N_i the number of included concepts and N_c the number of correct concepts.

The obtained results can be seen in Table 2. The values are in the range [0,1]. They are good in both methods for every measure, but the best results are obtained using the KeyWord Connection based approach. The best value of precision and F-measure corresponding to the ontology generated by the KeyWordConnection and the background documents from 2007. The other methods achieve better recall values but both precision values are poor because the number of generated relations in both ontologies is higher than the better one. These results illustrate the ability of fuzzy ontologies to represent background knowledge. In this regard, it is important to remark that we do not achieve better results by using a more complex method as the Fuzzy Multirelationship method.

Table 2 Experimental Results

Ontology	Precision	Recall	F - measure
Keyword Connection - 1997	0.35	**0.86**	0.49
Keyword Connection - 2007	**0.62**	0.68	**0.65**
Fuzzy Multi-Relationship - 2007	0.38	0.70	0.50
Relatedness Degree - 2007	0.44	0.79	0.56

4.1 Gold Standard Ontology Evaluation

In this section a gold standard ontology evaluation [22] process is presented. This process consists of comparing an ontology with another ontology that is deemed to be the benchmark. Precision has been used to perform this evaluative analysis.

In this work we have used standard measures implemented in WordNet::Similarity [23], which were contrasted with ontologies relations. The results of precision are shown in Table 3. The obtained value of precision is acceptable, however, the results still depends highly on the golden standard used, in this case, WordNet/ WordNetSimilarity. These results are not surprising. This may occurs due to several reasons. Firstly, the kind of non-taxonomic relationship generated by fuzzy ontology construction method is not contained in a golden standard, which is often organized in syntactic and taxonomic levels. Secondly, we generate ontologies that specifically comprise the climate change domain. On the other hand, golden standards are very generic and focus on the description of a large portion of the world, making it quite difficult to cover certain particular aspects, which are highly important in domain ontologies. For example, around 35% of terms generated for climate

Table 3 Precision Values for Wordnet::Similarity Measures

Source Documents	Einstein Product	Drastic Product
1997	0.81	0.81
2007	0.41	0.40

change domain are not present in WordNet/WordNetSimilarity database. Because of this, only the relations whose terms are in WordNet/WordNetSimilarity have been taken into account in the evaluation.

5 Conclusions and Future Work

The flexible nature of fuzzy ontologies can support a wide range of approaches to the information retrieval and text categorization problems. The use of fuzzy ontologies to represent background knowledge is proposed in this work and its application in a Internet Watch Web Application named PLINIO is discussed. We described and evaluated three different approaches to automatically construct fuzzy ontologies. Experiment results indicated that knowledge generated are a good summary of the background knowledge about climate change.

Future work includes updating the ontology-based background knowledge representation using the web documents crawled in the PLINIO projects (rss feeds, web pages, blogs, etc) about climate change. More experiments will be conducted with other collections of documents in order to verify the quality of the obtained ontologies. Further research is directed towards the task of improving the knowledge quality, using a pruning process to avoid concepts that have no significance.

Acknowledgements. This research was partially supported by the Spanish Science and Innocation Ministry (MEC) under TIN2010-20395 project and by the Regional Government of Castilla-La Mancha under PEIC09-0196-3018, POII10-0133-3516 (PLINIO) projects.

References

1. Sendhilkumar, S., Geetha, T.V.: Personalized ontology for web search personalization. In: COMPUTE 2008: Proceedings of the 1st Bangalore Annual Compute Conference, pp. 1–7. ACM, New York (2008)
2. Welty, C.: Ontology Research. AI Magazine 24(3), 11–12 (2003)
3. Lau, R.: Fuzzy Domain Ontology Discovery for Business Knowledge Management. IEEE Intelligent Informatics Bulletin 8(1), 29–41 (2007)
4. Deborah, L.: McGuinness and Frank van Harmelen. OWL Web Ontology Language Overview. Technical Report REC-owl-features-20040210, W3C (2004)
5. Solskinnsbakk, G., Gulla, J.A.: Combining ontological profiles with context in information retrieval. Data Knowl. Eng. 69(3), 251–260 (2010)

6. Tho, Q.T., Hui, S.C., Cao, T.H.: FOGA: A Fuzzy Ontology Generation Framework for Scholarly Semantic Web. In: Proceedings of the 2004 Knowledge Discovery and Ontologies Workshop (KDO 2004), Pisa, Italy (2004)
7. Ogawa, Y., Morita, T., Kobayashi, K.: A fuzzy document retrieval system using the keyword connection matrix and a learning method. Fuzzy Sets and Systems 39(2), 163–179 (1991)
8. Romero, F.P., Mateus, F.-S., Olivas, J.A., Jesus, S.-G.: PLINIO: Observatorio de Efectos del Cambio Climático basado en la extracción inteligente de Información en Internet. In: Actas del III Simposio sobre Lógica Fuzzy y Soft Computing, LFSC - CEDI 2010 (EUSFLAT), Valencia, Spain, pp. 385–392 (2010)
9. Hrebicek, J., Kubasek, M.: EnviWeb and Environmental Web Services: Case Study of an Environmental Web Portal, pp. 21–24. Springer, London (2004)
10. Kubasek, M.: Semantic web technology - ontology extraction from environmental web. In: 17th International Conference Informatics for Environmental Protection, The Information Society and Enlargement of the European Union, Cottbus, Germany, Metropolis, pp. 905–909 (2003)
11. Nathalie, A.-G., Mothe, J.: Ontologies as background knowledge to explore document collections. In: Seventh Triennial RIAO Conference: Coupling Approaches, Coupling Media and Coupling Languages for Information Retrieval, pp. 129–142 (2004)
12. Sabou, M., D'Aquin, M., Motta, E.: Exploring the Semantic Web as Background Knowledge for Ontology Matching. Journal on Data Semantics 11, 156–190 (2006)
13. Nikravesh, M.: Concept-based search and questionnaire systems. In: Nikravesh, M., Kacprzyk, J., Zadeh, L. (eds.) Forging New Frontiers: Fuzzy Pioneers I. Studies in Fuzziness and Soft Computing, vol. 217, pp. 193–215. Springer, Heidelberg (2007)
14. Lau, R.Y.K., Song, D., Li, Y., Cheung, T.C.H., Hao, J.-X.: Toward a Fuzzy Domain Ontology Extraction Method for Adaptive e-Learning. IEEE Transactions on Knowledge and Data Engineering 21(6), 800–813 (2009)
15. Calegari, S., Sanchez, E.: Object-fuzzy concept network: An enrichment of ontologies in semantic information retrieval. Journal of the American Society for Information Science and Technology 59(13), 1532–2890 (2008)
16. Emilio, F.-V., Jesús, S.-G., Olivas, J.A., De La Mata, J., Soto, A.: FOG: Arquitectura flexible para la generación automática de ontologías. In: ESTYLF 2010, Huelva, Spain (2010)
17. Horng, Y.-J., Chen, S.-M., Lee, C.-H.: Automatically Constructing Multi-Relationship Fuzzy Concept Networks for Document Retrieval. Applied Artificial Intelligence 17(4), 303–328 (2003)
18. Korfhage, R.R.: Information storage and retrieval. John Wiley & Sons, Chichester (1997)
19. David, A., Hull, D.A.: Stemming algorithms: A case study for detailed evaluation. J. Am. Soc. Inf. Sci. 47, 70–84 (1996)
20. Mateus, F.-S., Romero, F.P., Olivas, J.A., Braga, J.L.: A fuzzy ontology and user profiles approach to improve semantic information filtering. In: Proceedings of the 2009 Int. Conf. on Artificial Intelligence, ICAI 2009, pp. 849–854 (2009)
21. Mateus, F.-S., Romero, F.P., Menndez, V.H., Zapata, A., Prieto, M.E.: A fuzzy ontology approach to represent user profiles in e-learning environments. In: FUZZ-IEEE 2010 IEEE International Conference on Fuzzy Systems - WCCI 2010 IEEE World Congress on Computational Intelligence, Barcelona (Spain), pp. 161–168 (2010)

22. Ning, H., Shihan, D.: Structure-based ontology evaluation. In: ICEBE 2006: Proceedings of the IEEE International Conference on e-Business Engineering, pp. 132–137. IEEE Comp. Soc, USA (2006)
23. Pedersen, T., Patwardhan, S., Michelizzi, J.: Wordnet:similarity: measuring the relatedness of concepts. In: HLT-NAACL 2004: Demonstration Papers at HLT-NAACL 2004 on XX, pp. 38–41. Association for Computational Linguistics, Massachusetts (2004)

On the Semantics of Bipolarity and Fuzziness

J. Tinguaro Rodríguez, Camilo A. Franco, and Javier Montero

Abstract. This paper analyzes the relationship between fuzziness and bipolarity, notions which were devised to address different kinds of uncertainty: linguistic imprecision, in the former, and knowledge relevance and character or polarity, in the latter. Although different types of fuzziness and bipolarity have been defined, these relations are not always clear. This paper proposes the use of four-valued extensions to provide a formal method to rigorously define and compare the semantics and logical structure of diverse combinations of fuzziness and bipolarity types. As a result, this paper claims that these notions and their different types are independent and not semantically equivalent despite its possible formal equivalence.

1 Introduction

Fuzziness [39] and bipolarity [11] are two independent but complementary notions originally (and separately) devised to face the mathematical modelling of different features of natural languages and human reasoning. Though their influence (more than considerable in fields as decision theory [14,17,26,29] or machine learning [19]) has spread separately, in the last few decades both notions have started to appear together in many developments on these and other fields (see for instance [4,6,13,15,30, 31,38]), which comes to show its high relevance as a topic of research inside soft computing [20] and logics [36].

However, the relationships (and differences) between fuzziness and bipolarity are not always clear. In order to introduce our point, let us remind that, on one hand, *fuzziness* is concerned with the *imprecision* inherent to natural languages: many relevant predicates (i.e. words) P, as *good* or *young*, have ill-defined boundaries, and uncertainty arises regarding whether objects x of a universe of discourse X (e.g. decision *al ternatives* or *ages* of customers) fulfil them or not.

On the other hand, *bipolarity* is concerned with the *character* (or polarity) and relevance of information: it has become clear (see [7,25,28]) that human reasoning tends to analyze reality (e.g. a decision to be taken [23,29]) by checking separately

J. Tinguaro Rodríguez · Camilo A. Franco · Javier Montero
Faculty of Mathematics, Complutense University of Madrid,
Plaza de Ciencias 3 28040 Madrid, Spain
jtrodrig@mat.ucm.es

B. De Baets et al. (Eds.): Eurofuse 2011, AISC 107, pp. 193–205, 2011.
springerlink.com

both the positive and negative sides of the available information (e.g. an alternative could be *good* for certain criteria and *bad* for other set of criteria) in order to acquire a more expressive and relevant knowledge. Thus, reality is judged in terms of pairs of poles of reference *P*/*Q*, as *false*/*true* or *good*/*bad*, which organize and give relevance to the available information.

Moreover, different *types* of fuzziness [40] and bipolarity [12] have been studied and defined. While usual (type-1) fuzziness (F1) measures linguistic imprecision in a precise way (assigning a gradable but precise truth-value $\mu_P(x) \in [0,1]$ to the proposition "*x* fulfils *P*", thus modelling *P* as a *fuzzy set*), type-2 fuzziness (F2) enables such an imprecision to be measured *imprecisely* (since it assigns a fuzzy set of the truth scale [0,1] to "$x \in P$").

Similarly, while type-1 bipolarity (B1) relies on the idea that negative information is just the *negation* or complementation of the positive one, type-2 bipolarity (B2) allows the relation between poles to be not so simple (for example $bad \neq not\ \ good$), and thus evaluating the pair $(\mu_P(x), \mu_Q(x))$ could be necessary in order to capture all relevant information.

Notice that, as they try to address different kinds of uncertainty, fuzziness and bipolarity seem to be not necessarily related or interlinked: in principle a B2 formalism could be either an F1 or F2 (or even crisp!) model, and an F2 framework could be associated to either a B1 or B2 setting. Nevertheless, a commonly-used instance of type-2 fuzziness, *interval valued fuzzy sets* (IVFS, see [18]), actually devised as B1 objects, has been shown (see [8,9]) to be in certain sense equivalent to *Atanassov fuzzy sets* (AFS, see [1]), which however were originally devised as F1 and B2 objects. In fact, as a consequence of this formal equivalence, a bitter dispute (see [10] and [3]) raised between Atanassov and his followers, on one side, and an important part of the fuzzy community, on the other, about the exact meaning of AFS and their real relevance in the context of bipolarity.

The main objective of this paper is to shed some light on the relations between fuzziness and bipolarity from a different perspective, and try to lead the referred differences between Atanassov and his detractors, apparently not totally solved, to a definitive solution. For this aim, the notion of four-valued extension (that clearly resembles that of preference structure) is used in order to rigorously define and compare the semantics and underlying logical structure of each possible combination of fuzziness and bipolarity types 1 and 2. This will allow us to separate and distinguish IVFS from AFS in a practical way, and will enable us to show the independency of fuzziness and bipolarity. AFS in a practical way, and will enable us to show the independency of fuzziness and bipolarity. AFS in a practical way, and will enable us to show the independency of fuzziness and bipolarity.

This paper is organized as follows: the notions of type-1 and type-2 fuzziness are revised in Section 2, and those of type-1 and type-2 bipolarity will be revised in Section 3. Four-valued extensions are introduced and applied to the four possible combinations of bipolarity and fuzziness types in Section 4. Finally, some conclusions are shed in Section 5.

2 Type-1 and Type-2 Fuzziness

Since the first proposal of L.A. Zadeh in the middle-sixties of the last century (see [39]), fuzzy set theory and fuzzy logic have enabled an increasingly sophisticated mathematical treatment of the *imprecision* inherent to natural languages. As said above, the imprecision of a predicate P entails uncertainty about whether objects x of a universe of discourse X verify it or not. Fuzzy logic addresses this uncertainty by allowing the truth of the proposition "x verifies P" (i.e. "$x \in P$") to be evaluated in the interval [0,1] (rather than in the classical, binary *valuation space* $\{0,1\}$). Therefore, the crisp index function of P is generalized into a *membership function* $\mu_P : X \to [0,1]$ that specifies the degree up to which each object x verifies P. This enables objects to partially fulfil an imprecise predicate or, in other words, the *semantics* (or *use*, see [34]) of P on X is modeled as a (type-1) *fuzzy set* (T1FS) $P = \{(x, \mu_P(x)) \mid x \in X\}$.

Also, in response to some criticism raised about the possibility of obtaining totally precise membership-degrees $\mu_P(x) \in [0,1]$, L.A Zadeh introduced in [40] the notion of type-n fuzziness by allowing to measure the truth of "$x \in P$" by means of a type-n–1 fuzzy set of [0,1]. Particularly, if $F([0,1])$ denotes the set of all type-1 fuzzy sets of [0,1], a type-2 fuzzy set (T2FS) P is associated with a membership function $\psi_P : X \to F([0,1])$, in such a way that $\psi_P(x) : [0,1] \to [0,1]$ expresses the *plausibility-degree* of the proposition "x verifies P with truth-degree μ" for each truth-degree $\mu \in [0,1]$. Therefore, imprecision in the measurement of type-1 truth degrees $\mu_P(x)$ is allowed and, in general, higher types of fuzziness enable further imprecision to be introduced in truth degrees.

Perhaps the simplest and most used instance of type-2 fuzzy sets are interval-valued fuzzy sets (IVFS, see [18]), which assign to each object x and predicate P an interval $[\mu_L(x), \mu_U(x)]_P$ as (equally and totally) plausible values of the truth degree $\mu_P(x)$. Therefore, the valuation space of IVFS is the set

$$L^I = \{[a,b] \subseteq [0,1] \mid a \leq b\} \subset F([0,1]) .$$

Also, the wider an interval $[\mu_L(x), \mu_U(x)]$ is, the bigger the uncertainty associated to it, where its length $u(x) = \mu_U(x) - \mu_L(x)$ is usually taken as the *degree of uncertainty* inherent to such an evaluation. Thus, $u(x) = 0$ if there is not uncertainty about the degree up to which x verifies P, i.e. if $\mu_L(x) = \mu_P(x) = \mu_U(x)$.

Classic logical connectives such as *not*, *and*, *or* can be generalized by means of different fuzzy operators. The usual *negation* [22] for type-1 fuzzy sets is given by $n(\mu) = 1 - \mu$, so that $\mu_{\neg P}(x) = n(\mu_P(x))$, with $\neg P$ =*not-P*. Notice also that the usual negation defined over IVFS is $n_{L^I}([\mu_L(x), \mu_U(x)]_P) = [1 - \mu_U(x), 1 - \mu_L(x)]_{\neg P}$.

Also, t-norms and t-conorms [22,33] are usually taken as operators for conjunction and disjunction, respectively. Both connectives are related through the negation n, so that if T is a t-norm, then $S(\mu_P, \mu_Q) = n[T(n(\mu_P), n(\mu_Q))]$ is a t-conorm. For T1FS, common examples of so related operators are the Lukasiewicz t-norm and t-conorm, respectively given, for $a,b \in [0,1]$, by $a \odot b = \max\{a+b-1,0\}$, $a \oplus b = \min\{a+b,1\}$, or the minimum $a \wedge b = \min\{a,b\}$ and the maximum $a \vee b = \max\{a,b\}$. For IVFS, these operators are extended as follows:

$$[a,b] \odot_I [c,d] = [a \odot c, b \odot d], [a,b] \oplus_I [c,d] = [a \oplus c, b \oplus d],$$

$$[a,b] \wedge_I [c,d] = [a \wedge c, b \wedge d], [a,b] \vee_I [c,d] = [a \vee c, b \vee d].$$

Notice that fuzzy logic seems to underestimate the notion of negative information. If any, it assumes that the *falsehood* of "$x \in P$" is equal to the *truth* of "$x \notin P$", i.e. $\mu_{\neg P}(x)$, though this *falsehood* could be of different nature than negative information. Anyway, it is clear that fuzzy logic does not consider an extra, independent evaluation (being either a *falsehood* degree or a measure of negative information) together with the (precise or not) degrees $\mu_P(x)$ or $\psi_P(x)$.

Typical applications of fuzzy logic (today more or less covered below the term soft computing [20]) include, among others, intelligent control [24], decision theory [14] or machine learning [19]. Type-2 FS (specially IVFS) have also found extensive application in various fields (see for example [6]).

3 Type-1 and Type-2 Bipolarity

Although the idea of measuring independent positive and negative information has a psychological inspiration (see [25,28]) and has appeared separately and without a unitary label (and even not explicitly) in the scientific literature (specially that concerned with decision theory), quite recently the term *bipolarity* seems to have succeeded in becoming a widely accepted label for this rather general idea [11].

Basically, bipolarity assumes the existence of a pair of reference poles *P*/*Q*, as *false*/*true* or *good*/*bad*, which provide absolute landmarks that confer information its intrinsic positive and negative *character*. Information having neither positive nor negative character is therefore irrelevant or *neutral* in terms of such references. Thus, the poles *P*/*Q* organize and give relevance to the available information, so that positive information for one of the poles is taken as against the other.

If the relation between the poles P and Q is given by the complementation, i.e. $Q = not\, P = \neg P$ (e.g. bad=not good), then a single evaluation $\mu_P(x) \in [0,1]$ (or $\mu_P(x) \in L^I$) is enough to capture all the information that is relevant in terms of the polarity *P*/*Q*, since then $\mu_Q(x) = \mu_{\neg P}(x) = 1 - \mu_P(x)$ and therefore the negative information is just the negation of the positive one. This situation is usually referred to as *type-1 bipolarity* (B1, see [13]), and let us remark that fuzzy sets (both F1 and F2) are usually assumed to belong to this category as exposed in last section.

However, if the relation between *P* and *Q* is not so simple (e.g. $bad \neq not\ good$), then it is usually also necessary to evaluate $\mu_Q(x)$ in order to capture all the relevant information about "$x \in P$". Particularly, two different scales L_P^+ and L_P^- can be used to evaluate, respectively, $\mu_P(x)$ and $\mu_Q(x)$, in such a way that the evidence couple $(\mu^+(x), \mu^-(x))_P = (\mu_P(x), \mu_Q(x)) \in L_P^+ \times L_P^-$ measures the degree of positive and negative information regarding "$x \in P$". Besides, if the relation between the poles is assumed to be symmetric (though it is possible to remove this assumption, see [31]), then

$$(\mu^+(x), \mu^-(x))_Q = (\mu_Q(x), \mu_P(x)) \in L_Q^+ \times L_Q^- = L_P^- \times L_P^+ .$$

These bivariate evidences are typical of *type-2 bipolarity* (B2, see [13]), and note that, as *P* and *Q* are not necessarily complementary, B2 models in principle admits 4 possible cases or epistemic states regarding an object *x* and the pair *P*/*Q*:

- *Positive truth t*: *x* verifies *P* but does not verify *Q*, i.e., information is positive and not negative. The extreme (or crisp) representative of this state is given by the evidence couple $(\mu^+, \mu^-) = (1,0)$;
- *Negative truth* (or *positive falsehood*) *f*: *x* fulfils *Q* and does not verify *P*, i.e., information is negative and not-positive. Its extreme pair is $(\mu^+, \mu^-) = (0,1)$;
- *Irrelevance* or *neutrality i*: *x* neither fulfils *P* nor *Q*, and thus information is both not-positive and not-negative, so in the limit it is $(\mu^+, \mu^-) = (0,0)$;
- *Conflict k*: *x* simultaneously fulfil *P* and *Q*, so information is positive and negative, as in the crisp case $(\mu^+, \mu^-) = (1,1)$.

Let us remark that the third and fourth cases are not possible in B1 frameworks (as fuzzy logic), in which *P* and *Q* are tightly linked through the negation. If the relationship between the poles is somehow restricted (for example, by introducing a constraint in $L^+ \times L^-$), the third or the fourth case could cease to hold, but not simultaneously: if this is the case, then we are back in a B1 setting.

In relation with preference modelling and decision analysis, the notion of independent positive and negative information has also appeared in, among others, cumulative prospect theory (see [21,37]), outranking *concordance-discordance* ELECTRE methods (see [32]), Atanassov *intuitionistic* fuzzy sets (AFS, see [1]) and, more recently, in DDT logic [35]. Current applications of type-2 bipolarity along with fuzzy logic are promising and increasing (see [15,29,30,38])

It is interesting to examine in more detail the case of AFS, which assign to "$x \in P$" both a degree of membership $\mu_P(x)$ and an independent degree of *non-membership* $\mu_{\neg P}(x)$, such that $\mu_P(x), \mu_{\neg P}(x) \in [0,1]$ and $\mu_P(x) + \mu_{\neg P}(x) \leq 1$, i.e, $L^* = \{(x,y) \in [0,1]^2 \mid x + y \leq 1\} \subset [0,1]^2$ is the valuation space of AFS, and notice that the same scale $L^+ = L^- = [0,1]$ is used for both evaluations (though a constraint has been included). Note also that Atanassov implicitly treats a B1 framework (i.e. in which the relevant polarity is $P / \neg P$) as a B2 one, assuming that the truth of "$x \notin P$" can not be obtained from that of "$x \in P$", thus requiring an independent evaluation $\mu_{\neg P}(x) = \nu_P(x)$. As it is $\mu_P(x) + \nu_P(x) \leq 1$, it is assumed that the disjunction of P and its negation (in terms of the Lukasiewicz t-conorm) could not be a tautology, thus violating the law of the excluded middle ($P \vee \neg P$), that was Atanassov's aim as he wanted its model to be *intuitionistic*.

However, other fuzzy structures also violate such a law without considering an independent negation (for example, by taking the maximum t-conorm $\vee$ and $\mu_{\neg P}(x) = 1 - \mu_P(x)$). Thus, the exact meaning of AFS is rather obscure, showing the semantics of a B1 framework while formally seeming to be a B2 one. Moreover, as Atanassov and his followers failed to give a clear definition of AFS's underlying logical and semantical structure (see [27]), no solid reasons to separate AFS from IVFS were available when both formalisms were proven [8,9] to be equivalent (through the isomorphism $\Phi: L^* \rightarrow L^I$ given by

$$\Phi(\langle \mu_P(x), \mu_{\neg P}(x) \rangle) = [\mu_P(x), 1 - \mu_{\neg P}(x)]).$$

This equivalence triggered a strong controversy (see [3,10]) between Atanassov and his followers, which thought AFS were a valid B2 model (or at least a valid *intuitionistic* model) different from IVFS, and an important part of the fuzzy community, which instead thought the *intuitionistic* meaning of AFS was not clear at all and, since IVFS were formulated some years before than AFS, the relevance of the latter as bipolar objects should be reduced due to such equivalence (despite IVFS were originally conceived as B1 objects).

In our opinion, AFS' original semantics (whatever it may be) is not of B2 type, and therefore AFS are not a really relevant landmark in the field of bipolarity. However, if ν is interpreted as the membership function of a polarity Q such that $Q \neq \neg P$, i.e. $\nu_P(x) = \mu_Q(x)$, and the constraint $\mu_P(x) + \nu_P(x) \leq 1$ is maintained, then a B2 semantics (with a particular relation between poles P and Q) is easily obtained, and our opinion is that these objects (which will be referred to as *bipolar* AFS: BAFS) are not *semantically* equivalent to IVFS. In order to support this last claim, we will formally define the *semantics* of B2 logical objects, and will show that semantics of BAFS (and in general that of F1 and B2 formalisms) is different from that of IVFS (in general, from that of B1 and F2 models). Such a *semantical differentiation* will be introduced through the notion of *four-valued extension*.

4 Four-Valued Extensions

Given an evidence couple (μ^+, μ^-) that evaluates the positive and negative information regarding "$x \in P$", how can we assess (and quantify) from such evidence whether x can be identified with the positive pole P (in the sense of having positive character or verifying the *positive truth* epistemic state t) or not? In other words, what is the exact or relevant meaning or *semantics* of a given evidence couple? For example, the pair $(\mu^+, \mu^-) = (1,0)$ indicates that x has a fully positive character, and thus the positive truth of "$x \in P$" can be assumed in this case to be maximum, i.e., 1. This could lead to think that t coincides with the truth of "$x \in P$" (since $\mu^+(x) = \mu_P(x) = 1$) or the falsehood of Q (since $\neg\mu^-(x) = 1 - \mu_Q(x) = 1$).

However, the pairs (1,1) or (0,0) do not represent neither a prevalence of the positive pole over the negative one nor the reciprocal. Rather, they make it clear that in a B2 context, the truth of "$x \in P$" (i.e. $\mu_P(x)$) should not be directly identified with the *positive truth* epistemic state *t* resulting from conjointly evaluating positive and negative information. Moreover, as in a fuzzy (F1 or F2), type-2 bipolar setting an object *x* could be partially compatible with both poles *P* and *Q*, the associated evidence couple (μ^+, μ^-) could be partially compatible with more than one of the epistemic states *t*, *f*, *i*, *k* described in last section. Therefore, these states are also gradable or fuzzy in nature, and the notion of *4-valued extension* is useful to quantify these compatibilities (and thus the *semantics* of the evidence couple) under some reasonable assumptions (as in preference modelling [14], where the 4 values of the preference structure are the relevant, final items of the analysis).

Thus, supposing that $L^+ = L^- = L$, with each evidence couple $(\mu^+, \mu^-) \in L^+ \times L^-$ there is associated an *evidence matrix*

$$EM = \begin{bmatrix} t(\mu^+, \mu^-) & i(\mu^+, \mu^-) \\ k(\mu^+, \mu^-) & f(\mu^+, \mu^-) \end{bmatrix},$$

obtained through the formulae

$$t(\mu^+, \mu^-) = \mu^+ \wedge \neg\mu^-, \; f(\mu^+, \mu^-) = \neg\mu^+ \wedge \mu^- \tag{1}$$

$$i(\mu^+, \mu^-) = \neg\mu^+ \odot_W \neg\mu^-, \; k(\mu^+, \mu^-) = \mu^+ \odot_W \mu^-, \tag{2}$$

where $\neg\mu = n \circ \mu = 1 - \mu$. Note that $t, f, i, k \in L$, and therefore *EM* assesses the degree up to which each of these four epistemic states hold in the same scale in which both μ^+ and μ^- are measured. The 4-valued extension given in formulae (1)-(2) was first proposed for the case *L*=[0,1] in [29] (and have been further analyzed in [36]), in which it is proven that it is the *unique* continuous t-norm-based extension simultaneously verifying that the states *i* and *k* are mutually exclusive (since if *i*>0 then *k*=0 and *vice versa*) and $t + f + i + k = 1$. Notice that by exchanging the places of the Lukasiewicz t-norm $\odot$ and the minimum t-norm $\wedge$, a different extension is obtained (and in fact it is used in preference modelling, see [14]), in which the last equality also holds but now *t* and *f* are exclusive. Nevertheless, as we shall see, this last extension does not generalize classical truth and falsehood in the case in which $Q = \neg P$, while that given by (1)-(2) does.

Before we apply this extension to the different combinations of fuzziness and bipolarity types, we enunciate the following theorem, that states the properties verified by such an extension for the case $L^+ = L^- = L^I$, i.e., in a simultaneous B2 and F2 situation in which imprecision is allowed for the couple (μ^+, μ^-):

Theorem 1: Let $\mu^+ = [\mu_L^+, \mu_U^+]$ and $\mu^- = [\mu_L^-, \mu_U^-]$ ($0 \le \mu_L^- \le \mu_U^- \le 1$, $0 \le \mu_L^+ \le \mu_U^+ \le 1$) be the positive and negative interval-valued evaluations of a predicate "$x \in P$". Let also $t = [t_L, t_U]$, $f = [f_L, f_U]$, $i = [i_L, i_U]$ and $k = [k_L, k_U]$ be the 4-valued extension associated to the pair (μ^+, μ^-), obtained through the formulae (5)-(8). Then k_L and i_U as well as i_L and k_U are mutually exclusive and it holds that $t_L + f_L + i_U + k_U = t_U + f_U + i_L + k_L = 1 + m$, where $m = \min\{u^+, u^-, u_t, u_f, u_u, u_k\}$ and $u^+ = \mu_U^+ - \mu_L^+, u^- = \mu_U^- - \mu_L^-, u_t = t_U - t_L$, $u_f = f_U - f_L$, $u_i = i_U - i_L, u_k = k_U - k_L$ are the uncertainty degrees of each interval.

Proof: Let us denote $M_1 = t_L + f_L + i_U + k_U$, $M_2 = t_U + f_U + i_L + k_L$, $s^U = \mu_U^+ + \mu_U^-$, $s^L = \mu_L^+ + \mu_L^-$, $s_1 = \mu_L^+ + \mu_U^-$, $s_2 = \mu_U^+ + \mu_L^-$ and recall that

$$t_L = \mu_L^+ \wedge (1 - \mu_U^-), t_U = \mu_U^+ \wedge (1 - \mu_L^-), f_L = \mu_L^- \wedge (1 - \mu_U^+), f_U = \mu_U^- \wedge (1 - \mu_L^+)$$

$$i_L = (1 - \mu_U^+ - \mu_U^-) \vee 0, i_U = (1 - \mu_L^+ - \mu_L^-) \vee 0, k_L = (\mu_L^+ + \mu_L^- - 1) \vee 0, k_U = (\mu_U^+ + \mu_U^- - 1) \vee 0 .$$

Clearly, both k_L, i_U and i_L, k_U are mutually exclusive. Then notice that s_1 determines the relation between μ_L^+ and $1 - \mu_U^-$, since $s_1 < 1$ (resp. $s_1 \ge 1$) implies $\mu_L^+ < 1 - \mu_U^-$ ($\mu_L^+ \ge 1 - \mu_U^-$) and therefore $t_L = \mu_L^+$ and $f_U = \mu_U^-$ ($t_L = 1 - \mu_U^-$ and $f_U = 1 - \mu_L^+$). Similarly, s_2 determines the values t_U and f_L, and thus the 4 possible situations of s_1 and s_2 wrt 1 produce four possible cases for t_L, t_U, f_L, f_U.

On the other hand, s^U and s^L determine, respectively, the values i_L, k_U and i_U, k_L (e.g., $s^U < 1$ implies $i_L = 1 - s^U$ and $k_U = 0$), but only 3 cases are now possible since the case $s^U < 1, s^L \ge 1$ is excluded as it is $s^L \le s^U$. Moreover, from the 12 (4·3) cases in principle possible, 6 have to be removed since the constraints $0 \le s^L \le s_1, s_2 \le s^U \le 2$ and $s^L + s^U = s_1 + s_2$ also hold. Thus, this proof will consist on showing that the equality $M_1 = M_2 = 1 + m$ holds in each of these 6 cases:

1) $s_1, s_2, s^L, s^U < 1$: in this case it holds that $M_1 = \mu_L^+ + \mu_L^- + 1 - \mu_L^+ - \mu_L^- = 1$, $M_2 = \mu_U^+ + \mu_U^- + 1 - \mu_U^+ - \mu_U^- = 1$, $u_t = u^+, u_f = u^-, u_i = u^+ + u^-, u_k = 0$, hence m=0 and $M_1 = M_2 = 1 + m$.

2) $s_1, s_2, s^L < 1, s^U \ge 1$: now it is $M_1 = \mu_L^+ + \mu_L^- + 1 - \mu_L^+ - \mu_L^- + \mu_U^+ + \mu_U^- - 1 =$ $= \mu_U^+ + \mu_U^- = s^U$, $M_2 = \mu_U^+ + \mu_U^- = s^U$, $u_t = u^+, u_f = u^-, u_i = 1 - s^L, u_k = s^U - 1$, and since $u^+ - (s^U - 1) = 1 - s_1 > 0$, $u^- - (s^U - 1) = 1 - s_2 > 0$, $1 - s^L - (s^U - 1) =$ $= 2 - (s^L + s^U) = 2 - (s_1 + s_2) > 0$, then $m = u_k = s^U - 1$ and $M_1 = M_2 = 1 + m$.

3) $s_1, s^L < 1,\ s_2, s^U \geq 1$: it is $M_1 = \mu_L^+ + 1 - \mu_U^+ + 1 - \mu_L^+ - \mu_L^- + \mu_U^+ + \mu_U^- - 1 = 1 + \mu_U^- - \mu_L^- = 1 + u^-$, $M_2 = 1 - \mu_L^- + \mu_U^- = 1 + u^-$, $u_t = 1 - s^L, u_f = s^U - 1,\ u_i = s^U - 1$, $u_k = 1 - s^L$, and it also holds that $u^+ - u^- = s_2 - s_1 > 0$, $1 - s^L - u^- = 1 - s_1 > 0$, $s^U - 1 - u^- = s_2 - 1 \geq 0$, so $m = u^-$ and $M_1 = M_2 = 1 + m$ again.

4) $s_2, s^L < 1,\ s_1, s^U \geq 1$: it is $M_1 = 1 - \mu_U^- + \mu_L^- + 1 - \mu_L^+ - \mu_L^- + \mu_U^+ + \mu_U^- - 1 = 1 + \mu_U^+ - \mu_L^+ = 1 + u^+$, $M_2 = \mu_U^+ + 1 - \mu_L^+ = 1 + u^+$, $u_t = s^U - 1, u_f = 1 - s^L$, $u_i = 1 - s^L$, $u_k = s^U - 1$. As $u^- - u^+ = s_1 - s_2 > 0$, $1 - s^L - u^+ = 1 - s_2 > 0$, $s^U - 1 - u^+ = s_1 - 1 \geq 0$, it is $m = u^+$ and $M_1 = M_2 = 1 + m$.

5) $s^L < 1,\ s_1, s_2, s^U \geq 1$: it is $M_1 = 1 - \mu_U^- + 1 - \mu_U^+ + 1 - \mu_L^+ - \mu_L^- + \mu_U^+ + \mu_U^- - 1 = 2 - \mu_L^+ - \mu_L^- = 1 + (1 - s^L)$, $M_U = 1 - \mu_L^+ + 1 - \mu_L^- = 1 + (1 - s^L)$, $u_t = u^-, u_f = u^+$, $u_i = 1 - s^L$, $u_k = s^U - 1$, and also $u^+ - (1 - s^L) = s_1 - 1 \geq 0$, $u^- - (1 - s^L) = s_2 - 1 \geq 0$, $s^U - 1 - (1 - s^L) = s^U + s^L - 2 = s_1 + s_2 - 2 \geq 0$, thus $m = 1 - s^L$ and $M_1 = M_2 = 1 + m$.

6) $s_1, s_2, s^L, s^U \geq 1$: finally, it is $M_1 = 1 - \mu_U^- + 1 - \mu_U^+ + \mu_U^+ + \mu_U^- - 1 = 1$, $M_2 = 1 - \mu_L^+ + 1 - \mu_L^- + \mu_L^+ + \mu_L^- - 1 = 1$, $u_t = u^-, u_f = u^+, u_i = 0, u_k = u^+ + u^-$, so it follows $m = u_i = 0$ and $M_1 = M_2 = 1 + m$. ■

Next, Theorem 1 will be applied to different combinations of bipolarity and fuzziness types to illustrate its particular *semantics*:

Type-1 fuzziness and bipolarity (F1,B1)

In this case, $L=[0,1]$ and $Q = \neg P$, so it is $\mu^+ = \mu_P$ and $\mu^- = \mu_{\neg P} = n(\mu_P) = 1 - \mu_P$. This leads to $t = \mu^+$, $f = \mu^-$ and $k = i = 0$. Particularly, $m = 0$ as uncertainty regarding the actual values of μ^+, μ^- is not allowed. For example, let us suppose $x \in X$ such that $\mu_P(x) = 0.6$. Then we get the evidence couple $(\mu^+, \mu^-) = (0.6, 0.4)$ and the evidence matrix

$$EM_1 = \begin{bmatrix} 0.6 & 0 \\ 0 & 0.4 \end{bmatrix}.$$

Thus, the extension (4)-(5) generalizes fuzzy logic *truth* and *falsehood* (identified with μ_P and $\mu_{\neg P}$), and reflects the B1 semantics of this setting, that excludes *k* and *i*.

Type-2 fuzziness and type-1 bipolarity (F2,B1)

Suppose now that in the previous example, the value $\mu_P(x)$ is not precisely known, but rather it is known that such a value lies in the interval $[\mu_L(x), \mu_U(x)]_P$. Thus, it still is assumed $Q = \neg P$, but now $L = L^I$ and then $\mu^+ = [\mu_L, \mu_U]_P$ and $\mu^- = n_{L^I}([\mu_L, \mu_U]_P) = [1 - \mu_U, 1 - \mu_L]_{\neg P}$. Thus, it is again obtained $t = \mu^+$ and $f = \mu^-$

but now $k = i = [0,u]$ and m=u, with $u = \mu_U - \mu_L$ the uncertainty associated to μ_P. For example, if $\mu^+ = [0.5, 0.7]$ (and thus $\mu^- = [0.3, 0.5]$), the matrix

$$EM_2 = \begin{bmatrix} [0.5,0.7] & [0,0.2] \\ [0,0.2] & [0.3,0.5] \end{bmatrix}$$

is obtained, showing that now the evaluations of positive truth t and falsehood f are also uncertain, allowing i and k to be non-zero. However, if uncertainty is reduced, i.e. as u tends to 0, k and i tend to [0,0], approaching the case (F1,B1).

Type-1 fuzziness and type-2 bipolarity (F1,B2)

If negative information about "$x \in P$" is allowed to be independent from the negation n in the first case (F1,B1), then we obtain a B2 framework in which $Q \neq \neg P$, $\mu^+ = \mu_P$, $\mu^- = \mu_Q$ and $L^+ = L^- = L = [0,1]$. Thus, m=u=0, but t (f) can no more be identified with μ^+ (μ^-), and one of the states i,k can hold. For example, if $(\mu^+, \mu^-) = (0.6, 0.7)$, we get the matrix

$$EM_3 = \begin{bmatrix} 0.3 & 0 \\ 0.3 & 0.4 \end{bmatrix},$$

which shows that $t \neq \mu^+$, $f \neq \mu^-$ and informs that information is partially conflictive (k=0.3), thus entailing $t < \mu^+$ and $f < \mu^-$.

Similarly, notice that $(\mu^+, \mu^-) = (0.5, 0.3) \in L^*$ is the evidence couple associated to the interval $[0.5, 0.7] \in L^I$ of the previous case by means of the isomorphism $\Phi : L^* \to L^I$ (see Section 3), and it produces the evidence matrix

$$EM_4 = \begin{bmatrix} 0.5 & 0.2 \\ 0 & 0.3 \end{bmatrix}.$$

It is $EM_2 \neq EM_4$, and not only because EM_2 does introduce uncertainty while EM_4 does not. If uncertainty u is reduced, then it could be $\mu^+_{EM_2} = [0.5, 0.5]$, in such a way that $t_{EM_2} = [0.5, 0.5] = 0.5 = t_{EM_4}$, but then it is $f_{EM_2} = [0.5, 0.5] \neq 0.3 = f_{EM_4}$ and $i_{EM_2} = [0,0] \neq 0.2 = i_{EM_4}$ as $\mu^-_{EM_2} = n_{L^I}(\mu^+_{EM_2}) = [0.5, 0.5]$ due to the B1 setting of IVFS. However, BAFS allows μ^- to be independent from μ^+ since they exhibit a type-2 bipolarity, though the condition $\mu^+ + \mu^- \leq 1$ entails that k=0. Thus, it is clear that *semantics* of BAFS is different from that of IVFS, and their equivalence should be regarded as a formal coincidence between two scales that belong to different universes, since $[0,1] \subset L^I \subset F([0,1])$ while $L^* \subset [0,1]^2 \subset L^I \times L^I \subset F([0,1]) \times F([0,1])$.

Type-2 fuzziness and bipolarity (F2,B2)

Finally, suppose that uncertainty about μ^+, μ^- is allowed in a B2 setting, in such a way that an evidence couple $(\mu^+, \mu^-) \in L^I \times L^I$ is obtained. Notice that this case generalizes all the previous ones and corresponds to the hypothesis of Theorem 1. Thus, if $\mu^+ = [0.5, 0.7]$ and $\mu^- = [0.6, 0.9]$, we get the evidence matrix

$$EM_5 = \begin{bmatrix} [0.1, 0.4] & [0, 0] \\ [0.1, 0.6] & [0.3, 0.5] \end{bmatrix},$$

and notice that now each of its elements have a different associated uncertainty (and m=0) and that these results extend those of EM_1, EM_2 and EM_3. Note also that either u^+ or u^- can be reduced without affecting the other, and in the limit case in which $u^+ = u^- = 0$, a F1 and B2 setting is reached. Instances of B2 and F2 objects are Atanassov IVFS (AIVFS, see [2]), though the same argument of Section 3 applies in order to affirm that they can not be seen as B2 objects unless it is assumed the existence of a negative pole $Q \neq \neg P$ (leading to *bipolar* AIVFS).

5 Conclusions

The notions of fuzziness and bipolarity have been revised, and its relations analyzed. As these notions were developed to address different kinds of uncertainty, respectively *linguistic imprecision* and information *relevance* and *polarity*, both of them are in fact complementary and independent. An F2 object could be B2 or not, and reciprocally a B2 formalism could allow uncertainty (thus being also a F2 object) or not. Particularly, in this paper these differences have become formally evident, clarifying some aspects of the relations between B1,F2 formalisms and B2,F1 ones, and hopefully allowing to lead to an at least relative solution over the controversy between Atanassov's supporters and critics. As this paper implicitly shows, there is a difference between *intuitionism* (in the sense of using a *sub-additive* negation $n(\mu) \leq 1 - \mu$) and bipolarity. However, if $Q \neq \neg P$ is assumed, then BAFS and BAIVFS have been shown to be relevant B2 objects. Table 1 presents a summary of the cases and formalisms analyzed in this paper, which have been shown to exhibit a different *semantics* in terms of four-valued extensions, and thus in terms of its ability to model knowledge states.

Table 1 Combinations of bipolarity and fuzziness types-1 and 2

Bipolarity \ Fuzziness	Type-1 (F1)	Type-2 (F2)
Type-1 (B1)	T1FS	T2FS, IVFS
Type-2 (B2)	BAFS, BFS	BT2FS, BAIVFS

Further work is forthcoming in order to extend this work in several directions: *lattices* (see [5] and notice that all formalisms in Table 1 are special cases of *L-fuzzy sets* [16]) will be introduced in order to obtain a more rigorous and general logical and formal setting. The relations between the notions of bipolarity and intuitionism, the use of a sub-additive negation and dissimilarity operators [31] will be further explored (see [15]), as well as the possible use of a F2 formalism in order to combine fuzziness and other types of uncertainty theories.

Acknowledgments. This research has been partially supported by grant TIN2009-07901.

References

[1] Atanassov, K.T.: Intuitionistic Fuzzy-Sets. Fuzzy Set Syst., 20 (1) 87–96 (1986)

[2] Atanassov, K.T., Gargov, G.: Interval Valued Intuitionistic Fuzzy Sets. F Set Syst., 31(3) 343–349 (1989)

[3] Atanassov, K.T., Dubois, D., Gottwald, S., Hajek, P., Kacprzyk, J., Prade's Paper, H.: Terminological difficulties in fuzzy set theory - the case of Intuitionistic fuzzy sets. Fuzzy Set Syst. 156(3), 496–499 (2005)

[4] Beringer, J., Hullermeier, E.: Case-based learning in a bipolar possibilistic framework. Int.l J. Intell. Syst. 23(10), 1119–1134 (2008)

[5] Birkhoff, G.: Lattice theory, vol. 25. Amer. Math. Soc. Colloq. Publ, New York (1948)

[6] Bustince, H., Pagola, M., Barrenechea, E., Fernandez, J., Melo-Pinto, P., Couto, P., Tizhoosh, H.R., Montero, J.: Ignorance functions. An Application to the Calculation of the Threshold in Prostate Ultrasound Images, Fuzzy Set Syst. 161(1), 20–36 (2010)

[7] Cacioppo, J.T., Gardner, W.L., Berntson, G.G.: Beyond bipolar conceptualizations and measures: The case of attitudes and evaluative space. Pers Soc. Psychol. Rev. 1, 3–25 (1997)

[8] Cornelis, C., Atanassov, K., Kerre, E.: Intuitionistic fuzzy sets and interval-valued fuzzy sets: a critical comparison. In: Proc. Eusflat 2003, pp. 159–163 (2003)

[9] Deschrijver, G., Kerre, E.: On the relationship between some extensions of fuzzy set theory. Fuzzy Set Syst. 133, 227–235 (2004)

[10] Dubois, D., Gottwald, S., Hajek, P., Kacprzyk, J., Prade, H.: Terminological difficulties in fuzzy set theory - The case of Intuitionistic Fuzzy Sets. Fuzzy Set Syst. 156(3), 485–491 (2005)

[11] Dubois, D., Prade, H.: Special issue on bipolar representations of information and preference part 1A: Cognition and decision - Foreword. Int. J. Intell Syst. 23(8), 863–865 (2008)

[12] Dubois, D., Prade, H.: An introduction to bipolar representations of information and preference. Int. J. Intell Syst. 23(8), 866–877 (2008)

[13] Dubois, D., Prade, H.: An overview of the asymmetric bipolar representation of positive and negative information in possibility theory. Fuzzy Set Syst. 160(10), 1355–1366 (2009)

[14] Fodor, J., Roubens, M.: Fuzzy preference modelling and multicriteria decision support. Kluwer Academic, Dordrecht (1994)

[15] Franco, C.A., Montero, J., Rodriguez, J.T.: Aggregation Weights for a Fuzzy Preference-Aversion Model. In: Procs. of the WConSC 2011, pp. 199–206 (2011)

[16] Goguen, J.A.: L-fuzzy sets. J. Math. Anal. Appl. 18, 145–174 (1967)

[17] Grabisch, M., Labreuche, C.: A decade of application of the Choquet and Sugeno integrals in multi-criteria decision aid. A Quarterly Journal of Operations Research 6, 1–44 (2008)

[18] Grattan-Guinness, I.: Fuzzy membership mapped onto interval and many-valued quantities, Z. Z. Math. Logik Grundlag. Mathe. 22, 149–160 (1975)

[19] Hullermeier, E.: Fuzzy methods in machine learning and data mining: Status and prospects. Fuzzy Set Syst. 156(3), 387–406 (2005)

[20] Kacprzyk, J., Zadrozny, S.: Soft computing and web intelligence for supporting consensus reaching. Soft Computing 14(8), 833–846 (2010)

[21] Kahneman, D., Tversky, A.: Prospect theory: decision analysis under risk. Econom. 47, 263–291 (1979)

[22] Klir, G.J., Yuan, B.: Fuzzy Sets and Fuzzy Logic: Theory and Applications. Prentice-Hall, Englewood Cliffs (1995)

[23] Labreuche, C., Grabisch, M.: The representation of conditional relative importance between criteria. Annals of Operations Research 154, 93–122 (2007)

[24] Mamdani, E.H.: Application of Fuzzy Algorithms for the Control of a Dynamic Plant. Proc. IEEE 121, 1585–1588 (1974)

[25] Medin, D., Schwanenflugel, P.: Linear separability in classification learning. J. of Experimental Psychology: Human Learning & Memory 7, 355–368 (1981)

[26] Montero, J.: Arrow's theorem under fuzzy rationality. Behavioral Science 32, 267–273 (1987)

[27] Montero, J., Gómez, D., Bustince, H.: On the relevance of some families of fuzzy sets. Fuzzy Set Syst. 158(22), 2439–2442 (2007)

[28] Osgood, C.E., Suci, G.J., Tannenbaum, P.H.: The measurement of meaning. Chicago, UIP (1957)

[29] Öztürk, M., Tsoukiàs, A.: Modelling uncertain positive and negative reasons in decision aiding. Decision Support Systems 43(4), 1512–1526 (2007)

[30] Rodriguez, J.T., Vitoriano, B., Montero, J.: Rule-based classification by means of bipolar criteria. In: IEEE MCDM, pp. 197–204 (2011)

[31] Rodriguez, J.T., Franco, C.A., Vitoriano, B., Montero, J.: An axiomatic approach to the notion of semantic antagonism. In: Procs. of the IFSA-AFSS 2011, FT104-1/6 (2011)

[32] Roy, B.: Ranking and Choice in Pace of Multiple Points of View (Electre Method). Revue Francaise D Informatique De Recherche Operationnelle 2(8), 57–75 (1968)

[33] Schweizer, B., Sklar, A.: Probabilistic Metric Spaces. North–Holland, Elsevier, New York (1983)

[34] Trillas, E.: On the use of words and fuzzy sets. Information Sciences 176(11), 1463–1487 (2006)

[35] Tsoukiàs, A.: A first-order, four valued, weakly paraconsistent logic and its relation to rough sets semantics. Foundations of Computing and Decision Sciences 12, 85–108 (2002)

[36] Turunen, E., Ozturk, M., Tsoukiàs, A.: Paraconsistent semantics for Pavelka style fuzzy sentential logic. Fuzzy Set Syst. 161(14), 1926–1940 (2010)

[37] Tversky, A., Kahneman, D.: Advances in prospect theory: cumulative representation of uncertainty. Journal of Risk and Uncertainty 5, 297–323 (1992)

[38] Victor, P., Cornelis, C., Cock, M.D., Teredesai, A.: Trust- and distrust-based recommendations for controversial reviews. In: Proc. of WEBSCI (2009)

[39] Zadeh, L.A.: Fuzzy Sets. Information and Control 8(3), 338–353 (1965)

[40] Zadeh, L.A.: The Concept of a Linguistic Variable and Its Application to Approximate Reasoning. Information Sciences 9(1), 43–80 (1975)

Part IV

Decision Making

A Heterogeneous Evaluation Model for the Assessment of Sustainable Energy Policies

M. Espinilla, R. de Andrés, F.J. Martínez, and L. Martínez

Abstract. Decision makers are increasingly involved in complex real decisions that require multiple viewpoints. A specific case of this fact is the evaluation of sustainable policies related to environment and energy sectors. In this evaluation process, multiple experts are involved to assess a set of scenarios, according to multiple criteria that might have different nature. These evaluation processes aim to achieve an overall value for each scenario to obtain a ranking among them with the goal of identifying the best one. In this evaluation process a key issue is the treatment of experts' assessments for each criterion. Due to the uncertainty and vagueness in the judgments of the experts and the nature of the criteria, these assessments can be expressed in different information formats, generating an heterogeneous framework. There are diverse approaches to deal with this type of framework, the use of one approach or another could be crucial in the evaluation process, according to the necessities and requirements of the evaluation models regarding the expected results. In this contribution, we propose an evaluation model applied to energy policy selection based on the decision analysis which may use different approaches to deal with heterogeneous information. We present a comparative study of the proposed model using two different approaches to deal with heterogeneous information. Finally, we show the strengths and weaknesses of the evaluation model depending on the approach used to manage heterogeneous information

M. Espinilla · F.J. Martínez · L. Martínez
Department of Computer Sciences,
University of Jaén, Campus Las Lagunillas s/n, 23071. Jaén
e-mail: mestevezluis.martinez@ujaen.es,
fjmm0008@estudiante.ujaen.es

R. de Andrés
Department of Economic and Economic History,
University of Salamanca, Campus Miguel de Unamuno, 37007. Salamanca
e-mail: rocioac@usal.es

B. De Baets et al. (Eds.): Eurofuse 2011, AISC 107, pp. 209–220, 2011.
springerlink.com

1 Introduction

Energy consumption in general is increasing in our world because of the development of all countries. This fact implies energy resources are decreasing, resulting in a problem in the long-term energy scenario. In addition, extraction of energy resources generates other health and safety problems for the people and the environment.

It is thus common that international and national institutions in different areas such as economy, society and environment are quite concerned about this problem. Due to the fact, the development of sustainable energy policies has been a subject of wide-ranging. However, the main problem is that usually different policies are not compatible with current ones or unviable because of their costs. Hence, the evaluation of different policies is reason for discussion and debate for many circles (government, non-government and academic) [2], and is producing an increase in the research on sustainable development and its evaluation.

Evaluation processes can be carried out by using different methods. The use of decision approaches has been successfully applied to solve different evaluation problems in the literature [3, 5, 16, 19, 20]. From a simple resolution decision scheme (see Figure 1), the squared steps carry out an analysis process that allows to make decisions consistently, i.e., it helps to cope with difficult decisions. Such steps are called *decision analysis* and they are an appropriate approach for evaluation processes because it helps to analyze the alternatives, criteria, indicators of the items under study that is the objective of an evaluation process.

The evaluation of energy policies requires a study from multiple viewpoints. So, the scenarios are evaluated by multiple experts with different background, according to a set of criteria that may have different nature, either quantitative or qualitative. Moreover, judgments provided by experts can be vague or uncertain. Therefore, the experts can provide their assessments in different information formats, generating in the evaluation process a heterogeneous framework.

In the literature, we can find different approaches to treat with heterogeneous information [10, 15], each one with its own characteristics. The use of one approach or another is crucial in the evaluation process because it fixes the treatment of the information, the computation of experts' assessments and the output format of the overall values of the scenarios.

In this contribution, we propose an evaluation model for energy policies in long term scenarios based on decision analysis [4] defined in a heterogeneous framework where the experts' assessments can be expressed by numerical, interval-valued and linguistic terms. In order to manage such framework, we present a comparative study of the proposed evaluation model using two different approaches to deal with heterogeneous information. First, we use the approach presented in [10] that unifies the heterogeneous information in a linguistic domain. Second, we use the approach presented in [15] that keeps the format of information, computing the closeness to the Ideal Solution (IS) and the distance to the Negative Ideal Solution (NIS). Finally, we analyze the ranking obtained with both approaches and show the goodness and weaknesses of the evaluation model of energy policies, using both approaches.

To do so, this contribution is structured as follows: Section 2 outlines a scheme of decision analysis and reviews some concepts related to deal with heterogeneous information. Section 3 presents a heterogeneous evaluation model for sustainable energy policies. Section 4 shows a case study of the proposed model in Long-Term Scenarios using two approaches to deal with heterogeneous information. Finally, Section 5 points out concluding remarks.

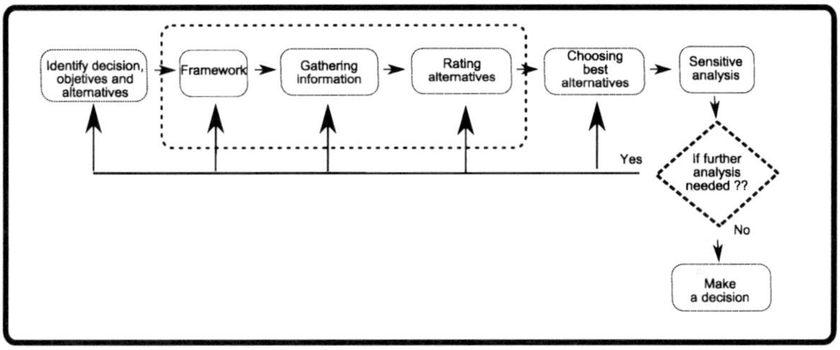

Fig. 1 Decision Analysis Scheme

2 Preliminaries

In this section, we outline the scheme of the decision analysis in which our evaluation model will be based and we briefly review the background to manage heterogeneous information that presents different concepts to understand the proposed evaluation model.

2.1 Decision Analysis

Decision Analysis is a discipline, which belongs to Decision Theory, whose purpose is to help decision makers to reach a consistent decision in a decision making problem. The evaluation process can be modeled as a type of decision making problems, we model the evaluation process as a Multi-Attribute Group Decision Making (MAGDM) problem where decision makers express their opinions about a set of alternatives by means of an utility vector. A classical decision analysis scheme consists of the following phases (see Figure 1 [4]):

- *Identify decision, objectives and alternatives of the problem.*
- *Framework:* It defines the structure of the problem, in our case modelled as a MAGDM [9], and the expression domains in which the preferences can be assessed, in our case as numerical, interval-valued and linguistic terms.
- *Gathering information:* Decision makers provide their information.

- *Rating alternatives:* This phase obtains a collective value for each alternative.
- *Choosing best alternatives:* It selects the solution from the set of alternatives (applying a choice degree [1, 18] to the collective values computed in the before phase).
- *Sensitive analysis:* The solution obtained is analyzed in order to know if it is good enough to make a decision, otherwise, go back initial phases to improve the the quality of the results.
- *Make a decision.*

The application of the decision analysis to an evaluation process does not imply all phases. The essential phases regarding an evaluation problem that will be used in our proposal are those dashed in a rectangle of Figure 1.

2.2 Heterogeneous Information Background

In this section, we review some necessary concepts related to heterogeneous information in order to understand our proposal.

2.2.1 Domains for Assessments

In evaluation processes, different criteria need to be evaluated, which can be of both of a quantitative and qualitative nature. Moreover, the knowledge provided by experts may be vague or uncertain. Many aspects of uncertainties clearly have a non-probabilistic character since they are related to imprecision and vagueness of meanings [17].

Quantitative aspects are assessed by means of numerical values [11]. However, some quantitative criteria may be difficult to qualify using precise values because they are vague. In this cases is suitable the use of interval values [12, 21]. Furthermore, there are criteria that involve subjectivity, vagueness and uncertainty that is not probabilistic in nature, but rather imprecise or vague. In this type of criteria, the fuzzy linguistic approach have been successfully applied [6, 22] by modeling them with linguistic terms information.

Our proposal will be use the approach introduced in [10], where the fuzzy linguistic approach provides the necessary tools for managing such a type of information. Therefore, next section reviews its main concepts.

Fuzzy Linguistic Approach

Many aspects of different activities in the real world cannot be assessed in a quantitative form, but rather in a qualitative one, i.e., with vague or imprecise knowledge. In that case, a better approach may be to use linguistic assessments instead of numerical values. The fuzzy linguistic approach represents qualitative aspects as linguistic values by means of linguistic variables [23].

In this approach, it is necessary to choose the appropriate linguistic descriptors for the term set and their semantics, there exist different possibilities (further description see [7]). One possibility of generating the linguistic term set consists of directly supplying the term set by considering all terms distributed on a linguistic term set on which a total order is defined [22]. For example, a seven-term set S, could be:

$$s_0 = None\ (N)\ s_1 = Very\_Low\ (VL)\quad s_2 = Low\ (L)\qquad s_3 = Medium\ (M)$$
$$s_4 = High\ (H)\ s_5 = Very\_High\ (VH)\ s_6 = Perfect\ (P)$$

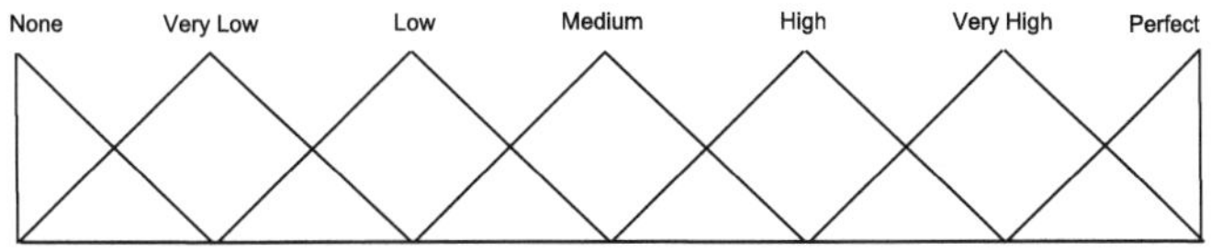

Fig. 2 Linguistic Terms Set

Usually, in these cases, it is required that in the linguistic term set there exist:

1. A negation operator: $Neg(s_i) = s_j$ such that $j = g - i$ ($g+1$ is the cardinality).
2. An order: $s_i \leq s_j \Longleftrightarrow i \leq j$. Therefore, there exists a min and a max operator.

The semantics of the terms are given by fuzzy numbers defined in the [0,1] interval, which are usually described by membership functions.

2-tuple Linguistic Representation Model

The 2-tuple fuzzy linguistic representation model represents the linguistic information by means of a 2-tuple, (s, α), where s is a linguistic label and α is a numerical value that represents the value of the symbolic translation.

Definition 1. *[8] Let β be the result of an aggregation of the indices of a set of labels assessed in a linguistic term set S, i.e., the result of a symbolic aggregation operation. $\beta \in [0, g]$, being $g+1$ the cardinality of S. Let $i = round(\beta)$ and $\alpha = \beta - i$ be two values, such that, $i \in [0, g]$ and $\alpha \in [-.5, .5)$ then α is called a Symbolic Translation.*

Based on the symbolic translation concept, a linguistic representation model which represents the linguistic information by means of 2-tuples (s_i, α), $s_i \in S$ and $\alpha \in [-.5, .5)$ was developed in [8]. This representation model has associated a computational model without loss of information that was presented in [8].

2.2.2 Approaches to Deal with Heterogeneous Information

In our proposal, we consider a heterogenous framework, in which the experts could use different information formats (numerical, interval-valued and linguistic terms) to provided their experts' assessments. In the literature, we find two different approaches to manage this type of framework [10, 15] that we review here.

Approach Based on Fusion of the Information

This approach was presented in [10] and is based on the unification of heterogeneous information into linguistic information in order to obtain linguistic results, that facilitate their comprehension. This approach consists in the following steps (graphically, Figure 3):

- **Choosing a domain to unify the linguistic information.** The heterogeneous information will be unified into a specific linguistic domain, called *Basic Linguistic Terms Set*, S_T, that is selected with the aim of keeping as much knowledge as possible.
- **Unification of the information into Fuzzy Sets into a term set (S_T).** Each numerical, interval-valued and linguistic value is transformed into a fuzzy set on the S_T, $F(S_T)$, by using different transformation functions according to the information format (see [10]).
- **Transformation of the information into Linguistic 2 tuples into a term set (S_T).** In order to simplify the computations and improve the understanding of the results, the fuzzy sets in the S_T are transform into linguistic 2-tuples in the S_T.
- **Aggregation Phase.** The 2-tuple computational model is used to make the processes of CW over the linguistic 2-tuples expressed in the term set S_T.

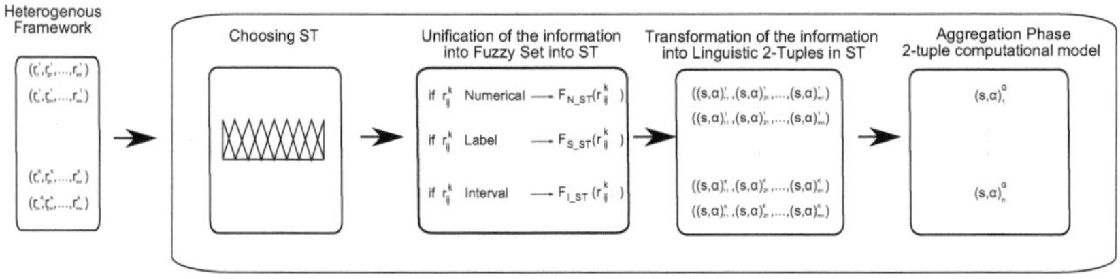

Fig. 3 Scheme of the approach based on Fusion of the Information

Approach Based on the Closeness to the IS and NIS.

Here, we review in short the approach presented in [15]. This approach keeps the different formats of information in order to compute a index ranking. To do so, this approach defines a model to compute the closeness to the Ideal Solution (IS) and the distance to the Negative Ideal Solution (NIS), using different distances, based on Minkowski [14], depending on the type of information. This approach defines a model that involves the following steps (graphically, Figure 4):

- **Computation and Normalization of the Ideal Solution (IS) and the Negative Ideal Solution (NIS) for each decision maker.** The computation of the IS, r^+, and NI, r^- depends on the information format.

- **Determine the weight for each attribute according to the group opinion**. Here, the weights of the attributes for the group are determined.
- **Calculate distances between alternatives and both the IS and the NIS for each decision maker.** The distances between each alternative and the IS are compute using distances based on the Minkowsky distance according to the information format.
- **Compute the relative closeness of the alternatives to the IS.** The relative closeness degrees of each alternative with respect the IS for each decision maker is calculated.
- **Establish the importance of each decision maker**. The importance of each decision maker is establishes.
- **Compute a ranking index based on the relative closeness of the alternative for the group.** The relative closeness degrees of alternatives with respect to the IS for the group is computed in order to obtain a ranking of the alternatives.

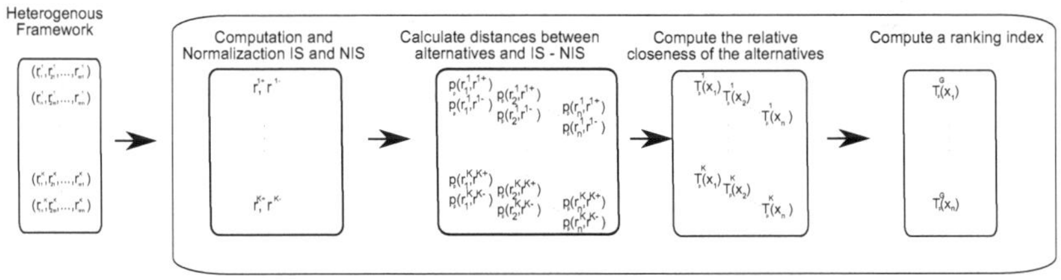

Fig. 4 Scheme of the approach based on the Closeness to the IS and NIS

3 An Energy Policy Evaluation Model in Heterogenous Framework

Our aim is to propose an evaluation model for energy policies in long term scenarios based on a decision analysis scheme, which may use different approaches to deal with heterogeneous information.

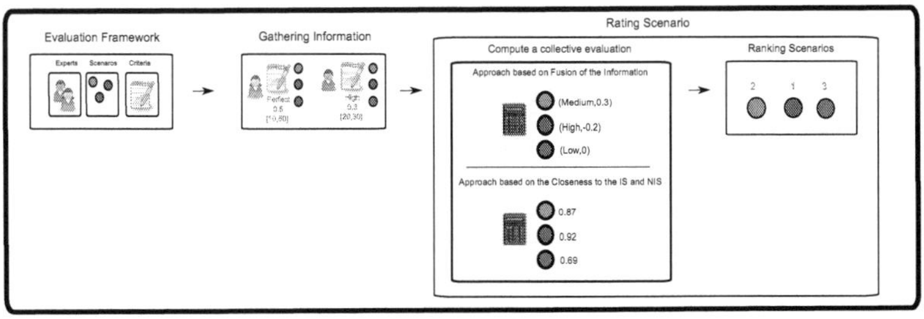

Fig. 5 Scheme of the Model

The decision analysis scheme for the evaluation model consists of the following phases (graphically, Figure 5) revised in Section 2.1:

- Evaluation Framework.
- Gathering Information.
- Rating Scenarios:
  - Step 1: Compute a Collective Evaluation for Each Scenario.
  - Step 2: Ranking Scenarios.

The following subsections present in detail phases of the above evaluation model.

3.1 Evaluation Framework

In this phase, the evaluation framework is defined to fix the problem structure. Hence, the evaluation framework will be as follows:

- Let $E = \{e_1, ..., e_K\}$ $(K \geq 2)$ be a set of experts.
- Let $S = \{S_1, S_2 ..., S_n\}$ $(n \geq 2)$ be a set of identified scenarios for evaluation by E.
- Let $C = \{c_1, c_2 ..., c_m\}$ $(m \geq 2)$ be a set of criteria that characterizes each scenario S_j.

Due to the uncertainty and vagueness in the judgments of the experts and the nature of the criteria, either quantitative or qualitative, we consider a heterogeneous information framework. So, we assume that each expert can use different information formats to assess each criterion.

3.2 Gathering Information

Once the framework has been defined to evaluate the different scenarios, the information must be obtained from the experts. The experts will provide their preferences by using utility vectors. Each expert, e_k provides his/her preferences of the scenario S_j by means of a utility vector: $P_j^k = \{r_{1j}^k, r_{2j}^k, ..., r_{mj}^k\}$ where r_{ij}^k is the preference provided to the criterion c_i of the scenario S_j by the expert e_k. The preferences of a criterion, c_i, are evaluated in an information format. So, the r_{jk}^i could be assessed as:

- Numerical values $N = \{v \,|v \in [0,1]\}$.
- Interval-values $I = \{[d,e] \mid d,e \in [0,1] \wedge d \leq e\}$.
- Linguistic values with different granularity $S = \{S = \{s_0, s_1, \ldots, s_h\} \mid h = 2 \cdot n + 1 \wedge n > 0 \wedge n \in \mathbb{N}$.

Another aspect to consider is that the criteria may be of benefit or cost, therefore preferences must be normalized according to the information format (see [15]).

3.3 Rating Scenarios

The evaluation process aims to rank the scenarios. So, this phase of the evaluation model computes a collective assessment for each scenario that will be used to obtain a ranking. This step has the following stages:

3.3.1 Stage 1: Compute a Collective Evaluation for Each Scenario

This stage of the evaluation model computes an overage assessment for each scenario that will be used to obtain a ranking. As already mentioned, our evaluation process is defined in a heterogeneous framework, therefore, we need an approach to deal with this type of framework. Choice of approach could be crucial in the evaluation model because it fixes the treatment of the information, the computation of experts' assessments and the output format of the overall values of the scenarios.

In our model, we have considered two different approaches to deal with heterogenous information, reviewed in subsection 2.2.2. On the one hand, the approach presented in [10] is based on the unification of heterogeneous information into linguistic information in order to obtain linguistic results. On the other hand, the approach presented in [15] that keeps the different formats of information in order to compute a index ranking. It is noteworthy that other approach that meets the needs and requirements of the evaluation model, regarding the expected results, can be included in the evaluation model.

So, in this phase of the evaluation model, we choose an approach for dealing with heterogeneous information and carry out its phases in order to compute an overall value for each scenario, u_j that will be used to obtain the ranking.

3.3.2 Stage 2: Ranking Scenarios

The final step in the evaluation process is to establish a ranking among scenarios with the purpose of identifying the best ones. The best scenario corresponds to the maximum collective evaluation, $max\{u_j, = 1,2,...,n\}$.

4 A Comparative Study in Long-Term Scenarios

In this section, we present a comparative study of the evaluation model presented previously in the problem of long-term scenarios, using the approaches revised in the subsection 2.2.2 to deal with heterogenous information.

4.1 Evaluation Framework

In this case study, the evaluation framework is composed by: 10 experts $E = \{e_1, e_2..., e_{10}\}$, who evaluate 8 scenarios $S = \{S_1, S_2..., S_8\} =$, where each scenario are involved 44 criteria $C = \{c_1, c_2..., c_{44}\}$. For further detail about scenarios and criteria see [13].

In this case, the information is defined in the interval $[0,100]$ because each expert gives two values to each criterion, worst and best values. In the interval, the value 0 represents the worst and the value 100 the best. Sometimes, experts could give the same values for the best and worst value, being the preference defined into a single numerical.

4.2 Gathering Information

In this phase, the information is gathered from the experts. Due to the great amount of information that we manage in this case study: (8 scenarios, 10 experts and 44 criteria), we have omitted the information gathered.

4.3 Rating Scenario

According to the evaluation model proposed in Section 3, this phase consists in the following stages:

4.3.1 Stage 1: Compute a Collective Evaluation for Each Scenario

The aim of this comparative study is analyze the results obtained by the approaches revised in the subsection 2.2.2. To do so, we computes a collective assessment for each scenario using both approaches.

It is noteworthy that in the approach based on fusion of the information the chosen domain for unifying information is shown in Figure 2. Moreover, in the approach based on the closeness to the IS and NIs, we use the Euclidean distance in order to compute the relative closeness degrees of the scenarios. In Table 1, we show the overall values for each scenario using both approaches.

Table 1 Comparative table of overall values using both approaches

Scenario	Fusion of the Information	Closeness to the IS and NIS
(S_1)	$(Low, -.384)$	0.117360
(S_2)	$(Low, -.482)$	0.000000
(S_3)	$(Low, -.325)$	0.153564
(S_4)	$(Low, -.285)$	0.249008
(S_5)	$(Low, -.079)$	0.567295
(S_6)	$(Low, -.180)$	0.410770
(S_7)	$(Low, .179)$	1.000000
(S_8)	$(Low, .135)$	0.954977

4.3.2 Stage 2: Ranking Scenarios

The ranking obtained in our comparative study using both approaches is the same, this fact shows that our proposal is consistent. $S_7 \succ S_8 \succ S_5 \succ S_6 \succ S_4 \succ S_3 \succ S_1 \succ S_2$.

4.4 Analyzing Results

This analysis points out the strengths and weaknesses of the evaluation model to use each approach, in order to choose the one more adequate in an evaluation process of sustainable energy policies.

The main strength of the evaluation model with the approach based on the fusion of the information is that results are verbalized, facilitating their comprehension. There is a slight loss of information when experts' assessments are unified in S_T. However, once the experts' assessments are expressed in linguistic 2 tuples, the computational processes are carried out in a precise way.

In the case of the evaluation model using the approach based on the closeness to the IS and NIS, the main advantage is that the computations are without loss of information. The main weakness is that results are not very suitable in those cases in which the experts need more than ranking of scenarios in order to understand the computational process and the result obtained.

5 Conclusions

In this contribution, we have proposed an evaluation model applied to the energy policy selection based on the decision analysis that can manage different types of information (numerical, interval-valued and linguistic). We have applied this model to a case study for evaluating long-term sustainable energy scenarios, using two different approaches to deal with heterogeneous information. We have analyzed the strengths and weaknesses of the model, using both approaches, in order to choose the most appropriate depending on the evaluation process.

References

1. Arrow, K.J.: Social Choice and Individual Values. Yale University Press, New Haven (1963)
2. Bilgen, S., Kele, S., Kaygusuz, A., Sari, A., Kaygusuz, K.: Global warming and renewable energy sources for sustainable development: A case study in turkey. Renewable and Sustainable Energy Reviews 12(2), 372–396 (1996); cited By 21 (since 1996)
3. Cheng-Kui Huang, T., Huang, C.-H.: An integrated decision model for evaluating educational web sites from the fuzzy subjective and objective perspectives. Computers and Education 55(2), 616–629 (1996); cited By 0 (since 1996)
4. Clemen, R.T.: Making Hard Decisions. An Introduction to Decision Analisys. Duxbury Press (1995)

5. de Andrés, R., García-Lapresta, J.L., Martínez, L.: A multi-granular linguistic model for management decision-making in performance appraisal. Soft Computing 14(1), 21–34 (1996); cited By 5 (since 1996)
6. Herrera, F., Herrera-Viedma, E.: Linguistic decision analysis: Steps for solving decision problems under lingusitic information. Fuzzy Sets and Systems 115, 67–82 (2000)
7. Herrera, F., Herrera-Viedma, E., Martínez, L.: A fusion approach for managing multi-granularity linguistic terms sets in decision making. Fuzzy Sets and Systems 114(1), 43–58 (2000)
8. Herrera, F., Martínez, L.: A 2-tuple fuzzy linguistic representation model for computing with words. IEEE Transactions on Fuzzy Systems 8(6), 746–752 (2000)
9. Herrera, F., Martínez, L.: A model based on linguistic 2-tuples for dealing with multi-granularity hierarchical linguistic contexts in multiexpert decision-making. IEEE Transactions on Systems, Man, and Cybernetics. Part B: Cybernetics 31(2), 227–234 (2001)
10. Herrera, F., Martínez, L., Sánchez, P.J.: Managing non-homogeneous information in group decision making. European Journal of Operational Research 166(1), 115–132 (2005)
11. Kacprzyk, J.: Group decision making with a fuzzy linguistic majority. Fuzzy Sets and Systems 18, 105–118 (1986)
12. Kundu, S.: Min-transitivity of fuzzy leftness relationship and its application to decision making. Fuzzy Sets and Systems 86, 357–367 (1997)
13. Laes, E.: Nuclear Energy and Sustainable Development. PhD thesis, Catholic University of Leuven, Leuven (October 2006)
14. Li, D.-F.: A fuzzy closeness approach to fuzzy multi-attribute decision making. Fuzzy Optimization and Decision Making 6(3), 237–254 (1996); cited By 12 (since 1996)
15. Li, D.-F., Huang, Z.-G., Chen, G.-H.: A systematic approach to heterogeneous multi-attribute group decision making. Computers & Industrial Engineering 59(4), 561–572 (2010)
16. Martínez, L.: Sensory evaluation based on linguistic decision analysis. International Journal of Aproximated Reasoning 44(2), 148–164 (2007)
17. Martínez, L., Liu, J., Yang, J.B., Herrera, F.: A multi-granular hierarchical linguistic model for design evaluation based on safety and cost analysis. International Journal of Intelligent Systems 20(12), 1161–1194 (2005)
18. Orlovski, S.A.: Decision-making with fuzzy preference relations. Fuzzy Sets and Systems 1(3), 155–167 (1978)
19. Sun, Y.-H., Ma, J., Fan, Z.-P., Wang, J.: A group decision support approach to evaluate experts for r&d project selection. IEEE Transactions on Engineering Management 55(1), 158–170 (1996); cited By 12 (since 1996)
20. Tchangani, A.P.: Evaluation model for multiattributes-multiagents decision making: Satisficing game approach. International Journal of Information Technology and Decision Making 8(1), 73–91 (1996); cited By 4 (since 1996)
21. Le Téno, J.F., Mareschal, B.: An interval version of PROMETHEE for the comparison of building products' design with ill-defined data on environmental quality. European Journal of Operational Research 109, 522–529 (1998)
22. Yager, R.R.: An approach to ordinal decision making. International Journal of Approximate Reasoning 12(3-4), 237–261 (1995)
23. Zadeh, L.A.: The concept of a linguistic variable and its applications to approximate reasoning. Information Sciences, Part I, II, I 8, 8, 9, 43–80, 199–249, 301–357 (1975)

A Model for B2B Supplier Selection

G. Campanella, R.A. Ribeiro, and L.R. Varela

Abstract. A supply chain is a set of geographically dispersed facilities that store and transform products, and that are connected by a transportation network. The main task of supply chain management is to design the supply chain so that a given set of objectives is achieved, for example by deciding the location and capacity of new production plants, or the location of warehouses. Since suppliers also play a key role in performance maximization, it is natural to integrate them in the supply chain as well [8]. For this reason, selection of potential suppliers has become a fundamental component of supply chain management; this is even more true in the globalized market of today. The problem of supplier selection can be easily understood as a multiple-criteria decision making (MCDM) problem: businesses express their preferences with respect to suppliers, which can then be ranked and selected. Doing so, however, does not take into account the temporal evolution of supplier performances, neither can it be easily applied when considering more than one customer. To overcome these problems, we introduce a model for supplier selection that extends the classic MCDM model by introducing feedback, and consider its application in the context of multiple customers by means of linear programming.

G. Campanella · R.A. Ribeiro
UNINOVA–CA3, Campus FCT–UNL, Caparica, Portugal
e-mail: gianluca@campanella.org, rar@uninova.pt

L.R. Varela
Departamento de Produção e Sistemas, Universidade do Minho, Portugal
e-mail: leonilde@dps.uminho.pt

B. De Baets et al. (Eds.): Eurofuse 2011, AISC 107, pp. 221–228, 2011.
springerlink.com

1 Introduction

In general, the aim of multiple-criteria decision making (MCDM) is to find the best compromise solution from a set of feasible alternatives assessed with respect to a predefined set of criteria. This type of problems is widespread in real-life situations, and many approaches have been proposed in the literature to deal with this static decision process, from utility methods to scoring and ranking ones [5, 13]. Numerous MCDM techniques, ranging from simple weighted averaging to complex mathematical programming models, have been applied to the supplier selection problem; for a detailed overview, the interested reader should refer to the recent work of Ho et al. [7].

However, all these models do not take into account the possibility that supplier performances may change over time, and that the set of available suppliers itself might also be altered (for example, because some suppliers went out of business, while others emerged in the market); also, they do not consider the possibility of a network of collaborating businesses.

The first limitation can be easily addressed by turning the problem into a dynamic decision making problem, considering feedback from previous steps; unfortunately, very few contributions to this interesting field can be found in the literature [9, 12]. Usually, dynamic MCDM belongs to spatial-temporal contexts, in that exploration of the problem might result in new alternatives being considered, others being discarded, and the set of criteria being similarly altered. The problem of supplier selection can be easily understood as a temporal MCDM problem: periodically, businesses express their preferences with respect to suppliers, which can then be ranked and selected. These decisions are usually complex, since they need to deal with criteria as diverse and competing as quality, service, reliability, organization, and other technical issues, which is why temporal aggregation becomes such an effective tool to deal with varying supplier performances.

Even considering a temporal MCDM model, however, does not take into account the possibility of planning for more than one customer, which is the motivation behind the introduction of a two-step Business to Business (B2B) model for supplier selection that incorporates:

(a) a dynamic decision making model [1] to handle one-to-many relationships (a single business facing several suppliers);
(b) a linear programming model to handle many-to-many relationships (a number of collaborating businesses facing several suppliers).

The rest of this paper is organized as follows. In Section 2, we briefly review the extensive literature on supplier selection problems. In Section 3, we introduce our new model, based on the classic multiple-criteria decision making one, in the case of a single business; this model is later extended (Section 4) to the case of many collaborating businesses that are served by the same set of suppliers.

2 Review of Related Work

Due to its criticality, many authors have focused on the problem of identifying and analyzing supplier selection criteria. Already in 1966, Dickson [6] examined different supplier selection strategies by means of questionnaires that were distributed among selected managers from the United States and Canada. Clearly, as companies become more and more dependent are on suppliers, outcomes of wrong decisions become more and more severe: for example, on-time delivery and material costs are both affected by careful selection of suppliers, especially in industries where raw material accounts for 70% or more of the total cost [14].

As outlined in the previous section, the problem of supplier selection can be easily understood as a multiple-criteria decision making problem: businesses express their preferences on suppliers, which are then ranked and selected. Many applications of classic MCDM models to the problem of supplier selection can be found in the literature, ranging from the work of Chan [2] (see also [3]), which is based on the well-known Analytic Hierarchy Process (AHP) of Saaty [10, 11], to advanced mathematical programming techniques Ho et al. [7]. Another approach that has found widespread use is Data Envelopment Analysis (DEA), originally developed by Charnes et al. [4]. Many authors have also considered integrated approaches that combine two or more of the described techniques, usually in a multiple-step process; the interested reader is again referred to the work of Ho et al. [7] for a more complete overview.

3 Supplier Selection Model for a Single Business

Let us first consider the case in which a single business has to select one or more suppliers to fulfill its needs. The business is considered to constantly place orders at the beginning of regularly spaced time intervals (for example, the first day of each month), to fulfill its needs for that period. When placing the order, it is faced with the problem of selecting which supplier or suppliers will handle it.

We consider time to be discretized, $\mathcal{T} = \{1, 2, \ldots\}$, and denote by $\mathcal{S}^{(t)}$ the set of n suppliers that are being considered at time $t \in \mathcal{T}$. Note that the number of suppliers needs not be constant along time, as they can be both removed (for example, because they went out of business) and added (for example, because new business opportunities opened up).

At the same time $t \in \mathcal{T}$, each supplier is also assumed to be assessed by the business according to a set $\mathcal{C}^{(t)}$ of m criteria that may also change over time (for example, because new information became available); possible criteria include reliability, speed and cost. Assessments are assumed to be expressed as a real number in the $[0, 1]$ interval, with the common understanding that lower values correspond to less satisfaction.

Let us now consider what happens at a fixed time $t \in \mathcal{T}$. Introducing an arbitrary ordering $\{c_1, \ldots, c_m\}$ for the set of criteria $\mathcal{C}$, we can define for each supplier $s \in \mathcal{S}$ the vector

$$\mathbf{a}_s^{(t)} = [a_1, \ldots, a_m] \tag{1}$$

whose entries $a_j \in [0, 1]$, $j = 1, \ldots, m$, denote the assessment of that supplier with respect to criteria c_j.

As in the classic multiple-criteria decision making approach, in order to evaluate each supplier globally (i.e., with respect to all criteria), let us introduce a vector-valued aggregation function

$$f : [0, 1]^m \to [0, 1] \tag{2}$$

mapping vectors of m criteria values $\mathbf{a}_s^{(t)}$, $s \in \mathcal{S}$, to a single value in the $[0, 1]$ interval that can be considered a score indicating how preferable the associated supplier is. In this context, it is common to associate a weight w_j, $j = 1, \ldots, m$, to each criterion, in order to express its importance relative to all others; these weights must satisfy the normalization condition $\sum_j w_j = 1$. In addition, we require that the function f satisfies the following properties for all $\mathbf{x}, \mathbf{y} \in [0, 1]^m$,

$$\begin{cases} f(\underbrace{0, 0, \ldots, 0}_{m \text{ times}}) = 0 \\ f(\underbrace{1, 1, \ldots, 1}_{m \text{ times}}) = 1 \end{cases} \qquad \text{(preservation of bounds),} \tag{3a}$$

$$\mathbf{x} \leq \mathbf{y} \Rightarrow f(\mathbf{x}) \leq f(\mathbf{y}) \qquad \text{(monotonicity).} \tag{3b}$$

Given the ratings

$$r_s^{(t)} = f\left(\mathbf{a}_s^{(t)}\right), \quad s \in \mathcal{S}, \tag{4}$$

suppliers may then be ordered and orders placed at the higher ranking ones.

In classic multiple-criteria decision making, the above procedure would simply be applied at each time $t \in \mathcal{T}$, without taking into account any historical information that might be available, as is done instead in the dynamic framework of Campanella and Ribeiro [1] that is depicted in Figure 1 and that we shall introduce next.

Let us start by defining the historical set as follows,

$$\mathcal{H}^{(0)} = \emptyset \quad \text{and} \quad \mathcal{H}^{(t)} \subseteq \mathcal{S}^{(t-1)} \cup \mathcal{H}^{(t-1)}, \ t \in \mathcal{T}, \tag{5}$$

meaning that, in the first iteration, no historical information will be available, and that, in later iterations, the historical set will contain information about some of the suppliers that were considered in the previous iteration, or came from the historical set at that time (i.e., from an iteration even longer away).

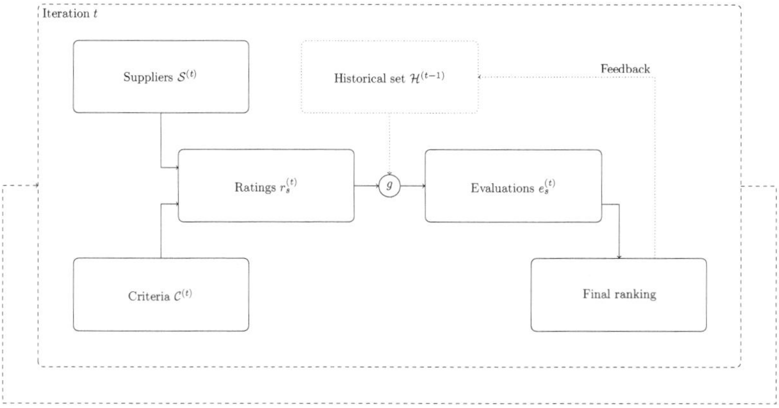

Fig. 1 Operations performed at each iteration t in the dynamic framework of Campanella and Ribeiro [1]: first, available suppliers and considered criteria are aggregated into ratings; then, by making use of the information stored in the historical set, a final ranking is produced; finally, the information in the historical set is updated and passed on to the next iteration.

Introducing a second associative aggregation function g, whose associativity is required to make sure that repeated pairwise application will yield the same result as application over vectors, we define the evaluation $e_s^{(t)} \in [0,1]$ of supplier $s \in \mathcal{S}$ at time $t \in \mathcal{T}$ as follows,

$$e_s^{(t)} = \begin{cases} r_s^{(t)} & s \in \mathcal{S}^{(t)} \setminus \mathcal{H}^{(t-1)} \\ g\left(e_s^{(t-1)}, r_s^{(t)}\right) & s \in \mathcal{S}^{(t)} \cap \mathcal{H}^{(t-1)} \\ e_s^{(t-1)} & s \in \mathcal{H}^{(t-1)} \setminus \mathcal{S}^{(t)} \end{cases}. \tag{6}$$

For each supplier $s \in \mathcal{S}^{(t)} \cup \mathcal{H}^{(t-1)}$, either belonging to the current set of available suppliers $\mathcal{S}^{(t)}$ or carried over from the previous one by means of the historical set $\mathcal{H}^{(t-1)}$, we have that:

1. if the supplier belongs only to the current set, but not to the historical set (first case), its evaluation is simply equal to its rating;
2. if the supplier belongs to both the current and the historical set of alternatives (second case), its evaluation is the aggregation (performed by the aggregation function g) of its evaluation in the previous iteration with its rating in the current one;
3. finally, if the supplier does not belong to the current set, but was carried over in the historical set (third case), its evaluation is also carried over from the previous iteration.

In the dynamic framework, evaluations (instead of ratings) are used to order and rank suppliers, thus taking into account satisfaction of criteria not only at

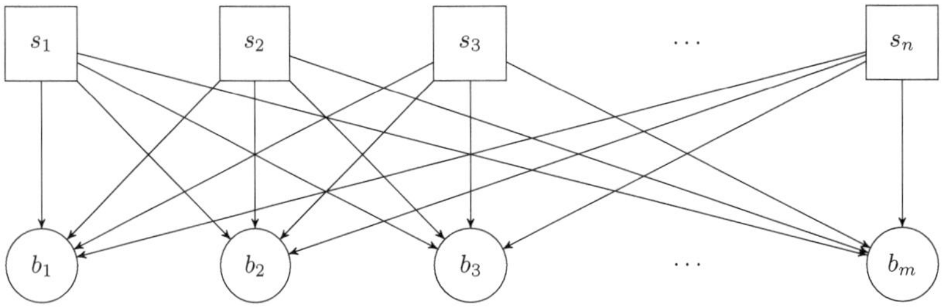

Fig. 2 Network of businesses and suppliers: each business b_j, $j = 1, \ldots, m$, depicted here as a circle, orders a certain quantity x_{ij} from supplier s_i, $i = 1, \ldots, n$, depicted here as a square.

the time of decision, but also at previous iterations. Choosing an appropriate aggregation function g, it is thus possible to reward suppliers that have had good scores in the past, and to penalize those that did not.

4 Supplier Selection Model for Multiple Businesses

Let us again examine what happens at a fixed time $t \in \mathcal{T}$, but this time considering that j businesses are collaboratively planning their orders to a set of n suppliers.

Each business b_j, $j = 1, \ldots, m$, is assumed to have computed evaluations for each supplier s_i, $i = 1, \ldots, n$, denoted $e_i^{(t)}$, using the previously described method. Furthermore, each business has a certain demand d_j, $j = 1, \ldots, m$, and each supplier a maximum capacity c_i, $i = 1, \ldots, n$. The variables of the problem are represented by the quantities x_{ij} that business b_j, $j = 1, \ldots, m$, shall order from supplier s_i, $i = 1, \ldots, n$. The situation is depicted in Figure 2.

The total satisfaction of business b_j, $j = 1, \ldots, m$ with respect to a certain allocation of orders to the suppliers can thus be written as follows,

$$\sigma_j = \sum_i e_i^{(t)} x_{ij}, \tag{7}$$

and we can define the following linear program that maximizes the satisfaction of all businesses, making sure all demands are met and no capacity is exceeded,

$$\begin{aligned} \max \quad & \sum_j \sigma_j \\ \text{s.t.} \quad & \sum_i x_{ij} = d_j \qquad & j = 1, \ldots, m, \\ & \sum_j x_{ij} \leq c_j \qquad & i = 1, \ldots, n. \end{aligned} \tag{8}$$

Note that similar linear programming models usually consider costs instead of satisfactions σ_j, $j = 1, \ldots, m$, and consequently minimize the objective function. Our approach, however, allows each business to take into account many more criteria that are then distilled into the individual evaluations $e_i^{(t)}$, $i = 1, \ldots, n$.

5 Conclusions

In this paper we have introduced:

(a) a supplier selection model for a single business, based on the dynamic MCDM model of Campanella and Ribeiro [1], that allows ranking and selection of suppliers from the point of view of a single business placing orders repeatedly over time;
(b) a B2B supplier selection model for multiple businesses, which extends the previous model by incorporating it into a linear program, that allows selection of the optimal quantities to be ordered from the suppliers by a network of collaborating businesses.

The first model overcomes the first problem of classic MCDM models, namely the fact that temporal evolution of supplier performances cannot be easily taken into account, while the second allows to consider a network of businesses and suppliers, instead of a single business and multiple suppliers.

We believe that these models have the potential for providing valuable decision support in supplier selection that can be exploited to improve the management of the supply chain. Moreover, they provide a unified method to assess supplier performances, and may thus also become an important part of quality management procedures.

Acknowledgements. This work has been partially funded by Fundação para a Ciência e a Tecnologia, Portugal, under contract CONT_DOUT/49/UNINOVA/ 0/902/1/2006.

References

[1] Campanella, G., Ribeiro, R.A.: A framework for dynamic multiple-criteria decision making. Decision Support Systems (in Press, 2011), doi:10.1016/j.dss.2011.05.003

[2] Chan, F.T.S.: Interactive selection model for supplier selection process: an analytical hierarchy process approach. International Journal of Production Research 41(15), 3549–3579 (2004)
[3] Chan, F.T.S., Kumar, N.: Global supplier development considering risk factors using fuzzy extended AHP-based approach. Omega 35(4), 417–431 (2007)
[4] Charnes, A., Cooper, W.W., Rhodes, E.: Measuring the efficiency of decision making units. European Journal of Operational Research 2(6), 429–444 (1978)
[5] Chen, S.-J., Hwang, C.L., Hwang, F.P.: Fuzzy Multiple Attribute Decision Making: Methods and Applications. Lecture Notes in Economics and Mathematical Systems, vol. 375. Springer, Heidelberg (1992)
[6] Dickson, G.W.: An Analysis of Vendor Selection Systems and Decisions. Journal of Purchasing 2(1), 5–17 (1966)
[7] Ho, W., Xu, X., Dey, P.K.: Multi-criteria decision making approaches for supplier evaluation and selection: A literature review. European Journal of Operational Research 202(1), 16–24 (2010)
[8] Kumar, M., Vrat, P., Shankar, R.: A fuzzy goal programming approach for vendor selection problem in a supply chain. Computers & Industrial Engineering 46(1), 69–85 (2004)
[9] Pais, T.C., Ribeiro, R.A., Devouassoux, Y., Reynaud, S.: Dynamic ranking algorithm for landing site selection. In: Magdalena, L., Ojeda-Aciego, M., Verdegay, J.L. (eds.) Proceedings of IPMU 2008, Málaga, Spain, pp. 608–613 (2008)
[10] Saaty, T.L.: Decision Making for Leaders: The Analytic Hierarchy Process for Decisions in a Complex World, 3rd edn. Analytic Hierarchy Process Series, vol. 2, pp. 978–96786. RWS Publications (1999); ISBN 978-0962031786
[11] Saaty, T.L.: Fundamentals of Decision Making and Priority Theory With the Analytic Hierarchy Process, 1st edn. Analytic Hierarchy Process Series, vol. 6, pp. 978–96762. RWS Publications (2031); ISBN 978-0962031762
[12] Townsend, J.T., Busemeyer, J.: Dynamic representation of decision-making. In: Port, R.F., van Gelder, T. (eds.) Mind as Motion: Explorations in the Dynamics of Cognition, pp. 101–120. MIT Press, Cambridge (1995)
[13] Triantaphyllou, E.: Multi-Criteria Decision Making Methods: A Comparative Study. Applied Optimization, vol. 44. Springer, Heidelberg (2000)
[14] Çebi, F., Bayraktar, D.: An integrated approach for supplier selection. Logistics Information Management 16(6), 395–400 (2003)

A Multicriteria Linguistic Decision Making Model Dealing with Comparative Terms

Rosa M. Rodríguez, Luis Martínez, and Francisco Herrera

Abstract. In this contribution our aim is to present a multicriteria linguistic decision making model in which experts might provide their assessments by using linguistic expressions based on comparative terms close to the expressions used by human beings in real world problems or single linguistic terms. To aggregate such a type of linguistic information two symbolic aggregation operators are introduced. Finally, an exploitation phase is proposed to build a preference relation among alternatives and then, a non-dominance choice degree is applied to obtain the solution set of alternatives.

Keywords: Linguistic decision making, hesitant fuzzy set, symbolic aggregation, linguistic expressions

1 Introduction

Decision Making is a common task for human beings in their daily lives and becomes a core research area in different fields such as planning, scheduling, engineering, medicine and so on. Decision problems are usually defined under context with imprecise, vague and uncertain information. Different approaches have been proposed to deal with such an uncertainty. One of them is the fuzzy linguistic approach [19] that has provided successful results on decision making problems under uncertainty [6, 9].

Rosa M. Rodríguez · Luis Martínez
Dept. of Computer Science, University of Jaén, 23071 - Jaén, Spain
e-mail: {rmrodrig,martin}@ujaen.es

Francisco Herrera
Dept. of Computer Science, University of Granada, 18071 - Granada, Spain
e-mail: herrera@decsai.ugr.es

B. De Baets et al. (Eds.): Eurofuse 2011, AISC 107, pp. 229–241, 2011.
springerlink.com

Often decision makers provide single linguistic terms to assess the alternatives or criteria of a decision making problem [1, 5, 17]. However, experts can prefer the use of linguistic expressions in natural language instead of a single linguistic term. Different approaches have been proposed in the literature to overcome this limitation and allow that decision makers can express their assessments by using more than one linguistic term [8, 13]. Tang and Zheng presented in [13] a linguistic model that dealt with linguistic expressions generated by applying logical connectives to the linguistic terms. Ma et al. introduced in [8] the concepts of determinacy and consistency of linguistic terms in multicriteria decision making problems and presented a model based on fuzzy-set in which decision makers could provide their assessments by using several linguistic terms together the reliability degree of each term.

In this contribution our aim is to introduce a multicriteria linguistic decision making model in which decision makers can provide their assessments by using linguistic expressions based on comparative terms or single linguistic terms. We will use the concept of hesitant fuzzy linguistic term set (HFLTS) [11] to model linguistic hesitant situations where decision makers hesitate among different linguistic values to provide their assessments. We propose a decision solving process with three phases: (i) a transformation phase that transforms the linguistic expressions and single linguistic terms into HFLTS, (ii) an aggregation phase that computes collective assessments for each alternative by using two symbolic aggregation operators and provides a linguistic interval, (iii) an exploitation phase that extracts a preference relation by means of a comparison between linguistic intervals and uses a non-dominance choice degree to obtain the solution set of alternatives of the decision making problem.

This contribution is organized as follows: Section 2 reviews briefly some preliminary concepts. Section 3 presents a multicriteria linguistic decision making model and defines two symbolic aggregation operators. An illustrative example is also introduced in this section, and finally, Section 4 points out some concluding remarks.

2 Preliminaries

The multicriteria linguistic decision making model proposed in this contribution represents the linguistic expressions by means of HFLTS which are based on the fuzzy linguistic approach [19] and hesitant fuzzy sets [15]. Therefore, in this section are reviewed some basic concepts about such approaches and the definition of the linguistic expressions based on comparative terms represented by HFLTS.

2.1 Fuzzy Linguistic Approach

In many real decision making problems is suitable the use of linguistic information due to the imprecision and vagueness of the framework in which are defined such problems. In such situations the fuzzy linguistic approach [19] that uses the fuzzy set theory [18] to manage the uncertainty and model such a type of information has provided successful results [1, 5, 7].

Zadeh [19] introduced the concept of linguistic variable as *a variable whose values are not numbers but words or sentences in a natural or artificial language*. A linguistic value is less precise than a number, but it is closer to the natural language used by human beings.

To deal with linguistic variables, it is necessary to choose the appropriate linguistic descriptors for the linguistic term sets and their semantics. To do so, there are different possibilities [19]. The choice of the linguistic descriptors can be carried out as follows:

- Supplying directly the term set by considering all the terms symmetrically distributed on a scale which has an order defined [17]. For example a set of seven terms S could be:

$$S = \{s_0 : nothing, s_1 : very\_low, s_2 : low, s_3 : medium, s_4 : high, s_5 : very\_high, s_6 : perfect\}$$

- Defining the linguistic term set by means of a context-free grammar, G, such that the linguistic terms are sentences generated by G [2, 3, 19]. A grammar G is a 4-tuple (V_N, V_T, I, P), where V_N is the set of non-terminal symbols, V_T is the set of terminals symbols, I is the starting symbol, and P the production rules defined in an extended Backus Naur Form [3]. Among the terminal symbols of G, we can find primary terms (e.g., low, medium), hedges (e.g., not, very), relations (e.g., lower than, higher than), conjunctions (e.g., and, but), and disjunctions (e.g., or).

To accomplish the definition of the semantics of the linguistic descriptors there also are different ways:

- A semantics based on membership functions and a semantic rule. It assumes that the meaning of each linguistic term is given by means of a fuzzy subset defined in the interval [0,1], which is described by membership functions [3]. This semantic approach is used when the linguistic descriptors are generated by means of a context-free grammar.
- A semantics based on an ordered structure of the linguistic term set that introduces the semantics from the structure defined over the linguistic term set. So, the users use an ordered linguistic term set to provide their assessments [14, 17].
- Mixed semantics that uses elements from the previous approaches.

In this contribution, we assume an ordered linguistic term set that requires the following operators:

1. Negation: $\text{Neg}(s_i) = s_j$ such that $j = g - i$ ($g + 1$ is the cardinality).
2. Maximization: $\max(s_i, s_j) = s_i \; if \; s_i \geq s_j$.
3. Minimization: $\min(s_i, s_j) = s_i \; if \; s_i \leq s_j$.

2.2 *Hesitant Fuzzy Sets*

Torra in [15] introduced the definition of hesitant fuzzy sets to fulfil the management of decision situations in quantitative contexts where the decision makers hesitate among different values to assess an alternative or criteria.

A hesitant fuzzy set is defined in terms of a function that returns a set of membership values for each element in the domain [15].

Definition 1 *[15]. Let X be a reference set, a hesitant fuzzy set on X is a function h that returns a subset of values in [0,1].*

$$h : X \rightarrow \{[0,1]\}$$

A hesitant fuzzy set can be also defined in terms of the union of their membership degree to a set of fuzzy sets.

Definition 2 *[15]. Let* $M = \{\mu_1, \mu_2, ..., \mu_n\}$ *be a set of n membership functions. The hesitant fuzzy set associated with M,* h_M*, is defined as:*

$$h_M : M \rightarrow \{[0,1]\}$$

$$h_M(x) = \bigcup_{\mu \in M} \{\mu(x)\}$$

In [15] were defined different hesitant fuzzy sets operations as union, intersection, complement and so on.

2.3 Hesitant Fuzzy Linguistic Term Set

Similarly to the hesitant fuzzy sets [15], where an expert might hesitate among several values to assess an alternative or criterion, in the qualitative context might happen that experts hesitate among several linguistic values. To manage such situations, it was proposed the concept of HFLTS [11] which is based on the fuzzy linguistic approach and hesitant fuzzy sets previously revised.

Definition 3 *[11]. Let S be a linguistic term set,* $S = \{s_0, \ldots, s_g\}$*, a HFLTS,* H_S*, is an ordered finite subset of consecutive linguistic terms of S.*

Two operators were defined to obtain the maximum and minimum bounds of a HFLTS.

Definition 4 *[11]. The upper bound,* H_{S^+}*, and lower bound,* H_{S^-}*, of the HFLTS,* H_S*, are defined as:*

- $H_{S^+} = max(s_i, s_j) = s_i$, *if* $s_i \geq s_j$; $s_i, s_j \in H_S$
- $H_{S^-} = min(s_i, s_j) = s_i$, *if* $s_i \leq s_j$; $s_i, s_j \in H_S$

The comparison between linguistic terms is necessary in many decision making problems and has been defined in different approaches. The comparison between HFLTS is not simple, therefore, it was introduced the concept of envelope of a HFLTS.

Definition 5 *[11]. The envelope of the HFLTS,* $env(H_S)$*, is a linguistic interval whose limits are obtained by means of upper bound (*max*) and lower bound (*min*):*

$$env(H_S) = [H_{S^-}, H_{S^+}], \;\; H_{S^-} <= H_{S^+}$$

More operations with HFLTS can be found in [11].

2.4 Elicitation of Linguistic Expressions

The main objective of the definition of HFLTS was to improve the flexibility of the elicitation of linguistic expressions when experts hesitate among several linguistic values to assess linguistic variables. Therefore, new linguistic expressions, ll, closer to human beings expressions are generated by a context-free grammar and then represented by HFLTS. The context-free grammar was defined as follows:

Definition 6 *[11]. Let* G_H *be a context-free grammar and* $S = \{s_0, \ldots, s_g\}$ *a linguistic term set. The elements of* $G_H = (V_N, V_T, I, P)$ *are defined as follows:*

$V_N = \{\langle primary\ term\rangle, \langle composite\ term\rangle, \langle unary\ relation\rangle, \langle binary\ relation\rangle, \langle conjunction\rangle\}$

$V_T = \{lower\ than, greater\ than, between, and, s_0, s_1, \ldots, s_g\}$

$I \in V_N$

The production rules are defined in an extended Backus Naur Form such that the brackets enclose optional elements and the symbol $|$ *indicate alternative elements [3]. For the context-free grammar,* G_H*, the production rules are the following ones:*

$P = \{I ::= \langle primary\ term\rangle | \langle composite\ term\rangle$

$\langle composite\ term\rangle ::= \langle unary\ relation\rangle\langle primary\ term\rangle | \langle binary\ relation\rangle$
$\langle primary\ term\rangle\langle conjunction\rangle\langle primary\ term\rangle$

$\langle primary\ term\rangle ::= s_0 | s_1 | \ldots | s_g$

$\langle unary\ relation\rangle ::= lower\ than | greater\ than$

$\langle binary\ relation\rangle ::= between$

$\langle conjunction\rangle ::= and\}$

It was also defined a transformation function, E_{G_H}, to obtain HFLTS from the linguistic expressions, ll, generated by the context-free grammar, G_H.

Definition 7 *[11]. Let* E_{G_H} *be a function that transforms linguistic expressions,* ll*, obtained by* G_H*, into HFLTS,* H_S*, where* S *is the linguistic term set used by* G_H*.*

$$E_{G_H} : ll \longrightarrow H_S$$

The linguistic expressions generated by using the production rules are transformed into HFLTS in different ways as follows:

- $E_{G_H}(s_i) = \{s_i / s_i \in S\}$
- $E_{G_H}(\text{less than } s_i) = \{s_j / s_j \in S \text{ and } s_j \leq s_i\}$
- $E_{G_H}(\text{greater than } s_i) = \{s_j / s_j \in S \text{ and } s_j \geq s_i\}$
- $E_{G_H}(\text{between } s_i \text{ and } s_j) = \{s_k / s_k \in S \text{ and } s_i \leq s_k \leq s_j\}$

3 Linguistic Decision Making Model Based on Comparative Terms

This section presents a multicriteria linguistic decision making model based on symbolic computational processes in which the assessments can be provided by using linguistic expressions based on comparative terms or single linguistic terms. To carry out the aggregation of such assessments are introduced two symbolic aggregation operators, *min_upper* and *max_lower*, that deal with HFLTS to obtain linguistic intervals as aggregated assessments. Finally, it is proposed an exploitation process that uses a preference relation to compare the intervals of the alternatives and then a non-dominance choice degree is applied to obtain the solution set of alternatives.

3.1 Multicriteria Linguistic Decision Making Problem

A multicriteria decision making problem consists of a finite set of alternatives, $X = \{x_1,\dots,x_n\}$, where each alternative is defined by means of a finite set of criteria, $C = \{c_1,\dots,c_m\}$ that can be assessed by using linguistic expressions based on the comparative terms or single linguistic terms.

Let $S = \{s_0,\dots,s_g\}$ be a linguistic term set and G_H a context-free grammar which produces the linguistic expressions, $ll(x_i,c_j)$, based on comparative terms to assess the criteria, $C = \{c_1,\dots,c_m\}$, for each alternative, $X = \{x_1,\dots,x_n\}$:

$$ll(x_i,c_j) = lower\ than\ s_k$$
$$ll(x_i,c_j) = greater\ than\ s_k$$
$$ll(x_i,c_j) = between\ s_k\ and\ s_l \quad \text{iff} \quad s_k < s_l$$

where $s_k, s_l \in S$, $i \in \{1,\dots,n\}$ and $j \in \{1,\dots,m\}$.

3.2 Multicriteria Linguistic Decision Making Model

The multicriteria linguistic decision making model proposed in this contribution is based on symbolic computational processes in which the assessments might be provided by using linguistic expressions generated by means of the context-free grammar, G_H, or single linguistic terms. To carry out the processes of computing with words, this decision making model uses linguistic intervals. The decision model consists mainly of three phases (see Fig. 1):

1. Transformation phase: the linguistic expressions and single linguistic terms are transformed into HFLTS by using the transformation function, E_{G_H}.
2. Aggregation phase: the assessments represented by HFLTS are aggregated by using symbolic aggregation operators that obtain a linguistic interval that represents the support of the set of criteria for each alternative.
3. Exploitation phase: the linguistic intervals obtained in the previous phase are used to build a preference relation between alternatives and a non-dominance choice degree is applied to obtain a solution set of alternatives.

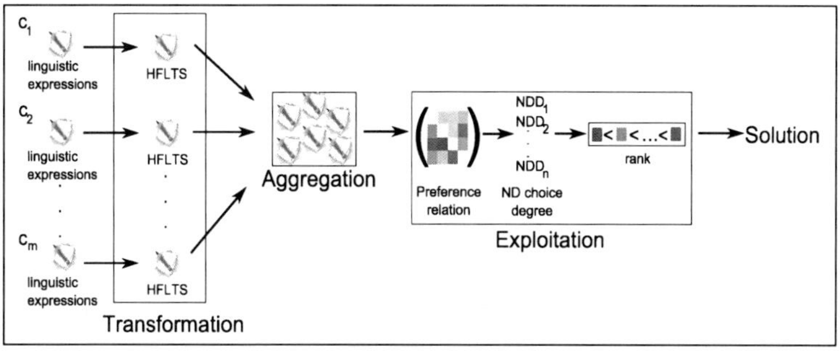

Fig. 1 Schema of the multicriteria linguistic decision making model

Next, these phases are explained in further detail and an illustrative example is presented.

1. *Transformation of the linguistic expressions into HFLTS*
   Our aim is to facilitate the expressiveness of the experts in linguistic decision making problems providing a linguistic modelling close to human beings that uses linguistic information either single linguistic terms or linguistic expressions generated by a context-free grammar, G_H. The linguistic expressions provided must be transformed into HFLTS to accomplish the aggregation process by means of the transformation function, E_{G_H}, introduced in Definition 7.
   Example: Let $X = \{x_1, x_2, x_3\}$ be a set of alternatives, $C = \{c_1, c_2, c_3\}$ a set of criteria defined for each alternative, and $S = \{s_0 : nothing(n), s_1 : very\ low(vl), s_2 : low(l), s_3 : medium(m), s_4 : high(h), s_5 : very\ high(vh), s_6 : perfect(p)\}$ the linguistic term set used by the context-free grammar, G_H, to generate the linguistic expressions. The assessments provided in such a problem are showed in Table 1.

Table 1 Assessments provided for the decision problem

		criteria		
	ll	c_1	c_2	c_3
alternatives	x_1	*between vl and m*	*between h and vh*	*h*
	x_2	*between l and m*	*m*	*lower than l*
	x_3	*greater than h*	*between vl and l*	*greater than h*

   The transformation of such expressions into HFLTS by means of the transformation function, E_{G_H}, are showed in Table 2.
2. *Aggregation of the assessments represented by HFLTS*
   Once the assessments are represented by HFLTS, it is necessary to fuse the set of criteria for each alternative by using symbolic aggregation operators. We shall define two operators, *min_upper* and *max_lower* that carry out the computations by using linguistic intervals. These operators allow to have two approximations,

Table 2 Assessments transformed into HFLTS

		criteria		
	$H_S^j(x_i)$	c_1	c_2	c_3
alternatives	x_1	$\{vl, l, m\}$	$\{h, vh\}$	$\{h\}$
	x_2	$\{l, m\}$	$\{m\}$	$\{n, vl, l\}$
	x_3	$\{h, vh, p\}$	$\{vl, l\}$	$\{h, vh, p\}$

the *pessimistic* one that selects the best of the inferior values of the HFLTS obtained in the previous phase, and the *optimistic* one that chooses the worst of the superior values of the HFLTS. The result of applying these operators is a linguistic interval associated to each alternative and represents the core information of the HFLTS aggregated.

- **Min_upper operator**
  It is a symbolic aggregation operator introduced to combine HFLTS. It obtains the worst of the superior linguistic terms and represents the right limit of a linguistic interval.

  Definition 8. *Let* $X = \{x_1, \ldots, x_n\}$ *be a set of alternatives,* $C = \{c_1, \ldots, c_m\}$ *a set of criteria,* $S = \{s_0, \ldots, s_g\}$ *be a linguistic term set and* $\{H_S^j(x_i)/i \in \{1, \ldots, n\}, j \in \{1, \ldots, m\}\}$ *a set of HFLTS, the min_upper operator consists of two steps:*
  a. Apply the upper bound, H_{S^+}*, for each HFLTS associated to each alternative:*

  $$H_{S^+}(x_i) = \{H_{S^+}^1(x_i), \ldots, H_{S^+}^m(x_i)\},\ i \in \{1, \ldots, n\}$$

  b. Obtain the minimum linguistic term for each alternative:

  $$H_{S^+_{min}}(x_i) = min\{H_{S^+}^j(x_i)\ /\ j \in \{1, \ldots, m\}\},\ i \in \{1, \ldots, n\}$$

  Example: The aggregation with such an operator is carried out as follows:
  a. Apply for each HFLTS the upper bound.
  b. Obtain for each alternative the minimum linguistic term of the set of criteria (see Table 3):

Table 3 Minimum linguistic term of the set of criteria

alternatives $/H_{S^+_{min}}(x_i)$		
x_1	x_2	x_3
$\{m\}$	$\{l\}$	$\{l\}$

- **Max_lower operator**
  This symbolic operator is also introduced to combine HFLTS. It is the opposite to the previous one because it obtains the best of the inferior linguistic terms and it represents the left limit of a linguistic interval.

  Definition 9. *Let* $X = \{x_1, \ldots, x_n\}$ *be a set of alternatives,* $C = \{c_1, \ldots, c_m\}$ *a set of criteria,* $S = \{s_0, \ldots, s_g\}$ *be a linguistic term set and* $\{H_S^j(x_i)/i \in \{1, \ldots, n\}, j \in \{1, \ldots, m\}\}$ *a set of HFLTS, the max_lower operator consists also of two steps:*
  a. Apply the lower bound for each HFLTS associated to each alternative:

$$H_{S^-}(x_i) = \{H_{S^-}^1(x_i), \ldots, H_{S^-}^m(x_i)\},\ i \in \{1, \ldots, n\}$$

  b. Obtain the maximum linguistic term for each alternative:

$$H_{S_{max}^-}(x_i) = max\{H_{S^-}^j(x_i)\ /\ j \in \{1, \ldots, m\}\},\ i \in \{1, \ldots, n\}$$

  Example: In the illustrative example the results obtained of applying the *max_lower* operator are the following ones:
  a. Apply the lower bound for each HFLTS.
  b. Obtain for each alternative the maximum linguistic term of the set of criteria (see Table 4):

Table 4 Maximum linguistic term of the set of criteria

alternatives/$H_{S_{max}^-}(x_i)$		
x_1	x_2	x_3
$\{h\}$	$\{m\}$	$\{h\}$

- The linguistic terms obtained from the previous aggregation operators are used to build a linguistic interval for each alternative that represents the core information of the aggregated HFLTS:

$$H'_{max}(x_i) = max\{H_{S_{min}^+}(x_i), H_{S_{max}^-}(x_i)\}$$

$$H'_{min}(x_i) = min\{H_{S_{min}^+}(x_i), H_{S_{max}^-}(x_i)\}$$

$$H'(x_i) = [H'_{min}(x_i), H'_{max}(x_i)]$$

  Example: Such linguistic intervals are showed in Table 5.

3. *Exploitation phase*
   Once the linguistic information has been aggregated, it is carried out the exploitation phase where the set of alternatives is ordered to select the best one/s according to the following steps:

Table 5 Linguistic intervals for the alternatives

alternatives/$H'(x_i)$		
x_1	x_2	x_3
$[m,h]$	$[l,m]$	$[l,h]$

a. Extraction of a preference relation. Right now the information regarding each alternative is expressed by a linguistic interval. Hence to order such alternatives, first it is built a binary preference relation [4, 10] between alternatives. This preference relation is obtained by adapting the method proposed by Wang et. al. in [16].

Definition 10 *[16]. Let $A = [a_1, a_2]$ and $B = [b_1, b_2]$ be two interval utilities, the preference degree of A over B (or $A > B$) is defined as:*

$$P(A > B) = \frac{max(0, a_2 - b_1) - max(0, a_1 - b_2)}{(a_2 - a_1) + (b_2 - b_1)}$$

and the preference degree of B over A (or $B > A$) as:

$$P(B > A) = \frac{max(0, b_2 - a_1) - max(0, b_1 - a_2)}{(a_2 - a_1) + (b_2 - b_1)}$$

It is obvious that $P(A < B) + P(B > A) = 1$ and $P(A > B) = P(B > A) = 0.5$ when $A = B$, i.e. $a_1 = b_1$ and $a_2 = b_2$.

Remark 1. *If the interval values are equal ($a_1 = a_2$, $b_1 = b_2$), they will be compared as crisp numbers. Therefore, if $a_1 > b_1$ then $P(A > B) = 1$, $P(A < B) = 0$, and if $a_1 = b_1$ then $P(A > B) = P(A < B) = 0.5$.*

This function is adapted to deal with linguistic intervals as follows:
$Ind(s_i) = i, s_i \in S = \{s_0, \ldots, s_g\}$

$A = [Ind(s_i), Ind(s_j)], s_i \leq s_j$

$B = [Ind(s_k), Ind(s_l)], s_k \leq s_l$

$P(A > B) = \frac{max(0, Ind(s_j) - Ind(s_k)) - max(0, Ind(s_i - Ind(s_l))}{(Ind(s_j) - Ind(s_i)) + (Ind(s_l) - Ind(s_k))}$

Therefore, the preference relation for the alternatives is obtained as follows:

Definition 11 *[16]. Let P_D be a preference relation,*

$$P_D = \begin{pmatrix} - & p_{12} & \cdots & p_{1n} \\ p_{21} & - & \cdots & p_{2n} \\ \vdots & \vdots & \ddots & \vdots \\ p_{n1} & p_{n2} & \cdots & - \end{pmatrix}$$

where

$$p_{ij} = P(x_i > x_j) = \frac{max(0, Ind(H'_{S+}(x_i)) - Ind(H'_{S-}(x_j))) - max(0, Ind(H'_{S-}(x_i)) - Ind(H'_{S+}(x_j)))}{(Ind(H'_{S+}(x_i)) - Ind(H'_{S-}(x_i))) - (Ind(H'_{S+}(x_j)) - Ind(H'_{S-}(x_j)))}$$

is the preference degree of the alternative x_i *over* x_j*;* $i, j \in \{1, \ldots, n\}$*;* $i \neq j$*;* $Ind(s_l) = l$ *(it provides the index associated to the label).*

b. Applying of a choice degree. For ranking alternatives from the preference relation, different choice functions can be applied [10, 12]. Here we propose the use of a non-dominance choice degree, *NDD*, that indicates the degree in which the alternative x_i is not dominated by the remaining alternatives. Its definition is given as:

Definition 12 *[10]. Let* $P = [p_{ij}]$ *be a preference relation defined over a set of alternatives* X*. For the alternative* x_i*, its non-dominance degree,* NDD_i*, is obtained as:*

$$NDD_i = min\{1 - p^S_{ji}, j \neq i\}$$

where $p^S_{ji} = max\{p_{ji} - p_{ij}, 0\}$ *represents the degree to which* x_i *is strictly dominated by* x_j*.*

c. Finally, we obtain the set of non-dominated alternatives as follows:

$$X^{ND} = \{x_i / x_i \in X, NDD_i = max_{x_j \in X}\{NDD_j\}\}$$

Example: Keep doing the example, the exploitation phase consists of the following steps:

a. Extract the preference degrees by using the definition introduced by Wang et al [16].

$$P_D = \begin{bmatrix} - & 1 & 0.667 \\ 0 & - & 0.333 \\ 0.333 & 0.667 & - \end{bmatrix}$$

b. We apply a non-dominance choice degree, NDD_i, to the preference relation:

$$P^S_D = \begin{bmatrix} - & 1 & 0.334 \\ 0 & - & 0 \\ 0 & 0.334 & - \end{bmatrix}$$

$NDD_1 = min\{(1-0), 1-0)\} = 1$
$NDD_2 = min\{(1-1), (1-0.334)\} = 0$
$NDD_3 = min\{1-0.334), (1-0)\} = 0.664$

c. Finally, the solution set of alternatives is, $\mathbf{X^{ND}} = \{\mathbf{x_1}\}$

4 Conclusions

In real world the decision making problems are usually defined under uncertainty where the information is vague and imprecise. In such problems might happen that a decision maker hesitate among several linguistic terms to provide his/her assessment and prefers using a linguistic expressions rather than a single linguistic term. In this contribution we have presented a multicriteria linguistic decision making model in which decision makers can provide their assessments by means of linguistic expressions based on comparative terms or single linguistic terms. To carry out the processes of computing with words the decision model manages such linguistic information by means of linguistic intervals and rank the alternatives by using two symbolic aggregation operators and a non-dominance choice degree.

Acknowledgements. This work is partially supported by the Research Project TIN-2009-08286, P08-TIC-3548 and FEDER funds.

References

1. Arfi, B.: Fuzzy decision making in politics: A linguistic fuzzy-set approach (LFSA). Political Analysis 13(1), 23–56 (2005)
2. Bonissone, P.P.: A fuzzy sets based linguistic approach: theory and applications. In: Gupta, M.M., Sanchez, E. (eds.) Approximate Reasoning in Decision Analysis, pp. 99–111. North-Holland Publishing Company, Amsterdam (1982)
3. Bordogna, G., Pasi, G.: A fuzzy linguistic approach generalizing boolean information retrieval: A model and its evaluation. J. of the American Society for Inf. Sci. 44, 70–82 (1993)
4. Fan, Z.P., Ma, J., Zhang, Q.: An approach to multiple attribute decision making based on fuzzy preference information alternatives. Fuzzy Sets and Systems 131(1), 101–106 (2002)
5. García-Lapresta, J.L.: A general class of simple majority decision rules based on linguistic opinions. Information Sciences 176(4), 352–365 (2006)
6. Herrera, F., Alonso, S., Chiclana, F., Herrera-Viedma, E.: Computing with words in decision making: Foundations, trends and prospects. Fuz. Opt. and Dec. Making 8(4), 337–364 (2009)
7. Huynh, V.N., Nakamori, Y.: A satisfactory-oriented approach to multi-expert decision-making under linguistic assessments. IEEE Trans. Systems, Man, and Cybernetics, SMC 35(2), 184–196 (2005)
8. Xu, Y., Ma, J., Ruan, D., Zhang, G.: A fuzzy-set approach to treat determinacy and consistency of linguistic terms in multi-criteria decision making. International Journal of Approximate Reasoning 44(2), 165–181 (2007)
9. Martínez, L., Ruan, D., Herrera, F.: Computing with words in decision support systems: An overview on models and applications. Int. J. of Comp. Inteligence Sys. 3(4), 382–395 (2010)
10. Orlovski, S.A.: Decision-making with fuzzy preference relations. Fuzzy Sets and Systems 1, 155–167 (1978)
11. Rodríguez, R. M., Martínez, L., Herrera, F.: Hesitant fuzzy linguistic term sets. Technical Report TR-3-2011, University of Jaén (2011), `http://sinbad2.ujaen.es/sinbad2/files/publicaciones/310.pdf`

12. Roubens, M.: Some properties of choice functions based on valued binary relations. European Journal of Operational Research 40, 309–321 (1989)
13. Tang, Y., Zheng, J.: Linguistic modelling based on semantic similarity relation among linguistic labels. Fuzzy Sets and Systems 157(12), 1662–1673 (2006)
14. Torra, V.: Negation functions based semantics for ordered linguistic labels. International Journal of Intelligent Systems 11, 975–988 (1996)
15. Torra, V.: Hesitant fuzzy sets. Int. Journal of Intelligent Systems 25(6), 529–539 (2010)
16. Wang, Y.M., Yang, J.B., Xu, D.L.: A preference aggregation method through the estimation of utility intervals. Computers and Operations Research 32, 2027–2049 (2005)
17. Yager, R.R.: An approach to ordinal decision making. Int. Journal of Approximate Reasoning 12(3-4), 237–261 (1995)
18. Zadeh, L.: Fuzzy sets. Information and Control 8, 338–353 (1965)
19. Zadeh, L.: The concept of a linguistic variable and its applications to approximate reasoning. Information Sciences, Part I, II, III (8,9), 43–80, 199–249, 301–357 (1975)

Construction of Interval-Valued Fuzzy Preference Relations Using Ignorance Functions: Interval-Valued Non Dominance Criterion

Edurne Barrenechea, Alberto Fernández, Francisco Herrera, and Humberto Bustince

Abstract. In this work we present a construction method for interval-valued fuzzy preference relations from a fuzzy preference relation and the representation of the lack of knowledge or ignorance that experts suffer when they define the membership values of the elements of that fuzzy preference relation. We also prove that, with this construction method, we obtain membership intervals for an element which length is equal to the ignorance associated with that element. We then propose a generalization of Orlovsky's non dominance method to solve decision making problems using interval-valued fuzzy preference relations.

1 Introduction

Fuzzy preference relations have been widely used to model preferences for decision making problems due to their high expressiveness and their effectiveness as a tool for modeling decision processes. In this case, the experts express their opinions using a difference scale $[0,1]$ (see [3, 8, 11, 14]). In some cases interval-valued fuzzy preference relations are used instead of fuzzy preference relations when the experts have a difficulty in expressing their preferences with exact numerical values

Edurne Barrenechea · Humberto Bustince
Dept. Automática y Computación, Universidad Pública de Navarra
e-mail: {edurne.barrenechea,bustince}@unavarra.es

Alberto Fernández
Dept. of Computer Science, University of Jaén
e-mail: alberto.fernandez@ujaen.es

Francisco Herrera
Dept. of Computer Science and Artificial Intelligence, Universidad de Granada
e-mail: herrera@decsai.ugr.es

B. De Baets et al. (Eds.): Eurofuse 2011, AISC 107, pp. 243–255, 2011.
springerlink.com

depending on the knowledge they have about the possible alternatives [2, 9, 22, 23]. Sometimes, experts suffer from a great lack of knowledge about the environment where the fuzzy decision making method is going to be applied.

Once the fuzzy preference relation (FPR) for a decision making problem is known, the goal is to improve using interval-valued fuzzy sets [19, 27] the solution that is obtained with common fuzzy methods, as Orlovsky's non dominance method [14].

To achieve this goal we will measure the ignorance (lack of knowledge) of the expert when providing the membership values of the elements of the FPR. We will do this using ignorance functions [4]. So each element will be associated with two values: the first value is given by the expert, and corresponds to the degree of membership of the element to the original FPR; the second value is calculated with the ignorance function and represents the lack of knowledge of the expert in the assignation of the first value. From these two values we will build an interval-valued fuzzy preference relation ($IVFPR$).

For the new $IVFPR$ we introduce a generalization of the non dominance method, which allows to recover the classical algorithm solution.

This work is organized as follows: In the next section we introduce the basic necessary concepts. In Section 3 we introduce the relationship between the concept of strict fuzzy preference relation given by Fodor and Roubens and the one given by Orlovsky. In Section 4 we consider a construction method of interval-valued fuzzy preference relations from fuzzy preference relations and ignorance functions. In Section 5, we propose a generalization of the non-dominance criterion proposed by Orlovsky to solve decision making problems. We finish with some conclusions and future lines of research.

2 Preliminaries

In fuzzy set theory, we know that a function $\mathbf{N}:[0,1]\to[0,1]$, with $\mathbf{N}(0)=1, \mathbf{N}(1)=0$ that is strictly decreasing and continuous, is called strict negation. If $\mathbf{N}$ is also involutive, then it is a strong negation.

Definition 1 *[26]. A fuzzy set $\tilde{A}$ on a finite universe U is a mapping $U\to[0,1]$.*

We will denote by $FS(U)$ the set of all the fuzzy sets on U.

Let us denote by $L([0,1])$ the set of all closed subintervals in $[0,1]$, that is,

$$L([0,1])=\{\mathbf{x}=[\underline{x},\overline{x}]|(\underline{x},\overline{x})\in[0,1]^2 \text{ and } \underline{x}\leq\overline{x}\}.$$

We also denote $0_L=[0,0]$, $1_L=[1,1]$ and the length of $\mathbf{x}\in L([0,1])$ as $W(\mathbf{x})=\overline{x}-\underline{x}$.

Definition 2 *[27]. An interval-valued fuzzy set A on a universe U is a mapping $A:U\to L([0,1])$.*

Note that the membership of each element $u_i \in U$ is given by $A(u_i) = [\underline{A}(u_i), \overline{A}(u_i)]$. We will denote by $IVFS(U)$ the set of all interval-valued fuzzy sets on U.

An interval-valued negation (IV negation) is a function $N_{IV} : L([0,1]) \to L([0,1])$ that is decreasing (with respect to the order: $\mathbf{x} \leq_L \mathbf{y}$ if and only $\underline{x} \leq \underline{y}$ and $\overline{x} \leq \overline{y}$) and with $N_{IV}(1_L) = 0_L$ and $N_{IV}(0_L) = 1_L$. If for all $\mathbf{x} \in L([0,1])$, $N_{IV}(N_{IV}(\mathbf{x})) = \mathbf{x}$, N_{IV} is said to be involutive.

Theorem 1 *[5]. A function $N_{IV} : L([0,1]) \to L([0,1])$ is an involutive IV negation if and only if there exists an involutive negation $\mathbf{N}$ such that*

$$N_{IV}(\mathbf{x}) = [\mathbf{N}(\overline{x}), \mathbf{N}(\underline{x})].$$

Throughout this paper we use involutive IV negations, N_{IV}, generated by the standard negation $\mathbf{N}(x) = 1 - x$ for all $x \in [0,1]$ in such a way that $N_{IV}(\mathbf{x}) = [\mathbf{N}(\overline{x}), \mathbf{N}(\underline{x})] = [1 - \overline{x}, 1 - \underline{x}]$.

A triangular norm (t-norm for short) $T : [0,1]^2 \to [0,1]$ is an associative, commutative, non-decreasing function such that $T(1,x) = x$ for all $x \in [0,1]$. A t-norm T is called *idempotent* if, $T(x,x) = x$ for all $x \in [0,1]$.

Three basic t-norms are the following: the minimum $T_M(x,y) = \min(x,y)$, the product $T_P(x,y) = x \cdot y$ and the Łukasiewicz $T_Ł(x,y) = \max(x + y - 1, 0)$.

In this paper we will also use the following relationship on $L([0,1])$ given in [24]: let $\mathbf{x}, \mathbf{y} \in L([0,1])$ and let $s(\mathbf{x}) = \underline{x} + \overline{x} - 1$ and $s(\mathbf{y}) = \underline{y} + \overline{y} - 1$ be the *scores* of $\mathbf{x}$ and $\mathbf{y}$ respectively. Let $h(\mathbf{x}) = 1 - (\overline{x} - \underline{x})$ and $h(\mathbf{y}) = 1 - (\overline{y} - \underline{y})$ be the *accuracy degrees* of $\mathbf{x}$ and $\mathbf{y}$ respectively. Then

1. If $s(\mathbf{x}) < s(\mathbf{y})$, then $\mathbf{x} < \mathbf{y}$;
2. If $s(\mathbf{x}) = s(\mathbf{y})$, then
   2.1 If $h(\mathbf{x}) = h(\mathbf{y})$, then $\mathbf{x} = \mathbf{y}$;
   2.2 if $h(\mathbf{x}) < h(\mathbf{y})$, then $\mathbf{x} < \mathbf{y}$.

The relation between the score function s and the accuracy function h is similar to the relation between the mean and the variance in statistics. Observe that any two intervals are comparable with this order relation. Moreover, it follows easily that 0_L is the smallest element in $L([0,1])$ and 1_L is the largest.

3 Fuzzy Binary Preference Relations and Interval-Valued Fuzzy Binary Preference Relations

First, we recall the concept of strict fuzzy binary preference relation given by Fodor and Roubens [7] and relate it with the definition given, for the same concept, by Orlovsky in [14]. Later, we recall the definition of interval-valued fuzzy binary preference relations and the reciprocity property.

3.1 Strict Fuzzy Binary Preference Relations

Let $R \in FR(X \times X)$ be a fuzzy preference relation over a set of alternatives $X = \{x_1, \ldots, x_n\}$ (see [1, 7, 9, 15]); for each pair of alternatives x_i and x_j, $R_{ij} = R(x_i, x_j)$ represents a degree of (weak) preference of x_i over x_j, namely the degree to which x_i is considered as least as good as x_j.

From a weak preference relation R, Fodor and Roubens [7] (see also [1, 18]) derive the following relation:

A *Strict preference* $P_{ij} = P(x_i, x_j)$ is a measure of strict preference of x_i over x_j, indicating that x_i is (weakly) preferred to x_j but x_j is not (weakly) preferred to x_i.

More specifically, Fodor and Roubens propose to express the above relation in terms of a t-norm T and a strict negation $\mathbf{N}$:

$$P_{ij} = T(R_{ij}, \mathbf{N}(R_{ji})) \text{ for all } i, j \in \{1, \ldots, n\}; \tag{1}$$

Fuzzy preference structures have been studied deeply as their axiomatic construction [7, 1, 14, 16, 17].

A fuzzy preference relation R satisfies the property of *reciprocity* if $R_{ij} + R_{ji} = 1$ for all $i, j \in \{1, \cdots, n\}$. In reciprocal preference relations is usual not to define the elements in the diagonal [11]. In a future, we will deepen about the fact of reciprocal relations avoiding the diagonal values how can affect to our models [12, 13].

Proposition 1. *Let R be a reciprocal fuzzy preference relation and $\mathbf{N}(x) = 1 - x$ for all $x \in [0, 1]$. Then,*

$$P_{ij} = R_{ij} \textit{ if and only if } T = T_M$$

for all $R_{ij} \in R$.

Orlovsky in [14] gives the following definition of *strict* fuzzy preference relation $R \in FR(X \times X)$:

$$R^s_{ij} = \begin{cases} R_{ij} - R_{ji} & \text{if } R_{ij} > R_{ji} \\ 0 & \text{otherwise} \end{cases}. \tag{2}$$

Next, we present the relationship between *strict* fuzzy preference relation given by Fodor and Roubens [7] P_{ij} and the one given by given by Orlovsky [14] R^s_{ij}.

Lemma 1. *If $T = T_{\L}$ and $\mathbf{N}(x) = 1 - x$ for all $x \in [0, 1]$, then*

$$P_{ij} = R^s_{ij}, \quad \textit{for all } i, j \in \{1, \ldots, n\}.$$

3.2 Interval-Valued Reciprocal Preference Relations

A first approach to add some flexibility to the uncertainty representation problem is by means of interval-valued fuzzy relations. An Interval-valued fuzzy binary relation $\mathbf{r}$ on X is defined as an interval-valued fuzzy subset of $X \times X$; that is,

$\mathbf{r}: X \times X \to L([0,1])$. The interval $\mathbf{r}(x_i,x_j) = \mathbf{r}_{ij}$ denotes the degree to which elements x_i and x_j are related in the relation $\mathbf{r}$ for all $x_i, x_j \in X$. By $IVFR(X \times X)$ we denote the set of all interval-valued fuzzy relations on $X \times X$.

Definition 3. *Let* $\mathbf{r} \in IVFR(X \times X)$. *We say that* $\mathbf{r}$ *satisfies the reciprocity property if for all* $\mathbf{r}_{ij}, \mathbf{r}_{ji} \in \mathbf{r}$ *the following identities hold:*

$$\begin{aligned} \underline{r}_{ij} + \overline{r}_{ji} &= 1 \\ \underline{r}_{ji} + \overline{r}_{ij} &= 1 \end{aligned} \tag{3}$$

In this work we use interval-valued fuzzy preference relations that satisfy the reciprocity property and such that the elements in their main diagonal are not defined.

4 Construction of Interval-Valued Fuzzy Preference Relations from Fuzzy Preference Relations and Weak Ignorance Functions

The goal of this section is to build interval-valued fuzzy preference relations arising from a fuzzy preference relation. For this purpose, we use the concept of weak ignorance function and a new construction method of intervals.

4.1 Weak Ignorance

The concept of ignorance functions is defined in order to quantify the lack of knowledge of an expert when he or she assigns a numerical value to the membership of an object to a given class and another numerical value for the membership of the same element to a different class.

Definition 4 *[4]. An ignorance function is a continuous mapping* $G_i : [0,1]^2 \to [0,1]$ *such that:*

(G_i1) $G_i(x,y) = G_i(y,x)$ *for all* $x,y \in [0,1]$;
(G_i2) $G_i(x,y) = 0$ *if and only if* $x = 1$ *or* $y = 1$;
(G_i3) *If* $x = 0.5$ *and* $y = 0.5$, *then* $G_i(x,y) = 1$;
(G_i4) G_i *is decreasing in* $[0.5,1]^2$;
(G_i5) G_i *is increasing in* $[0,0.5]^2$.

Observe that this definition implies that we have assumed that a value of 0.5 corresponds to complete lack of knowledge of the expert on the membership of an element to a class.

In order to build the interval-valued fuzzy sets the authors define in [20] a new function called weak ignorance for modeling the uncertainty associated with the definition of the membership functions. From this new concept, they represent the linguistic labels using by means of interval-valued fuzzy sets and present a natural extension of both the Fuzzy Reasoning Method (FRM) and the computation of the rule weight.

Proposition 2 *[20]. Let $G_i : [0,1]^2 \rightarrow [0,1]$ be an ignorance function. The mapping:*

$$\begin{aligned} &g : [0,1] \rightarrow [0,1] \textit{ given by} \\ &g(x) = G_i(x, 1-x) \end{aligned} \tag{4}$$

is a continuous function that satisfies:

(i) $g(x) = g(1-x)$ *for all* $x \in [0,1]$;
(ii) $g(x) = 0$ *if and only if* $x = 0$ *or* $x = 1$;
(iii) $g(0.5) = 1$.

Definition 5 *[20]. A continuous mapping $g : [0,1] \rightarrow [0,1]$ is called weak ignorance function if it satisfies the items $(i)-(iii)$ in Proposition 2.*

The name is due to the fact that they are only associated with an element, in the sense that they depend on a single variable, and not on two. We understand weak ignorance functions as a quantification of the lack of knowledge an expert suffers from when assigning a numerical value to the membership of an object to a given class (set).

Example 1. The function $g(x) = 2 \cdot \min(x, 1-x)$ for all $x \in [0,1]$, is a weak ignorance function.

4.2 Construction of Interval-Valued Fuzzy Preference Relations

One of the main goals of this work is to build an $IVFR$ arising from a FR, in such a way that for each element, the length of the interval that represents the membership to the new relation, is equal to the weak ignorance associated with the membership degree of the same element to the original fuzzy relation.

Proposition 3. *Let $R \in FR(X \times X)$. In the setting of Proposition 2 the following items hold:*

1. *The relation* $\mathbf{r}$ *given by*

$$\mathbf{r}_{ij} = \begin{cases} [(R^s_{ij} \cdot (1 - g(R_{ij})), R^s_{ij} \cdot (1 - g(R_{ij})) + g(R_{ij})] & \textit{if } R_{ij} > R_{ji} \\ [0, g(R_{ij})] & \textit{otherwise} \end{cases} \tag{5}$$

   for all i, j is an interval-valued fuzzy relation on $X \times X$;
2. $W(\mathbf{r}_{ij}) = g(R_{ij})$ *for all* $R_{ij} \in R$;
3. *If R satisfies the reciprocity property, then the interval-valued fuzzy preference relation* $\mathbf{r}$ *given by item* 1. *also satisfies it (in the sense of Eq. (3));*

5 An Approach to Multi-criteria Decision Making with Interval-Valued Fuzzy Preference Relations

In this section we propose a generalization of the non-dominance criterion proposed by Orlovsky. We always consider normalized fuzzy preference relations to satisfy the reciprocity property. In the algorithm we will use the construction method given in Proposition 4 (item 1.) and the concept of weak ignorance function.

Given a fuzzy preference relation $R^* \in FR(X \times X)$, to normalize such relation to $[0,1]$ we use Eq. (6), in such a way that for each element of the new relation it holds that $R_{ij} = 1 - R_{ji}$.

$$R_{ij} = \begin{cases} \frac{R^*_{ij}}{R^*_{ij}+R^*_{ji}} \text{ if } R^*_{ij}+R^*_{ji} \neq 0 \\ 0 \text{ othercase} \end{cases} \tag{6}$$

From the normalized fuzzy preference relation R we must extract a set of non-dominated alternatives as the solution of the decision making problem. Specifically, the maximal non-dominated elements of R are calculated by means of the following operations, according to the non-dominance criterion proposed by Orlovsky in [14]:

Step 1. Compute the fuzzy strict preference relation R^s as indicated in Eq. (2);

Step 2. Compute the non-dominance degree of each alternative, ND_i, in the following way:

$$ND_i = 1 - \max_j \{R^s_{ji}\} \tag{7}$$

This value represents the degree to which the alternative i is dominated by onc of thc rcmaining altcrnativcs.

Step 3. Take the alternative corresponding to the index of the maximal non-domınance degree:

$$Alternative(x_p) = \arg \max_{i=1,\cdots,n} \{ND_i\} \tag{8}$$

We must point out that it could happen that there exist two or more alternatives with the same degrees of membership to the set ND. In this case, the algorithm does not choose any of those alternatives. This fact has led many authors to propose other algorithms [9, 21, 25].

The main idea of our approach is to build an interval-valued fuzzy preference relation from the strict preference relation R^s (Step 1) of the non-dominance algorithm.

Given a fuzzy preference relation R^* (without defined elements in the main diagonal) and given a weak fuzzy ignorance function g in the sense of Proposition 2, the Non-dominance Interval-valued Algorithm (NDIVA) that we propose is the following:

(NDIVA1) Construct R normalizing the fuzzy preference relation $R^* \in FR(X \times X)$ by means of Eq. (6);

(NDIVA2) Compute the fuzzy strict preference relation R^s using Eq. (2);

(NDIVA3) Build the interval-valued fuzzy relation $\mathbf{r}$ using Eq. (5):

$$\mathbf{r}_{ij} = \begin{cases} [(R^s_{ij} \cdot (1 - g(R_{ij})), R^s_{ij} \cdot (1 - g(R_{ij})) + g(R_{ij})] & \text{if } R_{ij} > R_{ji} \\ [0, g(R_{ij})] & \text{otherwise} \end{cases} \tag{9}$$

(NDIVA4) Build the interval-valued fuzzy set:

$$\begin{aligned} ND_{IV} &= \{(x_j, ND_{IV}(x_j)) | x_j \in X\} \text{ where} \\ ND_{IV}(x_j) &= \mathbf{S}(\mathbf{r}_{ij}) = [\bigvee_{i=1}^{n} (\underline{r}_{ij}), \bigvee_{i=1}^{n} (\overline{r}_{ij})] \end{aligned} \tag{10}$$

(NDIVA5) Build the interval-valued fuzzy set:

$$N_{IV}(ND_{IV}) = \{(x_j, N_{IV}(ND_{IV}(x_j))) | x_j \in X\} \text{ where} \tag{11}$$

$$N_{IV}(ND_{IV})(x_j) = [1 - \bigvee_{i=1}^{n} (\overline{r}_{ij}), 1 - \bigvee_{i=1}^{n} (\underline{r}_{ij})] \tag{12}$$

(NDIVA6) Order the elements of $N_{IV}(ND_{IV})$ in a decreasing way in terms of accuracy and score functions.

(NDIVA7) If there exist several alternatives occupying the first place, take as solution the alternative with the biggest upper bound of its interval associated.

NDIVA Algorithm

Remark

- If for a majority of the elements $\mathbf{r}_{ij}$ given by Eq. (9) we have that $g(R_{ij}) \to 0$, i.e., if $R^s_{ij} = R_{ij} - R_{ji} \to 1$,then the resulting intervals have a very small length and it is reasonable to assume that the result obtained with the algorithm (NDIVA) is the same than the result obtained with the non-dominance algorithm.

  This is due to the fact that if $R_{ij} - R_{ji} \to 1$, then R_{ij} is very large and R_{ji} is very small; that is, the expert is very sure about the preference of alternative x_i against x_j. Moreover, in this case we have that the weak ignorance is very small. So, due to our construction method with Eq. (9) the intervals have a very small length.
- If for a majority of the elements $\mathbf{r}_{ij}$ given by Eq. (9) we have that $g(R_{ij}) \to 1$; i.e., if $R_{ij} \approx R_{ji} \approx 0.5$, then the (NDIVA) algorithm allows us to distinguish better than the non-dominance algorithm the alternative or alternatives that we must take as solution.

The reason for this is that if $R_{ij} \approx R_{ji} \approx 0.5$ and $R_{ij} + R_{ji} = 1$, by Definition 4 we have that $g(R_{ij}) = g(R_{ji}) \to 1$, in such a way that the product $(R_{ij} - R_{ji})(1 - g(R_{ij}))$ goes to zero faster than $R_{ij} - R_{ji}$. So, in the (NDIVA) algorithm, for the cases for which the expert shows a great indifference for choosing one alternative or the other, we *penalize* (diminish) even more the difference of his or her preferences, ins such a way that, when we negate intervals in step (NDIVA5) we strengthen even more the worst possible cases, and the intervals obtained with Eq. (9) have a very large length.

It is necessary to notice that Eq. (10) and (12) is a generalization of *Step*2. of the non-dominance algorithm when using interval-valued preference relations.

Example 2. Let $X = \{x_1, x_2, x_3, x_4\}$ be the set of alternatives. Consider the normalized fuzzy relation

$$R^* = \begin{pmatrix} - & 0.70 & 0.65 & 0.30 \\ 0.30 & - & 0.70 & 0.60 \\ 0.35 & 0.30 & - & 0.70 \\ 0.70 & 0.40 & 0.30 & - \end{pmatrix} \tag{13}$$

For this algorithm we consider the weak ignorance function $g(x) = 2 \cdot \min(x, 1-x)$.
(NDIVA1) Construct R. In this case $R = R^*$.
(NDIVA2) Transform R to R^s.

$$R^s = \begin{pmatrix} - & 0.40 & 0.30 & 0.00 \\ 0.00 & - & 0.40 & 0.20 \\ 0.00 & 0.00 & - & 0.40 \\ 0.40 & 0.00 & 0.00 & - \end{pmatrix} \tag{14}$$

(NDIVA3) Build the interval-valued fuzzy relation $\mathbf{r}$:

$$\mathbf{r} = \begin{pmatrix} - & [0.16,0.76] & [0.09,0.79] & [0.00,0.60] \\ [0.00,0.60] & - & [0.16,0.76] & [0.04,0.84] \\ [0.00,0.70] & [0.00,0.60] & - & [0.16,0.76] \\ [0.16,0.76] & [0.00,0.80] & [0.00,0.60] & - \end{pmatrix} \tag{15}$$

(NDIVA4) Build the interval-valued fuzzy set ND_{IV}:
$ND_{IV} = \{(x_1, [0.16, 0.76]), (x_2, [0.16, 0.80]), (x_3, [0.16, 0.79]), (x_4, [0.16, 0.84])\}$
(NDIVA5) Apply N_{IV} to the interval-valued fuzzy sets ND_{IV}:
$N_{IV}(ND_{IV}) =$
$\{(x_1, [0.24, 0.84]), (x_2, [0.20, 0.84]), (x_3, [0.21, 0.84]), (x_4, [0.16, 0.84])\}$
(NDIVA6) Order alternatives in a non-increasing way using the order relationship defined in terms of the *score and accuracy functions*.

$$x_1 \geq x_3 \geq x_2 \geq x_4.$$

5.1 Relationship between NDIVA Algorithm and Orlovsky's Algorithm

Next, we present a new operator P to associate a fuzzy set with each interval-valued fuzzy set satisfying a specific set of properties. If we apply this operator on the non dominance interval-valued fuzzy algorithm (Step $NDIVA6$) we recover the results given by Orlovsky's algorithm.

Definition 6. *A P operator is a mapping $L([0,1]) \to [0,1]$ given by:*

$$P(\mathbf{x}) = \begin{cases} \frac{\underline{x}}{1-W(\mathbf{X})} & \text{if } W(\mathbf{x}) \neq 1 \\ 0 & \text{if } W(\mathbf{x}) = 1 \end{cases} \tag{16}$$

Proposition 4. *Let $R \in FR(X \times X)$ and let $\mathbf{r} \in IVFR(X \times X)$ given by Eq. (5). In the setting of Proposition 2 the following items hold:*

1. $P(\mathbf{r}_{ij}) = R_{ij}$ for all $R_{ij} \in R$ and for all $g(R_{ij}) \neq 1$.
2. If $g(R_{ij}) = 1$ then, $P(\mathbf{r}_{ij}) = 0$.

Proposition 5. *Let P given in Eq. (16). The following properties hold:*

1. $\underline{x} \leq P(\mathbf{x}) \leq \overline{x}$ for all $\mathbf{x} \in L([0,1])$;
2. $P([x,x]) = x$ for all $x \in [0,1]$;
3. If $W(\mathbf{x}) \neq 1$, then $P(\mathbf{x}) + P(1-\mathbf{x}) = 1$;
4. If $W(\mathbf{x}) = 1$, then $P(\mathbf{x}) + P(1-\mathbf{x}) = 0$.

Remark. Notice that in step (NDIVA5) of Example 2 all the elements have membership intervals with the same upper bound (see item (2) of Proposition 7). So, if for relation R given by Eq. (13) we apply the non-dominance algorithm, all of the alternatives dominate with the same numerical value and we are not able of choosing the best one.

Proposition 6. *Let $A_{IV} \in IVFS(U)$ and let P the operator introduced in Definition 6. Then*

$$A = \{(u_i, P(A_{IV}(u_i)))|u_i \in U\} \tag{17}$$

is a fuzzy set on U.

Proposition 7. *The following items hold:*

1. Let P be the operator given in Definition 6. If in the (NDIVA) algorithm we replace (NDIVA4) by:
(NDFS4) Build the set

$$\begin{aligned} ND_{IVS} &= \{(x_j, ND_{IVS}(x_j))|x_j \in X\} \text{ where} \\ ND_{IVS}(x_j) &= [\bigvee_{i=1}^{n} P(\mathbf{r}_{ij}), \bigvee_{i=1}^{n} P(\mathbf{r}_{ij})], \end{aligned} \tag{18}$$

then we recover the non-dominance algorithm.

2. *If in step (NDIVA6) we reorder the elements of the set $N_{IV}(ND_{IV})$ in a non-increasing way with respect to the upper bounds of the intervals, then the alternative(s) which are solution(s) for this algorithm are the same than those of the non-dominance algorithm.*

Remark. Notice that in step (NDIVA5) of Example 2 all the elements have membership intervals with the same upper bound (see item (2) of Proposition 7). So, if for relation R given by Eq. (13) we apply the non-dominance algorithm, all of the alternatives dominate with the same numerical value and we are not able of choosing the best one.

6 Conclusions

In this paper we have presented an algorithm for decision making problems starting from a fuzzy preference relation. We use weak ignorance functions (in the sense of Proposition 2) to penalize indifference situations. That is, situations in which the preference of one alternative against the other is close to 0.5. We also represent the preference degree of a relation by means of intervals such that their lengths is equal to the weak ignorance of the expert when he or she assigns a specific value.

Finally, we define a new operator that allows us to associate each interval-valued fuzzy set to a fuzzy set. The analysis of this operator has allowed us to settle minimal conditions under which our first algorithm recovers the classical non dominance algorithm.

In future works we will apply NDIVA algorithm to classification problems as in [6, 10] and we will analyze the results in contrast with other approaches. We also consider necessary to study a construction method to generalize the one presented in this work. From this generalization we also propose new algorithms taking into account different criteria for selection of alternatives and different order relations between intervals.

Acknowledgements. This work has been partially supported by research grant TIN2010-15055 from the Government of Spain.

References

1. De Baets, B., Van de Walle, B., Kerre, E.: Fuzzy preference structures without incomparability. Fuzzy Sets and Systems 76 (3), 333–348 (1995)
2. Bilgiç, T.: Interval-valued preference structures. European Journal of Operational Research 105 (1), 162–183 (1998)
3. Chiclana, F., Herrera, F., Herrera-Viedma, E.: Integrating three representation models in fuzzy multipurpose decision making based on fuzzy preference relations. Fuzzy Sets and Systems 97, 33–48 (1998)

4. Bustince, H., Pagola, M., Barrenechea, E., Fernandez, J., Melo-Pinto, P., Couto, P., Tizhoosh, H.R., Montero, J.: Ignorance functions. An application to the calculation of the threshold in prostate ultrasound images. Fuzzy Sets and Systems 161 (1), 20–36 (2010)
5. Deschrijver, G., Cornelis, C., Kerre, E.E.: On the representation of intuitionistic fuzzy T-norms and T-conorms. IEEE Transactions on Fuzzy Systems 12(1), 45–61 (2004)
6. Fernández, A., Calderón, M., Barrenechea, E., Bustince, H., Herrera, F.: Solving multiclass problems with linguistic fuzzy rule based classification systems based on pairwise learning and preference relations. Fuzzy Sets and Systems 161(23), 3064–3080 (2010)
7. Fodor, J., Roubens, M.: Fuzzy Preference Modelling and Multicriteria Decision Support. Kluwer Academic Publishers, Dordrecht (1994)
8. González-Pachón, J., Gómez, D., Montero, J., Yáñez, J.: Searching for the dimension of valued preference relations. International Journal of Approximate Reasoning 33 (2), 133–157 (2003)
9. Herrera, F., Martínez, L., Sánchez, P.J.: Managing non-homogeneous information in group decision making. European Journal of Operational Research 166 (1), 115–132 (2005)
10. Hüllermeier, E., Brinker, K.: Learning valued preference structures for solving classification problems. Fuzzy Sets and Systems 159(18), 2337–2352 (2008)
11. Kacprzyk, J.: Group decision making with a fuzzy linguistic majority. Fuzzy Sets and Systems 18(2), 105–118 (1986)
12. Montero, F.J., Tejada, J.: Some problems on the definition of fuzzy preference relations. Fuzzy Sets and Systems 20, 45–53 (1986)
13. Montero, F.J., Tejada, J.: A necessary and sufficient condition for existence of Orlovsky's choice set. Fuzzy Sets and Systems 26, 121–125 (1988)
14. Orlovsky, S.A.: Decision-making with a fuzzy preference relation. Fuzzy Sets and Systems 1 (3), 155–167 (1978)
15. Ovchinnikov, S.V., Roubens, M.: On Strict Preference Relations. Fuzzy Sets and Systems 43 (3), 319–326 (1991)
16. Ovchinnikov, S.V., Ozernoy, V.M.: Using fuzzy binary relations for identifying noninferior decision alternatives. Fuzzy Sets and Systems 25 (1), 21–32 (1988)
17. Perny, P., Roy, B.: The use of fuzzy outranking relations in preference modelling. Fuzzy Sets and Systems 49, 33–53 (1992)
18. Roubens, M., Vincke, P.: Preference Modelling. Lecture Notes in Economics and Mathematical Systems, vol. 250. Springer, Berlin (1985)
19. Sambuc, J.: Function Φ-Flous, Application a l'aide au Diagnostic en Pathologie Thyroidienne. These de Doctorat en Medicine, Marseille (1975)
20. Sanz, J., Fernández, A., Bustince, H., Herrera, F.: A genetic tuning to improve the performance of fuzzy rule-based classification systems with interval-valued fuzzy sets: degree of ignorance and lateral position. International Journal of Approximate Reasoning (2011); doi:10.1016/j.ijar.2011.01.011
21. Szmidt, E., Kacprzyk, J.: Using intuitionistic fuzzy sets in group decision making. Control and Cybernetics 31, 1037–1053 (2002)
22. Szmidt, E., Kacprzyk, J.: A consensus-reaching process under intuitionistic fuzzy preference relations. International Journal of Intelligent Systems 18(7), 837–852 (2003)
23. Xu, Z.: On compatibility of interval fuzzy preference relations. Fuzzy Optimization and Decision Making 3 (3), 217–225 (2004)

24. Xu, Z., Yager, R.R.: Some geometric aggregation operators based on intuitionistic fuzzy sets. International Journal of General Systems 35, 417–433 (2006)
25. Xu, Z.: A method based on distance measure for interval-valued intuitionistic fuzzy group decision making. Information Sciences 180 (1), 181–190 (2010)
26. Zadeh, L.A.: Fuzzy sets. Information Control 8, 338–353 (1965)
27. Zadeh, L.A.: The concept of a linguistic variable and its application to approximate reasoning I. Information Sciences 8, 199–249 (1975)

Learning Valued Relations from Data

Willem Waegeman, Tapio Pahikkala, Antti Airola, Tapio Salakoski, and Bernard De Baets

Abstract. Driven by a large number of potential applications in areas like bioinformatics, information retrieval and social network analysis, the problem setting of inferring relations between pairs of data objects has recently been investigated quite intensively in the machine learning community. To this end, current approaches typically consider datasets containing crisp relations, so that standard classification methods can be adopted. However, relations between objects like similarities and preferences are in many real-world applications often expressed in a graded manner. A general kernel-based framework for learning relations from data is introduced here. It extends existing approaches because both crisp and valued relations are considered, and it unifies existing approaches because different types of valued relations can be modeled, including symmetric and reciprocal relations. This frame work establishes in this way important links between recent developments in fuzzy set theory and machine learning. Its usefulness is demonstrated on a case study in document retrieval.

1 Introduction

Relational data can be observed in many predictive modeling tasks, such as forecasting the winner in two-player computer games [1], predicting proteins that interact with other proteins in bioinformatics [2], retrieving documents that are similar to a target document in text mining [3], investigating the persons that are friends of each

Willem Waegeman · Bernard De Baets
Ghent University, KERMIT, Department of Applied Mathematics, Biometrics and Process Control, Coupure links 653, B-9000 Ghent
e-mail: {forname.surname}@ugent.be

Tapio Pahikkala · Antti Airola · Tapio Salakoski
University of Turku, Department of Information Technology and the Turku Centre for Computer Science, Joukahaisenkatu 3-5 B 20520 Turku
e-mail: {forname.surname}@utu.fi

B. De Baets et al. (Eds.): Eurofuse 2011, AISC 107, pp. 257–268, 2011.
springerlink.com

other on social network sites [4], etc. All these examples represent fields of application in which specific machine learning and data mining algorithms are successfully developed to infer relations from data; pairwise relations, to be more specific.

The typical learning scenario in such situations can be summarized as follows. Given a dataset of known relations between pairs of objects and a feature representation of these objects in terms of variables that might characterize the relations, the goal usually consists of inferring a statistical model that takes two objects as input and predicts whether the relation of interest occurs for these two objects. Moreover, since one aims to discover unknown relations, a good learning algorithm should be able to construct a predictive model that can generalize towards unseen data, i.e., pairs of objects for which at least one of the two objects was not used to construct the model. As a result of the transition from predictive models for single objects to pairs of objects, new advanced learning algorithms need to be developed, resulting in new challenges with regard to model construction, computational tractability and model assessment.

As relations between objects can be observed in many different forms, this general problem setting provides links to several subfields of machine learning, like statistical relational learning [5], graph mining [6], metric learning [7] and preference learning [8]. More specifically, from a graph-theoretic perspective, learning a relation can be formulated as learning edges in a graph where the nodes represent information of the data objects; from a metric learning perspective, the relation that we aim to learn should satisfy some well-defined properties like positive definiteness, transitivity or the triangle inequality; and from a preference learning perspective, the relation expresses a (degree of) preference in a pairwise comparison of data objects.

The topic of learning relations between objects is also closely related to recent developments in fuzzy set theory. This article will elaborate on these connections via two important contributions: (1) the extension of the typical setting of learning crisp relations to valued relations and (2) the inclusion of domain knowledge about relations into the inference process by explicit modeling of mathematical properties of these relations. For algorithmic simplicity, one can observe that many approaches only learn crisp relations, that is relations with only 0 and 1 as possible values, so that standard binary classifiers can be modified. Think in this context for example at inferring protein-protein interaction networks or metabolic networks in bioinformatics [2, 9].

However, not crisp but graded relations are observed in many real-world applications [10], resulting in a need for new algorithms that take graded relational information into account. Furthermore, the properties of valued relations have been investigated intensively in the recent fuzzy logic literature[1], and these properties are very useful to analyze and improve current algorithms. Using mathematical properties of valued relations, constraints can be imposed for incorporating domain knowledge in the learning process, to improve predictive performance or simply to

[1] Often the term fuzzy relation is used in the fuzzy set literature to refer to valued relations. However, fuzzy relations should be seen as a subclass of valued relations. For example, reciprocal relations should not be considered as fuzzy relations, because they often exhibit a probabilistic semantics rather than a fuzzy semantics.

guarantee that a relation with the right properties is learned. This is definitely the case for properties like transitivity when learning similarity relations and preference relations – see e.g. [11, 12], but even very basic properties like symmetry, antisymmetry or reciprocity already provide domain knowledge that can steer the learning process. For example, in social network analysis, the notion "person A being a friend of person B" should be considered as a symmetric relation, while the notion "person A wins from person B in a chess game" will be antisymmetric (or, equivalently, reciprocal). Nevertheless, many examples exist too where neither symmetry nor antisymmetry necessarily hold, like the notion "person A trusts person B".

In this paper we present a general kernel-based approach that unifies all the above cases into one general framework where domain knowledge can be easily specified by choosing a proper kernel and model structure, while different learning settings are distinguished by means of the loss function. From this perspective, one can make a subdivision between learning crisp relations, ordinal relations and $[0,1]$-valued relations. Furthermore, one can integrate in our framework different types of domain knowledge, by guaranteeing that certain properties are satisfied. Apart from the general case of arbitrary binary relations, we will specifically emphasize the prediction of reciprocal and symmetric relations.

2 General Framework

The framework that we propose strongly relies on graphs, where nodes represent the data objects that are studied and the edges represent the relations present in the training set. The weights on the edges characterize the values of known relations, while unconnected nodes indicate pairs of objects for which the unknown relation needs to be predicted.

Let us start with introducing some notations. We assume that the data is structured as a graph $G = (\mathscr{V},\mathscr{E},Q)$, where $\mathscr{V}$ corresponds to the set of nodes v and $\mathscr{E} \subseteq \mathscr{V}^2$ represents the set of edges e, for which training labels are provided in terms of relations. Moreover, these relations are represented by training weights y_e on the edges, generated from an unknown underlying relation $Q : \mathscr{V}^2 \rightarrow [0,1]$. Relations are required to take values in the interval $[0,1]$ because some properties that we need are historically defined for such relations, but an extension to real-valued relations $h : \mathscr{V}^2 \rightarrow \mathbb{R}$ can always be realized with a simple increasing mapping $\sigma : \mathbb{R} \rightarrow [0,1]$ such that

$$Q(v,v') = \sigma(h(v,v')) , \quad \forall (v,v') \in \mathscr{V}^2 . \tag{1}$$

Following the standard notations for kernel methods, we formulate our learning problem as the selection of a suitable function $h \in \mathscr{H}$, with $\mathscr{H}$ a certain hypothesis space, in particular a reproducing kernel Hilbert space (RKHS). More specifically, the RKHS supports in our case hypotheses $h : \mathscr{V}^2 \rightarrow \mathbb{R}$ denoted as

$$h(e) = \mathbf{w}^T \Phi(e) ,$$

with $\mathbf{w}$ a vector of parameters that needs to be estimated from training data, Φ a joint feature mapping for edges in the graph (see below) and $\mathbf{a}^T$ the transpose of a vector $\mathbf{a}$. Let us denote a training dataset of cardinality $q = |\mathscr{E}|$ as a set $T = \{(e, y_e) \mid e \in \mathscr{E}\}$ of input-label pairs, then we formally consider the following optimization problem, in which we select an appropriate hypothesis h from $\mathscr{H}$ for training data T:

$$\hat{h} = \text{argmin}_{h \in \mathscr{H}} \frac{1}{q} \sum_{e \in \mathscr{E}} \mathscr{L}(h(e), y_e) + \lambda \|h\|^2_{\mathscr{H}} \tag{2}$$

with $\mathscr{L}$ a given loss function, $\|\cdot\|^2_{\mathscr{H}}$ the traditional quadratic regularizer on the RKHS and $\lambda > 0$ a regularization parameter. According to the representer theorem [13], any minimizer $h \in \mathscr{H}$ of (2) admits a dual representation of the following form:

$$h(\bar{e}) = \mathbf{w}^T \Phi(\bar{e}) = \sum_{e \in \mathscr{E}} a_e K^{\Phi}(e, \bar{e}), \tag{3}$$

with $a_e \in \mathbb{R}$ dual parameters, K^{Φ} the kernel function associated with the RKHS and Φ the feature mapping corresponding to K^{Φ} and

$$\mathbf{w} = \sum_{e \in \mathscr{E}} a_e \Phi(e).$$

We will alternate several times between the primal and dual representation for h in the remainder of this article.

The primal representation as defined in (2) and its dual equivalent (3) yield an RKHS defined on edges in the graph. In addition, we will establish an RKHS defined on nodes, as every edge consists of a couple of nodes. Given an input space $\mathscr{V}$ and a kernel $K : \mathscr{V} \times \mathscr{V} \to \mathbb{R}$, the RKHS associated with K can be considered as the completion of

$$\left\{ f \in \mathbb{R}^{\mathscr{V}} \;\middle|\; f(v) = \sum_{i=1}^{m} \beta_i K(v, v_i) \right\},$$

in the norm

$$\|f\|_K = \sqrt{\sum_{i,j} \beta_i \beta_j K(v_i, v_j)},$$

where $\beta_i \in \mathbb{R}, m \in \mathbb{N}, v_i \in \mathscr{V}$.

As mentioned in the introduction, both crisp and valued relations can be handled by our framework. To make a subdivision between different cases, a loss function needs to be specified. For crisp relations, one can typically use the hinge loss, which is given by:

$$\mathscr{L}(h(e), y) = [1 - yh(e)]_+,$$

with $[\cdot]_+$ the positive part of the argument. Alternatively, one can opt to optimize a probabilistic loss function like the logistic loss:

$$\mathscr{L}(h(e),y) = \ln(1+\exp(-yh(e)))\,.$$

Conversely, if the observed relations in a given application are valued instead of crisp, other loss functions have to be considered. Further below, we will run experiments with a least-squares loss function:

$$\mathscr{L}(h(e),y) = (y_e - h(e))^2\,, \tag{4}$$

resulting in a regression type of learning setting. Alternatively, one could prefer to optimize a more robust regression loss like the ε-insensitive loss, in case outliers are expected in the training dataset.

So far, our framework does not differ from standard classification and regression algorithms. However, the specification of a more precise model structure for (2) offers a couple of new challenges. In the most general case, when no further restrictions on the underlying relation can be specified, the following Kronecker product feature mapping is proposed to express pairwise interactions between features of nodes:

$$\Phi(e) = \Phi(v,v') = \phi(v) \otimes \phi(v')\,,$$

where ϕ represents the feature mapping for individual nodes. Remark that in general the Kronecker product of two matrices $\mathbf{M}$ and $\mathbf{N}$ is defined as

$$\mathbf{M} \otimes \mathbf{N} = \begin{pmatrix} \mathbf{M}_{1,1}\mathbf{N} & \cdots & \mathbf{M}_{1,n}\mathbf{N} \\ \vdots & \ddots & \vdots \\ \mathbf{M}_{m,1}\mathbf{N} & \cdots & \mathbf{M}_{m,n}\mathbf{N} \end{pmatrix}.$$

As first shown in [14], the Kronecker product pairwise feature mapping yields the Kronecker product edge kernel (a.k.a. the tensor product pairwise kernel) in the dual representation:

$$K_{\otimes}^{\Phi}(e,\overline{e}) = K_{\otimes}^{\Phi}(v,v',\overline{v},\overline{v}') = K^{\phi}(v,\overline{v})K^{\phi}(v',\overline{v}')\,, \tag{5}$$

with K^{ϕ} the kernel corresponding to ϕ. With an appropriate choice for K^{ϕ}, such as the Gaussian RBF kernel, the kernel K^{Φ} generates a class $\mathscr{H}$ of universally approximating functions for learning any type of relation (formal proof omitted).

3 Special Relations

If no further information is available about the relation that underlies the data, one should definitely use the Kronecker product edge kernel. In this most general case, we allow that for any pair of nodes in the graph several edges can exist, in which an edge in one direction not necessarily imposes constraints on the edge in the opposite direction and multiple edges in the same direction can connect two nodes. This construction is required to allow repeated measurements. However, two important subclasses of relations deserve further attention: reciprocal relations and symmetric relations. Let us start with the former.

Definition 1. *A binary relation* $Q : \mathscr{V}^2 \to [0,1]$ *is called a reciprocal relation if for all* $(v,v') \in \mathscr{V}^2$ *it holds that* $Q(v,v') = 1 - Q(v',v)$.

Given the increasing transformation (1), every reciprocal relation $Q : \mathscr{V}^2 \to [0,1]$ can be rewritten as an antisymmetric relation $h : \mathscr{V}^2 \to \mathbb{R}$, formally defined as follows.

Definition 2. *A binary relation* $h : \mathscr{V}^2 \to \mathbb{R}$ *is called an antisymmetric relation if for all* $(v,v') \in \mathscr{V}^2$ *it holds that* $h(v,v') = -h(v',v)$.

For reciprocal and antisymmetric relations, every edge $e = (v,v')$ induces in the multi-graph that was defined above an unobserved invisible edge $e_R = (v',v)$ with appropriate weight in the opposite direction. Applications arise here in domains such as preference learning, game theory and bioinformatics for representing preference relations, choice probabilities, winning probabilities, gene regulation, etc. The weight on the edge defines the real direction of such an edge. If the weight on the edge $e = (v,v')$ is higher than 0.5, then the direction is from v to v', but when the weight is lower than 0.5, then the direction should be interpreted as inverted. If the relation is 3-valued as $Q : \mathscr{V}^2 \to \{0,1/2,1\}$, then we end up with a three-class ordinal regression setting instead of an ordinary regression setting. Interestingly, reciprocity can be easily incorporated in our framework.

Proposition 1. *Let* Ψ *be a feature mapping on* $\mathscr{V}^2$*, let* $\sigma : \mathbb{R} \to [0,1]$ *be an increasing mapping and let h be a hypothesis defined by (2), then the relation Q of type (1) is reciprocal if* Φ *is given by*

$$\Phi_R(e) = \Phi_R(v,v') = \Psi(v,v') - \Psi(v',v),$$

while σ *satisfies* $\sigma(1/2) = 0$ *and* $\sigma(x) = 1 - \sigma(-x)$ *for all* $x \in \mathbb{R}$.

The proof is immediate. In addition, one can easily show that reciprocity as domain knowledge can be enforced in the dual formulation. Let us in the least restrictive form now consider the Kronecker product for Ψ, then one obtains for Φ_R the kernel $K^{\Phi}_{\otimes R}$ given by

$$K^{\Phi}_{\otimes R}(e,\overline{e}) = 2\big(K^{\phi}(v,\overline{v})K^{\phi}(v',\overline{v}') - K^{\phi}(v,\overline{v}')K^{\phi}(v',\overline{v})\big). \tag{6}$$

Similar to the general case, one can show that this kernel can represent any type of reciprocal relation by means of universal approximation.

Symmetric relations form another important subclass of relations in our framework. As a specific type of symmetric relations, similarity relations constitute the underlying relation in many application domains where relations between objects need to be learned. Symmetric relations are formally defined as follows.

Definition 3. *A binary relation* $Q : \mathscr{V}^2 \rightarrow [0,1]$ *is called a symmetric relation if for all* $(v,v') \in \mathscr{V}^2$ *it holds that* $Q(v,v') = Q(v',v)$.

Definition 4. *A binary relation* $h : \mathscr{V}^2 \rightarrow \mathbb{R}$ *is called a symmetric relation if for all* $(v,v') \in \mathscr{V}^2$ *it holds that* $h(v,v') = h(v',v)$.

For symmetric relations, edges in the multi-graph introduced above become undirected. Applications arise in many domains and metric learning or learning similarity measures can be seen as special cases. If the relation is 2-valued as $Q : \mathscr{V}^2 \rightarrow \{0,1\}$, then we end up with a classification setting instead of a regression setting. Just like reciprocal relations, it turns out that symmetry can be easily incorporated in our framework.

Proposition 2. *Let* Ψ *be a feature mapping on* $\mathscr{V}^2$, *let* $\sigma : \mathbb{R} \rightarrow [0,1]$ *be an increasing mapping and let* h *be a hypothesis defined by (2), then the relation* Q *of type (1) is symmetric if* Φ *is given by*

$$\Phi_S(e) = \Phi_S(v,v') = \Psi(v,v') + \Psi(v',v) \,.$$

In addition, by using the mathematical properties of the Kronecker product, one obtains in the dual formulation an edge kernel that looks very similar to the one derived for reciprocal relations. Let us again consider the Kronecker product for Ψ, then one obtains for Φ_S the kernel $K^{\Phi}_{\otimes S}$ given by $K^{\Phi}_{\otimes S}(e,\overline{e}) =$

$$2\big(K^{\phi}(v,\overline{v})K^{\phi}(v',\overline{v}') + K^{\phi}(v,\overline{v}')K^{\phi}(v',\overline{v})\big) \,.$$

Thus, the substraction of kernels in the reciprocal case becomes an addition of kernels in the symmetric case. The above kernel has been used for predicting protein-protein interactions in bioinformatics [14]. Unlike many existing kernel-based methods for pairwise data, the models obtained with these kernels are able to represent any reciprocal or symmetric relation respectively, without imposing additional transitivity properties of the relations.

We also remark that for symmetry as well, one can prove that the Kronecker product edge kernel yields a model that is flexible enough to represent any type of underlying symmetric relation.

4 Relationships with Fuzzy Set Theory

The previous section revealed that specific Kronecker product edge kernels can be constructed for modeling reciprocal and symmetric relations, without requiring any further background about these relations. In this section we demonstrate that the Kronecker product edge kernels $K^{\Phi}_{\otimes}$, $K^{\Phi}_{\otimes R}$ and $K^{\Phi}_{\otimes S}$ are particularly useful for modeling intransitive relations, which occur in a lot of real-world scenarios, like game playing [15, 16], competition between bacteria [17, 18, 19, 20, 21, 22] and fungi [23], mating choice of lizards [24] and food choice of birds [25], to name just a few.

Despite the occurrence of intransitive relations in many domains, one has to admit that most applications are still characterized by relations that fulfill relatively strong transitivity requirements. For example, in decision making, preference modeling and social choice theory, one can argue that reciprocal relations like choice probabilities and preference judgments should satisfy certain transitivity properties, if they represent rational human decisions made after well-reasoned comparisons on objects [26, 27, 28]. For symmetric relations as well, transitivity plays an important role [29, 30], when modeling similarity relations, metrics, kernels, etc.

It is for this reason that transitivity properties have been studied extensively in fuzzy set theory and related fields. For reciprocal relations, one can distinguish the notions of stochastic transitivity [26], FG-transitivity [31] and the more general recent framework of cycle transitivity [32, 12]. For valued symmetric relations, the notion of T-transitivity has been put forward [33, 34]. In addition, several authors have shown that various forms of transitivity give rise to utility representable or numerically representable relations, also called fuzzy weak orders – see e.g. [26, 35, 36, 37, 38]. We will use the term ranking representability to establish a link with machine learning. We give a slightly specific definition that unifies reciprocal and symmetric relations.

Definition 5. *A reciprocal or symmetric relation $Q : \mathscr{V}^2 \rightarrow [0,1]$ is called ranking representable if there exists a ranking function $f : \mathscr{V} \rightarrow \mathbb{R}$ and an increasing mapping $\sigma : \mathbb{R} \rightarrow [0,1]$ such that for all pairs $(v,v') \in \mathscr{V}^2$ it respectively holds that*

1. $Q(v,v') = \sigma(f(v) - f(v'))$ (reciprocal case) ;
2. $Q(v,v') = \sigma(f(v) + f(v'))$ (symmetric case) .

The main idea is that ranking representable relations can be constructed from a utility function f. Ranking representable reciprocal relations correspond to directed acyclic graphs, and a unique ranking of the nodes in such graphs can be obtained with topological sorting algorithms.

Interestingly, ranking representability of reciprocal relations and symmetric relations can be easily achieved in our framework by simplifying the joint feature mapping Ψ. Let $\Psi(v,v') = \phi(v)$ such that K^{Φ} simplifies to

$$\begin{aligned} K^{\Phi}_{fR}(e,\bar{e}) &= K^{\phi}(v,\bar{v}) + K^{\phi}(v',\bar{v}') - K^{\phi}(v,\bar{v}') - K^{\phi}(v',\bar{v}) , \\ K^{\Phi}_{fS}(e,\bar{e}) &= K^{\phi}(v,\bar{v}) + K^{\phi}(v',\bar{v}') + K^{\phi}(v,\bar{v}') + K^{\phi}(v',\bar{v}) , \end{aligned}$$

when $\Phi(v,v') = \Phi_R(v,v')$ or $\Phi(v,v') = \Phi_S(v,v')$, respectively, then the following proposition holds.

Proposition 3. *The relation $Q : \mathscr{V}^2 \rightarrow [0,1]$ given by (1) and h defined by (2) with $K^{\Phi} = K^{\Phi}_{fR}$ (respectively $K^{\Phi} = K^{\Phi}_{fS}$) is a ranking representable reciprocal (respectively symmetric) relation.*

The proof directly follows from the fact that for this specific kernel, $h(v,v')$ can be respectively written as $f(v) - f(v')$ and $f(v) + f(v')$. The kernel K^{Φ}_{fR} has been initially introduced in [39] for ordinal regression and during the last decade it has been extensively used as main building block in many kernel-based ranking algorithms.

Since ranking representability of reciprocal relations implies strong stochastic transitivity of reciprocal relations, K^{Φ}_{fR} can represent this type of domain knowledge.

The notion of ranking representability is powerful for reciprocal relations, because the majority of reciprocal relations satisfy this property, but for symmetric relations it has a rather limited applicability. Ranking representability as defined above cannot represent relations that originate from an underlying metric or similarity measure. For such relations, one needs another connection with its roots in Euclidean metric spaces [29].

5 An Illustration in Document Retrieval

In the experiments, we test the ability of the pairwise kernels to model different relations, and the effect of enforcing prior knowledge about the properties of the learned relations. To this end, we train the regularized least-squares (RLS) algorithm to regress the relation values [40]. Extensive empirical results have been reported for reciprocal relations in [41], as a consequence we focus in this article on symmetric relations. To this end, we compare the ordinary and symmetric Kronecker kernels on a real-world data set based on newsgroups documents[2]. The data is sampled from 4 newsgroups: rec.autos, rec.sport.baseball, comp.sys.ibm.pc.hardware and comp.windows.x. The aim is to learn to predict the similarity of two documents as measured by the number of common words they share. The node features correspond to the number of occurrences of a word in a document. Unlike previously reported experiments, the feature representation is very high-dimensional and sparse, as there are more than 50000 possible features, the majority of which are zero for any given document. First, we sample separate training, validation and test sets each consisting of 1000 nodes. Second, we sample edges connecting the nodes in the training and validation set using exponentially growing sample sizes to measure the effect of sample size on the differences between the kernels. The sample size grid is $[100, 200, 400, \ldots, 102400]$. Again, we sample only edges with different starting and end nodes. When computing the test performance, we consider all the edges in the test set, except those starting and ending at the same node. We train the RLS algorithm using conjugate gradient optimization with early stopping [42], optimization is terminated once the MSE on the validation set has failed to decrease for 10 consecutive iterations. The mean predictor achieves around 145 MSE test performance on this data.

The results are presented in Figure 1. Even for 100 pairs the errors are for both kernels much lower than the mean predictor results, showing that the RLS algorithm succeeds with both kernels in learning the underlying relation. Increasing the training set size leads to a decrease in test error. Using the prior knowledge about the symmetry of the learned relation is clearly helpful. The symmetric kernel achieves for all sample sizes a lower error than the ordinary Kronecker kernel and the largest differences are observed for the smallest sample sizes. For 100 training instances, the error is almost halved by enforcing symmetry.

[2] Available at: `http://people.csail.mit.edu/jrennie/20Newsgroups/`

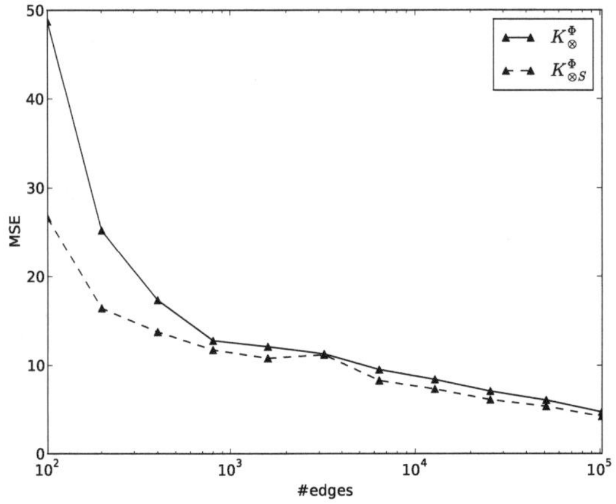

Fig. 1 The comparison of the ordinary Kronecker product pairwise kernel $K_{\otimes}^{\Phi}$ and the symmetric Kronecker product pairwise kernel $K_{\otimes S}^{\Phi}$ on the Newsgroups dataset. The mean squared error is shown as a function of the training set size.

6 Conclusion

A general kernel-based framework for learning various types of valued relations was presented in this article. This framework extends existing approaches for learning relations, because it can handle crisp and valued relations. A Kronecker product feature mapping was proposed for combining the features of pairs of objects that constitute a relation (edge level in a graph). In addition, we clarified that domain knowledge about the relation to be learned can be easily incorporated in our framework, such as reciprocity and symmetry properties. Experimental results on synthetic and real-world data clearly demonstrate that this domain knowledge really helps in improving the generalization performance. Moreover, important links with recent developments in fuzzy set theory and decision theory can be established, by looking at transitivity properties of relations.

Acknowledgments. W.W. is supported as a postdoc by the Research Foundation of Flanders (FWO Vlaanderen) and T.P. by the Academy of Finland (grant 134020).

References

1. Bowling, M., Fürnkranz, J., Graepel, T., Musick, R.: Machine learning and games. Machine learning 63(3), 211–215 (2006)
2. Yamanishi, Y., Vert, J. P., Kanehisa, M.: Protein network inference from multiple genomic data: a supervised approach. Bioinformatics 20, 1363–1370 (2004)
3. Yang, Y., Bansal, N., Dakka, W., Ipeirotis, P., Koudas, N., Papadias, D.: Query by document. In: Proceedings of the Second ACM International Conference on Web Search and Data Mining, Barcelona, Spain, pp. 34–43 (2009)

4. Taskar, B., Wong, M., Abbeel, P., Koller, D.: Link prediction in relational data. In: Advances in Neural Information Processing Systems (2004)
5. De Raedt, L.: Logical and Relational Learning. Springer, Heidelberg (2009)
6. Vert, J.-P., Yamanishi, Y.: Supervised graph inference. In: Advances in Neural Information Processing Systems, vol. 17 (2005)
7. Xing, E., et al.: Distance metric learning with application to clustering with side information. In: Advances in Neural Information Processing Systems, vol. 16, pp. 521–528 (2002)
8. Hüllermeier, E., Fürnkranz, J.: Preference Learning. Springer, Heidelberg (2010)
9. Geurts, P., Touleimat, N., Dutreix, M., d'Alché-Buc, F.: Inferring biological networks with output kernel trees. BMC Bioinformatics 8(2), S4 (2007)
10. Doignon, J.-P., Monjardet, B., Roubens, M., Vincke, P.: Biorder families, valued relations and preference modelling. Journal of Mathematical Psychology 3030, 435–480 (1986)
11. Switalski, Z.: Transitivity of fuzzy preference relations - an empirical study. Fuzzy Sets and Systems 118, 503–508 (2000)
12. De Baets, B., De Meyer, H., De Schuymer, B., Jenei, S.: Cyclic evaluation of transitivity of reciprocal relations. Social Choice and Welfare 26, 217–238 (2006)
13. Schölkopf, B., Smola, A.: Learning with Kernels, Support Vector Machines, Regularisation, Optimization and Beyond. The MIT Press, Cambridge (2002)
14. Ben-Hur, A., Noble, W.: Kernel methods for predicting protein-protein interactions. Bioinformatics 21(1), 38–46 (2005)
15. De Schuymer, B., De Meyer, H., De Baets, B., Jenei, S.: On the cycle-transitivity of the dice model. Theory and Decision 54, 164–185 (2003)
16. Fisher, L.: Rock, Paper, Scissors: Game Theory in Everyday Life. Basic Books, New York (2008)
17. Kerr, B., Riley, M., Feldman, M., Bohannan, B.: Local dispersal promotes biodiversity in a real-life game of rock-paper-scissors. Nature 418, 171–174 (2002)
18. Czárán, T., Hoekstra, R., Pagie, L.: Chemical warfare between microbes promotes biodiversity. Proceedings of the National Academy of Sciences 99(2), 786–790 (2002)
19. Nowak, M.: Biodiversity: Bacterial game dynamics. Nature 418, 138–139 (2002)
20. Kirkup, B., Riley, M.: Antibiotic-mediated antagonism leads to a bacterial game of rock-paper-scissors in vivo. Nature 428, 412–414 (2004)
21. Károlyi, G., Neufeld, Z., Scheuring, I.: Rock-scissors-paper game in a chaotic flow: The effect of dispersion on the cyclic competition of microorganisms. Journal of Theoretical Biology 236(1), 12–20 (2005)
22. Reichenbach, T., Mobilia, M., Frey, E.: Mobility promotes and jeopardizes biodiversity in rock-paper-scissors games. Nature 448, 1046–1049 (2007)
23. Boddy, L.: Interspecific combative interactions between wood-decaying basidiomycetes. FEMS Microbiology Ecology 31, 185–194 (2000)
24. Sinervo, S., Lively, C.: The rock-paper-scissors game and the evolution of alternative male strategies. Nature 340, 240–246 (1996)
25. Waite, T.: Intransitive preferences in hoarding gray jays (Perisoreus canadensis). Journal of Behavioural Ecology and Sociobiology 50, 116–121 (2001)
26. Luce, R., Suppes, P.: Preference, Utility and Subjective Probability. In: Handbook of Mathematical Psychology, pp. 249–410. Wiley, Chichester (1965)
27. Fishburn, P.: Nontransitive preferences in decision theory. Journal of Risk and Uncertainty 44, 113–134 (1991)
28. Tversky, A.: In: Shafir, E. (ed.) Preference, Belief and Similarity. MIT Press, Cambridge (1998)

29. Gower, J., Legendre, P.: Metric and Euclidean properties of dissimilarity coefficients. Journal of Classification 3, 5–48 (1986)
30. Jäkel, F., Schölkopf, B., Wichmann, F.: Similarity, kernels, and the triangle inequality. Journal of Mathematical Psychology 52(2), 297–303 (2008)
31. Switalski, Z.: General transitivity conditions for fuzzy reciprocal preference matrices. Fuzzy Sets and Systems 137, 85–100 (2003)
32. De Baets, B., De Meyer, H.: Transitivity frameworks for reciprocal relations: cycle-transitivity versus FG-transitivity. Fuzzy Sets and Systems 152, 249–270 (2005)
33. De Baets, B., Mesiar, R.: Metrics and T-equalities. Journal of Mathematical Analysis and Applications 267, 531–547 (2002)
34. Moser, B.: On representing and generating kernels by fuzzy equivalence relations. Journal of Machine Learning Research 7, 2603–2620 (2006)
35. Billot, A.: An existence theorem for fuzzy utility functions: A new elementary proof. Fuzzy Sets and Systems 74, 271–276 (1995)
36. Koppen, M.: Random Utility Representation of Binary Choice Probilities: Critical Graphs yielding Critical Necessary Conditions. Journal of Mathematical Psychology 39, 21–39 (1995)
37. Fono, L., Andjiga, N.: Utility function of fuzzy preferences on a countable set under max-*-transitivity. Social Choice and Welfare 28, 667–683 (2007)
38. Bodenhofer, U., De Baets, B., Fodor, J.: A compendium of fuzzy weak orders. Fuzzy Sets and Systems 158, 811–829 (2007)
39. Herbrich, R., Graepel, T., Obermayer, K.: Large margin rank boundaries for ordinal regression. In: Smola, A., Bartlett, P., Schölkopf, B., Schuurmans, D. (eds.) Advances in Large Margin Classifiers, pp. 115–132. MIT Press, Cambridge (2000)
40. Pahikkala, T., Tsivtsivadze, E., Airola, A., Järvinen, J., Boberg, J.: An efficient algorithm for learning to rank from preference graphs. Machine Learning 75(1), 129–165 (2009)
41. Pahikkala, T., Waegeman, W., Tsivtsivadze, E., Salakoski, T., De Baets, B.: Learning intransitive reciprocal relations with kernel methods. European Journal of Operational Research 206, 676–685 (2010)
42. Pahikkala, T., Waegeman, W., Airola, A., Salakoski, T., De Baets, B.: Conditional ranking on relational data. In: Balcázar, J.L., Bonchi, F., Gionis, A., Sebag, M. (eds.) ECML PKDD 2010. LNCS, vol. 6322, pp. 499–514. Springer, Heidelberg (2010)

Multicriteria Decision Making by Means of Interval-Valued Choquet Integrals

H. Bustince, J. Fernandez, J. Sanz, M. Galar, R. Mesiar, and A. Kolesárová

Abstract. In this work we propose a new multicriteria decision making algorithm for interval-valued fuzzy preference relations based on the use on a appropriate definition of interval-valued Choquet integrals. This algorithm allows to recover some of the best known usual fuzzy algorithms when the considered intervals are reduced to a single point. Since a key point in every decision making problem is that of the ordering, we propose a method to build orders based on the use of aggregation functions that, on one hand, allows to define several different total orders and, on the other hand, recovers some of the most commonly used total orders between intervals.

1 Introduction

Multicriteria decision-making (MCDM) problems are very common in everyday life. These kind of problems can be modeled as follows [6]: given a set of alternatives $\{A_1,\ldots,A_n\}$ and a set of criteria $\{x_1,\ldots,x_k\}$, choose the alternative that is the best one according to the criteria. A useful way of dealing with this situation is to

H. Bustince · J. Fernandez · J. Sanz · M. Galar
Dept. Automatica y Computacion, Universidad Publica de Navarra, Campus Arrosadia s/n, 31006 Pamplona, Spain
e-mail: {bustince, fcojavier.fernandez, joseantonio.sanz mikel.galar}@unavarra.es

R. Mesiar
Department of Mathematics, Faculty of Civil Engineering, Slovak University of Technology, 81368 Bratislava, Slovakia and UTIA CAS Prague, Czech Republik
e-mail: mesiar@math.sk

A. Kolesárová
Institute of Information Engineering, Automation and Mathematics, Slovak University of Technology, SK-812 37 Bratislava, Slovakia
e-mail: anna.kolesarova@stuba.sk

B. De Baets et al. (Eds.): Eurofuse 2011, AISC 107, pp. 269–278, 2011.
springerlink.com

provide a value that measures the degree to which alternative A_i fulfills criterium x_j. In this way, a MCDM problem can be expressed by means of a matrix (the MCDM matrix) where the entry at row i and column j denotes up to what extent alternative A_i satisfies criterium x_j. Once this MCDM matrix has been built, many algorithms are based on assigning a value to each of the alternatives, namely, to each of the rows in the MCDM matrix. Finally alternatives are ordered according to this score in order to choose the best one. Observe that with this approach a MCDM problem, once the MCDM matrix is given, is equivalent to, first, finding a way of aggregating the values in each row into a single one and second, to find an appropriate order for these aggregated values.

In order to deal with these situations, fuzzy sets theory has revealed itself as a very useful tool. Recall that for a fixed finite universe $U = \{u_1, \ldots, u_n\}$, a fuzzy subset F of U is given by its membership function $F : U \to [0,1]$ (we will not distinguish fuzzy subsets and the corresponding membership functions notations). In this case, entries in the MCDM matrix are normalized to $[0,1]$. Moreover, for the aggregation of the values in each row, it is advisable to introduce the expected value $E(F)$ of F. The original Zadeh's approach to this concept in [12] was based on a probability measure P on U, $P(u_i) = p_i$, and then $E(F) = \sum_{i=1}^{n} p_i F(u_i)$. But it is also possible to consider a more general approach based on a fuzzy measure and the Choquet integral [5, 4].

In recent years, interval-valued fuzzy sets in MCDM problems have attracted a wide interest of the scientific community [9, 11], since by providing an interval as a measure of the extent to which a criterium is satisfied by a given alternative, it is possible to take care of the uncertainty or lack of knowledge that an expert can suffer when determining that satisfaction[3]. Even more, Atanassov's intuitionistic fuzzy sets, which are mathematical isomorphic to interval-valued fuzzy sets can in some situations be considered as a natural setting to deal with this kind of problems, see. This extensions has brought the necessity to introduce the expected value also for these objects. Moreover, notice that, contrary to the real numbers case, there is not an universally accepted total order between intervals in $[0,1]$.

The aim of this paper is to discuss the use of interval-valued fuzzy sets in MCDM problems. To do so, we will introduce the expected value of interval–valued fuzzy sets based on the concept of discrete interval–valued Choquet integral [7]. The choice of the Choquet integral is due to the fact that in many of the most widely used methods for evaluation of the alternatives, as that of the weighted voting [10], the aggregation is done by means of a weighted mean, which is just a particular case of Choquet integral for symmetric measures. Moreover, since in any MCDM algorithm based on the use of interval-valued fuzzy sets, if the final score of alternatives is also determined by means of intervals, it is necessary to determine an appropriate ordering, we will also present a method of building different total orders by means of aggregation functions. We will leave for future works the analysis of which of these orders is the most appropriate one for a given MCDM problem.

The paper is organized as follows. In the next section, standard approach to the discrete interval–valued Choquet integral that arises from the concept of Aumann integral [1] is recalled. In Section 3, a method to obtain various linear orders on intervals are proposed. In Section 4 we present our specific approach to interval-valued Choquet integral. Section 5 is devoted to our MCDM algorithm. Finally, some conclusions are added.

2 Interval-Valued Choquet Integral

Recall that an interval–valued fuzzy set F is characterized by its membership function $F : U \to J([0,1])$, where

$$J([0,1]) = \{[a,b]|, 0 \leq a \leq b \leq 1\}$$

Generalization of reals into (closed real) intervals was forced by the development of computers (especially rounding problems) and it has lead into the interval arithmetics [8]. Recall, for example, that the summation in this case is given by

$$[a,b]+[c,d] = \{x+y|x \in [a,b], y \in [c,d]\} = [a+c, b+d].$$

A similar idea has lead Aumann [1] to introduce his integral of set–valued functions. Both concepts are of the same nature as Zadeh's extension principle [13] is, and in the framework of Choquet integral they appear in several works, see e.g. [7, 14]. This concept is defined as follows. Given a fuzzy measure

$$m : 2^U \to [0,1],\ m(\emptyset) = 0,\ m(U) = 1,\ m(A) \leq m(B) \text{ whenever } A \subseteq B \subseteq U$$

the (discrete) Choquet integral (or expectation) of F is defined as

$$E(F) = C_m(F) = \Sigma_{i=1}^{n} F(u_{\sigma(i)})(m(\{u_{\sigma(i)},\dots,u_{\sigma(n)}\}) - \\ -m(\{u_{\sigma(i+1)},\dots,u_{\sigma(n)}\})),$$

where $\sigma : \{1,\dots,n\} \to \{1,\dots,n\}$ is a permutation so that $F(u_{\sigma(1)}) \leq F(u_{\sigma(2)}) \leq \dots \leq F(u_{\sigma(n)})$, and $\{u_{\sigma(n+1)}, u_{\sigma(n)}\} = \emptyset$ by convention.

This discrete Choquet integral can be extended to the interval-valued setting

Definition 1. *Let $F : U \to J([0,1])$ be an interval–valued fuzzy set, and $m : 2^U \to [0,1]$ a fuzzy measure. Choquet integral–based expectation $\mathbf{C}_m(F)$ is given by*

$$\mathbf{C}_m(F) = \{C_m(f)|f : U \to [0,1],\ f(u_i) \in F(u_i)\} = \\ = [C_m(f_*), C_m(f^*)], \tag{1}$$

where $f_*, f^* : U \to [0,1]$ are given by $f_*(u_i) = a_i$ and $f^*(u_i) = b_i$, with $[a_i, b_i] = F(u_i)$.

Several properties of the discrete interval–valued Choquet integral $\mathbf{C}_m$ are discussed and introduced in [7, 14]. For example, this integral is comonotone additive,

$$\mathbf{C}_m(F+G) = \mathbf{C}_m(F) + \mathbf{C}_m(G)$$

whenever $F,G : U \to J([0,1])$ are such that $F(u_i)+G(u_i) \subseteq [0,1]$ for each $u_i \in U$, and F,G are comonotone, that is,

$$(f^*(u_i) - f^*(u_j))\,(g^*(u_i) - g^*(u_j)) \geq 0$$

and

$$(f_*(u_i) - f_*(u_j))\,(g_*(u_i) - g_*(u_j)) \geq 0$$

for all $u_i, u_j \in U$.

3 Total Orders for Intervals

Notice that the idea of a discrete Choquet integral C_m is based on a permutation $\sigma : \{1,\dots,n\} \to \{1,\dots,n\}$ forcing $F(u_{\sigma(1)}) \leq F(u_{\sigma(2)}) \leq \dots \leq F(u_{\sigma(n)})$. This idea can be adapted for the interval case only if there is a linear order $\preceq$ on $J([0,1])$. Several different orders have considered in the literature. For instance:

(i) Yager-Xu's order based on score and accuracy. $[a,b] \preceq M,G[c,d]$ if and only if $a+b < c+d$ or $a+b = c+d$ and $d-c \leq b-a$.
(ii)The lexicographical order with respect to the first (second) variable: $[a,b] \preceq_{P1} [c,d]$ ($[a,b] \preceq_{P2} [c,d]$) if and only if $a < c$ ($b < d$) or $a = c$ and $b \leq d$ ($b = d$ and $a \leq c$).

Observe that these two examples extend the usual partial order between intervals $[a,b] \leq [c,d]$ if and only if $a \leq c$ and $b \leq d$. Now we introduce a class of total orders that also extends this usual order.

Lemma 1. *Let $A,B : [0,1]^2 \to [0,1]$ be two aggregation functions [5] such that $A(x,y) = A(u,v)$ and $B(x,y) = B(u,v)$ can only happen if $(x,y) = (u,v)$. Define a relation $\preceq_{A,B}$ on $J([0,1])$ by*

$$[x,y] \preceq_{A,B} [u,v] \text{ whenever } A(x,y) < A(u,v)$$

or

$$A(x,y) = A(u,v) \text{ and } B(x,y) \leq B(u,v).$$

Then $\preceq_{A,B}$ is a linear order on $J([0,1])$ with the minimal element $\{0\} = [0,0]$, and the maximal element $\{1\} = [1,1]$.

The proof of this lemma is trivial and therefore omitted. Note that the linear order $\preceq_{A,B}$ always refines the standard partial order $\leq$ on intervals, $[x,y] \leq [u,v]$ whenever $x \leq u$ and $y \leq v$, i.e., $[x,y] \leq [u,v]$ implies $[x,y] \preceq_{A,B} [u,v]$.

With this definition, Yager and Xu's order based on score and accuracy functions can be recovered as the linear order $\preceq_{M,G}$ on $J([0,1])$, where M is the arithmetic mean and G is the geometric mean. Regarding the lexicographical order with respect to the first (second) variable, it can be seen as the linear order obtained by taking as A the projection with respect to the first (second) variable and as B the projection with respect to the second (first) variable.

On the other hand, for several couples $(A_1,B_1),(A_2,B_2),\ldots,$ the linear orders $\preceq_{A_1,B_1}, \preceq_{A_2,B_2}, \ldots,$ may coincide. However, then $\mathbf{C}_m^{A_1,B_1} = \mathbf{C}_m^{A_2,B_2}$. Consider, for example $\preceq_{Min,Max} \equiv \preceq_{P_1,P_1} \equiv \preceq_{P_1,B}$, where $P_1,P_2 : [0,1]^2 \to [0,1]$ are projections, $P_1(x,y) = x$, $P_2(x,y) = y$ and $B : [0,1]^2 \to [0,1]$ is an arbitrary cancellative aggregation function.

4 Discrete Interval-Valued (*A*,*B*)–Choquet Integrals

Let $\preceq_{A,B}$ be a total order defined as in 1. Discrete interval–valued (*A*,*B*)–Choquet integrals can be defined as follows.

Definition 2. *Let $F : U \to J([0,1])$ be an interval–valued fuzzy set, and $m : 2^U \to [0,1]$ a fuzzy measure. Under the constraints of Lemme 1, the (A, B)–Choquet integral $\mathbf{C}_m^{A,B}(F)$ is given by*

$$\mathbf{C}_m^{A,B}(F) = \tag{2}$$

$$= \sum_{i=1}^n F(u_{\sigma_{A,B}(i)})(m(\{u_{\sigma_{A,B}(i)},\ldots,u_{\sigma_{A,B}(n)}\}) - \tag{3}$$

$$-m(\{u_{\sigma_{A,B}(i+1)},\ldots,u_{\sigma_{A,B}(n)}\})), \tag{4}$$

where $\sigma_{A,B} : \{1,\ldots,n\} \to \{1,\ldots,n\}$ is a permutation such that $F(u_{\sigma_{A,B}(1)}) \preceq_{A,B} F(u_{\sigma_{A,B}(2)}) \preceq_{A,B} \cdots \preceq_{A,B} F(u_{\sigma_{A,B}(n)})$.

Observe that if $F(u_i) = [a_i, b_i]$, i = 1, ..., *n*, then 4 can be rewritten into

$$\mathbf{C}_m^{A,B}(F) = \tag{5}$$

$$= \Big[\sum_{i=1}^n a_{\sigma_{A,B}(i)} \cdot (m(\{u_{\sigma_{A,B}(i)},\ldots,u_{\sigma_{A,B}(n)}\}) - \tag{6}$$

$$-m(\{u_{\sigma_{A,B}(i+1)},\ldots,u_{\sigma_{A,B}(n)}\})), \tag{7}$$

$$\sum_{i=1}^n b_{\sigma_{A,B}(i)} \cdot (m(\{u_{\sigma_{A,B}(i)},\ldots,u_{\sigma_{A,B}(n)}\}) - \tag{8}$$

$$- m(\{u_{\sigma_{A,B}(i+1)},\ldots,u_{\sigma_{A,B}(n)}\}))\Big]. \tag{9}$$

The concept of an interval–valued (A,B)–Choquet integral $\mathbf{C}_m^{A,B}$ extends the standard discrete Choquet integral. Indeed if $F : U \to J([0,1])$ is singleton–valued, i.e., F is a fuzzy subset of U, then $C_m(F) = \mathbf{C}_m(F) = \mathbf{C}_m^{A,B}(F)$ independently of A, B.

It is also worth to remark that for a fixed $F : U \to J([0,1])$ such that f_* and f^* are comonotone, i.e., $(f_*(u_i) - f_*(u_j)) \cdot (f^*(u_i) - f^*(u_j)) \geq 0$ for all u_i, $u_j \in U$, for any A, B satisfying the constraints of Lemma 1, the discrete Choquet integrals introduced in (1) and (4) coincide, $\mathbf{C}_m(F) = \mathbf{C}_m^{A,B}(F)$. However, in general the integral $\mathbf{C}_m$ cannot be expressed in the form $\mathbf{C}_m^{A,B}$.

5 A Multicriteria Decision-Making Algorithm

In this section we apply all the previously developed concepts to a multicriteria decision making algorithm. This algorithm makes use of an interval-valued Choquet integral in order to evaluate each of the alternatives with respect to the given criteria. Finally, alternatives are ordered using one of the orders constructed by means of Lemma 1.

So assume that we are given the set of alternatives $\{A_1,\ldots,A_n\}$ and the set of criteria $\{x_1,\ldots,x_k\}$. Suppose all this information is merged in an interval-valued multicriteria decision making matrix as follows:

$$\begin{array}{c} \\ A_1 \\ A_2 \\ \\ A_n \end{array} \begin{array}{c} \begin{array}{ccc} x_1 & \cdots & x_k \end{array} \\ \begin{pmatrix} ([\underline{\mu_{A_1}}(x_1),\overline{\mu_{A_1}}(x_1)] & \cdots & [\underline{\mu_{A_1}}(x_k),\overline{\mu_{A_1}}(x_k)] \\ ([\underline{\mu_{A_2}}(x_1),\overline{\mu_{A_2}}(x_1)] & \cdots & [\underline{\mu_{A_2}}(x_k),\overline{\mu_{A_2}}(x_k)] \\ \cdots & \cdots & \\ ([\underline{\mu_{A_n}}(x_1),\overline{\mu_{A_n}}(x_1)] & \cdots & [\underline{\mu_{A_n}}(x_k),\overline{\mu_{A_n}}(x_k)] \end{pmatrix} \end{array}$$

where $[\underline{\mu_{A_i}}(x_j),\overline{\mu_{A_i}}(x_j)]$ denotes the degree to which alternative A_i satisfies criterion x_j. This means that we can understand each alternative as an interval-valued fuzzy set over the referential set of criteria and in such a way that each of the intervals $\mu(A_i)(x_j) = [\underline{\mu_{A_i}}(x_j),\overline{\mu_{A_i}}(x_j)]$ provides the membership value of criteria x_j to the interval valued fuzzy set A_i.

The algorithm that we propose is the following.

1. Fix a linear order $\preceq_{A,B}$ over the set $J([0,1])$.
2. Select a fuzzy measure m over the set of criteria $\{x_1,\ldots,x_k\}$.
3. FOR each row $i = 1,\ldots,n$ of the MCDM decision matrix DO

 3.1 Order the elements in increasing order $\{x_{(1)},\ldots,x_{x(k)}\}$ withe respect to their corresponding memberships to the set A_i and using the linear order chosen in Step 1.

 3.2 Take $\underline{C(A_i)} = 0$ and $\overline{C(A_i)} = 0$.

 3.3 FOR each $j = 1,\ldots k$ DO

  3.3.1 $\underline{C(A_i)} = \underline{C(A_i)} + \underline{\mu_{A_i}}(x_{(j)})(m(\{x_{(j)},\ldots,x_{(k)}) - m(\{x_{(j+1)},\ldots,x_{(k)})$

  3.3.2 $\overline{C(A_i)} = \overline{C(A_i)} + \overline{\mu_{A_i}}(x_{(j)}(m(\{x_{(j)},\ldots,x_{(k)}) - m(\{x_{(j+1)},\ldots,x_{(k)})$

 ENDFOR

 3.4 Take $\mathbf{C}(A_i) = [\underline{C(A_i)},\overline{C(A_i)}]$.

 ENDFOR

4. Choose as best alternative the one for which $\mathbf{C}(A_i)$ is the largest.

In this algorithm, steps 1 and 2 are crucial, since they determine its behaviour. With respect to the order, in the following examples we will see how it can affect the final decision, the choice of one order or another should be based on the general information we have about the way the different alternatives have been evaluated.

With respect to the measure, this is a complicate point. In future works we intend to carry on a deep study to determine which is the best measure for each particular problem. Observe that the measure should take into account the interrelationship between the different criteria, as well as the relative importance conceded to each of them.

Notice also that in case we are dealing with a pure (non interval-valued) MCDM the previous algorithm collapses into the fuzzy one that evaluates each of the rows by means of the corresponding measure-based Choquet integral. In particular, by choosing a symmetric measure, we evaluate each alternative with an OWA operator, which in particular proves that our algorithm extends the well known weighted voting method.

Now we present specific examples to show how this algorithm works.

Example 1. This example is taken from [11], who based it on the previous work [6]. Let's a set of four alternatives and three criteria with the following MCDM matrix:

$$\begin{array}{c} \\ A_1 \\ A_2 \\ A_3 \\ A_4 \end{array} \begin{array}{c} \begin{array}{ccc} \quad x_1 \quad & \quad x_2 \quad & \quad x_3 \quad \end{array} \\ \begin{pmatrix} ([0.45,0.65] & [0.50,0.70] & [0.20,0.45] \\ ([0.65,0.75] & [0.65,0.75] & [0.45,0.85] \\ ([0.45,0.65] & [0.45,0.65] & [0.45,0.80] \\ ([0.75,0.85] & [0.35,0.80] & [0.65,0.85] \end{pmatrix} \end{array}$$

We will take as linear order the following one: $[a,b] \preceq_{MG} [c,d]$ if and only if $a+b < c+d$ or $a+b = c+d$ and $ab \leq cd$. That is, we consider the linear order obtained by taking A as the arithmetic mean and B as the geometric mean. As we have already stated, this order is the same as the score and accuracy based order proposed by Yager and Xu.

Now we need to fix the fuzzy measure over the set of criteria that we are going to use. In this case, we start by using the following symmetric fuzzy measure:

$$\begin{gathered} m(\emptyset) = 0 \\ m(\{x_1\}) - m(\{x_2\}) = m(\{x_3\}) = \tfrac{1}{3} \\ m(\{x_1,x_2\}) = m(\{x_1,x_3\}) = m(\{x_2,x_3\}) = \tfrac{2}{3} \\ m(\{x_1,x_2,x_3\}) = 1 \end{gathered}$$

This m can be understood as a measure of how close the measured set is from the total set $\{x_1,x_2,x_3\}$, namely, how far the considered set is from the ideal situation of fulfilling all of the criteria. In this sense, this algorithm can be seen as connected to other algorithms that make use of fuzzy entropy. Nevertheless, this is a very coarse, simplified approach, and it is considered here only to illustrate the way the algorithm works.

If we proceed to order with respect to $\preceq_{MG}$ the rows of the MCDM matrix, we obtain the following. For A_1:

$$\{x_3,x_1,x_2\}$$

For A_2 we have:

$$\{x_3,x_1,x_2\}$$

For A_3:

$$\{x_1,x_2,x_3\}$$

and finally, for A_4, we arrive at:

$$\{x_2, x_3, x_1\}\ .$$

The corresponding calculations then provide that:

$$\begin{aligned}
\mathbf{C}(A_1) &= [0.38, 0.50] \\
\mathbf{C}(A_2) &= [0.58, 0.78] \\
\mathbf{C}(A_3) &= [0.45, 0.70] \\
\mathbf{C}(A_4) &= [0.58, 0.83]
\end{aligned}$$

So the final ordering of alternatives is A_4, A_2, A_3, A_1. This is not the same ordering obtained in [11], since the first and the second alternatives in that case are interchanged. Nevertheless, this can be explained by the fact that the approach in Ye's work is completely different, since it is based in the use of entropies and correlation, whereas in our case we are only based in aggregation function theory. Of course, here the choice of the measure has been crucial. In this particular case, also notice that our Choquet integral reduces to the arithmetic mean of the membership intervals under consideration.

Example 2. Let's consider now the following MCDM matrix, taken from [10]:

$$\begin{array}{c} \\ A_1 \\ A_2 \\ A_3 \\ A_4 \end{array}
\begin{array}{c} \quad x_1 \qquad\qquad x_2 \qquad\qquad x_3 \\
\begin{pmatrix}
([0.70, 0.70] & [0.80, 0.90] & [0.90, 0.90] \\
([0.60, 0.80] & [0.80, 0.80] & [0.80, 0.90] \\
([0.60, 0.90] & [0.50, 0.90] & [0.80, 0.80] \\
([0.40, 0.50] & [0.90, 0.90] & [0.40, 0.90]
\end{pmatrix}\end{array}$$

Ordering with respect to $\preceq_{MG}$ we obtain, for A_1 $\{x_1, x_2, x_3\}$; for A_2, $\{x_1, x_2, x_3\}$; for A_3, $\{x_2, x_1, x_3\}$; and for A_4, $\{x_1, x_3, x_2\}$. Let's also consider the fuzzy measure proposed in the same paper:

$$\begin{gathered}
m(\emptyset) = 0 \\
m(\{x_1\}) = m(\{x_2\}) = 0.4\ ;\ m(\{x_3\}) = 0.3 \\
m(\{x_1, x_2\}) = 0.6\ ;\ m(\{x_1, x_3\}) = m(\{x_2, x_3\}) = 0.8 \\
m(\{x_1, x_2, x_3\}) = 1\ ,
\end{gathered}$$

which is not symmetric. So if we carry on the corresponding calculations as in the previous case, we arrive at:

$$\begin{aligned}
\mathbf{C}(A_1) &= [0.81, 0.86] \\
\mathbf{C}(A_2) &= [0.76, 0.83] \\
\mathbf{C}(A_3) &= [0.64, 0.87] \\
\mathbf{C}(A_4) &= [0.60, 0.73]
\end{aligned}$$

So the final ordering of alternatives is A_1, A_2, A_3, A_4. This is the same ordering that is obtained in Xu's paper.

Let's consider now the order $\preceq_{P2}$. In this case, we have that the ordering of the criteria for each alternative is for A_1 $\{x_1, x_2, x_3\}$; for A_2, $\{x_1, x_2, x_3\}$; for A_3, $\{x_3, x_2, x_1\}$; and for A_4, $\{x_1, x_3, x_2\}$. The evaluation of each alternative by means of the interval-valued Choquet interval provides:

$$\begin{aligned} \mathbf{C}(A_1) &= [0.81, 0.86] \\ \mathbf{C}(A_2) &= [0.76, 0.83] \\ \mathbf{C}(A_3) &= [0.66, 0.86] \\ \mathbf{C}(A_4) &= [0.60, 0.82], \end{aligned}$$

So alternatives are finally ordered as follows: A_1, A_3, A_2, A_4. This is different from the ordering we have obtained before.

6 Conclusions and Future Research

We have introduced a new concept of a discrete interval–valued (A, B)–Choquet integral. This concept extends the classical one of discrete Choquet integral for the interval-valued setting, and collapses into the latter in the case of degenerate intervals.

We have also introduced a way of generating total orders in the set of intervals contained in the $[0, 1]$ interval. This orders contain as particular instances the lexicographical orders and the score and accuracy based order of Yager and Xu.

With these two concepts at hand, we have presented a MCDM algorithm that can be seen as a generalization of some well known fuzzy algorithms, as the weighted voting. We have provided several examples of how this algorithm works.

Of course, this algorithm is only a first approximation. In future works we intend to analize two points, namely, the choice of the measure and the choice of the order, that could be different and specific for each given problem.

Acknowledgements. The work on this paper was supported by grants P402/11/0378, APVV-0073–10 and VEGA 1/0080/10, and project TIN 2010-15055 from the Government of Spain.

References

1. Aumann, R.J.: Integrals of set–valued functions. J. Math. Anal. Appl. 12, 1–12 (1965)
2. Beliakov, G., Bustince, H., Goswami, D.P., Mukherjee, U.K., Pal, N.R.: On averaging operators for Atanassovs intuitionistic fuzzy sets. Inform. Sci. 181, 1116–1124 (2010)
3. Bustince, H., Barrenechea, E., Pagola, M., Fernandez, J.: Interval-valued fuzzy sets constructed from matrices: Application to edge detection. Fuzzy Sets Syst. 160, 1819–1840 (2009)
4. Choquet, G.: Theory of capacities. Ann. Inst. Fourier (Grenoble) 5, 131–292 (1953-1954)

5. Grabisch, M., Marichal, J.–L., Mesiar, R., Pap, E.: Aggregation functions. Cambridge University Press, Cambridge (2009)
6. Herrera, F., Herrera-Viedma, E.: Linguistic decision analysis:steps for solving decision problems under linguistic information. Fuzzy Sets and Systems 115, 67–82 (2000)
7. Jang, L.C.: Interval–valued Choquet integrals and their applications. J. Appl. Math. and Computing 16, 429–443 (2004)
8. Moore, R.E.: Interval Analysis. Prentice-Hall, Englewood Cliffs (1966)
9. Tan, C., Chen, X.: Intuitionistic fuzzy Choquet integral operator for multi-criteria decision making. Expert Systems with Appl. 37, 149–157 (2010)
10. Xu, Z.S.: Choquet integrals of weighted intuitionistic fuzzy information. Information Sciences 180, 726–736 (2010)
11. Ye, J.: Fuzzy decision-making method based on the weighted correlation coefficient under intuitionistic fuzzy environment. European Journal of Operational Research 205, 202–204 (2010)
12. Zadeh, L.A.: Probability measures of fuzzy events. J. Math. Anal. Appl. 23, 421–427 (1968)
13. Zadeh, L.A.: The concept of a linguistic variable and its applications to approximate reasoning. Inform. Sci. 8, Part I, 199–251, Part II pp. 301–357, Inform. Sci. 9, Part III pp. 43–80 (1975)
14. Zhang, D., Wang, Z.: On set–valued fuzzy integrals. Fuzzy Sets and Systems 56, 237–247 (1993)

Multiset Merging: The Majority Rule

Antoon Bronselaer, Guy De Tré, and Daan Van Britsom

Abstract. A well known problem that many sources of data nowadays cope with, is the problem of duplicate data. In general, we can represent a data source as a collection of objects. Deduplication then consists of two main problems: (a) finding duplicate objects and (b) processing those duplicate objects. This paper contributes to the study of the latter problem by investigating functions that map a multiset of objects to a single object. Such functions are called merge functions. We investigate the specific case where an object itself is a multiset. An interesting application of this case is the problem of multiple document summarization. Next to the basic definition of such merge functions, we focus on an important property borrowed from the (more general) field of information fusion: the majority rule.

1 Introduction

In the current climate of digitization and information processing, proper management of data rapidly gains interest. An important aspect of such data and information management, is the prevention of duplicate data. If such prevention is impossible or has been neglected, the challenging problem of data deduplication is encountered. This problem has been the subject of in-depth research ([1],[3],[5]), but most of these studies are limited to the problem of detection. However, in many practical situations, the *detection* of duplicate data is not sufficient as it must be completed with a strategy for the *processing* of duplicate data. This paper contributes to the latter problem by studying the merging of duplicate data in the specific case where a piece of data is modeled by a *multiset*. The choice for this specific case is far from arbitrary as it has interesting applications in the processing of duplicate textual data and duplicate graphs (e.g., XML documents). The most important results of this

Antoon Bronselaer · Guy De Tré · Daan Van Britsom
Department of Telecommunications and Information Processing, Ghent University, Sint-Pietersnieuwstraat 41, 9000 Ghent, Belgium
e-mail: {antoon.bronselaer,guy.detre,daan.vanbritsom}@ugent.be

B. De Baets et al. (Eds.): Eurofuse 2011, AISC 107, pp. 279–292, 2011.
springerlink.com

paper are the definition of merge functions that optimize a predefined quality measure on the one hand and the extent to which the majority rule holds for such merge functions on the other hand.

The remainder of this paper is structured as follows. In Section 2, we introduce some basic definitions regarding multisets and merge functions. In Section 3, we first cast two well known measures of quality from the field of information retrieval (precision and recall) to the field of multisets and we define an f-optimal merge function as a merge function that maximizes the harmonic mean of these two measures. In Section 4, we investigate the majority rule, which is an important property in the field of information fusion. Section 5 provides an illustrative example of the introduced concepts in the practical setting of multiple document summarization. Finally, in Section 6, we summarize the most important contributions of this paper.

2 Preliminaries

2.1 *Multisets*

We briefly recall some important definitions regarding multisets [8]. Informally, a multiset is an unordered collection in which elements can occur multiple times. Many definitions have been proposed, but within the scope of this paper, we adopt the functional definition of multisets.

Definition 1 (Multiset). *A multiset M over a universe U is defined by a function:*

$$M : U \rightarrow \mathbb{N}. \tag{1}$$

For each $u \in U$, $M(u)$ denotes the multiplicity of u in M. The set of all multisets drawn from a universe U is denoted $\mathscr{M}(U)$.

The j-cut of a multiset M is a regular set, denoted as M_j and given as:

$$M_j = \{u | u \in U \wedge M(u) \geq j\}. \tag{2}$$

Whenever we wish to assign an index $i \in \mathbb{N}$ to a multiset M, we use the notation $M_{(i)}$. The notation M_j is preserved for the j-cut of M. A multiset $M_{(1)}$ is a subset of a multiset $M_{(2)}$, denoted by $M_{(1)} \subseteq M_{(2)}$, if:

$$\forall u \in U : M_{(1)}(u) \leq M_{(2)}(u). \tag{3}$$

The cardinality of a multiset M is given by:

$$|M| = \sum_{u \in U} M(u). \tag{4}$$

Yager has defined a number of operators for multisets [8] which cast into the functional notation as follows:

$$\left(M_{(1)} \cup M_{(2)}\right)(u) = \max\left(M_{(1)}(u), M_{(2)}(u)\right) \tag{5}$$

$$\left(M_{(1)} \cap M_{(2)}\right)(u) = \min\left(M_{(1)}(u), M_{(2)}(u)\right) \tag{6}$$

$$\left(M_{(1)} \oplus M_{(2)}\right)(u) = M_{(1)}(u) + M_{(2)}(u). \tag{7}$$

The $\in$-operator applies for multisets as follows:

$$u \in M \Leftrightarrow M(u) > 0. \tag{8}$$

2.2 Merge Functions

An explorative study of information fusion in the specific context of duplicate data is performed in [2] and has led to the definition of a merge function.

Definition 2 (Merge function). *A merge function over a universe U is defined by a function:*

$$\varpi : \mathcal{M}(U) \to U. \tag{9}$$

Several additional properties for merge functions have been discussed in [2]. One of these properties is called the *majority rule* (see also [6, 9]), which is generalized in this paper to the z-majority rule.

Definition 3 (z-majority rule). *A merge function ϖ over a universe U satisfies the z-majority rule, with $z \in [0.5, 1]$, if for any $M \in \mathcal{M}(U)$ we have that:*

$$\left(\exists v \in M : M(v) > z \cdot |M|\right) \Rightarrow \omega(M) = v. \tag{10}$$

2.3 Triangular Norms

Triangular norms or t-norms for short, are introduced by Schweizer and Sklar in the context of probabilistic metric spaces [4]. We provide the following definition.

Definition 4 (t-norm). *A triangular norm (or t-norm) T is an increasing, associative and commutative $[0,1]^2 \to [0,1]$ mapping that satisfies:*

$$\forall x \in [0,1] : \mathrm{T}(x,1) = x. \tag{11}$$

Some important examples of t-norms include the minimum operator:

$$\mathrm{T}_{\mathbf{M}}(x,y) = \min(x,y), \tag{12}$$

the algebraic product:

$$\mathrm{T}_{\mathbf{P}}(x,y) = x \cdot y, \tag{13}$$

the Łukasiewics t-norm

$$\mathrm{T_L}(x,y) = \max(x+y-1,0), \tag{14}$$

and the drastic product:

$$\mathrm{T_D}(x,y) = \begin{cases} \min(x,y) & \textbf{if} \quad \max(x,y) = 1 \\ 0 & \textbf{else} \end{cases}. \tag{15}$$

3 f-Optimal Merge Functions

3.1 Basic Notations

As mentioned in the introduction, we limit the focus of this paper to the case of merge functions for multisets. This means that we only consider merge functions of the following type:

$$\varpi : \mathscr{M}(\mathscr{M}(U)) \to \mathscr{M}(U). \tag{16}$$

In order to avoid confusion, we shall introduce some basic notations for the remainder of this paper. A multiset over the ground universe U will be called a *source* and will be denoted as S. A multiset of sources, i.e. a multiset over the universe $\mathscr{M}(U)$, will be denoted as M. We shall assume that there are n *distinct* and *non-empty* sources, which allows us to write $M = \{S_{(1)},...,S_{(n)}\}$. For each source S, $M(S)$ is called the *source multiplicity* and it denotes the number of times the source S occurs in M. If we consider a merge function for multisets ϖ, then $\varpi(M)$ is a multiset over U. We say that $\varpi(M)$ is a *solution* and we denote a general solution as $\mathscr{S}$.

Example 1. *Let us consider a finite universe $U = \{a,b,c,d\}$ and let us consider the multiset $M = \{S_{(1)},S_{(1)},S_{(2)},S_{(3)}\}$ such that:*

$$S_{(1)} = \{a,a,b,c\}$$
$$S_{(2)} = \{b,b,b,c\}$$
$$S_{(3)} = \{a,c,d,d\}.$$

In this example, source $S_{(1)}$ has a multiplicity of 2, while other sources have a multiplicity of 1. An example of a merge function is $\varpi = \cup$. We then have:

$$\varpi(M) = \mathscr{S} = \bigcup_{S \in M} (S)$$
$$= \{a,a,b,b,b,c,d,d\}.$$

Note that with $\varpi = \cup$, the z-majority rule is never satisfied for any $z \in [0.5,1]$.

3.2 Measures of Quality

In order to define merge functions that provide an optimal behavior, we must first define measures of quality. For this purpose, we borrow three well known measures from the field of information retrieval: precision, recall and the f-value [7]. In order to cast these measures to the scope of this paper, we first define *local* measures of quality.

Definition 5 (Local precision). *Given a multiset of sources* $M = \{S_{(1)},...,S_{(n)}\}$*, the local precision of an element u is defined by:*

$$p^* : U \times \mathbb{N} \to [0,1] : (u,j) \mapsto p^*(u,j|M) \tag{17}$$

such that:

$$p^*(u,j|M) = \frac{1}{|M|} \sum_{S \in M \wedge S(u) \geq j} M(S). \tag{18}$$

The local precision measures the *correctness* of putting an element u with multiplicity j in the solution of a merge function, given that M represents the multiset of sources. Within the scope of Example 1, we can see that $S_{(1)}(b) < 2$, $S_{(2)}(b) \geq 2$ and $S_{(3)}(b) < 2$. Also $M(S_{(1)}) = 2$, $M(S_{(2)}) = 1$, $M(S_{(3)}) = 1$ and $|M| = 4$. We thus find that $p^*(b,2|M) = \frac{1}{4}(1) = 0.25$.

Definition 6 (Local recall). *Given a multiset of sources* $M = \{S_{(1)},...,S_{(n)}\}$*, the local recall of an element u is defined by:*

$$r^* : U \times \mathbb{N} \to [0,1] : (u,j) \mapsto r^*(u,j|M) \tag{19}$$

such that:

$$r^*(u,j|M) = \frac{1}{|M|} \sum_{S \in M \wedge S(u) \leq j} M(S). \tag{20}$$

The local recall measures the *completeness* of putting an element u with multiplicity j in the solution of a merge function, given that M represents the multiset of sources. Within the scope of Example 1, we can see that $S_{(1)}(a) > 1$, $S_{(2)}(a) \leq 1$ and $S_{(3)}(a) \leq 1$. We thus find that $r^*(a,1|M) = \frac{1}{4}(1+1) = 0.5$.

It can be easily seen that p^* (resp. r^*) is a decreasing (resp. increasing) function in terms of the second argument. Moreover, the following property can be identified.

Property 1. *For each* $j \in \mathbb{N}$*, local precision and local recall satisfy:*

$$\forall u \in U : r^*(u,j|M) = 1 - p^*(u,j+1|M). \tag{21}$$

We are now able to define the precision and recall of an arbitrary solution.

Definition 7 (Precision). *Given a multiset of sources $M = \left\{S_{(1)}, ..., S_{(n)}\right\}$, the precision of a solution is defined by:*

$$p : \mathscr{M}(U) \rightarrow [0,1] : \mathscr{S} \mapsto p(\mathscr{S}|M) \tag{22}$$

such that:

$$p(\mathscr{S}|M) = \underset{u \in U}{\mathrm{T}} \left(p^*(u, \mathscr{S}(u)|M) \right) \tag{23}$$

where T *is a triangular norm.*

Definition 8 (Recall). *Given a multiset of sources $M = \left\{S_{(1)}, ..., S_{(n)}\right\}$, the recall of a solution is defined by:*

$$r : \mathscr{M}(U) \rightarrow [0,1] : \mathscr{S} \mapsto r(\mathscr{S}|M) \tag{24}$$

such that:

$$r(\mathscr{S}|M) = \underset{u \in U}{\mathrm{T}} \left(r^*(u, \mathscr{S}(u)|M) \right) \tag{25}$$

where T *is a triangular norm.*

These definitions of precision and recall simply state that a solution is complete or correct, to the extent that *all* elements in the solution are complete or correct. The quantification "*all*" is hereby characterized by means of a triangular norm. A balance between precision and recall is usually expressed as the f-value.

Definition 9 (f-value). *Given a multiset of sources $M = \left\{S_{(1)}, ..., S_{(n)}\right\}$, the f-value of a solution is defined as the harmonic mean of precision and recall, i.e.:*

$$f : \mathscr{M}(U) \rightarrow [0,1] : \mathscr{S} \mapsto f(\mathscr{S}|M) \tag{26}$$

such that:

$$f(\mathscr{S}|M) = \frac{2 \cdot r(\mathscr{S}|M) \cdot p(\mathscr{S}|M)}{r(\mathscr{S}|M) + p(\mathscr{S}|M)}. \tag{27}$$

Note that for any solution $\mathscr{S}$, $f(\mathscr{S}|M) \neq 0$ if and only if both the precision and recall of that solution differ from zero. Precision and recall as defined here, satisfy some interesting properties. For example, for $M = \left\{S_{(1)}, ..., S_{(n)}\right\}$, we can see that:

$$\left(\mathscr{S} \supset \bigcup_{S \in M} S\right) \Rightarrow p(\mathscr{S}|M) = 0 \tag{28}$$

and

$$\left(\mathscr{S} \subset \bigcap_{S \in M} S\right) \Rightarrow r(\mathscr{S}|M) = 0. \tag{29}$$

As a result, the f-value of a solution that is a subset of the source intersection or a superset of the source union is always zero. For these reasons, we call the source intersection the *lower solution* (denoted by $\underline{\mathscr{S}}$) and the source union the *upper solution* (denoted by $\overline{\mathscr{S}}$). Moreover, we have that:

$$p(\underline{\mathscr{S}}|M) = 1 \tag{30}$$
$$r(\overline{\mathscr{S}}|M) = 1. \tag{31}$$

Definition 10 (f-optimal merge function). *A merge function ϖ over $\mathscr{M}(U)$ is f-optimal if it satisfies for any $M \in \mathscr{M}(\mathscr{M}(U))$:*

$$\varpi(M) = \arg \max_{\mathscr{S} \in \mathscr{M}(U)} f(\mathscr{S}|M) \tag{32}$$

constrained by:

$$\left(\max_{\mathscr{S} \in \mathscr{M}(U)} f(\mathscr{S}|M) = 0 \right) \Rightarrow \varpi(M) = \emptyset. \tag{33}$$

Note that an optimal solution $\mathscr{S}$ must satisfy:

$$\underline{\mathscr{S}} \subseteq \mathscr{S} \subseteq \overline{\mathscr{S}}. \tag{34}$$

For such solutions, we can provide a more strict definition of precision and recall:

$$p(\mathscr{S}|M) = \underset{u \in \overline{\mathscr{S}}}{\mathrm{T}} \left(p^*(u, \mathscr{S}(u)|M) \right)$$
$$r(\mathscr{S}|M) = \underset{u \in \overline{\mathscr{S}}}{\mathrm{T}} \left(r^*(u, \mathscr{S}(u)|M) \right)$$

where we now only consider elements that belong to the upper solution, i.e. elements that occur in at least one source. Note that the result of an f-optimal merge function is not unique. In case of ties, we assume that a consistent tie-breaking mechanism is at hand, but we shall not treat this mechanism within the scope of this paper. We say that a merge function ϖ is f-optimal under T if ϖ maximizes the f-value where precision and recall are calculated by means of T.

4 The Majority Rule

We shall now investigate the extent to which a z-majority rule is valid under a t-norm T. We can see that the following property holds for *any* merge function.

Property 2. *If a merge function ϖ satisfies the z_1-majority rule, then for any $z \in [z_1, 1[$ it also satisfies the z-majority rule.*

In the remainder of this paper, we shall assume that the multiset of sources $M = \{S_{(1)},...,S_{(n)}\}$ contains a source $S_{(m)}$ that satisfies:

$$M\left(S_{(m)}\right) > z \cdot |M|. \tag{35}$$

It is said that $S_{(m)}$ has a z-majority. We can see that the 1-majority rule can never be satisfied, because it is impossible that a source satisfies $M(S) > |M|$. We thus only study the case where $z \in [0.5, 1[$. We can see by means of some counterexamples that the strongest majority rule, i.e. for $z = 0.5$, does not hold in general.

Example 2. *Assume $U = \{a,b,c,d\}$ and assume that M contains two sources as follows:*

$$S_{(1)} = \{a,a,b,b,c,d\}$$
$$S_{(2)} = \{a,a,a,b,b,b\}$$

with $M\left(S_{(1)}\right) = 5$ and $M\left(S_{(2)}\right) = 4$. Under these conditions, we see that $S_{(1)}$ has a 0.5-majority. Assume that we calculate precision and recall by means of $\mathrm{T}_{\mathbf{M}}$. We have that:

$$p^*\left(a, S_{(1)}(a)|M\right) = 1$$
$$p^*\left(b, S_{(1)}(b)|M\right) = 1$$
$$p^*\left(c, S_{(1)}(c)|M\right) = 5/9$$
$$p^*\left(d, S_{(1)}(d)|M\right) = 5/9$$

and:

$$r^*\left(a, S_{(1)}(a)|M\right) = 5/9$$
$$r^*\left(b, S_{(1)}(b)|M\right) = 5/9$$
$$r^*\left(c, S_{(1)}(c)|M\right) = 1$$
$$r^*\left(d, S_{(1)}(d)|M\right) = 1$$

which means that both precision and recall of $S_{(1)}$ equal $5/9$. As such:

$$f\left(S_{(1)}|M\right) = 5/9. \tag{36}$$

If we consider an alternative solution $\mathscr{S} = \{a,a,b,b\}$, we find that:

$$p^*\left(a, \mathscr{S}(a)|M\right) = 1$$
$$p^*\left(b, \mathscr{S}(b)|M\right) = 1$$
$$p^*\left(c, \mathscr{S}(c)|M\right) = 1$$
$$p^*\left(d, \mathscr{S}(d)|M\right) = 1$$

and:

$$r^*(a, \mathscr{S}(a)|M) = 5/9$$
$$r^*(b, \mathscr{S}(b)|M) = 5/9$$
$$r^*(c, \mathscr{S}(c)|M) = 4/9$$
$$r^*(d, \mathscr{S}(d)|M) = 4/9$$

leading to a precision of 1 *and a recall of* $4/9$. *As such:*

$$f(\mathscr{S}|M) = 8/13. \tag{37}$$

We thus find that, although $S_{(1)}$ *has a* 0.5*-majority, it does not maximize the* f*-value. Similar examples can be found for* $T_{\mathbf{P}}$, $T_{\mathbf{L}}$ *and* $T_{\mathbf{D}}$.

An interesting question is thus whether, for any f-optimal merge function, there always exists at least one z such that the z-majority rule is satisfied. Let us therefore first show an interesting property of local precision and local recall in the case of a z-majority.

Property 3. *Assume a multiset of sources* $M = \{S_{(1)}, ..., S_{(n)}\}$ *and let* $S_{(m)}$ *have a* z*-majority. We then have for any* $u \in \overline{\mathscr{S}}$:

$$p^*(u, S_{(m)}(u)|M) > z$$
$$r^*(u, S_{(m)}(u)|M) > z$$

and for any $j \in \mathbb{N}$ *different from* m:

$$\bigvee \begin{pmatrix} p^*(u, S_{(j)}(u)|M) < 1 - z \\ r^*(u, S_{(j)}(u)|M) < 1 - z \end{pmatrix}. \tag{38}$$

Based on this property, we prove the following interesting theorem.

Theorem 1. *For any continuous t-norm* T, *there always exists a* $z \in [0.5, 1[$ *such that an* f*-optimal merge function under* T, *satisfies the* z*-majority rule.*

Proof. Let us denote $k = |(\overline{\mathscr{S}})_1|$, which means that there are k distinct elements that occur in an least one source. Taking into account Property 3, we then have that:

$$\min\left(p\left(S_{(m)}|M\right), r\left(S_{(m)}|M\right)\right) > \mathop{\mathrm{T}}_{i=1}^{k} z. \tag{39}$$

Due to the fact that the f-value is increasing in terms of both precision and recall, we thus have that:

$$f\left(S_{(m)}|M\right) > \mathop{\mathrm{T}}_{i=1}^{k} z. \tag{40}$$

In addition, we also have for any solution $\mathscr{S} \neq S_{(m)}$ that:

$$\min\left(p\left(\mathscr{S}|M\right), r\left(\mathscr{S}|M\right)\right) < 1 - z. \tag{41}$$

As such, an upper limit for the maximal f-value of an alternative solution $\mathscr{S}$ is given by:

$$\frac{2 \cdot (1 - z)}{2 - z}. \tag{42}$$

The z-majority rule is then satisfied for any z that satisfies the following inequality:

$$\mathop{\mathrm{T}}_{i=1}^{k} z > \frac{2 \cdot (1 - z)}{2 - z}. \tag{43}$$

Due to the fact that T is continuous and increasing and the fact that the right hand side is a continuous, decreasing function g that satisfies $g(1) = 0$, we can see that there always exists a $z \neq 1$, such that the above inequality is satisfied.

If we instantiate T with some well known continuous t-norms (Section 2.3), we can solve the inequality and provide z-majority rules for merge functions under those t-norms. For example, in the case of $\mathrm{T}_{\mathbf{M}}$ (in fact any non Archimedean t-norm), we get the inequality:

$$z > \frac{2 \cdot (1 - z)}{2 - z} \tag{44}$$

which can be solved easily and leads to the constraint $z > 2 - \sqrt{2}$. We see that if T is non Archimedean, the obtained majority rule does not depend on the number of elements in the upper solution, which is in fact the number of distinct elements that occur in the sources. This does not hold for Archimedean t-norms. Figure 1 shows

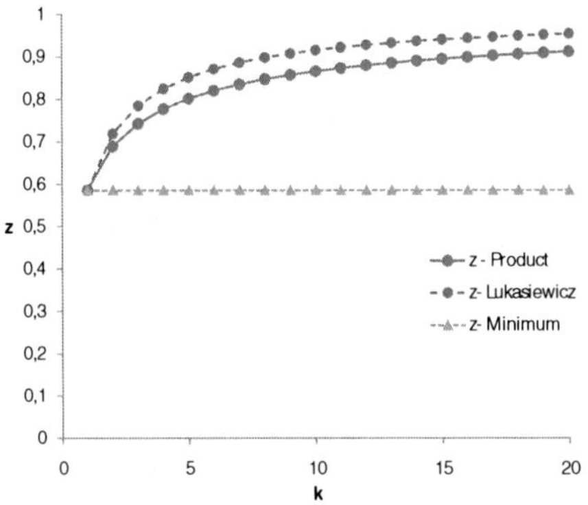

Fig. 1 z-majority rules under $\mathrm{T}_{\mathbf{M}}$, $\mathrm{T}_{\mathbf{P}}$ and $\mathrm{T}_{\mathbf{L}}$ in function of $k = |(\overline{\mathscr{S}})_1|$

the values of z that are obtained by solving the corresponding inequality in the case of T_M, T_P and T_L for different values of $k = |(\mathscr{S})_1|$. Note that, in the case of T_L, $\frac{k-1}{k}$ gives a strict lower bound for z because the left hand side of inequality (43) must be strictly greater than 0. We can also see that the case of T_D would lead to the limit of $z = 1$. We have noticed that the 1-majority rule can never be satisfied because a source in M can not have a multiplicity strictly larger than $|M|$. However, it can be *equal* to $|M|$. In that case, there is only one distinct source and the result of an f-optimal merge function will be equal to that source. This stems with the fact that an f-optimal merge function is *idempotent* under any t-norm (even non continuous t-norms).

It is important to observe that inequality (43) does not always provide the most *strict* majority rule. It is very well possible that for a combination of T and k, there exists a z' smaller than the solution of inequality (43) such that the z'-majority rule is satisfied. We can for example formulate the following theorem.

Theorem 2. *An f-optimal merge function under any* T*, satisfies the* 0.5*-majority rule if* $|(\mathscr{S})_1| = 1$.

Proof. . Assume $M = \{S_{(1)}, \ldots, S_{(n)}\}$ with $(\mathscr{S})_1 = \{a\}$ and let $S_{(m)}$ have a 0.5-majority. We can see that for any solution $\mathscr{S}$ and for any T, we have that:

$$p(\mathscr{S}|M) = p^*(a, \mathscr{S}(a)|M)$$
$$r(\mathscr{S}|M) = r^*(a, \mathscr{S}(a)|M).$$

As such, the f-value for a solution $\mathscr{S}$ is given by:

$$f(\mathscr{S}|M) = \frac{2 \cdot p^*(a, \mathscr{S}(a)|M) \cdot r^*(a, \mathscr{S}(a)|M)}{p^*(a, \mathscr{S}(a)|M) + r^*(a, \mathscr{S}(a)|M)}. \tag{45}$$

Let us now consider a solution $\mathscr{S} \neq S_{(m)}$. For such a solution, we identify two cases.

Case 1: $\mathscr{S}(a) > S_{(m)}(a)$. In this case, with:

$$x = p^*(a, \mathscr{S}(a)|M) \tag{46}$$

we have that $x < 0.5$, due to the 0.5-majority of $S_{(m)}$. In addition, the definition of p^* implies:

$$p^*\left(a, S_{(m)}(a)|M\right) > 0.5 + x \tag{47}$$

and Property 3 implies:

$$r^*\left(a, S_{(m)}(a)|M\right) > 0.5. \tag{48}$$

We can thus say that:

$$f\left(S_{(m)}|M\right) > \frac{0.5 + x}{1 + x} \tag{49}$$

and

$$f(\mathscr{S}|M) \leq \frac{2 \cdot x}{1 + x}. \tag{50}$$

We thus have that $f\left(S_{(m)}|M\right) > f(\mathscr{S}|M)$ if and only if:

$$0.5 + x \geq 2 \cdot x \tag{51}$$

or equivalently:

$$0.5 \geq x \tag{52}$$

which is true.

Case 2: $\mathscr{S}(a) < S_{(m)}(a)$. The second case can be proven similarly, but now with $x = r^*(a, \mathscr{S}(a)|M)$.

The case of Theorem 2 signifies the case where all sources in M contain the same value to a different extent. We could imagine this as a voting process where each source votes about the importance of a. In this specific case, we thus have that the 0.5-majority rule always holds, which is indeed a stronger rule than the one obtained by solving (43). As such, it is possible that in other specific cases, stronger majority rules can be found, but these stronger rules rely upon specific constraints or specific properties of T.

The z-majority rule as defined in this paper requires that a z-majority of a source $S_{(m)}$ leads to the equality $\varpi(M) = S_{(m)}$. However, in some cases it might be interesting to weaken the rule with respect to the required relation between $S_{(m)}$ and $\varpi(M)$. It can for example be proven that a 0.5-majority results in a weaker relation (weaker then =) between $S_{(m)}$ and $\varpi(M)$, provided that T is continuous and strict. This statement is formalized by the following theorem.

Theorem 3. *Assume a multiset of sources* $M = \left\{S_{(1)}, \ldots, S_{(n)}\right\}$ *and let* $S_{(m)}$ *be a source that has a* 0.5*-majority. Then:*

$$\exists u \in \overline{\mathscr{S}} : \varpi(M)(u) = S_{(m)}(u) \tag{53}$$

always holds if ϖ *is* f*-optimal under a strict and continuous t-norm* T.

Proof. The proof follows from the fact that, for any strict and continuous t-norm T, we have that:

$$\mathop{\mathrm{T}}_{i=1}^{k} x > \mathop{\mathrm{T}}_{i=1}^{k} 1 - x \tag{54}$$

if $x > 0.5$.

5 Multiple Document Summarization

As mentioned before, a possible application of multiset merging lies in the field of multiple document summarization. Hereby, a set of documents dealing with the same topic needs to be summarized into one document. An important step in the automated construction of such a summary, is the selection of content. If each document d is transformed into a multiset of words S (i.e. each document is regarded as a source), then the introduced mechanism can be applied. As an example, consider a multiset M of 27 articles and interviews concerning President Barack Obama's recovery plan for the United States of America, the so called economic stimulus bill. The transformation of documents into multiset is obtained by a standard tokenization (whitespace as delimiter), followed by the elimination of stopwords. This results in a universe of 2859 words. The f-optimal solution under T_M is given by ($f = 0.5185$):

$\varpi(M)$ = *[obama:6, house:4, bill:4, economic:3, stimulus:3, president:3, senate:3, plan:3, tax:2, jobs:2, republicans:2, state:1, conference:1, congress:1, version:1, washington:1, government:1, time:1, money:1, country:1, tuesday:1, support:1, work:1, billion:1, final:1, american:1, republican:1, monday:1, democrats:1, federal:1, people:1, economy:1, news:1, 000:1, package:1, secretary:1, spending:1, create:1, crisis:1, public:1, barack:1, week:1, white:1, cuts:1].*

This multiset provides us with the most important keywords that occur in the sources and can be used as a basis in the formulation of a readable summary. Majority rules as investigated in this paper can contribute to the faster construction of $\varpi(M)$ by reducing the search space in which the optimal solution must be searched.

6 Conclusion

In this paper, we have contributed to the study of merge functions for duplicate objects. More specific, the case where each object is a multiset is investigated. Applications can be found in multiple document summarization and merging of duplicate graphs. Our study of merge functions for multisets has led to the definition of f-optimal merge functions. The result of such a merge function is a multiset that maximizes the harmonic mean of precision (the correctness of the result) and recall (the completeness of the result). Both precision and recall rely on local measures and combine the results of those local measures by means of a t-norm T. We have mentioned some properties of f-optimal merge functions and we have shown that there always exists a z-majority rule if T is continuous. We have provided a general inequality for finding such z-majority rules, but is proven that this inequality does not always provide the most strict majority rule under T.

Acknowledgements. This work is supported by the Flemish Fund for Scientific Research (FWO-Vlaanderen).

References

1. Elmagarmid, A., Ipeirotis, P., Verykios, V.: Duplicate record detection: A survey. IEEE Transactions on Knowledge and Data Engineering 19(1), 1–16 (2007)
2. Bronselaer, A., De Tré, G.: Aspects of object merging. In: Proceedings of the NAFIPS Conference, Toronto, Canada, pp. 27–32 (2010)
3. Bronselaer, A., De Tré, G.: Properties of possibilistic string comparison. IEEE Transactions on Fuzzy Systems 18(2), 312–325 (2010)

4. Schweizer, B., Sklar, A.: Probabilistic metric spaces. Elsevier, Amsterdam (1983)
5. Fellegi, I., Sunter, A.: A theory for record linkage. American Statistical Association Journal 64(328), 1183–1210 (1969)
6. Lin, J., Mendelzon, A.: Knowledge base merging by majority. In: Dynamic Worlds: From the Frame Problem to Knowledge Management. Kluwer, Dordrecht (1994)
7. Ricardo, B.-Y., Berthier, R.-N.: Modern information retrieval. ACM Press, New York (1999)
8. Yager, R.: On the theory of bags. International Journal of General Systems 13(1), 23–27 (1986)
9. Konieczny, S., Pérez, R.: Merging information under constraints: a logical framework. Journal of Logic and Computation 12(1), 111–120 (2002)

Penalty Fuzzy Function for Derivative-Free Optimization

J. Matias, P. Mestre, A. Correia, P. Couto, C. Serodio, and P. Melo-Pinto

Abstract. Penalty and Barrier methods are normally used to solve Nonlinear Optimization Constrained Problems. The problems appear in areas such as engineering and are often characterized by the fact that involved functions (objective and constraints) are non-smooth and/or their derivatives are not know. This means that optimization methods based on derivatives cannot be used. A Java based API was implemented, including only derivative-free optimization methods, to solve both constrained and unconstrained problems, which includes Penalty and Barriers methods. In this work a new penalty function, based on Fuzzy Logic, is presented. This function imposes a progressive penalization to solutions that violate the constraints. This means that the function imposes a low penalization when the violation of the constraints is low and a heavy penalization when the violation is high. The value of the penalization is not known in beforehand, it is the outcome of a fuzzy inference engine. Numerical results comparing the proposed function with two of the classic penalty/barrier functions are presented. Regarding the presented results one can conclude that the proposed penalty function besides being very robust also exhibits a very good performance.

1 Introduction

In most scientific areas, optimization problems must be addressd. For some of these problems it is not possible to determine its objective function because it is a very complex task,or due to the costs involved, etc. In some cases the objective function might be non-smooth, have many local minima or be non differentiable. As a

J. Matias · P. Mestre · P. Couto · C. Serodio · P. Melo-Pinto
CM-UTAD, Apartado 1013, 5001-801 Vila Real, Portugal
e-mail: {j_matias,pmestre,pcouto,cserodio,pmelo}@utad.pt

A. Correia
CM-UTAD and CIICESI/ESTGF/IPP, 4610-156 Felgueiras, Portugal
e-mail: aic@estgf.ipp.pt

B. De Baets et al. (Eds.): Eurofuse 2011, AISC 107, pp. 293–301, 2011.
springerlink.com

consequence derivative based methods cannot be used to solve these problems, as presented in [10]. In such problems a possible solution to cope with these issues is to use direct search methods that do not use derivatives or approximations to them. For further details see [15], [14] and [13].

Possible Optimization problems that can be of two types: unconstrained optimization problems or constrained optimization problems. Usually constrained optimization problems can be presented in the form of (1):

$$\begin{array}{rl} \min\limits_{x \in \mathbb{R}^n} & f(x) \\ \text{subject to} & c_i(x) = 0, i \in \mathscr{E} \\ & c_i(x) \leq 0, i \in \mathscr{I} \end{array} \tag{1}$$

where:

- $f : \mathbb{R}^n \rightarrow \mathbb{R}$ is the objective function;
- $c_i(x) = 0$, $i \in \mathscr{E}$, with $\mathscr{E} = \{1,2,...,t\}$, define the problem equality constraints;
- $c_i(x) \leq 0$, $i \in \mathscr{I}$, with $\mathscr{I} = \{t+1,t+2,...,m\}$, represent the inequality constraints;
- $\Omega = \{x \in \mathbb{R}^n : c_i = 0, i \in \mathscr{E} \wedge c_i(x) \leq 0, i \in \mathscr{I}\}$ is the set of all feasible points, i.e., the feasible region.

One of the strategies to solve such problems is to transform them into unconstrained problems and then solve them using the methods/algorithms that usually are used to solve unconstrained problems. For this purpose, penalty/barrier functions can be used. These methods only need information about value of the objective and the constraints functions at some points, and only use this information comparing these values to find the next iteration.

In this paper a new approach to penalty/barrier functions is presented. A new penalty function, based on Fuzzy Logic was developed and it was tested and compared with some of the classic penalty/barrier functions normally used to solve optimization problems. The objective of this new function is to do a progressive penalization of functions that violate the constraints.

2 Penalty and Barrier Functions

2.1 The Process

Penalty and Barrier Methods, generally presented in Fig. 1, are built by two processes:

- External Process (EP) - where a succession of Unconstrained Optimization Problems is created;
- Internal Process (IP) - where the Unconstrained Optimization Problems are solved.

with the objective to solve the original Constrained Optimization Problem presented in (1).

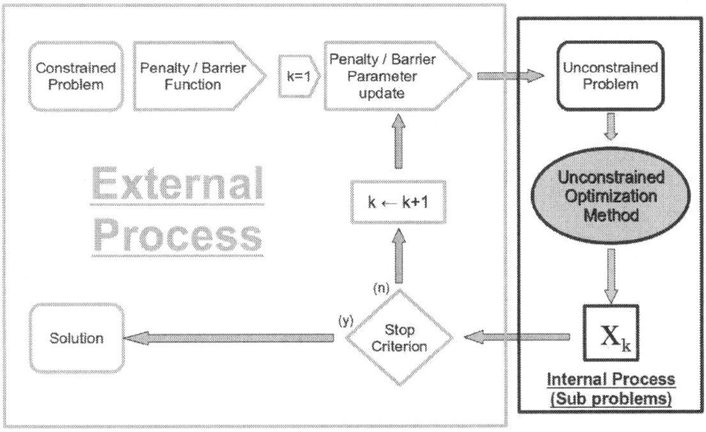

Fig. 1 Penalty and Barrier Methods - General Process

Using Penalty or Barrier functions a new objective function, Φ_k, is constructed, using information about the initial objective function, f, and the constraints functions. Therefore it is created a succession of Unconstrained Optimization Problems that depend on a positive parameter, r_k (Penalty/Barrier parameter) which solutions $x^*(r_k)$ converge to the solution of the initial problems x^* (*External Process*).

These Unconstrained Optimization Problems are then solved using Direct Search Methods (*Internal Process*), where the problem to be solved at each iteration k, is:

$$\min_{x_k \in \mathbb{R}^n} \Phi(x_k, r_k) = \min_{x_k \in \mathbb{R}^n} (f(x_k) + r_k p(x)) \tag{2}$$

where p is a function that penalizes (penalty) or refuses (barrier) points that violates the constraints.

In these methods optimality and feasibility are treated together.

In this work the following Penalty/Barrier functions were implemented and compared:

- Extreme Barrier Function (EB);
- ℓ_1 Penalty Function (ℓ_1);
- Fuzzy Penalty Function.

Barrier methods are more adequate for solving problems where a feasible initial point is know and when a feasible solution is needed. Penalty methods can be used with infeasible initial points, and the found solution can be infeasible. The fuzzy penalty function uses a different approach, seeking the potential of this theory and combining its strength with the no need for derivatives.

2.2 Extreme Barrier Function

Other authors such as Audet et. al., [1, 2, 4, 7, 6, 5] have used Extreme Barrier Function with Pattern Search Methods, which is defined by:

$$\Phi(x) = b_\Omega(x) = \begin{cases} f(x) & \text{if } x \in \Omega \\ +\infty & \text{if } x \notin \Omega \end{cases} \tag{3}$$

A new version of this method was presented by Audet et. al. [6]. This version admits infeasible initial points, that violate the inequality constraints but do not violate the equality constraints. However, this method was not implemented in this work.

2.3 Penalty Function ℓ_1

The penalty method ℓ_1 was presented initially by Pietrzykowski [16]. Although it was presented in 1969 it has been studied and used by many authors, for example, Gould et. al. in [12] and Byrd et. al. in [9] and it is also the basis for many other penalty methods proposed in the literature.

For the problem (1), its penalty is given by 4:

$$\min_{x\in\mathbb{R}^n} \ell_1^{(k)}(x,\mu) \tag{4}$$

with the ℓ_1 penalty function:

$$\ell_1^{(k)}(x,\mu) = f(x) + \mu \sum_{i=1}^{t} |c_i(x)| + \mu \sum_{i=t+1}^{m} \max[c_i(x), 0], \tag{5}$$

where $\mu \to +\infty$.

2.4 Fuzzy Penalty Function

The main objective of using Fuzzy Logic to determine the value of the penalty to apply to a constraint violation, is to create a progressive penalty function, having a good behaviour when used either with equality and inequality constraints. This means that functions with a low violation of the constraints will have a low penalty and the functions with higher violation values, will have a high penalty. Also the progression between a low penalty value and a high penalty value must be progressive.

First, the value of the constraint violation is calculated and then it is transformed into a grade of membership, i.e. the fuzzification is done. For this, a set of membership functions, presented in Fig. 2, are used. The violation of the constraint is then classified as 'Low Violation', 'Medium Violation', 'High Violation' or 'Very High Violation'.

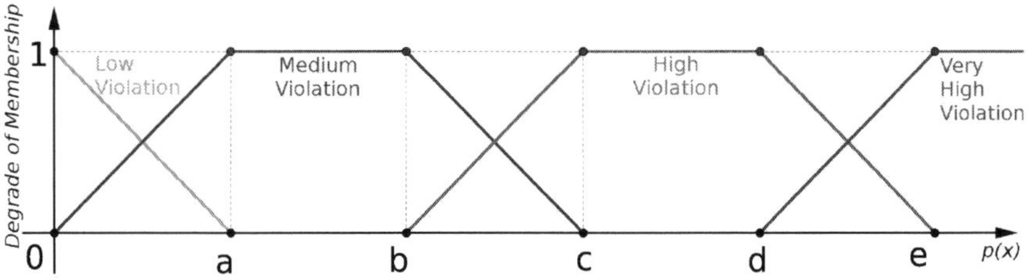

Fig. 2 Membership functions used

After this step, the penalty to be applied to the function is calculated using the fuzzy inference engine. In this step the following simple IF THEN rules are used:

- IF the violation is 'Low Violation' THEN the penalty is 'Low';
- IF the violation is 'Medium Violation' THEN the penalty is 'Medium';
- IF the violation is 'High Violation' THEN the penalty is 'High';
- IF the violation is 'Very High Violation' THEN the penalty is 'Very High';

To 'Very High', 'High', 'Medium' and 'Low' are assigned different weights, W_1, W_2, W_3 and W_4, such that $W_1 > W_2 > W_3 > W_4$. These values, and the values for a, b, c, d and e were chosen empirically. Also the number and the shape of the membership functions were chosen empirically.

The above presented steps are repeated for each constraint of the problem. If no violation occurs, the penalty will be zero, otherwise the penalty is calculated. The final value returned to the optimization method is the summation of all the calculated penalty values.

3 Derivative-Free Methods

Penalty and Barrier Methods can be used solve constrained problems by first transforming these problems into a sequence of unconstrained problems, of the form of (6). These problems are then solved, in the Internal Process, using methods typically used to solve Unconstrained Optimization Problems.

$$\min_{x \in \mathbf{R}^n} \Phi(x) \tag{6}$$

where, $\Phi : \mathbb{R}^n \rightarrow \mathbb{R}$ is the objective function.

To solve this type of problems, some derivative-free optimization algorithms were implemented. The implemented methods were:

- Opportunistic Coordinated search method;
- Hooke and Jeeves method;
- A version of Audet et. al. method;
- Nelder-Mead method;
- A Convergent Simplex method.

These methods are well known and can be found in the literature. The first three are Pattern Search Methods or Directional Direct-Search Methods (described, for example, in *Chapter 7 - Directional Direct-Search Methods*, by Conn et. al. [10]). These algorithms determine possible points using fixed search directions during the iterative process: starting at an iteration x_k, the next iteration will be found in a pattern or grid of points, in the fixed directions, at a distance s_k, said step size.

The last two are Simplex Methods (described, for example, in *Chapter 8 - Simplicial Direct-Search Methods*, by Conn et. al. [10]). These methods are characterized by constructing an initial simplex and change the directions of search in each iteration, using reflection, expansion and contraction movements and shrunk steps.

4 Numerical Results

4.1 Test Problems

A set of test problems was selected to assess the performance of the optimization methods combined with be penalty/barrier functions. These problems where selected from [3], the PA problem, from the CUTE collection [8], problems C801 and C802 and from the Schittkowski collection [17]. From the Schittkowski collection, the following fifteen problems were selected: S224; S225; S226; S227; S228; S231; S233; S234; S249; S264; S270; S323; S324; S325 and S326.

The choice of these eighteen test problems was not made in accordance with any special requirement, they were only used to assess the performance of the implemented methods, with special attention to the comparison between the classic functions and the proposed fuzzy penalty function.

4.2 Parameters and Return Values

External/Internal Process Parameters and Stopping Criterion:[1mm]

All the parameters used for the optimization methods are the same as presented in [11], except for the fuzzy penalty function which are: $W_1 = 100000$, $W_2 = 1000$, $W_3 = 10$, $W_4 = 0.01$, $a = 1$, $b = 2$, $c = 3$, $d = 4$ and $e = 5$.

The parameters of the stopping criteria are those presented in [11] with a tolerance of 10^{-5}

Return Values: The implemented penalty/barrier methods return several results, namely:

- Number of EP iterations - k;
- Number of Penalty/Barrier function evaluations - $nEvals$
- Last iteration - x_k;
- Value of the Penalty/Barrier function at the last iteration - $\Phi(x_k)$;
- Best feasible solutions found (if any) - x_{kf};
- Value of the objective function at the best feasible solution found - $f(x_{kf})$;

- Iteration where the best feasible solution was found - kf
- Best infeasible solutions found (if any) - x_{ki};
- Value of the objective function at the best infeasible solution found - $f(x_{ki})$;
- Value of the Penalty/Barrier function value at the best infeasible solution - $\Phi(x_{ki})$;
- Iteration were was found the best infeasible solution - ki
- Constraint violation value at the best infeasible solution, $V = \Phi(x_{ki}) - f(x_{ki})$.

These parameters were stored, for all the above presented problems, using all the five optimization methods combined with the three penalty/barrier functions. From the analysis of these results, the comparative tables presented in the results section were built.

4.3 Results

In Table 1 values reflecting the comparison of the number of admissible approaches to the solution, the number of solutions and the number of times that the method had the best performance in terms of calculations of the objective function, for the three penalty/barrier functions are presented. The values in this table are absolute values, the relative values for these tests are presented in Table 2.

Also a comparison of the performance of the used internal methods was extracted form the collected data, taking into consideration the same variables. The results of this comparison are presented presented in Table 3 and Table 4.

Analysing the obtained numerical results, from the experimental point of view, problem S325 from the Schittkowski collection, has not been solved by any combination of internal method versus penalty / barrier function;

Observing the Tables 1 and 2, one can conclude that, regarding the penalty/barrier functions:

- For most problems, even when convergence was not obtained, it was possible to obtain an admissible approximation;
- Good results were obtained for admissible approaches;
- Penalty fuzzy function has proved to be the most robust and the most efficient. It had the higher number of admissible approaches; the higher number of solutions; the higher number of times that the objective function was calculated less number of times.

For the internal methods, analysing Tables 3 and 4, it can be concluded that:

- Considering the number of obtained admissible approaches, the best internal methods were the Opportunistic Coordinate Search and the Hooke and Jeeves methods;
- Nelder Mead method obtained more solutions;
- The Convergent Simplex method was the more efficient, since it is the one with the fewer objective function calculations. Nevertheless, it only converged in 79.63% of tests.

Table 1 Obtained results by penalty/barrier function (absolute values)

Function	N. of Admissible Approaches [a]	N. of Solutions	N. of Less objective function calculations
Extreme Barrier	78	66	32
Penalty ℓ_1	71	67	28
Fuzzy Penalty	90	76	39

[a] A total of 90 tests.

Table 2 Obtained results by penalty/barrier function (relative values %)

Function	N. of Admissible Approaches	N. of Solutions	N. of Less objective function calculations
Extreme Barrier	86.7	73.3	35.6
Penalty ℓ_1	78.9	74.4	31.1
Fuzzy Penalty	100.0	84.4	43.3

Table 3 Obtained results by internal method (absolute values)

Internal Method	N. of Admissible Approaches [b]	N. of Solutions	N. of Less objective function calculations
Opportunistic Coordinate Search	49	39	21
Hooke and Jeeves	49	39	24
A version of Audet et. al.	48	43	20
Nelder-Mead	48	45	20
A Convergent Simplex	45	43	14

[b] A total of 54 tests.

Table 4 Obtained results by internal method (relative values %)

Internal Method	N. of Admissible Approaches [b]	N. of Solutions	N. of Less objective function calculations
Opportunistic Coordinate Search	90.74	72.22	38.89
Hooke and Jeeves	90.74	72.22	44.44
A version of Audet et. al.	88.89	79.63	37.04
Nelder-Mead	88.89	83.33	37.04
A Convergent Simplex	83.33	79.63	25.93

5 Conclusions and Future Work

In this work a novel approach to the penalty/barrier functions was presented. The proposed penalty function is based on Fuzzy Logic and its objective was to obtain a progressive penalty function. This penalty function was included in a Java-based

API presented in [11], built to solve both constrained and unconstrained optimization methods, without the used of derivatives.

The proposed function was compared with two of the most used penalty/barrier functions and the results show that the proposed method has a very good performance, for the used test problems. This means that this is a valid penalty function. Also, due to its characteristics, its association with direct search methods seems appropriate and efficient.

References

1. Audet, C.: Convergence results for pattern search algorithms are tight. Optimization and Engineering 2(5), 101–122 (2004)
2. Audet, C., Bchard, V., Digabel, S.L.: Nonsmooth optimization through mesh adaptive direct search and variable neighborhood search. J. Global Opt. (41), 299–318 (2008)
3. Audet, C., Dennis, J.: A pattern search filter method for nonlinear programming without derivatives. SIAM Journal on Optimization 5(14), 980–1010 (2004)
4. Audet, C., Dennis Jr., J.E.: Analysis of generalized pattern searches. SIAM Journal on Optimization 13(3), 889–903 (2002)
5. Audet, C., Dennis Jr., J.E.: Mesh adaptive direct search algorithms for constrained optimization. SIAM Journal on Optimization (17), 188–217 (2006)
6. Audet, C., Dennis, Jr., J.E.: A mads algorithm with a progressive barrier for derivative-free nonlinear programming. Tech. Rep. G-2007-37, Les Cahiers du GERAD, cole Polytechnique de Montral (2007)
7. Audet, C., Dennis Jr, J.E., Digabel, S.L.: Globalization strategies for mesh adaptative direct search.Tech. Rep. G-2008-74, Les Cahiers du GERAD, cole Polytechnique de Montral (2008)
8. Bongartz, I., Conn, A., Gould, N., Toint, P.: Cute: Constrained and unconstrained testing environment. ACM Transactions and Mathematical Software (21), 123–160 (1995)
9. Byrd, R.H., Nocedal, J., Waltz, R.A.: Steering exact penalty methods for nonlinear programming. Optimization Methods & Software 23(2), 197–213 (2008)
10. Conn, A.R., Scheinberg, K., Vicente, L.N.: Introduction to Derivative-Free Optimization. MPS-SIAM Series on Optimization. SIAM, Philadelphia (2009)
11. Correia, A., Matias, J., Mestre, P., Serdio, C.: Direct-search penalty/barrier methods. In: World Congress on Engineering 2010. Lecture Notes in Engineering and Computer Science, vol. 3, pp. 1729–1734. IAENG, London (2010)
12. Gould, N.I.M., Orban, D., Toint, P.L.: An interior-point l_1-penalty method for nonlinear optimization. Tech. rep., Rutherford Appleton Laboratory Chilton (2003)
13. Hooke, R., Jeeves, T.: Direct search solution of numerical and statistical problems. Journal of the Association for Computing Machinery 8(2), 212–229 (1961)
14. Kolda, T., Lewis, R., Torczon, V.: Optimization by direct search: New perspectives on some classical and modern methods. SIAM Review 45, 385–482 (2003)
15. Lewis, R., Torczon, V., Trosset, M.: Direct search methods: Then and now. J. Comput. Appl. Math. (124), 191–207 (2000)
16. Pietrzykowski, T.: An exact potential method for constrained maxima. SIAM Journal on Numerical Analysis 6(2), 299–304 (1969)
17. Schittkowski, K.: More Test Examples for Nonlinear Programming Codes, Economics and Mathematical Systems. Springer, Heidelberg (1987)

Part V

Applications

A PCA-Fuzzy Clustering Algorithm for Contours Analysis

Paulo Salgado and Getúlio Igrejas

Abstract. Principal component analysis (PCA) is a usefully tool for data compression and information extraction. It is often utilized in point cloud processing as it provides an efficient method to approximate local point properties through the examination of the local neighborhoods. This process does sometimes suffer from the assumption that the neighborhood contains only a single surface, when it may contain curved surface or multiple discrete surface entities, as well as relating the properties from PCA to real world attributes. This paper will present a new method that joins the fuzzy clustering algorithm with a local sliding PCA analysis to identify the non-linear relations and to obtain morphological information of the data. The proposed PCA-Fuzzy algorithm is performed on the neighborhood of the cluster center and normal approximations in order to estimate a tangent surface and the radius of the curvature that characterizes the trend and curvature of the data points or contour regions.

1 Introduction

One of the major objectives of data analysis is the extraction and instructive representation of the relevant information contained in the data. In cases of practical interest data are given by high-dimensional vectors corrupted by noise. Dimension reduction and elimination of noise is the essential step in analyzing the data Principal component analysis. PCA is one of the most prominent tool in this process.

PCA is designed to transform the original variables into new uncorrelated variables (axes) called the principal components, that are linear combinations of the

Paulo Salgado
Universidade de Trás-os-Montes e Alto Douro, Quinta de Prados, 5000 Vila Real
e-mail: psal@utad.pt

Getúlio Igrejas
Instituto Politécnico de Bragança, Campus de Sta. Apolónia, 5300 Bragança
e-mail: igrejas@ipb.pt

B. De Baets et al. (Eds.): Eurofuse 2011, AISC 107, pp. 305–311, 2011.
springerlink.com

original variables. The new axes lie along the directions of maximum variance. Principal component analysis as any other multivariate statistical method is sensitive to outliers, missing data, and poor linear correlation between variables due to poorly distributed variables. As a result, the classical principal components may describe the shape of the majority data incorrectly.

By uncovering the principal components of the data distribution, PCA creates a lower dimensional subspace which contains the relevant information of the data. Although highly successful in typical cases, PCA suffers from the drawback of being a linear method. However real world data manifolds besides of being nonlinear often are corrupted by noise and embedded into high dimensional spaces.

It is therefore necessary to apply robust methods that are resistant to possible outliers [5]. In this order, during the last decades, two robust approaches have been developed. The first is based on the eigenvectors of a robust covariance matrix such as the MCD-estimator or S-estimators of location and shape [10, 11], and is limited to relatively low dimensional data. The second approach is based on projection pursuit and can handle high-dimensional data [1].In this paper we discuss and apply a robust fuzzy PCA algorithm (FPCA). The efficiency of the new algorithm is illustrated on a data set concerning the open curved and a close circle clouds of points.

Point cloud processing relies on the analysis and examination of different observed attributes such as position, intensity and colour. These estimated attributes, which include curvature, surface normal and geometric surface properties, are of great importance in point cloud-processing procedures such as for surface classification and segmentation.

Often these attributes are estimated with the use of Principal Component Analysis (PCA) performed on the local neighborhoods of points, as it efficiently retrieves the local properties of a neighborhood. Some common uses have included approximating the normal direction [7], fitting first order planar surfaces [14], approximating surface curvature [9], defining the tensors for tensor voting [13] and providing local point coordinate systems [2].

There are two problems that can occur when using PCA. The first is that a neighborhood may contain multiple discrete surface entities. For most attributes, they are calculated under the assumption that there is only one surface structured. The effect of multiple surfaces can cause biases in the attributes, *e.g.* the surface normal approximation. While decreasing the size of the neighborhood may help in reducing the probability that a neighborhood contains more than one sampled surface, the neighborhood needs to be of sufficient size in order to reduce the effect of random errors and noise.

The other problem is how to relate the PCA results to the surface attributes. Often information can be lost, such as the approximation of curvature through surface variation where the comparable level of curvature is indicated, but there is no directional component or unit of measurement associated with the approximation.

The proposed method is a first step to elaborate a new automatic contour tracking system, *EdgeTrack*, for the ultrasound image sequences. The noise and unrelated high–contrast edges in ultrasound images make it very difficult to automatically detect the correct membrane surfaces. In our tracking system, a novel active contour model is developed. Unlike the classical active contour models which only use gradient of the image as the image force, the proposed model incorporates the edge gradient and intensity information in local regions around each snake element. The contour orientation is also taken into account so that any unnecessary edges in ultrasound images will be discarded.

It is the aim of this paper to present an iterative method of adjusting the neighborhood (clustering region) to remove effects of multiple surface entities. In addition, a formula between the eigenvalues of the PCA and the surface properties such as the radius and direction of local curvature will be presented. From this formula, the maximum and minimum curvature directions can be also calculated in a closed form solution and this information provides a novel and detailed information of the neighborhood of the point of interest.

2 Fuzzy Principal Component Analysis

Suppose observations $\boldsymbol{x}_k \in \Re^p$, $k = 1,2,3,\ldots,n$. The input data point cannot be directly used in fuzzy principal component analysis, so the first thing is to fuzzify the data set and get memberships and centroids. One method is that we can run FCM algorithm on training input data set until termination, to obtain membership matrix U and centroid V. But FCM is sensitive to noises or outliers [6]. Another method is shown as follows, and can reduce the effect of noises or outliers.

2.1 *Fuzzy Clustering*

Fuzzy clustering is an important tool to identify the structure in data. In general, a fuzzy clustering algorithm with objective function can be formulated as follows: let $X = \{\boldsymbol{x}_i,\ldots,\boldsymbol{x}_k,\ldots,\boldsymbol{x}_n\} \in \Re^p$ be a finite set of feature vectors, where n is the number of objects (measurements) and p is the number of the original variables, x. $S = \{\mathbf{S}_i,\ldots,\mathbf{S}_c\}$ is a s–tuple object prototypes, each one charactering one of the c clusters composing the cluster substructure of the data set; a partition of X into s–fuzzy clusters will be performed by minimizing the objective function:

$$J(U,S) = \sum_{i=1}^{c}\sum_{k=1}^{n} (u_{ik})^m d^2(\boldsymbol{x}_k, S_i)\,, \tag{1}$$

where $U = \{u_1,\ldots,u_c\}$ is the fuzzy partition, $u_{ik} = u_i(\boldsymbol{x}_k) \in [0,1]$ represents the membership degree of feature point $\boldsymbol{x}_k$ to cluster u_i, $d(\boldsymbol{x}_k, S_i) = d_{ik}$ is the distance from a feature point $\boldsymbol{x}_k$ to the prototype of the ith cluster.

The optimal fuzzy set will be determined by using an iterative method where J is successively minimized with respect to U and S. Supposing that S is given, the minimum of the function $J(\cdot,S)$ is obtained for:

$$u_{ik} = \frac{1}{\sum_{j=1}^{c} \left(\frac{d_{ik}}{d_{jk}} \right)^{\frac{1}{m-1}}} , \tag{2}$$

A cluster can have different shapes, depending on the choice of prototypes. The calculation of the membership values is dependent on the definition of the distance measure. According to the choice of prototypes and the definition of the distance measure, different fuzzy clustering algorithms are obtained. If the prototype of a cluster is a point the cluster center it will give spherical clusters, if the prototype is a line, it will give tubular clusters, and so on. In view of the linear form of the consequence part in linear fuzzy models, an obvious choice of fuzzy clustering was the Generalized fuzzy n-means algorithm [8], in which linear or planar clusters are allowed as prototypes to be sought.

In this paper a new type of prototype of a cluster is proposed, which is characterized by a cluster center point and a sliding vector, a tangential vector of a surface (general curved). This vector is the main principal component vector of a sliding PCA method of cluster data. The algorithm permits the determination of the u_{ik} values that best describe the fuzzy set U and the relation with its nonlinear (curved) prototype, which is a natural extension of the fuzzy PCA algorithm.

2.2 The Fuzzy PCA Clustering Algorithm

The Fuzzy clustering algorithm separates the set of all data points X (in analysis) into c fuzzy subsets. The membership function (Eq.2) gives the membership degree of each point $\boldsymbol{x}_k$ to cluster i . Let X_i be the subset with the associated membership $U_i = \{u_i(\boldsymbol{x}_k), \boldsymbol{x}_k \in X\}$. For simplicity we will omit the index i. We can obtain the following centroid for FPCA from the centroid of FCM:

$$\boldsymbol{v} = \frac{1}{\rho} \sum_{k=1}^{n} u_k^m \boldsymbol{x}_k , \tag{3}$$

where $\rho = \sum_{k=1}^{n} u_k^m$ with m a fuzzifier parameter. We can obtain the following fuzzy covariance matrix for FPCA from the fuzzy covariance matrix:

$$C = \frac{1}{\rho} \sum_{k=1}^{n} u_{ki}^m (\boldsymbol{x}_k - \boldsymbol{v}) (\boldsymbol{x}_k - \boldsymbol{v})^T , \tag{4}$$

FPCA diagonalizes the fuzzy covariance matrix C_{fpca} and finds the eigenvectors $\boldsymbol{e}$ and the corresponding eigenvalues λ. To do this, we can solve the following eigenequation:

$$\lambda \boldsymbol{e} = C_{fpca} \boldsymbol{e} , \tag{5}$$

By transforming an input data point, $\boldsymbol{x}_k$ we get:

$$y_k = \boldsymbol{e}^T (\boldsymbol{x}_k - \boldsymbol{v}) , \tag{6}$$

In Eq.(6), the eigenvectors $\boldsymbol{e}$ constitute an orthogonal coordinate system, and y_k can be seen as the coordinates in the the orthogonal base. Instead of using all the eigenvectors of the fuzzy covariance matrix, we may represent the data in terms of only a few basis vectors of the orthogonal basis. For a local neighborhood in $\Re^3$ of a point cloud, $\boldsymbol{e}_0$ approximates the local surface normal, with $\boldsymbol{e}_1$ and $\boldsymbol{e}_2$ approximating the tangential plane through $\boldsymbol{v}$. This result is equivalent to a first order least squares plane fit.

2.3 The Sliding FPCA Algorithm

In the PCA Fuzzy clustering the static PCA analysis in each cluster is not enough to go with the non-linear distribution geometrical data. In contrast, we propose a sliding analysis PCA method where the main components of the correlation matrix are sliding in data domain. Now, the principal component vector $\boldsymbol{e}_0(s)$ changes its direction, following the maximum variance of data, function of the sliding s variable. In this process the center of the data $\boldsymbol{v}(s)$, here designed as sliding cluster center (SCC), and the local density of probability is dragging by $\boldsymbol{e}_0(s)$.

The covariance matrix is a function of the sliding variable s:

$$C_i(s) = \frac{1}{\rho}\sum_{k=1}^{n} u_{ik}^{m}\varphi(\boldsymbol{x}_k - \boldsymbol{v}(s), \boldsymbol{e}_0)(\boldsymbol{x}_k - \boldsymbol{v}(s))(\boldsymbol{x}_k - \boldsymbol{v}(s))^T, \tag{7}$$

where $\varphi(\boldsymbol{x}_k - \boldsymbol{v}(s), \boldsymbol{e}_0)$ is the directional probability distribution of the data, *i.e.* a local probability distribution projected on the principal component vector $\boldsymbol{e}_0$, from the local center $\boldsymbol{v}(s)$:

$$\varphi_k(s) = \frac{1}{1 + e^{\lambda(\boldsymbol{x}_k - \boldsymbol{v}(s))^T \boldsymbol{e}_0(s)}}, \tag{8}$$

Eq. (7) is a local directional covariance matrix of local data centered in $\boldsymbol{v}$. In other words the projection of a data point $\boldsymbol{x}_k$ is given by its closest point on the principal component axis (in direction of $\boldsymbol{e}_0$). The projections agree with the center of gravity of the projectors. In this sense one may say that the principal component is running through the middle of its data points. $\boldsymbol{e}_0$ is also the velocity vector of the center of the sliding cluster center:

$$\boldsymbol{e}_0(s) = \frac{\delta \boldsymbol{v}(s)}{\delta s}, \tag{9}$$

This definition can be carried over to the nonlinear case. Hence a principal curve is defined as running through the center of gravities of the points projecting to it.

After the sliding process, SCC is recalculated by local mean gravity center:

$$\boldsymbol{v}(s) = \frac{1}{\rho(s)}\sum_{k=1}^{n} u_k^m \varphi_k(s)\boldsymbol{x}_k, \tag{10}$$

where $\rho(s) = \sum_{k=1}^{n} u_k^m \varphi_k(s)$. Resuming, the fuzzy set, in this case, may be characterized by a linear prototype, denoted $S(u, \boldsymbol{v}, \boldsymbol{e}_0)$, where $\boldsymbol{v}$ is the sliding cluster center of the class and $\boldsymbol{e}$, with $\|e\| = 1$, is the main direction. This line is named the first principal component for the set, and its direction is given by the unit eigenvector $\boldsymbol{e}$ associated with the largest eigenvalue λ_{max} of sliding covariance matrix (7), which is a slight generalization for fuzzy sets of the classical covariance matrix.

2.4 *Practical Application*

The present algorithm will be used to extract nonlinear data model so that non or multiply connected data structures can be represented without any prior knowledge of the number of components and their respective structure. The algorithm presented above rests on finding the data topology automatically, by finding the sliding center and tangential principal component vector and by combining the sliding PCA method with the fuzzy clustering algorithm.

The algorithm tries to adapt the center v and vector $\boldsymbol{e}_0$ to the chain of the data points as close as possible under the smoothness constraint. Two examples are shown is Figure 1a and Figure 1b. The first data set is an open cloud of points and the second case is a circular distribution. In both case, the algorithm has efficiently found the topology of the data distribution.

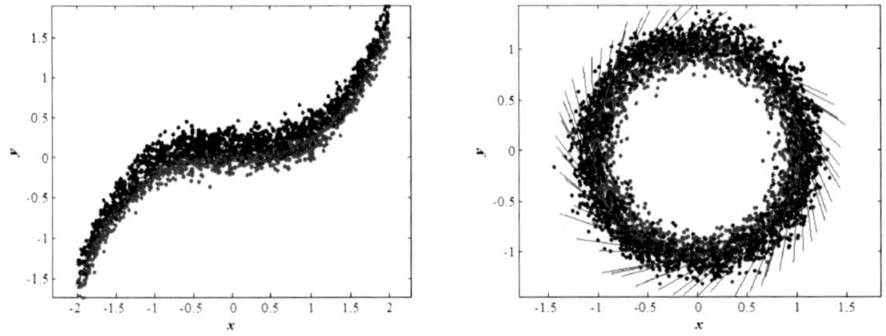

(a) Open cloud of points. **(b)** Closed cloud of points.

Fig. 1 The sliding cluster center $\boldsymbol{v}$ and the tangential principal component vector $\boldsymbol{e}_0$ (red lines).

3 Conclusions

A fuzzy principal component analysis method for robust estimation of principal components has been applied in this paper. The efficiency of the new algorithm was illustrated on estimation of contours (trends and curvature) of data sets.

The Sliding FPCA method achieved better results mainly because it is more compressible than classical PCA, *i.e.* the first fuzzy principal component accounts for significantly more of the variance than their classical counterparts, estimating the non-linear relations on the data. Moreover, the principal component vector is now a sliding tangential vector that follows the curvature (topology) of the clouds of points. The advantage of the proposed algorithm is the result of the combination of the fuzzy clustering strategy, useful for the reparation of the data set in a characteristic pattern region, with the sliding PCA method, which is capable to follow non-linear relations of data cluster data points.

References

1. Crouxa, C., Ruiz-Gazen, A.: High breakdown estimators for principal components: the projection-pursuit approach revisited. Journal of Multivariate Analysis 95, 206–226 (2005)
2. Daniels, J.D., Ha, L., Ochotta, T., Silva, C.T.: Robust smooth feature extraction from point clouds. In: IEEE International Conference on Shape Modeling and Applications 2007 (SMI 2007), Lyon, France, pp. 123–136 (2007)
3. Hubert, M., Rousseeuw, P.J., Verboven, S.: Chemom. Intell. Lab. Syst. 60, 101–111 (2002)
4. Hubert, M., Engelen, S.: Bioinformatics 20, 1728–1736 (2004)
5. Kafadar, K.: The influence of John Tukey's work in robust methods for chemometrics and environmetrics, Chemom. Chemom. Intell. Lab. Syst. 60, 127–134 (2002)
6. Lin, C.-F., Wang, S.-D.: Training algorithms for fuzzy support vector machines with noisy data. Pattern Recognition Letters 1656, 1647–1656 (2004)
7. Mitra, M.J., Nguyen, A., Guibas, L.: Estimating surface normals in noisy point cloud data. International Journal of Computational Geometry and Applications 14(4,5), 261–276 (2004)
8. Luukka, P.: Classification based on fuzzy robust PCA algorithms and similarity classifier. Expert Systems with Applications 36, 7463–7468 (2009)
9. Pauly, M., Gross, M., Kobbelt, L.P.: Efficient simplification of point-sampled surfaces. In: VIS 2002: Proceedings of the Conference on Visualization 2002, pp. 163–170. IEEE Computer Society, Boston (2002)
10. Rousseeuw, P.J., Leroy, A.: Robust Regression and Outlier Detection. Wiley, New York (1987)
11. Rousseeuw, P.J., Hubert, M.: Robust statistics for outlier detection. Wiley Interdisciplinary Reviews: Data Mining and Knowledge Discovery 1(1), 73–79 (2011)
12. Sârbu, C., Pop, H.F.: Fuzzy soft-computing methods and their applications in chemistry. In: Lipkowitz, K.B., Boyd, D.B., Cundari, T.R. (eds.) Reviews in Computational Chemistry, ch. 5, pp. 249–332. Wiley–VCH, Weinheim (2004)
13. Tong, W.-S., Tang, C.-K., Mordohai, P., Medioni, G.: First order augmentation to tensor voting for boundary inference and multiscale analysis in 3D. IEEE Transactions on Pattern Analysis and Machine Intelligence 26(5), 294–611 (2004)
14. Weingarten, J., Gruener, G., Siegwart, R.: A fast and robust 3d feature extraction algorithm for structured environment reconstruction. In: Proceedings of 11th International Conference on Advanced Robotics (ICAR), Portugal (2003)

Color Image Magnification with Interval-Valued Fuzzy Sets

Aranzazu Jurio, Miguel Pagola, Humberto Bustince, and Gleb Beliakov

Abstract. In this work we present a simple magnification algorithm for color images. It uses Interval-Valued Fuzzy Sets in such a way that every pixel has an interval membership constructed from its original intensity and its neighbourhood's one. Based on that interval membership, a block is created for each pixel, so this is a block expansion method.

1 Introduction

Image magnification [16, 13, 10, 11] is used in many applications nowadays. For instance, to upload images to a web page or to show images in devices such as mobile phones, PDAs or screens. Some of these devices have very limited memory, so they need to use simple image magnification algorithms.

In the literature, there exist several techniques for image magnification, whether based on one image or based on several ones. The most frecuently used methods working with a single image are based on interpolation [1, 8]. Common algorithms such as nearest neighbour or bilinear interpolation are computationally simple, but they suffer from smudge problems, especially in the areas containing edges. Nevertheless, linear approximations are the most popular ones based on their low computational cost, even providing worse results than cubic interpolation or splines.

Methods working with several images are used to enlarge individual images in learning frameworks [6, 9], but their main application is the magnification of video sequences [12, 14].

Aranzazu Jurio · Miguel Pagola · Humberto Bustince
Universidad Publica de Navarra, Pamplona (Spain)
e-mail: {aranzazu.jurio,miguel.pagola,bustince@}unavarra.es

Gleb Beliakov
Deakin University, Burwood (Australia)
e-mail: gleb@deakin.edu.au

B. De Baets et al. (Eds.): Eurofuse 2011, AISC 107, pp. 313–324, 2011.
springerlink.com

Our approach is based only in one image, where the magnified image is obtained by joining different constructed blocks. To create each block we use interval-valued fuzzy theory. This theory has been widely used in image processing to solve problems like edge detection [3], filtering [2] or segmentation [15, 4, 7].

In this sense, we present a new method to associate an interval-valued fuzzy set (IVFS) to the original image. The interval membership of each pixel represents its original intensity and its neighbourhoods' one, being the length of that membership a measure of the variation of intensities in the neighbourhood of that pixel. The block associated with each pixel is constructed using this interval, maintaining the intensity of the original pixel in the center of the block and filling in the rest using the relation between that pixel and its neighbours.

This work is organized as follows: is Section 2 we recall some preliminary definitions. In Section 3 we show the construction method of IVFSs. The magnification algorithm is described in detail in Section 4. We finish this work with some illustrative examples in Section 5 and conclusions in Section 6.

2 Preliminaries

Let us denote by $L([0,1])$ the set of all closed subintervals in $[0,1]$, that is,

$$L([0,1]) = \{\mathbf{x} = [\underline{x},\overline{x}] | (\underline{x},\overline{x}) \in [0,1]^2 \text{ and } \underline{x} \leq \overline{x}\}.$$

$L([0,1])$ is a lattice with respect to the relation $\leq_L$, which is defined in the following way. Given $\mathbf{x},\mathbf{y} \in L([0,1])$,

$$\mathbf{x} \leq_L \mathbf{y} \text{ if and only if } \underline{x} \leq \underline{y} \text{ and } \overline{x} \leq \overline{y}.$$

The relation above is transitive, antisymmetric and it expresses the fact that $\mathbf{x}$ strongly links to $\mathbf{y}$, so that $(L([0,1]), \leq_L)$ is a complete lattice, where the smallest element is $0_L = [0,0]$, and the largest is $1_L = [1,1]$.

Definition 1. *An interval-valued fuzzy set A on the universe $U \neq \emptyset$ is a mapping $A : U \to L([0,1])$.*

We denote by $IVFSs(U)$ the set of all IVFSs on U. Similarly, $FSs(U)$ is the set of all fuzzy sets on U.

From now on, we denote by $W([\underline{x},\overline{x}])$ the length of the interval $[\underline{x},\overline{x}]$; that is, $W([\underline{x},\overline{x}]) = \overline{x} - \underline{x}$.

Definition 2. *Let $\alpha \in [0,1]$. The operator $K_\alpha : L([0,1]) \to [0,1]$ is defined as a convex combination of the bounds of its argument, i.e.*

$$K_\alpha(\mathbf{x}) = \underline{x} + \alpha(\overline{x} - \underline{x})$$

for all $\mathbf{x} \in L([0,1])$.

Clearly, the following properties hold:

1. $K_0(\mathbf{x}) = \underline{x}$ for all $\mathbf{x} \in L([0,1])$,
2. $K_1(\mathbf{x}) = \overline{x}$ for all $\mathbf{x} \in L([0,1])$,
3. $K_\alpha(\mathbf{x}) = K_\alpha([K_0(\mathbf{x}), K_1(\mathbf{x})]) = K_0(\mathbf{x}) + \alpha(K_1(\mathbf{x}) - K_0(\mathbf{x}))$ for all $\mathbf{x} \in L([0,1])$.

Let $A \in IVFSs(U)$ and $\alpha \in [0,1]$. Then, we denote by $K_\alpha(A)$ the fuzzy set

$$K_\alpha(A) = \{u_i, K_\alpha(A(u_i))|u_i \in U\}.$$

Proposition 1. *For all $\alpha, \beta \in [0,1]$ and $A, B \in IVFSs(U)$, it is verified that*

(a)If $\alpha \leq \beta$, then $K_\alpha(A) \leq K_\beta(A)$.
(b)If $A \leq_L B$ then $K_\alpha(A) \leq K_\alpha(B)$.

where $\leq$ is Zadeh's order relation.

3 Construction of Interval-Valued Fuzzy Sets

In this section we propose a method to associate an image with an IVFS. We demand two properties to this method: the first one is that the intensity of every pixel in the original image must belong to the interval membership associated with it. The second one is that the length of each interval membership must depend on the intensities of the original pixel and its neighbours'. In this sense, we represent the variation of the intensities around each pixel, adjusted by a scaling factor (δ), by the length of the interval.

Proposition 2. *The mapping $F : [0,1]^2 \times [0,1] \to L([0,1])$ given by*

$$F(x,y,\delta) = [\underline{F}(x,y,\delta), \overline{F}(x,y,\delta)]$$

where

$$\underline{F}(x,y,\delta) = x(1-\delta y)$$
$$\overline{F}(x,y,\delta) = x(1-\delta y) + \delta y$$

satisfies that:

1. *$\underline{F}(x,y,\delta) \leq x \leq \overline{F}(x,y,\delta)$ for all $x \in [0,1]$;*
2. *$F(x,0,\delta) = [x,x]$;*
3. *$F(0,y,\delta) = [0,\delta y]$;*
4. *$F(x,y,0) = [x,x]$;*
5. *$W(F(x,y,\delta)) = \delta y$.*
6. *If $y_1 \leq y_2$ then $W(F(x,y_1,\delta)) \leq W(F(x,y_2,\delta))$ for all $x, \delta \in [0,1]$;*

Theorem 1. *Let $A_F \in FSs(U)$ and let $\omega, \delta : U \to [0,1]$ be two mappings. Then*

$$A = \{(u_i, A(u_i) = F(\mu_{A_F}(u_i), \omega(u_i), \delta(u_i)))|u_i \in U\}$$

is an Interval-Valued Fuzzy Set.

Corollary 1. *In the setting of Theorem 1, if for every* $u_i \in U$ *we take* $\delta(u_i) = 1$ *then*

$$\omega(u_i) = W(F(\mu_{A_F}(u_i), \omega(u_i), 1)).$$

Notice that under the conditions of Corollary 1 the set A is given as follows:

$$A = \{(u_i, \mu_{A_F}(u_i)(1-\omega(u_i)), \\ \mu_{A_F}(u_i)(1-\omega(u_i)) + \omega(u_i)) | u_i \in U\}$$

Example 1. Let $U = \{u_1, u_2, u_3, u_4\}$ and let $A_F \in FSs(U)$ given by

$$A_F = \{(u_1, 0.3), (u_2, 1), (u_3, 0.5), (u_4, 0.8)\}$$

and $\omega(u_i) = 0.3, \delta(u_i) = 1$ for all $u_i \in U$. By Corollary 1 we obtain the following Interval-Valued Fuzzy Set:

$$A = \{(u_1, [0.21, 0.51]), (u_2, [0.7, 1.00]), \\ (u_3, [0.35, 0.65]), (u_4, [0.56, 0.86])\}$$

4 Magnification Algorithm

In this section we propose a color image magnification algorithm based on block expansion, that uses IVFSs and the K_α operators.

We consider a color image Q in the RGB space as a $N \times M \times 3$ matrix. Each coordinate of the pixels in the image Q is denoted by (i, j, k). The normalized intensity of the pixel located at (i, j, k) is represented as q_{ijk}, with $0 \leq q_{ijk} \leq 1$ for each $(i, j, k) \in Q$.

The purpose of our algorithm is, given an image Q of dimension $N \times M \times 3$, to magnify it $n \times m$ times; that is, to build a new image of dimension $N' \times M' \times 3$ with $N' = n \times N$, $M' = m \times M$, $n, m \in \mathbb{N}$ with $n \leq N$ and $m \leq M$. We denote $(n \times m)$ as magnification factor.

Algorithm 1 presents the method we propose.

We explain the steps of this algorithm by means of an example. Given an image in Figure 1 of dimension $5 \times 5 \times 3$, we want to build a magnified image of dimension $15 \times 15 \times 3$ (magnification factor=(3×3)).

Step 1. Take $\delta \in [0, 1]$

In the example we take the middle value, $\delta = 0.5$.

Step 2.1 Fix a grid V of dimension $m \times n \times 1$ centered at each pixel.

INPUT: Q original image, $(n \times m)$ magnification factor.

1. Take $\delta \in [0,1]$.
2. FOR each pixel in each channel (i,j,k) DO

   2.1 Fix a grid V of dimension $n \times m \times 1$ centered at (i,j,k).
   2.2 Calculate W as the difference between the largest and the smallest intensities of the pixels in V.
   2.3 Build the interval $F(q_{ijk}, W, \delta)$.
   2.4 Build a block V' equal to V.
   2.5 FOR each element (r,s) of V' DO

   $$q_{rs} := K_{q_{rs}}(F(q_{ijk}, W, \delta)).$$

   ENDFOR
   2.6 Put the block V' in the magnified image.

   ENDFOR

Algorithm 1

0.60	0.65	0.68	0.70	0.70
0.60	0.68	0.70	0.72	0.73
0.70	0.69	0.69	0.73	0.75
0.15	0.14	0.15	0.17	0.19
0.13	0.16	0.12	0.15	0.21

(a)

0.20	0.20	0.25	0.25	0.30
0.20	0.21	0.20	0.20	0.30
0.22	0.20	0.25	0.60	0.65
0.60	0.65	0.62	0.65	0.70
0.65	0.70	0.70	0.73	0.75

(b)

0.79	0.83	0.85	0.82	0.86
0.81	0.82	0.79	0.83	0.81
0.80	0.81	0.50	0.83	0.84
0.52	0.50	0.51	0.51	0.54
0.50	0.51	0.53	0.52	0.53

(c)

Fig. 1 Example: original color image. (a) The R channel of the image. (b) The G channel of the image. (c) The B channel of the image.

This grid V represents the neighborhood that is used to build the interval. In the example, for pixel $(2,3,1)$ (marked in dark gray in Figure 2), we fix a grid of dimension 3×3 around it (in light gray).

Step 2.2. Calculate W as the difference between the largest and the smallest of the intensities of the pixels in V

For pixel $(2,3,1)$, we calculate W as:

$$\begin{aligned} W =& \max(0.65,0.68,0.70,0.68,0.70,0.72,0.69,0.69,0.73) - \\ & \min(0.65,0.68,0.70,0.68,0.70,0.72,0.69,0.69,0.73) = \\ & = 0.73 - 0.65 = 0.08 \end{aligned}$$

0.60	0.65	0.68	0.70	0.70
0.60	0.68	0.70	0.72	0.73
0.70	0.69	0.69	0.73	0.75
0.15	0.14	0.15	0.17	0.19
0.13	0.16	0.12	0.15	0.21

(a)

0.20	0.20	0.25	0.25	0.30
0.20	0.21	0.20	0.20	0.30
0.22	0.20	0.25	0.60	0.65
0.60	0.65	0.62	0.65	0.70
0.65	0.70	0.70	0.73	0.75

(b)

0.79	0.83	0.85	0.82	0.86
0.81	0.82	0.79	0.83	0.81
0.80	0.81	0.50	0.83	0.84
0.52	0.50	0.51	0.51	0.54
0.50	0.51	0.53	0.52	0.53

(c)

Fig. 2 Example: Grid V in original image for pixel $(2,3,1)$.

W is the maximum length of the interval associated with the pixel. The final length is calculated scaling it by the factor δ chosen in Step 1.

Step 2.3. Build interval $F(q_{ijk}, W, \delta)$

We associate to each pixel an interval of length $\delta \cdot W$ by the method explained in Section 3:

$$F(q_{ijk}, \delta \cdot W) = [q_{ijk}(1 - \delta \cdot W), q_{ijk}(1 - \delta \cdot W) + \delta \cdot W].$$

In the example, the interval associated to pixel $(2,3,1)$ is given by:

$$F(0.7, 0.08, 0.5) = [0.7(1 - 0.08 \cdot 0.5), 0.7(1 - 0.08 \cdot 0.5) + 0.08 \cdot 0.5] = [0.672, 0.712]$$

Step 2.4. Build a block V' equal to V

In the example, this new block is shown in Figure 3.

0.65	0.68	0.70
0.68	0.70	0.72
0.69	0.69	0.73

Fig. 3 Original V' block for pixel $(2,3)$

Step 2.5. Calculate $K_{q_{rs}}(F(q_{ijk}, W, \delta))$ for each pixel

Next, we expand the pixel (i, j, k) in image Q over the new block V'. In the example, the pixel $(2,3,1)$ is expanded as shown in Figure 4.

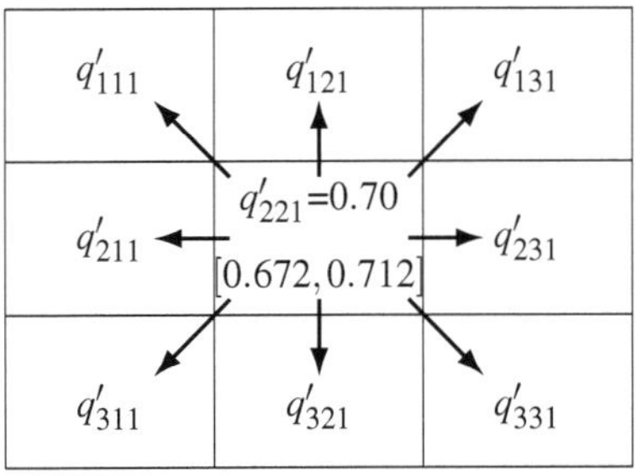

Fig. 4 Expanded block for pixel q_{231}

To keep the value of the original pixel at the center of the new block, we use the result obtained in Proposition 3.

Proposition 3. *In the settings of Proposition 1, if we take $\alpha = x$, then*

$$K_x(F(x,y,\delta)) = x$$

for all $x, y, \delta \in [0,1]$.

Proof. $K_x(F(x,y,\delta)) = K_x([x(1-\delta y), x(1-\delta y)+\delta y]) = x(1-\delta y) + x \cdot W(F(x,y \cdot \delta))$.

This proposition states that if we take α as the intensity of the pixel, we recover that same intensity from the constructed interval. In the case of pixel $(2,3,1)$ of the example we have

$$0.7 = q'_{221} = K_{q'_{221}}([0.672, 0.712]) = 0.672 + q'_{221}0.04 = 0.7.$$

We apply this method to fill in all the other pixels in the block. In this way, from Proposition 3 we take as α for each pixel, the value of that pixel in the grid V' :

- $\alpha = q'_{111}$. Then
  $q'_{111} = 0.672 + 0.65 \cdot 0.04 = 0.698$
- $\alpha = q'_{121}$. Then
  $q'_{121} = 0.672 + 0.68 \cdot 0.04 = 0.6992$
- $\cdots$
- $\alpha = q'_{331}$. Then
  $q'_{331} = 0.672 + 0.73 \cdot 0.04 = 0.7012$

In Figure 5 we show the expanded block for pixel $(2,3,1)$ in the example.

Once each of the pixels has been expanded, we join all the blocks (Step 2.6) to create the magnified image. This process can be seen in Figure 6.

0.6980	0.6992	0.7000
0.6992	0.7000	0.7008
0.6996	0.6996	0.7012

Fig. 5 Numerical expanded block for pixel q_{231}

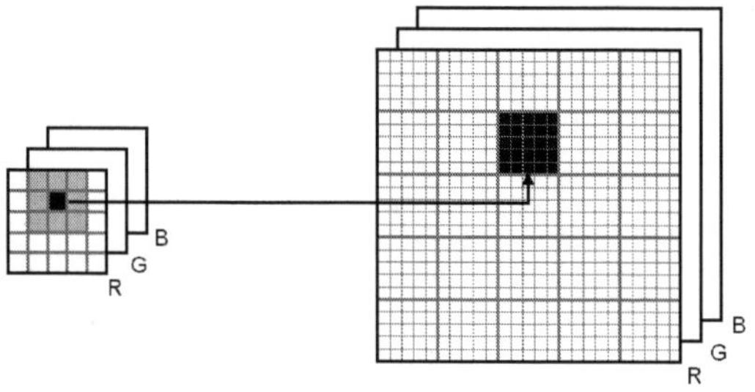

Fig. 6 Construction of the magnified image by joining all the created blocks

5 Illustrative Examples

In this section we show some illustrative examples of the proposed algorithm. To evaluate the quality of the results we use the following steps:

- We start from color images of 510×510 and we reduce them to 170×170 using the reduction algorithm proposed in [4].
- By means of Algorithm 1 we magnify the reduced images to a 510×510 size.
- Finally, we compare the obtained images with the original ones, using PSNR.

An scheme of this process is shown in Figure 7.

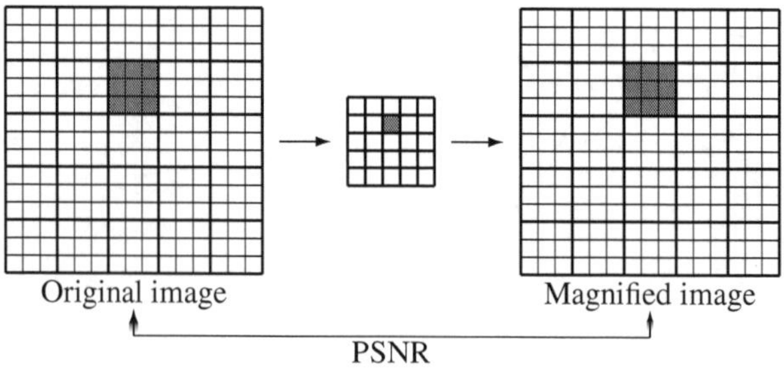

Fig. 7 Scheme to check the algorithm

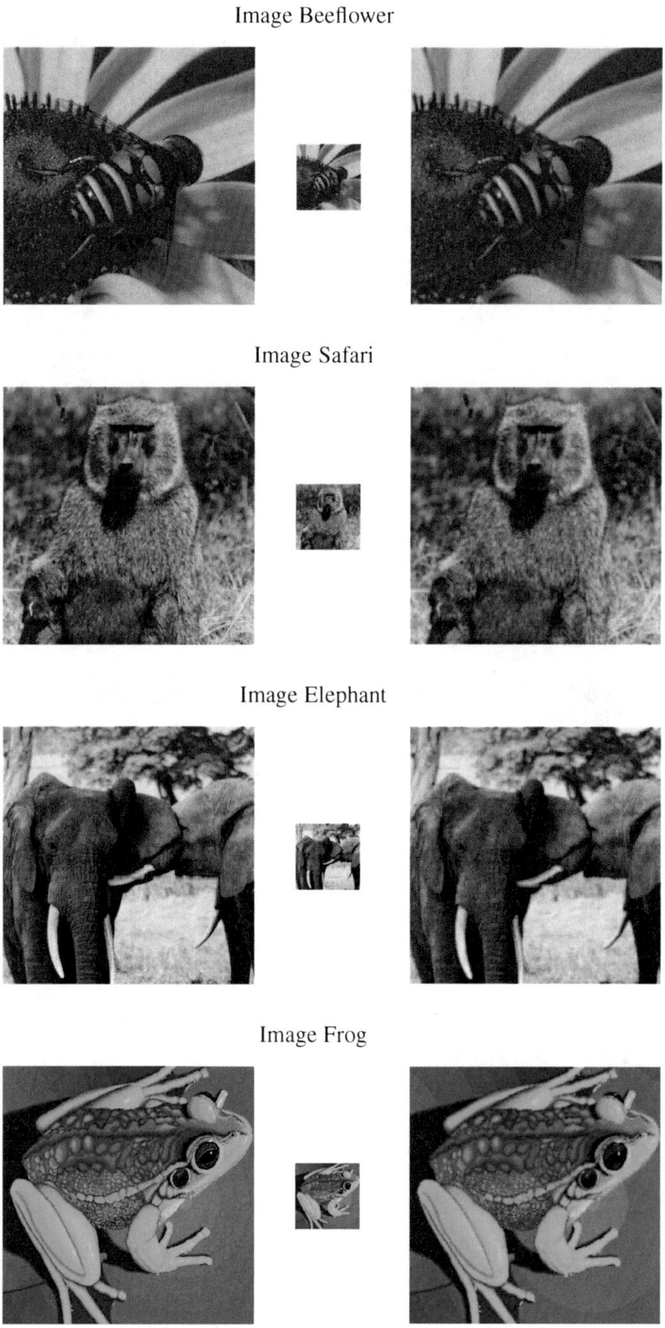

Fig. 8 Original images and reductions

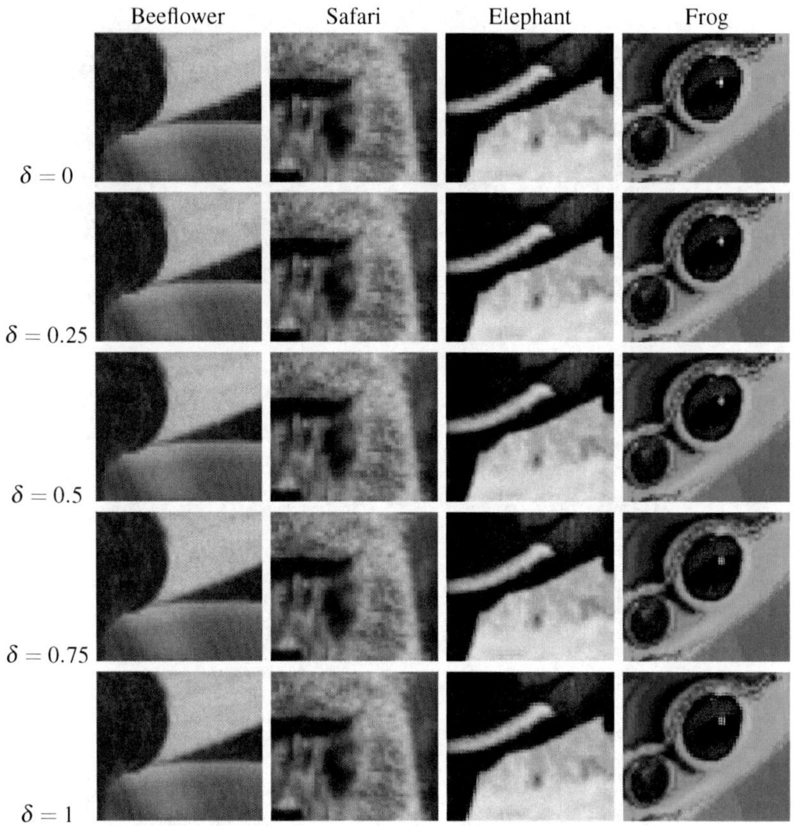

Fig. 9 Reconstructed images with different values of the parameter δ

Algorithm 1 has a parameter δ that is a scaling factor. This parameter can range between 0 and 1. In Figure 8 it is shown the original images Beeflower, Elephant, Frog and Safari, their reduced versions and the magnified images obtained by Algorithm 1 with $\delta = 0.5$ (middle value).

As we have said, depending on the value of δ parameter, the obtained results vary. When $\delta = 0$, we know by Proposition 2 that $F(x,y,0) = [x,x]$ for all $x,y \in [0,1]$, and we also know that $K_\alpha([x,x]) = x$ for all $\alpha \in [0,1]$. In this sense, when $\alpha = 0$ we build blocks in which all the elements take the value of the central pixel, and we loose information from the neighbourhood. But when δ increases, the length of the interval associated with each pixel increases too, so the range in which the intensities of pixels in each reconstructed block vary is bigger. If we take the biggest possible value ($\delta = 1$) every block intensities vary in a too big range, so there are some blurring problems. In this sense, an intermediate δ value allows to balance these two problems: jagging artifacts and blurring. In Figure 9 we show four cropped images magnified with five different δ values ($\delta = 0$, $\delta = 0.25$, $\delta = 0.5$, $\delta = 0.75$ and $\delta = 1$).

Table 1 Comparison of the magnified images with the original one

	$\delta=0$	$\delta=0.25$	$\delta=0.5$	$\delta=0.75$	$\delta=1$
Beeflower	29.3964	29.8177	30.0912	30.1858	30.0895
Safari	28.9410	29.2879	29.5676	29.7636	29.8645
Elephant	27.3838	28.0579	28.4589	28.4995	28.1698
Frog	27.7417	28.1376	28.3393	28.3182	28.0774

To compare the obtained images with the original ones we use the PSNR measure (see Table 1). We observe that the same conclusion is obtained: the best results are got with intermediate values of δ parameter.

6 Conclusions

In this work we have introduced a new magnification algorithm of color images. It is based on block expansion and it is characterized by its simplicity. This method uses interval-valued fuzzy sets to keep the information of the neighbourhood of each pixel. Besides, it maintains the original intensities with K_α operators. The parametrization used in the algorithm allows to adapt it in order to look for the optimal set-up for each image, balancing the solutions with jagging artifacts and the solutions with blurring.

Acknowledgements. This paper has been partially supported by the National Science Foundation of Spain, reference TIN2010-15055 and by the Research Services of the Universidad Publica de Navarra.

References

1. Amanatiadis, A., Andreadis, I.: A survey on evaluation methods for image interpolation. Measurement Science & Technology 20, 104015 (2009)
2. Bigand, A., Colot, O.: Fuzzy filter based on interval-valued fuzzy sets for image filtering. Fuzzy Sets and Systems 161, 96–117 (2010)
3. Bustince, H., Barrenechea, E., Pagola, M., Fernandez, J.: Interval-valued fuzzy sets constructed from matrices: Application to edge detection. Fuzzy Sets and Systems 160, 1819–1840 (2009)
4. Bustince, H., Paternain, D., De Baets, B., Calvo, T., Fodor, J., Mesiar, R., Montero, J., Pradera, A.: Two methods for image compression/reconstruction using OWA operators. In: Recent Developments in the Ordered Weighted Averaging Operators: Theory and Practice. Springer, Heidelberg (2011)
5. Bustince, H., Barrenechea, E., Pagola, M., Fernandez, J., Sanz, J.: Comment on: Image thresholding using type II fuzzy sets. Importance of this method. Pattern Recognition 43, 3188–3192 (2010)
6. Gajjar, P.P., Joshi, M.V.: New Learning Based Super-Resolution: Use of DWT and IGMRF Prior. IEEE Transactions on Image Processing 19, 1201–1213 (2010)

7. Jurio, A., Pagola, M., Paternain, D., Lopez-Molina, C., Melo-Pinto, P.: Interval-valued restricted equivalence functions applied on Clustering Techniques. In: 13th International Fuzzy Systems Association World Congress and 6th European Society for Fuzzy Logic and Technology Conference, Portugal (2009)
8. Karabassis, E., Spetsakis, M.E.: An analysis of image interpolation, defferentiation, and reduction using local polynomial. Graphical Models and Image Processing 57, 183–196 (1995)
9. Ni, K.S., Nguyen, T.Q.: Image Superresolution Using Support Vector Regression. IEEE Transactions on Image Processing 16, 1596–1610 (2007)
10. Park, S.C., Park, M.K., Kang, M.G.: Super-resolution image reconstruction: A technical overview. IEEE Signal Processing Magazine 20, 21–36 (2003)
11. Perfilieva, I.: Fuzzy transforms: Theory and applications. Fuzzy Sets and Systems 157, 993–1023 (2006)
12. Protter, M., Elad, M.: Super Resolution With Probabilistic Motion Estimation. IEEE Transactions on Image Processing 18, 1899–1904 (2009)
13. Qiu, G.: Interresolution Look-up Table for Improved Spatial Magnification of Image. Journal of Visual Communication and Image Representation 11, 360–373 (2000)
14. Takeda, H., Milanfar, P., Protter, M., Elad, M.: Super-Resolution Without Explicit Subpixel Motion Estimation. IEEE Transactions on Image Processing 18, 1958–1975 (2009)
15. Tizhoosh, H.R.: Image thresholding using type II fuzzy sets. Pattern Recognition 38, 2363–2372 (2005)
16. Unser, M., Aldroubi, A., Eden, M.: Enlargement and reduction of digital images with minimum loss of information. IEEE Transactions on Image Processing 4, 247–258 (1995)

Edge Detection on Interval-Valued Images

C. Lopez-Molina, B. De Baets, E. Barrenechea, and H. Bustince

Abstract. A digital image is an approximation of some real situation, and carries some uncertainty. In this work we model the ambiguity related to the brightness by associating an interval with each pixel, instead of a scalar brightness value. Then we adapt the Sobel method for edge detection to the new conditions of the image, leading to a representation of the edges in the shape of an interval-valued fuzzy set. To conclude, we illustrate the performance of the method and perform a qualitative comparison with the classical Sobel method on grayscale images.

1 Introduction

Any discretization process is based on sampling continuous facts, and hence misses part of the initial information. Moreover, the data can be contaminated in many different ways. In the case of a digital image, different factors are to be taken into account, either inherent to the model (as the limited number of tones) or alien to it (as noise or broken cells in the sensor). Hence, we can never have a full certainty about the tone of a pixel. In the field of edge detection, the problems originating from this uncertainty manifest themselves clearly, since sometimes not even two humans can reach an agreement on where the boundary between two objects is.

Fuzzy logic has been used in many different fields. Applications have been published, especially in those tasks where factors as ambiguity and partial or contradictory information arise. When considering classical fuzzy sets (FS), the membership of each element to a set is expressed by a degree in $[0,1]$. Subsequently, many

C. Lopez-Molina · E. Barrenechea · H. Bustince
Dpto. Automatica y Computacion, Universidad Publica de Navarra, Spain
e-mail: carlos.lopez@unavarra.es

B. De Baets
Department of Applied Mathematics, Biometrics and Process Control,
Ghent University, Belgium
e-mail: bernard.debaets@ugent.be

B. De Baets et al. (Eds.): Eurofuse 2011, AISC 107, pp. 325–337, 2011.
springerlink.com

extensions have been proposed in order to model the fuzziness, giving rise to different *extensions* or *generalizations* of the fuzzy sets. Among these extensions, one of the most popular ones has been the interval-valued fuzzy sets (IVFS), which represent membership by means of an interval $[a,b]$, $a,b \in [0,1]$.

Due to the nature of digital images, fuzzy logic appears as a natural option for handling the uncertainty. In fact, in the image processing field, edge detection has been one of the tasks where fuzzy logic has been most prolific. It has been used for different purposes at almost any step of the process, from the very interpretation of the image [1, 2] to the reconstruction of the edges once they have been characterized [3]. Authors have experimented with a wide variety of techniques based on fuzzy logic, including fuzzy inference systems [4, 5], fuzzy morphology [6], and fuzzy peer groups [7]. However, most of the applications make use of FS, partly because of the relatively new development of the FS extensions.

In this work we introduce and justify an interval-based representation of images, designed for better managing their inherent ambiguity. Then, we present the extension of the classical Sobel method [8] for its application on interval-valued images, leading to a representation of the edges based on IVFS. We also show how to turn the IVFS representation of the edges into a binary edge image.

The remainder of this work is organized as follows. Section 2 is devoted to the analysis and modeling of the ambiguity in digital images. Section 3 introduces the extension of the Sobel method for edge detection to interval-valued images. To conclude, we include experimental tests in Section 4 and draw some brief conclusions in Section 5.

2 Images and Ambiguity Representation

2.1 Image Representation

In this work we consider an image to be a matrix of M rows and N columns, with $P = \{1,\dots,M\} \times \{1,\dots,N\}$ the set of their positions. In an image I, the value of a pixel at a position $(x,y) \in P$, $I_{(x,y)}$ is a value in $\{0,\dots,255\}$. Moreover, we denote as $n(x,y) \subset P$ to the set of positions in a 3×3 neighborhood centered at (x,y).

2.2 Images and Inherent Ambiguity

Digital images are the result of a discretization of the reality. That is, a discrete, sampled version of a continuous fact. Hence, there are different sources of uncertainty and ambiguity to be considered when performing image processing tasks. Most of those sources are contextual, in the sense that they could be present (or not) in an image, depending on the situation the image was registered at. Some examples of those sources of contamination are noise, excessive illumination or shading. However, there is also some uncertainty embedded in the very nature of digital images:

the measurement error. The image discretizes the reality in two different facets, spatial and tonal, each of them producing a measurement error:

- *Spatial error:* Surfaces and objects are continuous in reality. However, they become discrete in the pixel representation of an image. Hence, due to the way the information is stored, we might be mislocating one object by 1 position in any direction.
- *Tonal error:* Images are stored using a finite number of tones. There are usually 2^8 tones in a greyscale image, or 2^{24} in a RGB one. However, even using finer-detail coding, there is always a limit in the tonal precision. Hence, the measure error associated to the tone of the pixel is as much as $\pm$ 1 tone.

In fact, both errors are derived from the imprecision of the discrete measure used in each dimension (spatial and tonal). A scalar representation of the pixel brightness is not sufficient for representing both errors. Hence, we will use an interval-valued representation of the image brightness. The construction will be straight, driven from the analysis of the measurement errors performed above. From an image I, we will generate an interval-valued (IV) image W so that the value at each pixel (x,y) is obtained as

$$W_{(x,y)} = \left[\max(0, \min_{(x',y') \in n(x,y)} I_{(x',y')} - 1),\ \min(255, \max_{(x',y') \in n(x,y)} I_{(x',y')} + 1)\right]. \quad (1)$$

That is, we assign to each position in the image an interval encompassing all of the brightness values in a 3×3 neighborhood (assuming the spatial error), modified by $\pm$ 1 tone (because of the tonal error). Fig. 1 includes as an example the Lena image, togheter with the upper and lower bound of its interval-valued representation.

(a) Original Image (b) IV representation

Fig. 1 The Lena image and the lower and upper bounds of its interval valued representation.

We have to bear in mind that the construction of IV images is quite robust against some specific types of noise (as Gaussian one), but also quite sensitive to some other kinds of contamination (as salt-and-pepper). However, dealing with external sources of contamination is not the main goal of the methodology, since we only intend to capture those factors intrinsically linked to the image representation.

The IV images provide a realistic interpretation of the image, since the measurement error is something we should not ignore. However, there are very few procedures in the image processing field able to deal with such a representation of the images [9]. We devote the rest of this paper to generalize the Sobel method for edge detection on IV images.

3 The Sobel Method and Its Application on Interval-Valued Images

3.1 The Sobel Method

The Sobel method for edge detection is based on filtering an image with 2 operators, each of them in charge of estimating the intensity change along one of the axis (horizontal and vertical). The estimation of the change along each axis is later used for generating the so-called *gradients*. Gradients are vectors representing the strength and direction of the intensity changes at each position of the image.

Horizontal Operator

-1		1
-2		2
-1		1

Vertical Operator

1	2	1
-1	-2	-1

Fig. 2 The Sobel operators for edge detection.

The Sobel operators (included in Fig. 2) act as discrete convolution operators, producing an approximation of each of the components of the gradient at each point. Some considerations on the search for gradients in discrete environments can be found in [10, 11]. For example, the normalized estimation of the horizontal intensity change at a position of the image ($H_{(x,y)} \in [-1,1]$), is obtained as

$$H_{(x,y)} = \frac{1}{4 \cdot 255} \sum \begin{pmatrix} -1\ 0\ 1 \\ -2\ 0\ 2 \\ -1\ 0\ 1 \end{pmatrix} * \begin{pmatrix} I_{(x-1,y+1)} & I_{(x,y+1)} & I_{(x+1,y+1)} \\ I_{(x-1,y)} & I_{(x,y)} & I_{(x+1,y)} \\ I_{(x-1,y-1)} & I_{(x,y-1)} & I_{(x+1,y-1)} \end{pmatrix}, \quad (2)$$

where $*$ represents the convolution operator. The vertical estimation ($V_{(x,y)}$) is calculated analogously.

Then, the gradient at a given position (x,y) is constructed as $\mathbf{G}_{(x,y)} = (H_{(x,y)}, V_{(x,y)})$. Once the information is taken as a gradient, many different techniques become applicable for turning the edges into a binary representation. Fig. 3 includes an example of the application of the Sobel method on the Lena image. The gradient magnitudes have been used to generate an intermediate representation of the edges, in the shape of a FS. In order to do so, we have selected the normalized Euclidean norm $||\mathbf{G}_{(x,y)}|| = \frac{1}{\sqrt{2}}\sqrt{H^2_{(x,y)} + V^2_{(x,y)}}$. (See [12] for further considerations on other options for combining the gradient components). To conclude, the conversion into binary edges has been carried out using the non-maximum suppression (NMS) [13] and the Rosin method for thresholding [14].

The Sobel method [8], as well as the Prewitt method [15], has had great impact in the literature. They are considered the pioneer methods in the derivative-based approach to edge detection [11, 16, 17], that has provided some of the most recognized methods in the literature, such as the Canny [13] or Marr-Hildreth [18] methods.

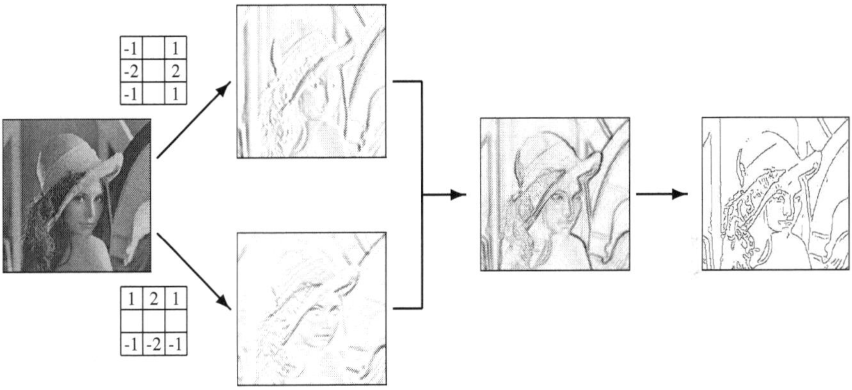

Fig. 3 Schematic representation of the Sobel edge detection method

3.2 Application to Interval-Valued Images

Once we have IV images, there are different options to apply the Sobel operators. For example, we could think of applying the detector individually on the upper and lower bounds of the intervals, as they are be themselves scalar images. However, this collides with the main idea behind the IV representation of the images. Hence, we will apply the classical interval operations [19]:

- Sum of intervals: $[\underline{a},\overline{a}]+[\underline{b},\overline{b}]=[\underline{a}+\underline{b},\overline{a}+\overline{b}]$;
- Difference of intervals: $[\underline{a},\overline{a}]-[\underline{b},\overline{b}]=[\underline{a}-\overline{b},\overline{a}-\underline{b}]$;
- Product of a positive scalar s and an interval: $s\cdot[\underline{a},\overline{a}]=[s\cdot\underline{a},s\cdot\overline{a}]$

In this way, both the horizontal and vertical intensity changes are expressed as intervals. More specifically, they are subintervals of $[-1,1]$. As a consequence, the gradient estimated at the position p will no longer be a vector in the $[-1,1]^2$ space, but an area, as illustrated in Fig. 4(a). This area can be seen as a projection of the uncertainty about the intensity of the pixels in the image (and the object boundaries). We are unsure about the initial data, so the uncertainty is propagated when measuring local features (in this case, the gradient). Following this interpretation of the image brightness, we understand that the gradient is located somewhere in the gray area in Fig. 4(a), while its exact position remains unknown.

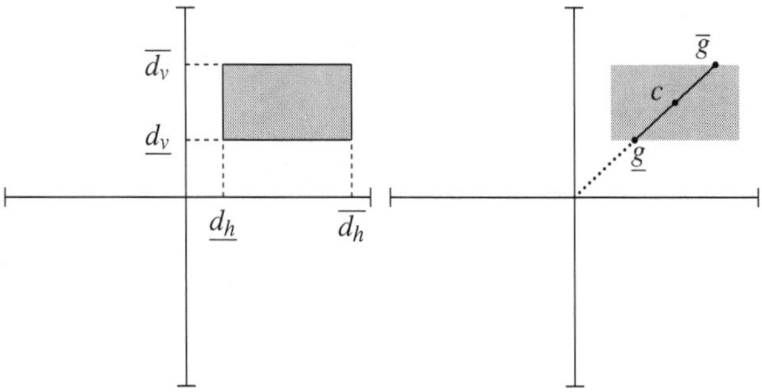

(a) Area-based interpretation (b) Interval-valued gradient

Fig. 4 Area gradient

The area-based interpretation has a complicated application. Moreover, our goal to obtain of binary edges, following the constraints by Canny [13]. The first step in binarization of the edges is the conversion of the area into a segment. There are different ways of performing the geometric interpretation of the area in Fig. 4(a). We propose to first calculate the center of gravity of the area,

$$c = \left[\frac{\underline{d_h} + \overline{d_h}}{2}, \frac{\underline{d_v} + \overline{d_v}}{2} \right] \tag{3}$$

Then, the candidates for being the gradient are those points in the intersection of the area and the straight line passing through the origin c. Note that, in case c coincides with the origin, we have a null gradient. This segment is delineated by $\underline{g}$ (the closest point to the origin) and $\overline{g}$ (the one furthest away), as illustrated in Fig. 4(b).

In order to follow the Sobel method as closely as possible, we have to generate a FS representing the edges. However, in our case, instead of creating directly a classical FS, we will first use an IVFS. The interval-valued membership degree will be determined by $[||\underline{g}||, ||\overline{g}||]$, where d stands for the normalized Euclidean norm, as before. Before using the classical procedures for binarization, the IVFS representation of the edges will be turned into a FS. To do so we will use the K_α operators, defined as $K_\alpha([\underline{a}, \overline{a}]) = (\underline{a} + \alpha(\overline{a} - \underline{a}))$. In this way, $\alpha \in [0,1]$ will measure our optimism with respect to the magnitude of the gradients. Once the edges are modeled in the shape of a FS, the processing is as it was in Fig. 3. We will first use NMS for thinning the edges, then the Rosin method for binarization. In this way we obtain an edges representation satisfying the Canny constraints [13].

This method for edge detection, as illustrated in Fig. 5, will be called the Interval-valued Sobel (IV-Sobel) method.

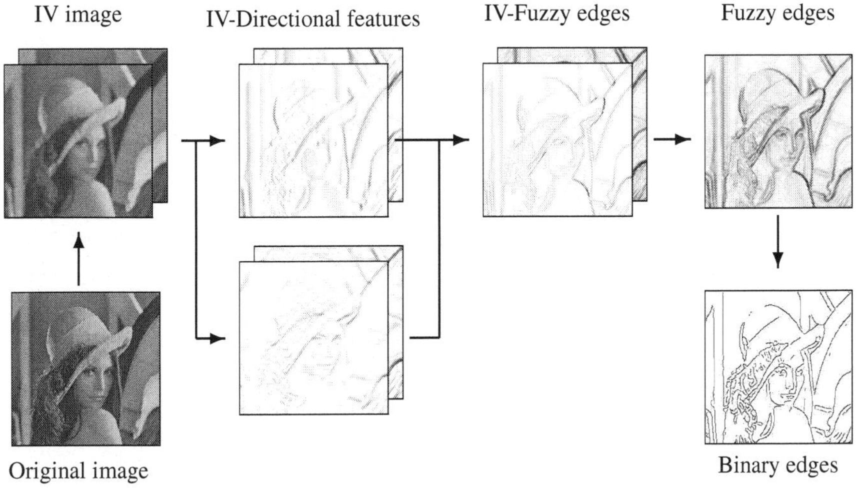

Fig. 5 Visual schema of the IV-Sobel edge detection method

In order to illustrate the performance of the proposed method, we use the Lena image in Fig. 1(a). First, we have regularized the image with a Gaussian filter with $\sigma = 1.80$. At the moment of creating a FS representation of the edges we have experimented with different values of α. The results are included in Fig. 6. Note that the upper and lower bounds of the IVFS are the FSs generated with $\alpha = 0.0$ and $\alpha = 1.0$, respectively. In this figure we observe how the increase of the optimism (represented by an increase of α) gives rise to the selection of a larger number of edge points. However, it remains unclear whether that is bad or good news, since it has both positive and negative consequences. For example, in Fig. 6 we can observe how the hat silhouette is almost completed using large values of α, but they also imply a lot of false positive detections, as those in the hair region.

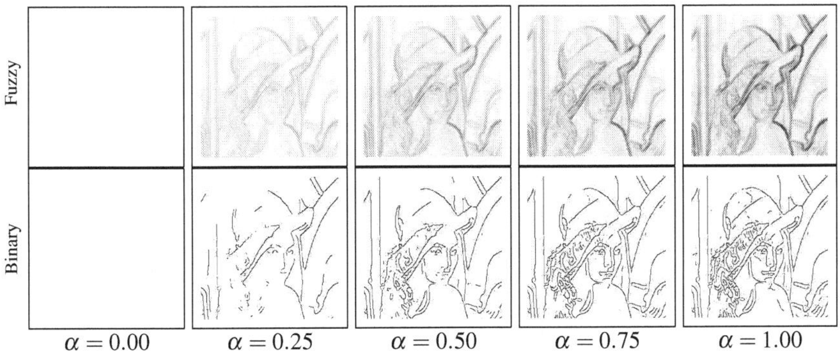

Fig. 6 Fuzzy and binary edge images obtained using different α values for the conversion of the IVFSs into FSs.

4 Experimental Results

4.1 Aim of the Experiment

In this experiment we want to compare the performance of the classical Sobel method with that of the IV-Sobel method. In order to do so, we have run both of the algorithms on the same images, under similar conditions, and then quantified the results.

4.2 Experimental Dataset

In the experiments we have used the Berkeley Segmentation Dataset (BSDS) [20], containing 300 images. The images are provided in grayscale with 256 tones and have a resolution of 481×321 (or 321×481) pixels. In addition, each of them comes along with 5 to 10 hand-made segmentation solutions. As those solutions are provided in the shape of region boundaries, we have used them as ground truth for the quantification of the quality of the edge detection results.

4.3 Comparison Method

Baddeley's Delta Metric (BDM) is a measure initially designed for the comparison of binary sets [21], but can also be used to compare binary images. It intends to measure the similarity of two subsets of *featured* points on the same referential. As the edges of an image are usually displayed as a binary image (following the Canny constraints [13]), this measure can be used for our purposes.

Let B_1 and B_2 be two binary images and let M,N and P be as defined in Section 2. Given a value $1 < k < \infty$ the k-BDM between the images B_1 and B_2 (denoted $\Delta^k(B_1,B_2)$) is defined as:

$$\Delta^k(B_1,B_2) = \left[\frac{1}{|P|} \sum_{p \in P} |w(d(p,B_1)) - w(d(p,B_2))|^k \right]^{\frac{1}{k}}, \tag{4}$$

where $d(p,B_i)$ represents a distance from the position p to the closest *featured* point of the image B_i and $w : [0,\infty] \to [0,\infty]$ is a concave, increasing function used for weighing. Note that we assume that both images contain at one or more edge pixels. In our experiments, we use the Euclidean distance in the computation of d. Hence, $d(p,B_i)$ stands for the minimum Euclidean distance from the position p to an edge point of B_i. As for the other settings, we use $w(x) = x$ [22, 23]. and $k = 2$, as in [23].

The score of an edge detection method on an image is the average distance of its result to the set of hand-made solutions associated with that image. Analogously, the score on a dataset is the average on its images.

4.4 Algorithm

Algorithm 1 lists all of the steps of the procedure. The proposed algorithm could be modified in order to obtain better results. For example, single-threshold binarization techniques are usually outperformed by double-threshold techniques (as *hysteresis*), specially on low contrast regions. Also the Gaussian smoothing could be substituted by edge preserving techniques [24]. However, the aim of this experiment is not to obtain the best possible results, but to perform a comparison between the Sobel and the IV-Sobel methods.

Data: An image I, an optimism indicator α
Result: A binary edge image B
begin
- Smoothen I with a Gaussian mask with standard deviation $\sigma_1 = 1.0$;
- Create the IV version of the image, I_w;
- Calculate the IV gradient estimations H_w and V_w;
- Using the technique in Fig. 4, generate the IV fuzzy edges E_w;
- Create the fuzzy edges as $E = K_\alpha(E_w)$;
- Obtain a thin version of fuzzy edge image (T) using NMS [13];
- Obtain a binary edge image B from T using Rosin method [14];

end

Algorithm 1. Algorithm of the IV-Sobel method with user-selected value α.

4.5 Results

We have executed the algorithm on the 100 images of the BSDS *test* set [20], using different fixed $\alpha \in \{0, 0.025, \ldots, 1\}$. Fig. 7 displays the average performance on the image set, and its comparison with the performance obtained by the classical Sobel method. In addition, we display the average number of pixels classified as edges for each value of α ($|B_\alpha|$) and its increase.

We observe how the results by the Sobel method can be greatly improved by the IV-Sobel one. First, there is an initial range of values of α producing very bad results, but then the performance rapidly increases. Then, when α tends to 1, the performance of the Sobel and IV-Sobel methods becomes similar. Moreover, we notice that the best results (average BDM) coincide with those values of α producing largest increments of $|B_\alpha|$.

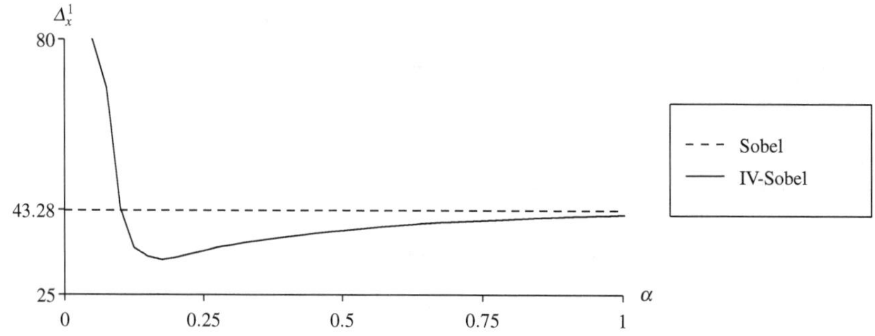

(a) Average BDM

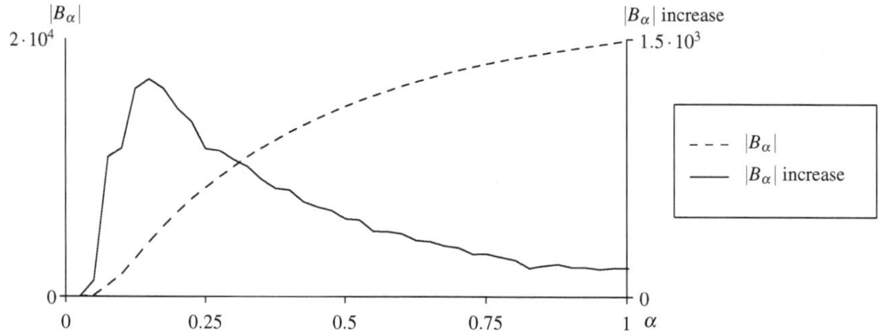

(b) Average number of edge pixels

Fig. 7 Results obtained by the Sobel and IV-Sobel methods on the BSDS test set

In order to confirm these facts, we have repeated the same experiment on the *train* dataset (Fig. 8). Again, we obtain similar results both for the performance (Δ) and the increase of $|B_\alpha|$.

Hence, we can conclude that a proper way to determine α is by experimentally testing a set of them, then selecting the one producing the largest increment of $|B_\alpha|$. The IV-Sobel method is as included in Algorithm 2. Obviously, in this way the first $\alpha = 0$ will never be selected, since its result cannot be compared with any previous B_α.

5 Conclusions

We have analyzed the role of the measurement error in digital images, proposing an interval-valued representation of the image to overcome it. Then, we have proposed a generalization of the Sobel method for edge detection applied on interval-valued images, displaying an example of its use and comparing its performance to the classical Sobel method. We have shown how the performance of the Sobel method can be improved by using the IV representation for ambiguity modeling.

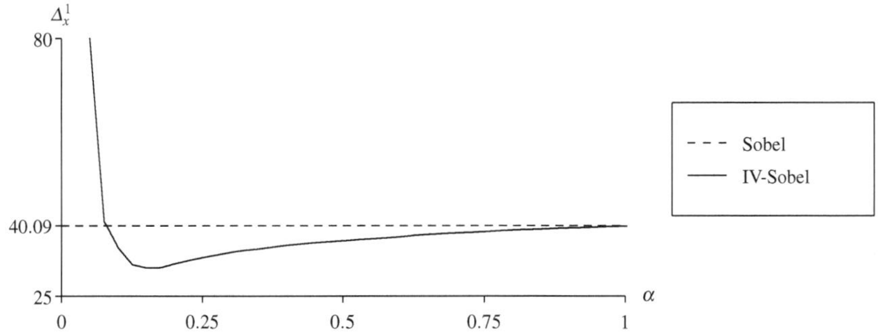

(a) Average BDM

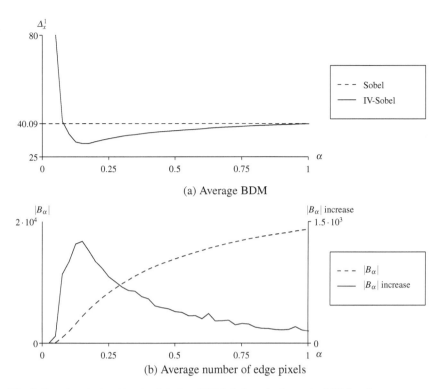

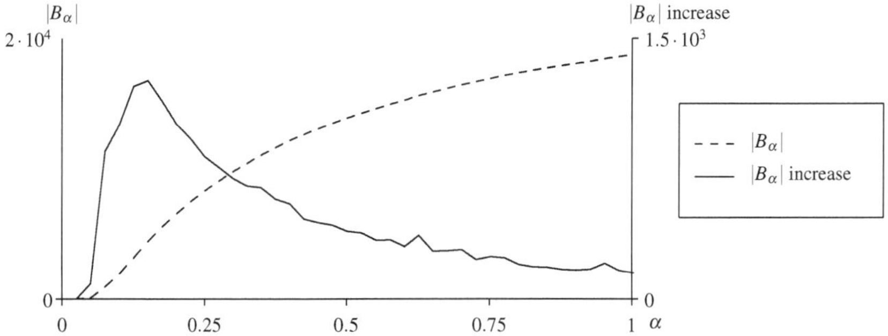

(b) Average number of edge pixels

Fig. 8 Results obtained by the Sobel and IV-Sobel methods on the BSDS train set

Data: An image I, a optimism indicator step $\delta\alpha$
Result: A binary edge image B
begin
  Smoothen I with a Gaussian mask with standard deviation $\sigma_1 = 1.0$;
  Create the IV version of the image, I_w;
  Calculate the IV gradient estimations H_w and V_w;
  Using the technique in Fig. 4, generate the IV fuzzy edges E_w;
  Calculate the candidate α, $A \leftarrow \{0, \delta\alpha, 2 \cdot \delta\alpha, \ldots, 1\}$;
  foreach $\alpha \in A$ **do**
    Create the fuzzy edges as $E_\alpha = K_\alpha(E_w)$;
    Obtain a thin version of fuzzy edge image (T_α) using NMS [13];
    Obtain a binary edge image B_α from T_α using Rosin method [14];
  $B \leftarrow$ The B_α producing a larger increment of $|B_\alpha|$;
end

Algorithm 2. Final algorithm of the IV-Sobel method with unsupervised selection of the value α.

References

1. Jacquey, F., Comby, F., Strauss, O.: Fuzzy edge detection for omnidirectional images. Fuzzy Sets and Systems 159(15), 1991–2010 (2008)
2. Pal, S.K., King, R.A.: On edge detection of x-ray images using fuzzy sets. IEEE Trans. on Pattern Analysis and Machine Intelligence 5(1), 69–77 (1983)
3. Law, T., Itoh, H., Seki, H.: Image filtering, edge detection, and edge tracing using fuzzy reasoning. IEEE Trans. on Pattern Analysis and Machine Intelligence 18(5), 481–491 (1996)
4. Hu, L., Cheng, H.D., Zhang, M.: A high performance edge detector based on fuzzy inference rules. Information Sciences 177(21), 4768–4784 (2007)
5. Russo, F.: Edge detection in noisy images using fuzzy reasoning. In: Proceedings of the Instrumentation and Measurement Technology Conference, vol. 1, pp. 369–372 (1998)
6. Jiang, J.-A., Chuang, C.-L., Lu, Y.-L., Fahn, C.-S.: Mathematical-morphology-based edge detectors for detection of thin edges in low-contrast regions. IET Image Processing 1(3), 269–277 (2007)
7. Morillas, S., Gregori, V., Hervas, A.: Fuzzy peer groups for reducing mixed Gaussian-impulse noise from color images. IEEE Trans. on Image Processing 18(7), 1452–1466 (2009)
8. Sobel, I., Feldman, G.: A 3x3 isotropic gradient operator for image processing. Presented at a talk at the Stanford Artificial Intelligence Project (1968)
9. Galar, M., Fernandez, J., Beliakov, G., Bustince, H.: Interval-valued fuzzy sets applied to stereo matching of color images. IEEE Trans. on Image Processing 20(7), 1949–1961 (2011)
10. Canny, J.: Finding edges and lines in images. Technical report, Massachussets Institute of Technology, Cambridge, MA, USA (1983)
11. Torre, V., Poggio, T.: On edge detection. IEEE Trans. on Pattern Analysis and Machine Intelligence 8,147–163 (1984)
12. Lopez-Molina, C., Fernandez, J., Jurio, A., Galar, M., Pagola, M., De Baets, B.: On the use of quasi-arithmetic means for the generation of edge detection blending functions. In: Proceedings of the IEEE International Conference on Fuzzy Systems (2010)
13. Canny, J.: A computational approach to edge detection. IEEE Trans. on Pattern Analysis and Machine Intelligence 8(6), 679–698 (1986)
14. Rosin, P.L.: Unimodal thresholding. Pattern Recognition 34(11), 2083–2096 (2001)
15. Prewitt, J.M.S.: Object enhancement and extraction, Picture Processing and Psychopictorics, pp. 75–149. Academic Press, London (1970)
16. Basu, M.: Gaussian-based edge-detection methods- A survey. IEEE Trans. on Systems, Man, and Cybernetics, Part C: Applications and Reviews 32(3), 252–260 (2002)
17. Papari, G., Petkov, N.: Edge and line oriented contour detection: State of the art. Image and Vision Computing 29(2-3), 79–103 (2011)
18. Marr, D., Hildreth, E.: Theory of edge detection. Proceedings of the Royal Society of London 207(1167), 187–217 (1980)
19. Moore, R.: Interval Analysis. Prentince-Hall, Englewood Cliffs (1996)
20. Martin, D., Fowlkes, C., Tal, D., Malik, J.: A database of human segmented natural images and its application to evaluating segmentation algorithms and measuring ecological statistics. In: Proceedings of the 8th International Conference on Computer Vision, vol. 2, pp. 416–423 (2001)
21. Baddeley, A.J.: Errors in binary images and an L^p version of the Hausdorff metric. Nieuw Archief voor Wiskunde 10, 157–183 (1992)

22. Lopez-Molina, C., Bustince, H., Fernandez, J., Couto, P., De Baets, B.: A gravitational approach to edge detection based on triangular norms. Pattern Recognition 43(11), 3730–3741 (2010)
23. Medina-Carnicer, R., Madrid-Cuevas, F.J., Carmona-Poyato, A., Muñoz-Salinas, R.: On candidates selection for hysteresis thresholds in edge detection. Pattern Recognition 42(7), 1284–1296 (2009)
24. Weickert, J., ter Haar Romeny, B.M., Viergever, M.A.: Efficient and reliable schemes for nonlinear diffusion filtering. IEEE Trans. on Image Processing 7(3), 398–410 (1998)

Histograms for Fuzzy Color Spaces

J. Chamorro-Martínez, D. Sánchez, J.M. Soto-Hidalgo, and P. Martínez-Jiménez*

Abstract. In this paper we introduce two kinds of fuzzy histograms on the basis of fuzzy colors in a fuzzy color space and the notion of gradual number by Dubois and Prade. Fuzzy color spaces are a collection of fuzzy sets providing a suitable, conceptual quantization with soft boundaries of crisp color spaces. Gradual numbers assign numbers to values of a relevance scale, typically [0,1]. Contrary to convex fuzzy subsets of numbers (called fuzzy numbers, but corresponding to fuzzy intervals as an assignment of intervals to values of [0,1]), they provide a more precise representation of the cardinality of a fuzzy set. Histograms based on gradual numbers are particularly well-suited for serving as input to another process. On the contrary, they are not the best choice when showing the information to a human user. For this second case, linguistic labels represented by fuzzy numbers are a better alternative, so we define linguistic histograms as an assignment of linguistic labels to each fuzzy color. We provide a way to calculate linguistic histograms based on the compatibility between gradual numbers and linguistic labels. We illustrate our proposals with some examples.

J. Chamorro-Martínez · D. Sánchez · P. Martínez-Jiménez
Dept. Computer Science and A.I., University of Granada, Spain
e-mail: {jesus,daniel,pedromartinez}@decsai.ugr.es

D. Sánchez
European Centre for Soft Computing, Mieres, Spain
e-mail: daniel.sanchezf@softcomputing.es

J.M. Soto-Hidalgo
Department of Computer Architecture, Electronics and Electronic Technology,
University of Córdoba, Spain
e-mail: jmsoto@uco.es

* This work has been supported by the Spanish Science and Innovation Ministry under the project TIN2009-08296.

B. De Baets et al. (Eds.): Eurofuse 2011, AISC 107, pp. 339–350, 2011.
springerlink.com

1 Introduction

Histograms are the basis of many techniques for image restoration, enhancement, segmentation, retrieval, etc. In principle, a color histogram is defined as a function $h(\mathbf{c_k}) = n_k$ where $\mathbf{c_k} = [x, y, z]$ is a color and n_k is the number of pixels in the image having the color $\mathbf{c_k}$. It is common to normalize a histogram by dividing each of its values by the total number of pixels, obtaining the frequency of occurrence of a color $\mathbf{c_k}$.

This simple approach has the drawback that a color space is not representative of the collection of colors we can distinguish and identify. In addition, the values n_k use to be very low because there are many colors in a crisp color space, and it is easy to find small color variations in real images. Since in practice many of the colors $\mathbf{c_k}$ are indistinguishable for us, a solution is to use an histogram defined on groups of indistinguishable colors C_k, in which the associated number of pixels is $h(C_k) = \sum_{\mathbf{c_k} \in C_K} n_k$. The collection of groups of colors $C_1 \ldots C_n$ defines a partition (quantization) of the color space employed.

Many approaches for defining the quantization of the color space are available in the literature. In this work we are interested in fuzzy approaches [10, 12, 14, 21, 24, 19, 20], in which each group of colors is in fact a fuzzy subset of colors, called a *fuzzy color*, and the whole collection of colors defining a fuzzy partition of the color space is called a *fuzzy color space*. This approach has the advantage that it is able to represent the fact that indistinguishability is a fuzzy, gradual concept for we humans, i.e. for us, colors are indistinguishable to a certain degree. Crisp boundaries inherent to crisp quantization are counterintuitive for us. In addition, both the (fuzzy) quantization and the histograms are less sensitive to small variations of the boundaries

In the literature there are several proposals which define histograms over a set of fuzzy colors [4, 9, 15, 16]. One drawback of most of these proposals is that they work only with intensities. In addition, the counting of fuzzy colors is performed by using the sigma-count (i.e., the sum of membership degrees). However, the sigma-count is not a suitable measure of cardinality in many applications, as it has been recognized by several authors [8]. Proposals based on the sigma-count summarize the counting into a single number, so they do not represent the fuzziness of the count itself.

In [1], we introduced an approach to fuzzy histograms based on the concept of fuzzy natural number introduced in [3]. This approach represented a kind of compromise between precision and understandability. In this work we introduce two approaches, a numerical one intended to provide a more accurate representation of cardinalities on the basis of gradual numbers, and a linguistic one obtained from the previous one, intended for improving understandability at the cost of precision.

The rest of the paper is organized as follows: in section 2 we recall the notion of fuzzy color space. Different approaches to fuzzy cardinality and their suitability for histogram representation on fuzzy color spaces are discussed in section 3. We introduce our new histograms and some examples in section 4. Finally, the main conclusions and future works are summarized in section 5.

2 Fuzzy Color Spaces

In order to represent the semantic compatibility between crisp colors and linguistic color terms, in [20] we introduce the following definitions of fuzzy color and fuzzy color space on a generic crisp color space XYZ with domain of components being D_X, D_Y, and D_Z:

Definition 1. *[20] A fuzzy color $\widetilde{C}$ is a linguistic label whose semantics is represented in a color space XYZ by a normalized fuzzy subset of $D_X \times D_Y \times D_Z$.*

Definition 2. *[20] A fuzzy color space $\widetilde{XYZ}$ is a set of fuzzy colors $\widetilde{C}_1,\ldots,\widetilde{C}_m$ that define a fuzzy partition of $D_X \times D_Y \times D_Z$, i.e., that satisfies:*

1. *$\bigcup_{\{1,\ldots,m\}} sup(\widetilde{C}_i) = XYZ$, i.e., the union of the support of the $\widetilde{C}_i$ covers the whole space.*
2. *$ker(\widetilde{C}_i) \cap ker(\widetilde{C}_j) = \emptyset\ \forall i \neq j$, i.e., the kernels of the $\widetilde{C}_i$ and $\widetilde{C}_j$ are pairwise disjoint.*
3. *$\forall i \in \{1,\ldots,m\}\ \exists \mathbf{c} \in XYZ$ such that $\widetilde{C}_i(\mathbf{c}) = 1$, i.e., there is at least one object fully representative of the fuzzy color $\widetilde{C}_i$.*

Condition 3 is always verified by definition of fuzzy color. Condition 1 implies $\forall \mathbf{c} \in XYZ\ \exists i \in \{1,\ldots,m\}$ such that $\widetilde{C}_i(\mathbf{c}) > 0$. Conditions 2 and 3 imply $\widetilde{C}_i \not\subseteq \widetilde{C}_j$ $\forall i \neq j$.

In [20] we proposed several fuzzy color spaces using color names provided by the well-known ISCC-NBS system [11]. ISCC-NBS provides several color sets in the form of sets of pairs (linguistic term, crisp color). Using the methodology introduced in [20], we calculate for each color set a fuzzy color space on the basis of a Voronoi diagram of the crisp color space, calculated using the crisp colors of the set of pairs considered. The Voronoi diagram is a crisp partition corresponding to the 0.5-cut of the fuzzy colors. The kernel and support of each fuzzy color is obtained as a scaling with parameters α and β respectively, with $\alpha < 1 < \beta$, and guaranteeing the conditions in definition 2. The membership functions of the fuzzy colors are obtained on the basis of distances in the crisp color space. For more details see [20].

In [20], we have obtained three fuzzy color spaces on the basis of the sets of color names Basic (13 colors), Extended (31 colors) and Complete (267 colors) in the RGB color space. For instance, the Basic set has color names corresponding to ten basic color terms (pink, red, orange, yellow, brown, olive, green, blue, violet, purple), and 3 achromatic ones (white, gray, and black). The corresponding representative crisp colors are shown in figure 1, together with a rough view of the core, the alpha-cuts of level 0.5, and the support of some fuzzy colors in the fuzzy color space obtained from ISCC-NBS Basic in [20]. We are not showing examples of the fuzzy color spaces for the sets Extended and Complete because of the lack of space.

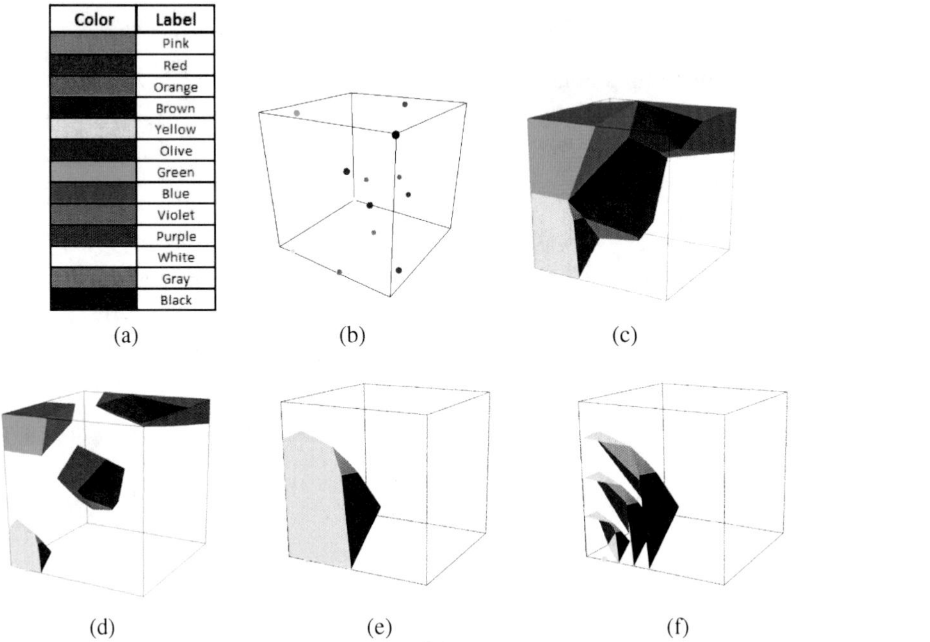

Fig. 1 Part of the RGB fuzzy color space obtained in [20] from the ISCC-NBS Basic set of colors. (a) ISCC-NBS Basic set of colors (representative crisp color and color name). (b) Situation of the representative crisp colors in the RGB color space. (c) Volumes of colors in the 0.5-cut for the fuzzy colors *yellow*, *blue*, *green*, and *gray* obtained from the Voronoi diagram in the RGB cube. (d) Volumes of colors in the kernel of the same fuzzy colors. (e) Volume of colors in the support of the fuzzy color *yellow*. (f) Superimposed views of part of the surfaces of the volumes of colors in the kernel (most internal), 0.5-cut (middle) and support (most external) for the fuzzy color *yellow*.

3 Cardinality of Fuzzy Sets

In order to define histograms on the basis of fuzzy color spaces, we need to cope with the problem of cardinality of fuzzy sets, since the set of pixels painted in a certain fuzzy color is in fact a fuzzy set whose cardinality must appear in the histogram. In this section we recall the most employed approaches for solving the problem of cardinality of fuzzy sets, and we discuss on their advantages and disadvantages, particularly in the definition of histograms.

3.1 Scalar Cardinalities

In these approaches, the cardinality of a fuzzy set is a crisp number, either real or integer [22]. The most employed cardinality for fuzzy sets is the scalar sigma-count, defined for any fuzzy set $F : X \to [0,1]$ as

$$sc(F) = \sum_{x \in X} F(x) \tag{1}$$

However, the sigma-count is not a suitable measure of cardinality in many applications. In fact, *sc* was introduced as a measure of *energy* of a fuzzy set [13]. Sigma-count is counterintuitive since *sc* is not a natural number in general. Even if the result is a natural number, we can obtain the same result in very different situations, thus loosing information. This is the case if we consider two situations like having 100 pixels compatible with a color $\widetilde{C}$ to a degree 0.1, and having 10 pixels compatible with $\widetilde{C}$ to a degree 1; in both cases, the result of the sigma count applied to the fuzzy set of pixels compatible with $\widetilde{C}$ is 10. Some alternatives consider in the addition only values above a certain threshold, but this does not solve other problems, as we shall see.

3.2 Fuzzy Numbers

The next approach in popularity is to consider that the cardinality of a fuzzy set is a fuzzy number, i.e., a normalized, convex subset of the real line or the nonnegative integers [5, 22]. However, in [2] we showed that in some cases this is counterintuitive. Consider for example the fuzzy set given by $A = 1/x_1 + 0.5/x_2 + 0.5/x_3$. The cardinality of A could be one (because x_1 belongs to A for sure) or, if we relax our criterion to accept elements in A, the cardinality could be three (accepting x_2 and x_3 belong to A as well). However, the cardinality cannot be two, since if $x_2 \in A$ then $x_3 \in A$ and vice versa. This way, the cardinality is not convex. In addition, this example illustrates that the sigma-count is not always a good measure since $sc(A) = 2$.

Several authors have related this issue to the idea that the possible cardinalities of a fuzzy set are the cardinalities of its α-cuts, since these are the possible crisp representatives of the fuzzy set, and hence that should be the support of the fuzzy cardinality [2]. In our previous example the possible cardinalities of A are 1 or 3 since its possible α-cuts are $\{x_1\}$ and $\{x_1, x_2, x_3\}$. Several alternative proposals that comply with this idea are available [23, 2]. In [1] we introduced histograms on fuzzy color spaces in which the cardinality associated to each fuzzy color was a fuzzy subset of the naturals calculated by using method *ED*, proposed in [2]. This fuzzy set is not convex in general. We will see an example later.

3.3 Gradual Numbers

In [7], Dubois and Prade introduced the ideas of gradual element and gradual number as a way to represent fuzzy quantities. Gradual numbers assign numbers to values of a relevance scale, typically [0,1]. The cardinality of a fuzzy set can be represented by a gradual number in which the cardinality of the α-cut of the fuzzy set is assigned to α.

Following the notation in [17], a gradual (real) number is a pair $(\Lambda, \mathscr{R})$ where $\Lambda \subset (0,1]$ is finite, and $\mathscr{R} : (0,1] \to \mathbb{R}$. Let $\mathbb{R}_{RL}$ be the set of gradual real numbers. Operations are extended as follows:

Definition 3. *Let $f : \mathbb{R}^n \to \mathbb{R}$ and let $R_1 \ldots R_n$ be gradual numbers. Then $f(R_1, \ldots, R_n)$ is a gradual number with*

$$\Lambda_{f(R_1,\ldots,R_n)} = \bigcup_{1 \leq i \leq n} \Lambda_{R_i} \tag{2}$$

and, $\forall \alpha \in \Lambda_{f(R_1,\ldots,R_n)}$

$$\mathscr{R}_{f(R_1,\ldots,R_n)}(\alpha) = f(\mathscr{R}_{R_1}(\alpha), \ldots, \mathscr{R}_{R_n}(\alpha)) \tag{3}$$

Gradual numbers offer several advantages. First, they don't introduce imprecision in the cardinality, since each α-cut is assigned a crisp number. On the contrary, the α-cut of a fuzzy number is an interval where the cardinality is assumed to be, and hence it is an imprecise representation of the cardinality. Hence, the so-called fuzzy numbers are in fact fuzzy intervals [6]. Another advantage with respect to fuzzy numbers is that gradual numbers have the same algebraic structure as ordinary numbers, whist fuzzy numbers satisfy the properties of interval arithmetic only. In addition, gradual numbers do not increase imprecision of the representation with operations, even to the extent that operations on gradual numbers may yield crisp numbers. In particular, it is always the case that for any gradual number a, $a - a = 0$, $a/a = 1$ (provided $a \neq 0$), etc. They are hence a precise and very useful representation when the cardinality is to be employed in any calculation.

However, gradual numbers are not as intuitive as fuzzy numbers when the cardinality is to be expressed to a human user. In particular, it is difficult to provide a linguistic label describing precisely a gradual number. Another disadvantage is related to the space needed for representing a gradual number, and the time needed for arithmetic calculations, both being proportional to the number of α-cuts employed. In practice, only a finite number of cuts is necessary, corresponding to levels in the representation of the gradual number, see [17].

3.4 Fuzzy Cardinality and Histograms on Fuzzy Color Spaces

Summarizing, we conclude the following:

- Scalar cardinalities are not accurate representations of the cardinality. They can be seen as summaries of the real cardinality, either by discarding the cardinality of all but one of the α-cuts, or by providing the center of gravity, like in the case of the sigma-count (this interpretation is given in [2]).
- Fuzzy numbers represent restrictions, linguistic concepts on numbers, but are not well suited for providing accurate representations, since the latter are non-convex in general [2]. Their arithmetics is an arithmetic of restrictions (fuzzy intervals), that is not appropriate for representing cardinalities that are to be employed in further calculations. On the contrary, they are the best choice for summarizing the cardinality into a linguistic term to be provided to human users (losing information and accuracy, but less than scalar cardinalities).

- Gradual numbers are the most accurate representation of cardinality, and the best choice for representing cardinalities to be used in further calculations. They are not the best choice for giving results to the user.

In [1] we employed non-convex fuzzy subsets of the non-negative integers for defining fuzzy histograms. This approach represents a compromise between the advantages and disadvantages of using fuzzy numbers and gradual numbers. In this paper, we propose to define two kinds of histograms: one based on gradual numbers, to be used in practice by the computer for image processing, and an approximation of this histogram by means of linguistic labels (fuzzy numbers) in order to give information to the user. These are introduced and illustrated in the next section.

4 New Histograms on Fuzzy Color Spaces

4.1 Gradual Histogram

Definition 4. *A gradual color histogram is a function h_G that assigns a gradual number to every fuzzy color in a fuzzy color space $\widetilde{XYZ}$.*

The gradual number corresponding to every fuzzy color $\widetilde{C}$ is obtained by assigning to α the cardinality of the α-cut of the fuzzy subset of pixels with color $\widetilde{C}$. Since the fuzzy subset of pixels with color $\widetilde{C}$ is finite, in practice we need only a finite number of cuts for representing the gradual number. However, this number may be large, so in some cases it may be interesting to consider just a fixed collection of cuts with equidistant values of α. See [17] for further discussion on that.

Let us remark that the gradual integer number can be easily transformed into a gradual rational number by dividing the number associated to each level α by the total number of pixels in the image. This is convenient when we want the histogram to represent proportions instead of absolute values. We shall see examples in section 4.3.

Finally, usual image processing operations performed on crisp histograms can be extended directly to gradual histograms by using definition 3, i.e., by applying the operation in each level.

4.2 Linguistic Histogram

Let $\widetilde{XYZ}$ be a fuzzy color space. Let us consider a collection of fuzzy linguistic quantifiers $S_Q = \{Q_1 \dots Q_k\}$ provided by the user, like *Around 10%* or *Most*, represented by appropriate fuzzy subsets of the unit interval, and *Between 50 and 100*, represented by appropriate fuzzy subsets of the non-negative integers.

Definition 5. *A linguistic color histogram is a function h_L that assigns a linguistic quantifier from S_Q to every fuzzy color in $\widetilde{XYZ}$.*

We consider user-defined quantifiers in order to improve understandability, since accuracy of the linguistic approach is always worst than that of the gradual approach.

In order to obtain the quantifier for every fuzzy color $\widetilde{C}$, we take the gradual number associated to $\widetilde{C}$ in the gradual color histogram (an accurate representation of cardinality) and we calculate the compatibility between this gradual number and every quantifier in S_Q. The quantifier that yields maximum compatibility is then chosen to represent the linguistic amount or proportion of pixels having color in the linguistic histogram. The compatibility is calculated by evaluating the accomplishment degree of the quantified statement Q_i *of pixels in the image are painted in* $\widetilde{C}$, by using the method introduced in [18] as follows: let (Λ, ρ) be the gradual number representing the cardinality of the fuzzy subset of pixels with color $\widetilde{C}$. Let $\Lambda = \{\alpha_1, \ldots, \alpha_r\}$ with $1 = \alpha_1 > \alpha_2 > \cdots > \alpha_r > \alpha_{r+1} = 0$, with $r > 0$. Then the evaluation of Q_i *of pixels in the image are painted in* $\widetilde{C}$ is a number in $[0,1]$ given by equation 4

$$\sum_{\alpha_i \in \Lambda} (\alpha_i - \alpha_{i+1}) \times Q(\rho(\alpha_i)) \tag{4}$$

4.3 A First Synthetic Example

Our first example, taken from [17], allows us to illustrate the differences between the different approaches to cardinality of fuzzy sets discussed in section 3. The fuzzy color space employed here, described in [17], is not that of section 2, but this is unimportant for our purpose in this first example. Figure 2 shows two images, the first one containing eight colors in the kernel of different fuzzy colors. Let $\widetilde{C_1}, \ldots, \widetilde{C_8}$ denote these colors, from left to right and top to bottom (see [17] for a definition of the membership functions). In the second one we have four colors that are compatible to a degree 0.5 with two of the eight fuzzy colors in the first image.

Table 1 shows the (relative) cardinality of the fuzzy set of pixels painted in each fuzzy color in both images, using different approaches. The result is the same for all the fuzzy colors in both images by the way they have been defined. In the case of image A, since the eight crisp colors employed are in the kernel of fuzzy colors,

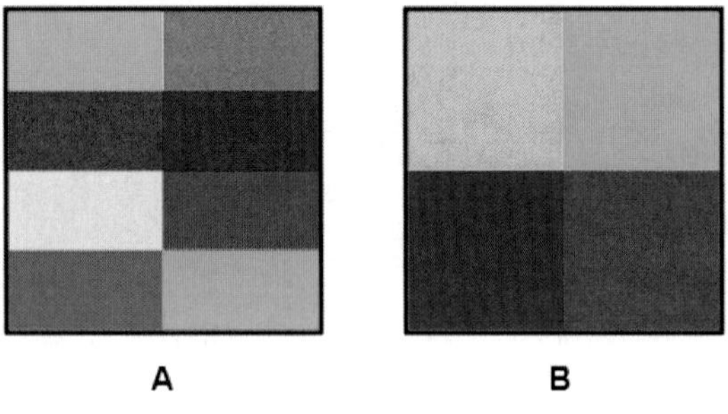

Fig. 2 Two synthetic images

the sets of pixels painted in every fuzzy color are crisp, and hence a crisp cardinality is obtained. In particular, the fuzzy number for image A is the same despite the method employed for calculating it. In the gradual approach, crisp results are represented by the fact that only the level $\alpha = 1$ is necessary, since all the levels are assigned the same cardinality (in this case, 1/8). In the case of the quantifier, any quantifier having 1/8 in its core will give a result of 1 in the evaluation of the quantified sentence. We assume here that the user has predefined a collection of quantifiers of the form *Around x %*, with $x \in [0,1]$ corresponding to some kind of fuzzy partition of the unit interval. Under this assumption, *Around 12.5%* is a triangular fuzzy number/quantifier with kernel 1/8 with appropriate specificity for the user. Of course, in this particular case, it is clear that a quantifier *Exactly 12.5%* would have been a better choice.

Table 1 Cardinalities (same for all fuzzy colors $\widetilde{C_1},\ldots,\widetilde{C_8}$) in the Color Histograms obtained for images A and B in figure 2 by using different approaches to cardinality

Quantifier	Around 12.5%	Around 0% or Around 25%
Gradual	$1 \mapsto 1/8$	$1 \mapsto 0$ $0.5 \mapsto 1/4$
Fuzzy number	1/(1/8)	$1/0 + \sum_{0<x\leq 1/4} 0.5/x$
Scalar	1/8	1/8
	Image A	Image B

In the case of image B, the first thing to highlight is the fact that the scalar cardinality yields the same result than that of image A. However, the perception of the colors that appear in the image and the corresponding frequencies is very different in both images, the only possible frequencies being intuitively 1/8 in image A, and 0 and 1/4 in image B. This example illustrates how the sigma-count is not an accurate measure in general. Notice also that using a threshold below 1/2 does not solve the problem.

The fuzzy cardinality provided is the one that would be provided by most existing methods, and represents the notion of approximately *Between 0 and 1/4*. Though a suitable information for a human user, it is not an accurate representation, since there is no chance that the cardinality is other than 0 or 1/4. However, notice that it yields a different result than that for image A, and hence it is more accurate than the sigma-count. As a final remark, the fuzzy cardinality defined in [2] and employed in [17] gives a fuzzy set $0.5/0 + 0.5/(1/4)$ which is not a fuzzy number since it is neither normalized nor convex. This is more accurate than a fuzzy number and, though good enough to be easily understandable by a human, this is not always the case, as the fuzzy set can have a large support. Hence, it is a kind of compromise between accuracy and understandability.

The gradual approach gives also a different result for both images, in this case representing accurately the only possible cardinalities in both images, associated to different levels. In addition, this representation is very easy to use for further calculations, since we have a crisp number in each level, and any kind of calculation to be employed is performed independently in each level following definition 3. The results obtained in each level can later be summarized using techniques similar to those employed here to provide a quantifier from a gradual number.

Finally, the evaluation of quantified sentences would provide for image B a fulfilment degree of 0.5 for both triangular quantifiers *Around 0%* and *Around 25%*. This is in consonance with our intuition about the possible cardinalities in image B. Here, two different options are to choose one of them on the basis of other application or user-specific information (for example, whether we prefer to be conservative or not with respect to the amount of colors), or to give both of them. Notice that this case is rather unfrequent, and it is motivated by the fact that very specific, synthetic images with particular memberships have been employed.

4.4 A Real Example

Let us consider the real image in figure 3. We are not giving fuzzy and gradual results here because of lack of space, being also little informative in a real case to the user. We have considered a collection of ten triangular relative quantifiers defining a fuzzy partition in Ruspini's sense, with kernels being the percentages 0, 1, 2, 3, 4, 12, 25, 50, 75, and 100. Let us denote by QP_x the quantifier with kernel x. Table 2 shows the (approximate) value obtained by evaluation of the quantified sentences QP_x *of the pixels are painted in color* $\widetilde{C}$ for every fuzzy color $\widetilde{C}$ in the fuzzy color space associated to the ISCC-NBS Basic set of colors explained in section 2. The last column contains the values of the linguistic histogram.

Let us remark that the addition of the result of the evaluation of quantified sentences is not necessarily 1, since these are not frequencies but values of compatibility between cardinality and quantifiers. Similarly, the addition of the quantities indicated in the linguistic histogram is not expected to be 100% in general, since these are fuzzy sets around the value. Specificity also play a role here, for example, the collection of quantifiers employed here "jumps" from 4% to 12%, hence the

Fig. 3 Parrot image

Table 2 Evaluation of quantified sentences and linguistic color histogram for figure 3

Fuzzy Color $\widetilde{C}$	**Quantifiers** QP_0	QP_1	QP_2	QP_3	QP_4	QP_{12}	QP_{25}	QP_{50}	QP_{75}	QP_{100}	**Linguistic Histogram** $h_L(\widetilde{C})$
Pink	**0.47**	0.29	0.14	0.08	0	0	0	0	0	0	Around 0%
Red	0.20	**0.25**	0.16	0.15	0.21	0	0	0	0	0	Around 1%
Orange	0	0.03	0.11	0.06	0.31	**0.43**	0.03	0	0	0	Around 12%
Brown	0.12	0.17	0.10	0.09	0.22	**0.25**	0.01	0	0	0	Around 12%
Yellow	0	0	0.13	0.14	**0.39**	0.32	0	0	0	0	Around 12%
Olive	0	0	0	0	0.02	0.19	0.35	**0.41**	0.01	0	Around 50%
Green	0.10	0.12	0.13	0.15	**0.16**	0.13	0.05	0	0	0	Around 4%
Blue	0.25	**0.26**	0.22	0.23	0.02	0	0	0	0	0	Around 1%
Violet	**0.47**	0.42	0.09	0	0	0	0	0	0	0	Around 0%
Purple	**0.73**	0.16	0.08	0.02	0	0	0	0	0	0	Around 0%
White	**0.85**	0.08	0.03	0	0	0	0	0	0	0	Around 0%
Gray	0.03	0.21	0.13	0.10	0.18	**0.26**	0.05	0	0	0	Around 12%
Black	0	0	0	0	0	0.06	**0.62**	0.31	0	0	Around 25%

latter quantifier is much less specific than those between 0-4%. In practice, we are just working on the basis of the quantifiers the user is interested in, his/her vocabulary, and the results may vary depending on the number and definition of quantifiers, and the fuzzy color space employed. However, we think that the results are compatible with what we can see in the image. Finally, choosing more than one quantifier when the difference in the accomplishment degree is very low can be an interesting alternative in some cases, like in the case of the fuzzy color Green.

5 Conclusions and Future Work

Different color histograms of images, based on different approaches to the cardinality of fuzzy sets, may be applied depending on the use they are intended. We propose to use gradual histograms when the result is to be employed as the basis for further calculation in the setting of image processing and analysis, whilst we consider linguistic histograms more appropriate when the information is to be given to the user. As future work we plan to use these histograms in developing applications for linguistic description of images and image information retrieval.

References

1. Chamorro-Martínez, J., Sánchez, D., Soto-Hidalgo, J.M.: A novel histogram definition for fuzzy color spaces. In: Proceedings IEEE WCCI 2008, pp. 2149–2156 (2008)
2. Delgado, M., Martín-Bautista, M.J., Sánchez, D., Vila, M.A.: A probabilistic definition of a nonconvex fuzzy cardinality. Fuzzy Sets and Systems 126(2), 41–54 (2002)
3. Delgado, M., Martín-Bautista, M.J., Sánchez, D., Vila, M.A.: Fuzzy integers: Representation and arithmetic. In: Proceedings of IFSA 2005 (2005)

4. Doulamis, A., Doulamis, N.: Fuzzy histograms for efficient visual content representation: application to content-based image retrieval. In: IEEE International Conference on Multimedia and Expo., pp. 893–896 (August 2001)
5. Dubois, D., Prade, H.: Fuzzy cardinality and the modeling of imprecise quantification. Fuzzy Sets and Systems 16, 199–230 (1985)
6. Dubois, D., Prade, H.: Fuzzy intervals versus fuzzy numbers: Is there a missing concept in fuzzy set theory? In: Linz Seminar 2005 Abstracts, pp. 45–46 (2005)
7. Dubois, D., Prade, H.: Gradual elements in a fuzzy set. Soft Computing 12, 165–175 (2008)
8. Dubois, D., Prade, H., Sudkamp, T.: A discussion of indices for the evaluation of fuzzy associations in relational databases. In: De Baets, B., Kaynak, O., Bilgiç, T. (eds.) IFSA 2003. LNCS, vol. 2715, pp. 111–118. Springer, Heidelberg (2003)
9. Han, J., Kai-Kuang: Fuzzy color histogram and its use in color image retrieval. IEEE Transactions on Image Processing 11(8), 944–952 (2002)
10. Hildebrand, L., Fathi, M.: Knowledge-based fuzzy color processing. IEEE. Tran. on Systems, Man and Cybernetics. Part C 34(4), 499–505 (2004)
11. Kelly, K.L., Judd, D.B.: Color: universal language and dictionary of names. National Bureau of Standards (USA) (440) (1976)
12. Louverdis, G., Andreadis, I., Tsalides, P.: New fuzzy model for morphological colour image processing. In: IEEE Proc. Vis. Image Signal Proc., vol. 149, pp. 129–139 (2002)
13. De Luca, A., Termini, S.: A definition of a nonprobabilistic entropy in the setting of fuzzy sets theory. Information and Control 20, 301–312 (1972)
14. Mitsuishi, T., Kayaki, N., Saigusa, K.: Color construction using dual fuzzy system. In: IEEE Int. Sym. Comp. Intelligence for Measurement Sys. and Appl., pp. 136–139 (2003)
15. Romani, S., Sobrebilla, P., Montseny, E.: Obtaining the relevant colors of an image through stability-based fuzzy color histograms. In: IEEE International Conference on Fuzzy Systems, St. Louis, Missouri (USA), vol. 2, pp. 914–919 (May 2003)
16. Runkler, T.A.: Fuzzy histograms and fuzzy chi-squared tests for independence. In: IEEE International Conference on Fuzzy Systems, vol. 3, pp. 1361–1366 (2004)
17. Sánchez, D., Delgado, M., Vila, M.A.: RL-numbers: An alternative to fuzzy numbers for the representation of imprecise quantities. In: Proc. Fuzz-IEEE 2008, pp. 2058–2065 (2008)
18. Sánchez, D., Delgado, M., Vila, M.A.: Fuzzy quantification using restriction levels. In: Di Gesù, V., Pal, S.K., Petrosino, A. (eds.) WILF 2009. LNCS, vol. 5571, pp. 28–35. Springer, Heidelberg (2009)
19. Seaborn, M., Hepplewhite, L., Stonham, J.: Fuzzy colour category map for the measurement of colour similarity and dissimilarity. Pattern Recognition 38(4), 165–177 (2005)
20. Soto-Hidalgo, J.M., Chamorro-Martínez, J., Sánchez, D.: A new approach for defining a fuzzy color space. In: Proceedings IEEE WCCI 2010, pp. 292–297 (2010)
21. Sugano, N.: Color-naming system using fuzzy set theoretical approach. In: IEEE Int. Conf. on Fuzzy Systems, pp. 81–84 (2001)
22. Wygralak, M.: Cardinalities of Fuzzy Sets. Springer, Heidelberg (2003)
23. Zadeh, L.A.: A theory of approximate reasoning. Machine Intelligence 9, 149–194 (1979)
24. Zhu, H., Zhang, H., Yu, Y.: Deep into color names: Matching color descriptions by their fuzzy semantics. In: Euzenat, J., Domingue, J. (eds.) AIMSA 2006. LNCS (LNAI), vol. 4183, pp. 138–149. Springer, Heidelberg (2006)

Image Reduction Using Fuzzy Quantifiers

D. Paternain, C. Lopez-Molina, H. Bustince, R. Mesiar, and G. Beliakov

Abstract. In this work we propose an image reduction algorith based on local reduction operators. We analyze the construction of weak local reduction operators by means of aggregation functions and we analyze the effect of several aggregation functions in image reduction with original and noisy images.

1 Introduction

Image resampling is a topic extensively used and studied in image processing. The two main processes are image magnification and image reduction [14]. In this work we focus on image reduction. Image reduction consists in diminishing the resolution of the image while keeping as much information of the original image as possible. Image reduction can be used to reduce the storage cost of images [10, 15] or to accelerate computations on an image [11].

In the literature there exists many methods for image reduction. We can divide these algorithms in two groups. In the first group, the image to be reduced is considered in a global way. This means that every pixel in the image is used to build a new pixel of the reduced image. In the second group, the image is divided in small

D. Paternain · C. Lopez-Molina · H. Bustince
Departamento de Automatica y Computacion, Universidad Publica de Navarra, Pamplona (Spain)
e-mail: {daniel.paternain,carlos.lopez,bustince}@unavarra.es

R. Mesiar
Department of Mathematics and Descriptive Geometry, Slovak University of Technology, Bratislava (Slovakia) and Institute of Information Theory and Automation, Czech Academy of Sciences, Prague (Czech Republic)
e-mail: mesiar@math.sk

G. Beliakov
School of Information Technology, Deakin University, Burwood (Australia)
e-mail: gleb@deakin.edu.au

B. De Baets et al. (Eds.): Eurofuse 2011, AISC 107, pp. 351–362, 2011.
springerlink.com

pieces or blocks and each block is then treated independently. The reduced image is made by the composition of the results of the algorithm for each block [9]. In this work we focus on the second group. We consider that treating each block independently allows to design simple reduction algorithms (as we only work with a small set of pixels). Moreover, as the algorithm acts locally on the imagen, we can develope algorithms with better features as keeping some properties of the image (edges or textures) or reducing the amount of noise in the image.

The objective of this work is to develope an image reduction algorithm such that, for each block in the imagen, we obtain a single pixel in the reduced image that represents all the pixels in the original block, and hence, such that keeps as much information as possible.

To solve this problem, we present weak local reduction operators. These operators take a block of the image and return a single value satisfying certain conditions. We propose the construction of these operators by means of aggregation functions, since these functions and their properties have been widely studied [1, 4, 5, 7]. We also want to analyze the effect of each local reduction operator in image reduction and the stability of the operators under the presence of noise.

Although measuring the effectiveness of an image reduction algorithm can be very difficult (since there is not a unique way), in this work we follow one of the most used strategy in the literature. We reconstruct a reduced image and then we compare with the original image. In this way, the more similar the reconstructed image with the original, the better the reduction operator.

The remainder of the work is organized as follows. In Sect. 2 we briefly introduce some theoretical concepts. In Sect. 3 we present the definition of weak local reductions operators and we present our image reduction algorithm. In Sec. 4 we study the construction of reduction operators by means of aggregation functions. Finally, we show some experimental results as well as some brief conclusions.

2 Preliminaries

We start by recalling some concepts that will be used along this work.

Definition 1. *An aggregation function of dimension n (n-ary aggregation function) is a non-decreasing mapping* $M : [0,1]^n \to [0,1]$ *such that* $M(0,\dots,0) = 0$ *and* $M(1,\dots,1) = 1$.

Definition 2. *Let* $M : [0,1]^n \to [0,1]$ *be a n-ary aggregation function.*

(i) M is said to be idempotent if $M(x,\dots,x) = x$ *for any* $x \in [0,1]$.
(ii) M is said to be homogeneous if $M(\lambda x_1,\dots,\lambda x_n) = \lambda M(x_1,\dots,x_n)$ *for any* $\lambda \in [0,1]$ *and for any* $(x_1,\dots,x_n) \in [0,1]^n$.
(iii) M is said to be shift-invariant if $M(x_1 + r,\dots,x_n + r) = M(x_1,\dots,x_n) + r$ *for all* $r > 0$ *such that* $0 \leq x_i + r \leq 1$ *for any* $i = 1,\dots,n$.

A complete characterization for shift-invariantness and homogeneity of aggregation functions can be seen in [12, 13].

We know that a triangular norm (t-norm for short) $T: [0,1]^2 \longrightarrow [0,1]$ is an associative, commutative, non-decreasing function such that $T(1,x) = x$ for all $x \in [0,1]$. A basic t-norm is the minimum ($T_M(x,y) = \wedge(x,y)$). Analogously, a triangular conorm (t-conorm for short) $S: [0,1]^2 \longrightarrow [0,1]$ is an associative, commutative, non-decreasing function such that $S(0,x) = x$ for all $x \in [0,1]$. A basic t-conorm is the maximum ($S_M(x,y) = \vee(x,y)$).

3 Weak Local Reduction Operators and Image Reduction Algorithm

We consider an image of $n \times m$ pixels as a set of $n \times m$ elements arranged in rows and columns. Hence we consider an image as a $n \times m$ matrix. Each element of the matrix has a value in $[0,1]$ that will be calculated by normalizing the intensity of the corresponding pixel in the image (by dividing by $L-1$, being L the number of gray levels). In this work we use this notation:

- $\mathscr{M}_{n\times m}$ is the set of all matrices of dimension $n \times m$ over $[0,1]$.
- Each element of a matrix $A \in \mathscr{M}_{n\times m}$ is denoted by a_{ij} with $i \in \{1,\ldots,n\}$, $j \in \{1,\ldots,m\}$.
- Let $A,B \in \mathscr{M}_{n\times m}$. We say that $A \leq B$ if for all $i \in \{0,\ldots,n\}, j \in \{0,\ldots,m\}$ the inequality $a_{ij} \leq b_{ij}$ holds.
- Let $A \in \mathscr{M}_{n\times m}$ and $c \in [0,1]$. $A = c$ denotes that $a_{ij} = c$ for all $i \in \{1,\ldots,n\}, j \in \{1,\ldots,m\}$. In this case, we will say that A is constant matrix or a flat image.

Definition 3. *A weak local reduction operator WO_{LR} is a mapping $WO_{LR}: \mathscr{M}_{n\times m} \to [0,1]$ that satisfies*

(WOLR1) For all $A,B \in \mathscr{M}_{n\times m}$, if $A \leq B$, then $WO_{LR}(A) \leq WO_{LR}(B)$.
(WOLR2) If $A = c$ then $WO_{LR}(A) = c$.

Remark. We call our operators weak local reduction operators since we demand the minimum number of properties that, in our opinion, a local reduction operator must fulfill.

Definition 4. *We say that a weak reduction operator WO_{LR} is:*

(WOLR3) homogeneous if $WO_{LR}(\lambda A) = \lambda \cdot WO_{LR}(A)$ for all $A \in \mathscr{M}_{n\times m}$ and $\lambda \in [0,1]$
(WOLR4) stable under translation (shift-invariant) if $WO_{LR}(A+r) = WO_{LR}(A) + r$ for all $A \in \mathscr{M}_{n\times m}$ and $r \in [0,1]$ such that $0 \leq a_{ij} + r \leq 1$ whenever $i \in \{1,\ldots,n\}, j \in \{1,\ldots,m\}$

From these definitions, we now present our image reduction algorithm that uses weak local reduction operators. Given an image $A \in \mathscr{M}_{n\times m}$ and a reduction block size $n' \times m'$ (with $n' \leq n$ and $m' \leq m$), the algorithm is the following:

(1) Choose a weak local reduction operator.
(2) Divide the image A into disjoint blocks of dimension $n' \times m'$. If n or m are not multiple of n' or m' respectively, we suppress the smallest number of rows and/or columns in A that ensures that these conditions hold.
(3) FOR each block in A DO

  (3.1) Apply the weak local reduction operator to the elements of the block.
  (3.2) Place the pixel obtained in the reduced image as in Fig. 1.

  END FOR

Algorithm 1

En Fig. 1 we show the scheme of Algorithm 1. We take an image $A \in \mathcal{M}_{9\times 9}$ and $n' = m' = 3$. When applying Algorithm 1 we obtain a reduced image $A' \in \mathcal{M}_{3\times 3}$.

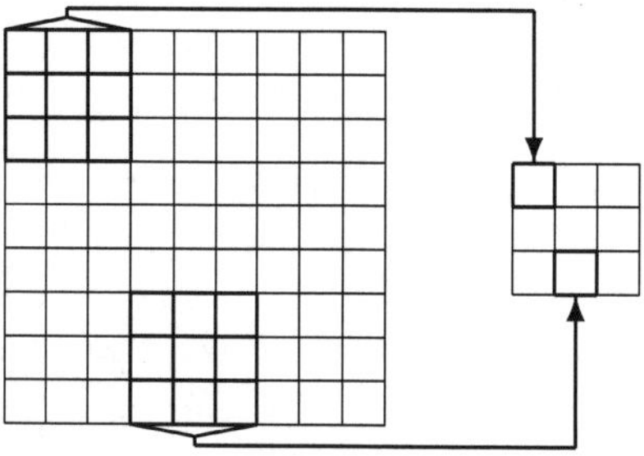

Fig. 1 Example of reduction of a matrix

4 Weak Local Reduction Operators and Aggregation Functions

In this section we study the construction of weak local reduction operators by means of some families of aggregation functions.

Proposition 1. *Let M be an idempotent aggregation function. The operator defined by*

$$WO_{LR}(A) = M(a_{11}, a_{12}, \ldots, a_{1m}, \ldots, a_{n1}, \ldots, a_{nm})$$

is a weak local reduction operator for all $A \in \mathcal{M}_{n\times m}$.

4.1 Minimum and Maximum

We know that the t-norm minimum $T = T_M$ is the only idempotent t-norm. Analogously, the t-conorm maximum $S = S_M$ is the only idempotent t-conorm. Moreover, observe that T_M and S_M are lower and upper bounds respectively of weak local reduction operators.

In Fig. 2 we apply to image (a) the weak local reduction operators constructed from T_M obtaining image (c). In the same way, we apply the weak local reduction operator constructed from S_M obtaining image (d). Observe that, with these weak local reduction operators we get that $image(c) \leq image(d)$. Now, we add salt and pepper noise to original image (image (b)) and we repeat the same experiment obtaining images (e) and (f). Observe that these two operators (minimum and maximum) are not good local reduction operators. If we take the minimum over a block with noise we always obtain the value 0. Analogously, if we consider the maximum and apply it to a block with noise, we always recover the value 1. In this way we lose all information about the elements in the block that have not been altered by noise. This fact leads us to study other aggregation functions that take into account all the pixels in the block.

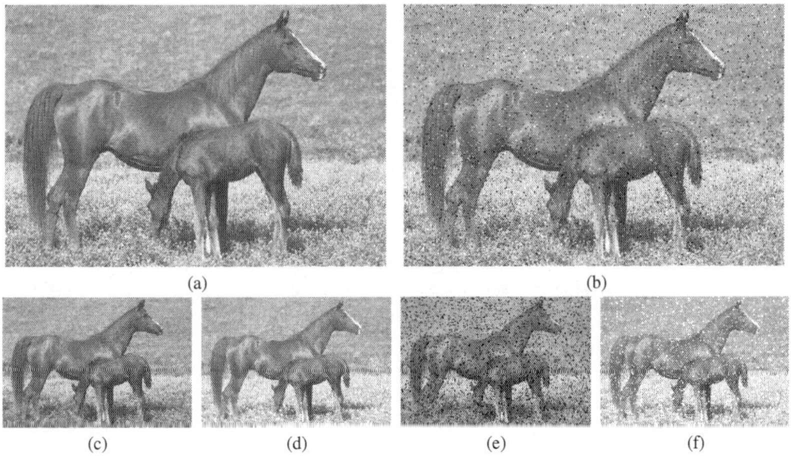

Fig. 2 Reduction using weak local reduction operators constructed from minimum and maximum

4.2 Quasi-arithmetic Means

Definition 5. *Let $g : [0,1] \to [-\infty,\infty]$ be a continuous and strictly monotone function. A quasi-arithmetic mean is a mapping $M_g : [0,1]^n \to [0,1]$ defined as*

$$M_g(x_1,\ldots,x_n) = g^{-1}\left(\sum_{i=1}^{n} g(x_i)\right)$$

Proposition 2. *Let $M_g : [0,1]^{n\cdot m} \to [0,1]$ be a quasi-arithmetic mean. The operator defined as*

$$WO_{LR}(A) = g^{-1}\left(\sum_{i=1}^{n}\sum_{j=1}^{m} g(a_{ij})\right)$$

is a weak local reduction operator for all $A \in \mathcal{M}_{n\times m}$.

Next, we analyze wether a wak local reduction operator constructed from a quasi-arithmetic mean satisfies properties $(WOLR3)$ and $(WOLR4)$.

Proposition 3. *A weak local reduction operator built from a quasi-arithmetic satisfies $(WOLR3)$ if and only if*

$$WO_{LR}(A) = \left(\prod_{i=1}^{n}\prod_{j=1}^{m} a_{ij}\right)^{\frac{1}{n\cdot m}} \quad or$$

$$WO_{LR}(A) = \left(\sum_{i=1}^{n}\sum_{j=1}^{m} \frac{a_{ij}^{\alpha}}{n\cdot m}\right)^{\frac{1}{\alpha}} \quad with\ \alpha \neq 0$$

for all $A \in \mathcal{M}_{n\times m}$.

Proof. See page 118 of [7] □

Proposition 4. *A weak local reduction operator built from a quasi-arithmetic mean satisfies $(WOLR4)$ if and only if*

$$WO_{LR}(A) = \frac{1}{n\cdot m}\sum_{i=1}^{n}\sum_{j=1}^{m} a_{ij} \quad or$$

$$WO_{LR}(A) = \frac{1}{\alpha}\log\left(\sum_{i=1}^{n}\sum_{j=1}^{m} \frac{e^{\alpha a_{ij}}}{n\cdot m}\right) \quad with\ \alpha \neq 0$$

for all $A \in \mathcal{M}_{n\times m}$

Proof. See page 118 of [7] □

In Fig. 3 we apply two weak local reduction operators constructed from the following quasi-arithmetic means:

(1) from the arithmetic mean. This weak local reduction operator satisfies properties $(WOLR3)$ and $(WOLR4)$ and is given by

$$WO_{LR}(A) = \frac{1}{n\cdot m}\sum_{i=1}^{n}\sum_{j=1}^{m} a_{ij}$$

(2) from the geometric mean. This weak local reduction operators satisfies property $(WOLR3)$ and is given by

$$WO_{LR}(A) = \sqrt[n\cdot m]{\prod_{i=1}^{n}\prod_{j=1}^{m} a_{ij}}$$

In Figure 3 we apply these weak local reduction operators.From the original image (a), we apply the operators constructed from the arihtmetic mean and geometric mean obtaining images (c) and (d) respectively. We repeat the experiment to analyze the reaction of the operators to the noise. We add salt and pepper noise to the original image obtaining image (b). We apply the operators obtaining images (e) and (f). Observe that image (f) obtained by means of the geometric mean is not a good weak local reduction operator for images with impulsive noise. This is due to the fact that if there is a pixel in the block with a value of 0, then the result of the operator is always 0 (so we do not take into account all the pixels in the block). In image (e) we see that the outliers introduced by impulsive noise also affects the result. However, we visually check that the arithmetic mean is a better reduction operators than the geometric mean for this type of noise.

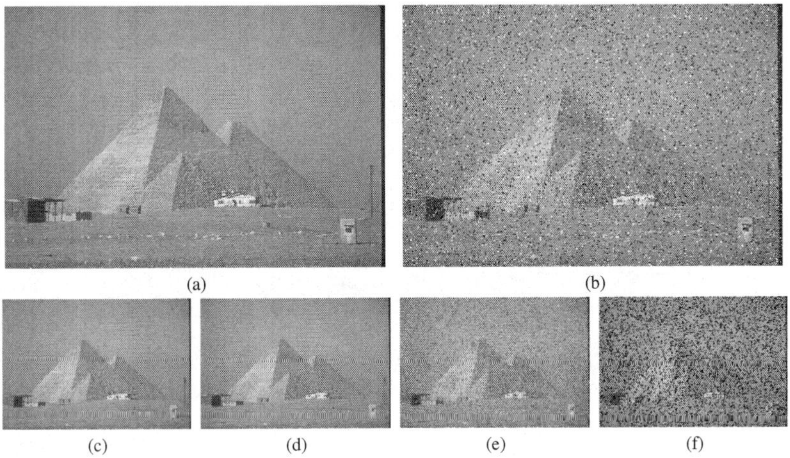

Fig. 3 Reduction using weak local reduction operators constructed from arithmetic and geometric mean

4.3 OWA Operators

Definition 6. *A mapping $F : [0,1]^n \longrightarrow [0,1]$ is called an OWA operator of dimension n if there exists a weighting vector $W = (w_1, w_2, \dots, w_n) \in [0,1]^n$ with $\sum_i w_i = 1$ and such that*

$$F(a_1, a_2, \dots, a_n) = \sum_{j=1}^{n} w_j b_j$$

with b_j the $j-th$ largest element of the sample $(a_1, \dots, a_n)$.

Remark: Observe that we can obtain well known aggregation functions from OWA operators, as the minimum, maximum, arithmetic mean or median.

Proposition 5. *Let F be an OWA operator of dimension $n \cdot m$. Then the operator given by*

$$WO_{LR}(A) = F(a_{11}, a_{12}, \ldots, a_{1m}, \ldots, a_{n1}, \ldots, a_{nm})$$

is a weak local reduction operator that satisfies $(WOLR3)$ *and* $(WOLR4)$ *for all* $A \in \mathcal{M}_{n\times m}$.

In this subsection we focus on OWA operators constructed using fuzzy quantifiers. These operators are used to model human expressions as, for example, "at least one half", "the largest part of", "a lot of" and, in general, expressions that represent the amount of items satisfying certain property. The weights of an OWA operator associated with these expression are given by

$$w_k = Q\left(\frac{k}{n}\right) - Q\left(\frac{k-1}{n}\right) \text{ with } k = 1, \ldots, n$$

where Q is a fuzzy quantifier as, for instance,

$$Q(r) = \begin{cases} 0 & \text{if } r < a \\ \frac{r-a}{b-a} & \text{if } a \leq r \leq b \\ 1 & \text{if } r > b \end{cases}$$

En Fig. 4 we apply weak local reduction operators from OWA operators and fuzzy quantifiers. We study the following situations:

(1) "at least one half" operator taking $a = 0$ and $b = 0.5$.
(2) "the largest possible amount" taking $a = 0.5$ and $b = 1$.
(3) "the largest part of" taking $a = 0.3$ and $b = 0.8$.

Observe that in general, the three operators act in a similar way with the original image. However, they act different under the presence of noise. We can see that the operator given by the expression "the largest possible amount" is less sensitive to impulsive noise. In the other way, we see that the operator given by "at least one half" erases black pixels but it is affected by white pixels. The operator given by "the largest part of" acts in an opposite way, as this operator is affected by black pixels.

5 Experimental Results

In Sect. 4 we have studied several constructions of weak local reduction operators. In this section we want to evaluate the reduced images. However, there is not a unique way of determining the best image and hence the best weak local reduction operator. In this we work we will use the following process:

1. Fix one image magnificaction algorithm;
2. Apply one weak local reduction operator to an original image;
3. Apply the magnification algorithm to the reduced image;

Fig. 4 Reduction using weak local reduction operators constructed from OWA operators and fuzzy quantifiers

4. Compare the reconstructed image with the original;
5. Take as the best image the image being more similar to the original.

We know that in the literature there exist many methods for image magnification. In this work we will use the algorithm given in [8] since it gets very good results and the computations cost of the algorithm is very low. To compare images, we will use fuzzy measures (since we can consider an image to be a fuzzy set [3]). In [2] an in depth study of such indices is carried out. Specifically we use the similarity measure based on contrast de-enhancement. With our notation, this index is given by:

$$S(A,B) = \frac{1}{n \times m} \sum_{i=1}^{n} \sum_{j=1}^{m} 1 - |a_{ij} - b_{ij}|.$$

In these experiments we are going to use the following weak local reduction operators: minimum (*min*), geometric mean (*geom*), arithmetica mean (*arith*), "at least one half" (*OWA1*), "the largest possible amount" (*OWA2*), "the largest part of" (*OWA3*) and the maximum (*max*). We are going to reduce (and then reconstruct) the set of three images showed along this work: horse, bird and pyramid. In Algorithm 1 we take $n' = m' = 2$. If the original images are of dimension 320×480, the reduced are 160×240.

In Table 1 we show the result of the comparison by means of the similarity index between original images and the reconstructed images. Observe that the results are good and similar. The best result is obtained with the operator "the largest possible amount". Results are very similar if we take the geometric mean and arithmetic mean. Results are worse if we take minimum, maximum or the operators "at least one half" and "the largest part of".

Table 1 Comparison between reconstructed and original images

	Image (a)	Image (b)	Image (c)
Min	0.9502	0.9835	0.9807
Geom	0.9658	**0.9878**	0.9864
Arith	0.9658	0.9876	0.9865
OWA1	0.9581	0.9852	0.9840
OWA2	**0.9663**	**0.9878**	**0.9866**
OWA3	0.9592	0.9860	0.9840
Max	0.9498	0.9830	0.9813

To analyze the reaction of the operators to impulsive noise, we modify original images by adding salt and pepper noise (with a probability of 5% and 10%). In Tables 2 and 3 we show the similarity index between original images with noise (5% in Table 2 and 10% in Table 3) and the reconstructed images. In these conditions we see that the best result is now obtained with the OWA operator "the largest possible amount". Although this operator is affected by low values, it dismisses all possible high values (aggregating only low values). Moreover, we observe that the operators constructed from minimum, geometric mean, "the largest part of" and maximum are very sensitive to this type of noise.

Table 2 Comparison between reconstructed and original images with noise (5%)

	Image (a)	Image (b)	Image (c)
Min	0.8953	0.9249	0.9240
Geom	0.9106	0.9275	0.9279
Arith	0.9360	0.9528	0.9528
OWA1	0.9286	0.9524	0.9504
OWA2	**0.9417**	**0.9605**	**0.9600**
OWA3	0.9202	0.9445	0.9440
Max	0.9123	0.9411	0.9372

A comparison between results is shown in Fig. 5. For each reduction operator we show the average of the similarity index in the test images. The first column of each operator corresponds to reconstruction of original images. The second and third column corresponds to images with salt and pepper noise with a probability of 5% and 10% respectively.

Table 3 Comparison between reconstructed and original images with noise (10%)

	Image (a)	Image (b)	Image (c)
Min	0.8043	0.8288	0.8314
Geom	0.8205	0.8317	0.8354
Arith	0.8837	0.8935	0.8969
OWA1	0.8746	0.8935	0.8914
OWA2	**0.8946**	**0.9091**	**0.9104**
OWA3	0.8530	0.8721	0.8745
Max	0.8470	0.8697	0.8639

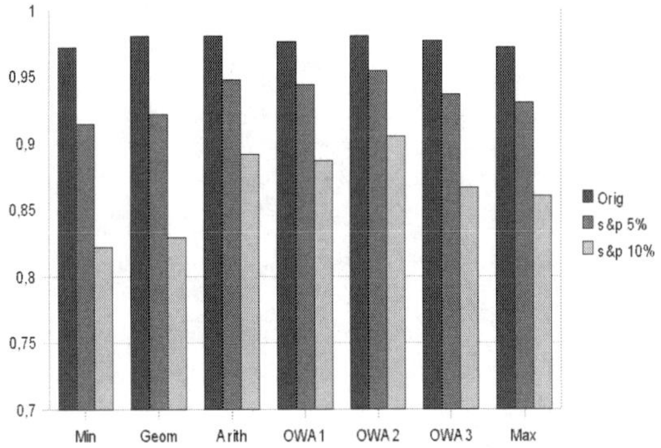

Fig. 5 Reduction using weak local reduction operators constructed from OWA operators and fuzzy quantifiers

6 Conclusions

In this work we have defined weak local reduction operators and the minimum properties these operators must fulfill. We have studied the construction of these operators by means of idempotent aggregation functions. We have focused on well known averaging functions, specially OWA operators.

From our definitions, we have proposed an image reduction algorithm that acts locally on small pieces of the image. We have compared some reduction operators by magnifying the reduced image and comparing with the original. We have also studied the reaction of different reduction operators to impulsive noise. We have seen that all the operators studied are sensitive to this noise, although there are differences between some of them. Specifically we have checked that the best reduction operator is the one constructed from the OWA operator the "largest possible amount".

As future work, we want to stablish new ways of determining the best reduction operator and to analyze other situations as, for example, other types of noise or preserving of features.

Acknowledgements. This work has been partially supported by the National Science Foundation of Spain, reference TIN2010-15055, by the Research Services of the Universidad Publica de Navarra and by grants P402/11/0378, APVV-0073-10 and VEGA 1/0080/10.

References

1. Beliakov, G., Pradera, A., Calvo, T.: Aggregation Functions: A Guide for Practitioners. Studies in Fuzziness and Soft Computing, vol. 221 (2007)
2. Bustince, H., Pagola, M., Barrenechea, E.: Construction of fuzzy indices from DI-subsethood measures: Application to the global comparison of images. Information Sciences 177, 906–929 (2007)
3. Bustince, H., Barrenechea, E., Pagola, M., Fernandez, J.: Interval-valued fuzzy sets constructed from matrices: Application to edge detection. Fuzzy Sets and Systems 160, 1819–1840 (2009)
4. Bustince, H., Calvo, T., De Baets, B., Fodor, J., Mesiar, R., Montero, J., Paternain, D., Pradera, A.: A class of aggregation functions encompassing two-dimensional OWA operators. Information Sciences 180, 1977–1989 (2010)
5. Calvo, T., Beliakov, G.: Aggregation functions based on penalties. Fuzzy sets and Systems 161, 1420–1436 (2010)
6. Chaira, T., Ray, A.K.: Fuzzy measures for color image retrieval. Fuzzy Sets and Systems 150, 545–560 (2005)
7. Fodor, J., Roubens, M.: Fuzzy Preference Modelling and Multicriteria Decision Support. Kluwer Academic Publishers, Dordrecht (1994)
8. Jurio, A., Pagola, M., Mesiar, R., Beliakov, G., Bustince, H.: Image magnification using interval information. IEEE Transactions on Image Processing (to appear)
9. Loia, V., Sessa, S.: Fuzzy relation equations for coding/decoding processes of images and videos. Information Sciences 171, 145–172 (2005)
10. Di Martino, F., Loia, V., Sessa, S.: A segmentation method for image compressed by fuzzy transform. Fuzzy Sets and Systems 161, 56–74 (2010)
11. Perfilieva, I.: Fuzzy Transforms and Their Applications to Image Compression. In: Bloch, I., Petrosino, A., Tettamanzi, A.G.B. (eds.) WILF 2005. LNCS (LNAI), vol. 3849, pp. 19–31. Springer, Heidelberg (2006)
12. Rückschlossová, T.: Aggregation operators and invariantness. PhD thesis, Slovak University of Technology, Bratislava, Slovakia (June 2003)
13. Rückschlossová, T., Rückschloss, R.: Homogeneous aggregation operators. Kybernetika (Prague) 42(3), 279–286 (2006)
14. Unser, M., Aldroubi, A., Eden, M.: Enlargement or reduction of digital images with minimum loss of information. IEEE Transactions on Image Processing 4, 247–258 (1995)
15. Xiang, S., Nie, F., Zhang, C.: Learning a Mahalanobis distance metric for data clustering and classification. Pattern Recognition 41, 3600–3612 (2008)

Indoor Location Using Fingerprinting and Fuzzy Logic

Pedro Mestre, Luís Coutinho, Luís Reigoto, João Matias, Aldina Correia, Pedro Couto, and Carlos Serodio

Abstract. Indoor location systems cannot rely on technologies such as GPS (Global Positioning System) to determine the position of a mobile terminal, because its signals are blocked by obstacles such as walls, ceilings, roofs, etc. In such environments the use of alternative techniques, such as the use of wireless networks, should be considered. The location estimation is made by measuring and analysing one of the parameters of the wireless signal, usually the received power. One of the techniques used to estimate the locations using wireless networks is fingerprinting. This technique comprises two phases: in the first phase data is collected from the scenario and stored in a database; the second phase consists in determining the location of the mobile node by comparing the data collected from the wireless transceiver with the data previously stored in the database. In this paper an approach for localisation using fingerprinting based on Fuzzy Logic and pattern searching is presented. The performance of the proposed approach is compared with the performance of classic methods, and it presents an improvement between 10.24% and 49.43%, depending on the mobile node and the Fuzzy Logic parameters.

Pedro Mestre · Pedro Couto · Carlos Serodio
CITAB - UTAD, Apartado 1013, 5001-801 Vila Real, Portugal
e-mail: {pmestre,pcouto,cserodio}@utad.pt

Luís Coutinho · Luís Reigoto
UTAD, Apartado 1013, 5001-801 Vila Real, Portugal
e-mail: luis_coutinho_86@hotmail.com, luisreigoto@gmail.com

João Matias
CM-UTAD, Apartado 1013, 5001-801 Vila Real, Portugal
e-mail: j_matias@utad.pt

Aldina Correia
CM-UTAD and CIICESI/ESTGF/IPP, 4610-156 Felgueiras, Portugal
e-mail: aic@estgf.ipp.pt

B. De Baets et al. (Eds.): Eurofuse 2011, AISC 107, pp. 363–374, 2011.
springerlink.com

1 Introduction

While in outdoor location technologies such as GPS can be successfully used, their use in indoor environments is seriously compromised by obstacles (floors, ceilings, walls, etc...). Alternative techniques must be used for indoor location systems. Some of these alternatives include the use of ultrasonic waves [9, 20], proximity sensors (e.g. pressure sensors [15]), infra-red [19], RFID (Radio Frequency Identification) [18] and wireless communications networks [17, 7, 16].

In this work a new approach to fingerprinting-based location, which is a Scene Analysis location technique, is presented. Fingerprinting consists in collecting information about the electromagnetic radiation from several sources (called the references) and store it in a database. The location of wireless nodes is then estimated by comparing the wireless signals detected by the mobile node and the values stored in the database.

Unlike the traditional approach to fingerprinting, in the presented approach, the distance between the node and the references is not calculated. Instead of searching for the nearest point(s), using the distance in the signal domain, it is made a pattern search. Points of the spatial domain exhibiting a pattern similar to that of the signals captured by the mobile node are taken in consideration to determine the current location.

Considering that fuzzy set theory [21] has worked in the treatment of models that present ambiguity and highly noisy data, this theory in an interesting alternative for indoor location, in order to obtain an accurate location estimation [11, 12, 21].

Moreover, uncertainly is present in almost every measurement, estimation or decision process and, therefore, fuzzy techniques have been widely used in almost any of these processes. Being Fuzzy Logic considered the adequate tool to deal with uncertainly, in this work, an approach for localisation using fingerprinting based on Fuzzy Logic and pattern searching is presented.

Tests were made in a real life scenario at the University of Trás-os-Montes and Alto Douro. The scenario consisted in two classrooms and a corridor connecting them. A laptop computer to collected the data used to generate the Fingerprint Map (FM) and two mobile phones to test the Location Estimation Algorithms (LEA), were used. IEEE 802.11 was used as the wireless communications technology, because it is one of the most popular and used technologies for Wireless Local Area Networks (WLAN).

2 Location Using Fingerprinting

Methodologies used to do the location can be classified as [10]: Triangulation, Proximity and Scene Analysis. Fingerprinting falls in this last classification, i.e., it is a Scene Analysis technique in which a given scene is observed and analysed to determine the location of nodes.

Fingerprinting comprises two different phases [13], also called the off-line phase, in which the data to build the fingerprinting map is collected and a second phase, called on-line phase, in which the location of the mobile node is made. This location is based on the information collected from the wireless transceiver and the previously stored data.

In the first phase, for every point belonging to the spatial domain, a given property of the wireless signal received from the several references, usually the Received Signal Strength (RSS) value, are stored into a database, called the Fingerprinting Map (FM). In a IEEE 802.11 network the references are the Access Points (AP) used to access to the wireless infrastructure. The values of RSS from the various references at the various points of the spatial domain, from an N dimensional space (N is the number of APs) called the signal domain.

During the on-line phase, the location of the mobile node is done based on the comparison between the values acquired from the wireless transceiver and the information that was previously stored in the FM. Although any property of the wireless signal can be used in this estimation, in this work the RSS value was used.

After acquiring the needed parameters from the wireless transceiver, the next step is to determine the distance, in the signal domain, between the current location and all the references that are part of the FM. The distance can be calculated using the Euclidean distance (Eq. 1), considering the RSS values as the coordinates of the point in the domain:

$$d_j = \sqrt{\sum_{i=1}^{n} (P_{ri} - P_{FM\,j,i})^2} \tag{1}$$

were d_j is the distance between the current location (point j) and the reference i, n is the number of dimensions of the signal domain, P_{ri} is the power received from reference i and $P_{FM\,j,i}$ the value of the power of reference i registered in the FM for the point of the spatial domain j.

The classical Location Estimation Algorithms, used to estimate de location of the node are:

- Nearest Neighbour, which considers that the coordinates (in the spatial domain) of the point belonging the FM which has the shortest distance (in the signal domain) to the current point, are the current coordinates of the node.
- k-Nearest Neighbour, on which it is considered that the current coordinates of the node are the average of the coordinates (in the spatial domain) of the coordinates of the k nearest points (in the signal domain).
- Weighted k-Nearest Neighbour, which is similar to the above LEA. In this case a weighted average is used.

In this paper a new approach, based on Fuzzy Logic, is presented in the next section. The performance of the proposed LEA is then compared with the performance of the above mentioned classic methods.

3 Location Using Fingerprinting and Fuzzy Logic

Although the classic methods are based on the distance (in the signal domain) between the coordinates of the current location and the nodes that belong to the FM, the approach proposed in this work is based on pattern search.

One of the problems associated with the use of the distance is related with the fact that the received power depends on the terminal type and the attenuation due to obstacles. The power that arrives to receiver is given by Eq. 2:

$$P_r = P_t - PL + G_t + G_r \quad (2)$$

where P_r is the received power in dBm, P_t is the transmitted power in dBm, PL is the total Path Loss (db) and G_t and G_r are the gains of the the transmitting and receiving antenna (in dB or dBi).

Since different types of terminals have different antennas and therefore different antenna gains, the received signal will not be the same for different types of mobile terminals. Another problem is related with the attenuation due to obstacles. The PL in Eq. 2 is dependent both on the free space losses and the attenuation due to obstacles between the transmitting and the receiving antennas. This means that for the same terminal different RSS values for the same references can be obtained at the same point, e.g. a cell phone will have different RSS values when placed on a table or inside a pocket.

To overcome these problems the proposed approach, instead of being based on the distance, is based on the pattern formed by the received power from the several references at each point.

3.1 Building the Patterns

Let us consider the coordinates of point i as a function of the received power (Pr) from the n references used in the scenario : $P_i = f(Pr_1, Pr_2, ..., Pr_n)$. Since the received power at each point is given by Eq. 2, then for each point of the spatial domain the pattern P_i will be different. To determine the location of the node, the pattern formed by the received power is then matched with the patterns stored in the FM.

However, the pattern to be stored in the FM cannot be directly obtained through the pattern formed by the received power, as above explained. Instead, a new pattern is generated and stored in the FM. As depicted in Fig. 1, to the RSS values obtained at a given point it is subtracted a reference value. The result of this operation is then matched against a set of membership functions which will classify it as 'Similar', 'Slightly Different' (positive or negative) and 'Different' (positive or negative).

After this step each point is assigned a weight, which is calculated according to the following set of IF-THEN rules:

- IF the distance is 'Different (Negative)' THEN the weight of the point is 'high negative';

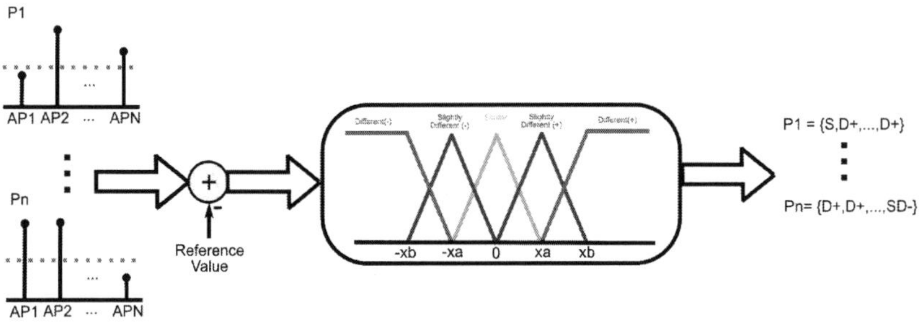

Fig. 1 Generation of the patterns for the points to be used in the FM and to do the location, using Fuzzy inference

- IF the distance is 'Slightly Different (Negative)' THEN the weight of the point is 'medium negative';
- IF the distance is 'Similar' THEN the weight of the point is 'low';
- IF the distance is 'Slightly Different (Positive)' THEN the weight of the point is 'medium positive';
- IF the distance is 'Different (Positive)' THEN the weight of the point is 'high positive';

Two different approaches were considered determine the reference RSSS value:

- RSS value of a predetermined reference – the RSS value of a predetermined reference is considered as the RSS reference value.

While the shape and number of the membership functions and weights for 'high negative','medium negative', 'low', 'medium positive' and 'high positive' were chosen empirically, the values for x_a and x_b were tuned using Nonlinear Optimization Methods as presented in subsection 3.3.

In Fig. 1 AP1, AP2,.. APN correspond to the Access Point 1, 2, ..., N.

3.2 Locating the Mobile Node

At the on-line phase, the location of the node is made by searching the FM for patterns that are similar to the pattern obtained at the current location. Although in the classic methods the number of points used to determine the actual coordinates is well known in beforehand, in this approach these parameters are determined by Fuzzy Logic based reasoning.

For each point belonging to the FM, it is determined the degree of similarity with the pattern obtained at the current location. The first step is to determine the pattern for the current location, as described above. After, using Eq. 3, the degree of similarity is calculated:

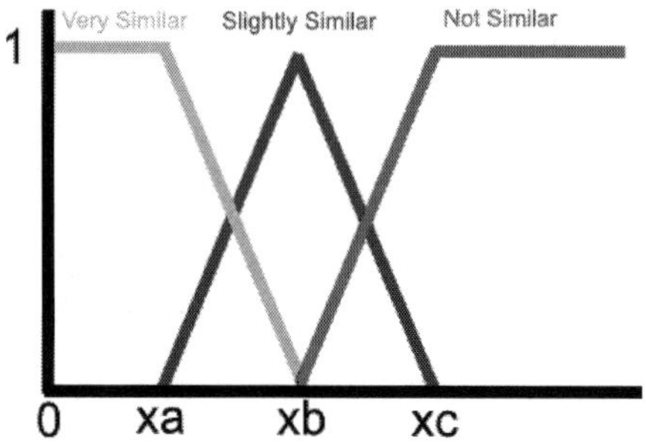

Fig. 2 Membership functions used in the on-line phase

$$S_j = \sum_{i=1}^{n} (|P_{j,i} - P_{FMi}|) \tag{3}$$

where n is the number of references, $P_{j,i}$ the value of the pattern of point j for the reference i and P_{FMi} is the value of the pattern in the FM for the point i.

These values can then be classified as 'Very Similar', 'Slightly Similar' and 'Not Similar'. The weight of point j in the final average, used to determine the current coordinates, depends on the grade of membership of S_j to the membership functions presented in Fig. 2. After determining the grade of membership, the following set of IF THEN rules are used:

- IF the value is 'Very Similar' THEN the point weight is 'high';
- IF the value is 'Slightly Similar' THEN point the weight is 'medium';
- IF the value is 'Not Similar' THEN point the weight is 'low';

After determining the final weight for each point of the FM, the weighted average of all the coordinates in the spatial domain of the points belonging to FM are considered the coordinates of the current location.

While the shape and number of membership functions were determined empirically, the values for x_a, x_b and x_c, of Fig. 2, and the weights for 'high', 'medium' and 'low' were determined as described the following subsection.

3.3 Tuning of the Fuzzy Logic Parameters

To tune the parameters of the membership functions used in the location systems as well as the weights of the rules used in the on-line phase, Direct Search Methods were used.

These methods are used in situations where it is not possible, or it is too difficult, to analytically determine the objective function of the problem to be solved. These methods are also very useful when the derivatives of the objective function are not known or cannot be determined.

Two types of Optimization Problems may arise: unconstrained and constrained. Constrained problems have the form:

$$\begin{aligned} \min_{x \in \mathbb{R}^n} \ & f(x) \\ s.t. \ & c_i(x) = 0, i \in \mathscr{E} \\ & c_i(x) \leq 0, i \in \mathscr{I} \end{aligned} \tag{4}$$

where:

- $f : \mathbb{R}^n \rightarrow \mathbb{R}$ is the *objective function*;
- $c_i(x) = 0$, $i \in \mathscr{E}$, with $\mathscr{E} = \{1, 2, ..., t\}$, define the problem *equality constraints*;
- $c_i(x) \leq 0$, $i \in \mathscr{I}$, with $\mathscr{I} = \{t+1, t+2, ..., m\}$, represent the *inequality constraints*;
- $\Omega = \{x \in \mathbb{R}^n : c_i = 0, i \in \mathscr{E} \wedge c_i(x) \leq 0, i \in \mathscr{I}\}$ is the set of all feasible points, i.e., the *feasible region*.

To solve this type of problems, Penalty and Barrier Methods can be used. These methods have been created to solve constrained problems, by solving a specially chosen sequence of unconstrained problems, of the form of Eq. 5. These unconstrained problems are then solved using the methods that are typically used to solve unconstrained problems (called the Internal Process).

$$\min_{x \in \mathbb{R}^n} f(x) \tag{5}$$

where:

- $f : \mathbb{R}^n \rightarrow \mathbb{R}$ is the *objective function*.

The new set of objective functions, Φ, contains information about the initial objective function, f, and the constrains of the problem. The new problem is defined by:

$$\Phi(x_k, r_k) : \min_{x_k \in \mathbb{R}^n} f(x_k) + r_k p(x) \tag{6}$$

where p is a function that penalizes (penalty) or refuses (barrier) points that violates the constraints and r_k is a positive parameter.

From the classic Penalty/Barrier functions and methods to solve unconstrained problems (used in the Internal Process), Extreme Barrier and Nelder-Mead methods were chosen, respectively.

The External Barrier Function, is widely used with Direct Search Methods with feasible point (for example by Audet et. al., [1, 2, 3, 4, 5, 6]), and it is defined by:

$$\Phi(x) = \begin{cases} f(x) & \text{se } x \in \Omega \\ +\infty & \text{se } x \notin \Omega \end{cases} \tag{7}$$

Conn et. al. in [8], (Chapter 8 - *Simplicial direct-search methods*) for example describes the Nelder-Mead method. It is characterized by starting from an initial simplex and modifying the search directions at the end of each iteration. It uses movements of reflection, expansion and contraction to the inside and the outside, together with the shrunk step towards the best vertex.

The localisation algorithm can be seen as an Optimization Problem, on which the parameters of the membership functions and the respective weights are the problem

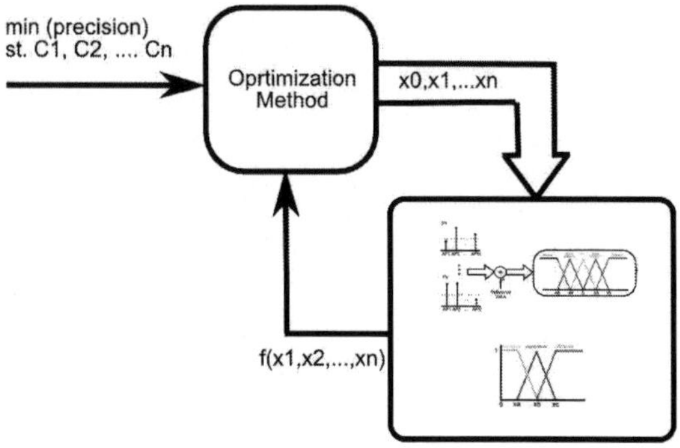

Fig. 3 The optimization method is used to tune the Fuzzy Logic parameters, by minimizing the value of the precision.

variables (Fig. 3). At each iteration of the optimization method, data from the FM and test data collected from mobile nodes are loaded, the patterns are created and the location estimation is made for all the points of the spatial domain under test. In this case the objective function ($f(x)$) to be minimized is the precision of the location system. As a result, the optimization method returns the values for the parameters for which the lower precision was obtained.

In this work x_0,x_1 are x_a and x_b from the first set of membership functions; x_2, x_3 and x_4 are parameters x_a, x_b and x_c from the second set of membership functions; x_5, x_6 and x_7 are respectively the values for the weights 'low', 'medium' and 'high' used at the on-line phase.

4 Tests and Results

A real life scenario consisting in a two classrooms and a corridor connecting them (Fig. 4), at the University of Trás-os-Montes and Alto Douro was used to test the proposed LEA.

Data to build the FM were collected using a laptop computer and data used for location estimation tests were collected using two different mobile phones: a HTC Desire (Mobile Phone 1) and a Sony Ericsson Xperia X10 mini (Mobile Phone 2). Data for the FM and to test the LEA were collected at the coordinates marked in Fig. 4. A distance of $2,5m$ was used between two consecutive points. For each point, 20 samples of the RSS value, for all the APs in the scene were taken. This data was then used to analyse the performance of the various LEA.

Also the location of the five Access Points (AP) used as reference are marked on the map. In this test, CISCO Aironet 1200 Access Points, with $6.5dBi$ patch antennas, were used. These APs were selected because they are the same model used in the wireless network of our University Campus.

After collecting the data, three sets of tests were made. The first consisted in testing the classic methods and the other two were made to evaluate the performance of the proposed LEA. For all these tests, the FM was generated using the data

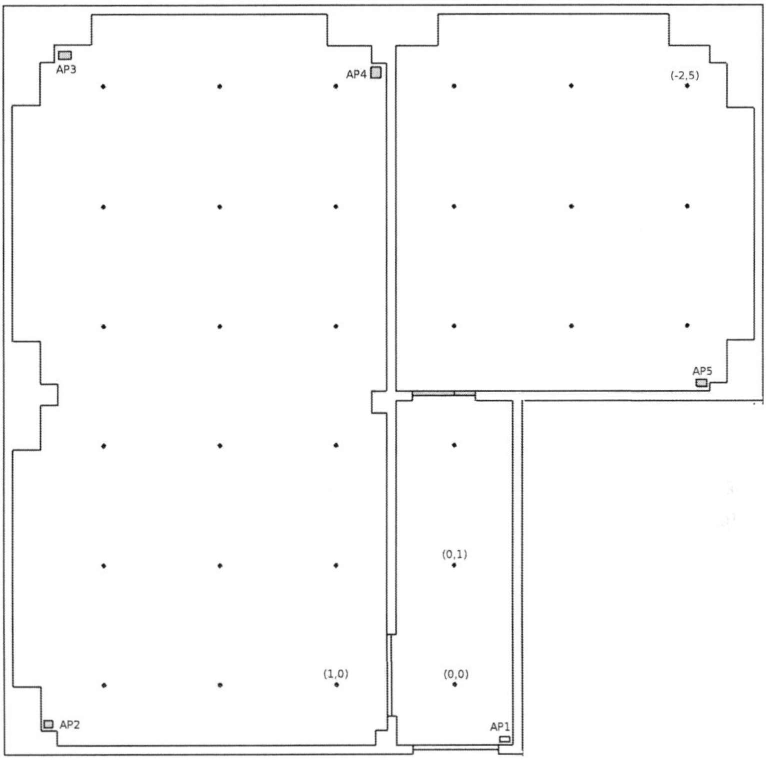

Fig. 4 Map of the testing scenario

collected using the laptop computer and the location estimation using the data from the mobile phones.

In Table 1 the values for the precision, in *m*, using the classic methods are presented. In these tests the value of *k* and the weights are as in [14].

Table 1 Precision values obtained with the classic methods

LEA	Mob. Phone 1 (table)	Mob. Phone 1 (hand)	Mob. Phone 2 (hand)
Nearest Neighbour	6.19*m*	7.33*m*	5.00*m*
k-Nearest Neighbour	4.86*m*	6.29*m*	3.81*m*
W. k-Nearest Neighbour	5.56*m*	6.85*m*	4.32*m*

The second and the third set of tests consisted in tuning the fuzzy parameters, as above mentioned, using the data acquired by the three mobile phones. In these tests, two different values were used as the reference RSS value to generate the pattern (as explained in 3.1). Table 2 presents the data from the tests made with the pattern generated using the average RSS value as reference. In the table, Calibration 1 corresponds to the tuning of the Fuzzy Logic parameters using the data from the Mobile Phone 1 placed on the table; Calibration 2 corresponds to the mobile phone 1 held by the user; Calibration 3 corresponds to the Mobile Phone 2.

Table 2 Precision values obtained using the LEA based on Fuzzy Logic and the average RSS to generate the patterns

LEA	Mob. Phone 1 (table)	Mob. Phone 1 (hand)	Mob. Phone 2 (hand)
Calibration 1	$3.17m$	$4.00m$	$3.32m$
Calibration 2	$3.13m$	$3.97m$	$3.42m$
Calibration 3	$3.31m$	$4.16m$	$3.25m$

The last set of tests were made using the value of RSS from AP1 as the reference value to generate the patterns stored in the FM and to do the pattern search (Table 3).

Table 3 Precision values obtained using the LEA based on Fuzzy Logic and the RSS value from AP1 as reference value to generate the patterns.

LEA	Mob. Phone 1 (table)	Mob. Phone 1 (hand)	Mob. Phone 2 (hand)
Calibration 1	$3.27m$	$4.13m$	$3.42m$
Calibration 2	$3.30m$	$4.10m$	$3.55m$
Calibration 3	$3.53m$	$4.48m$	$3.48m$

5 Discussion and Conclusion

When comparing only the results obtained using Fuzzy Logic, it is noticeable that the use of the average RSS as reference value leads to a slightly better precision than when using the RSS value of one of Access Points. This is due to the fact that the average will act as a dynamic threshold adapted to all the RSS values and not only one.

When the comparison is made between the classic methods and the method based on Fuzzy Logic it is noticeable that there is a great improvement in the precision of the system. Table 4 presents the values for the precision improvements, considering the best and the worse cases.

Table 4 Maximum and minimum improvement to the location precision

Terminal	Best case	Worse case
Mobile Phone 1 (table) 1	49.43%	31.89%
Mobile Phone 1 (hand)	45.84%	33.86%
Mobile Phone 2	35.00%	10.24%

Improvements on the location precision vary from 10.24% to 49.43%. The results are dependent on the type of mobile terminal. Mobile Phone 2 had worse results, however it is noticeable an improvement on the location precision. To be noticed that this mobile phone never detected all the five references at the same time, it only detected four.

As a conclusion, the LEA presented in this work was successfully tested and overcome the performance of the classic LEA, so it can be used for indoor location. Also the use of Nonlinear Optimization Methods to tune the Fuzzy parameters was very helpful to achieve the best performance of the location system.

References

1. Audet, C.: Convergence results for pattern search algorithms are tight. Optimization and Engineering 2(5), 101–122 (2004)
2. Audet, C., Bchard, V., Digabel, S.L.: Nonsmooth optimization through mesh adaptive direct search and variable neighborhood search. J. Global Opt. (41), 299–318 (2008)
3. Audet, C., Dennis Jr., J.E.: Analysis of generalized pattern searches. SIAM Journal on Optimization 13(3), 889–903 (2002)
4. Audet, C., Dennis Jr., J.E.: Mesh adaptive direct search algorithms for constrained optimization. SIAM Journal on Optimization (17), 188–217 (2006)
5. Audet, C., Dennis Jr., J.E.: A mads algorithm with a progressive barrier for derivative-free nonlinear programming. Tech. Rep. G-2007-37, Les Cahiers du GERAD, cole Polytechnique de Montral (2007)
6. Audet, C., Dennis Jr., J.E., Digabel, S.L.: Globalization strategies for mesh adaptative direct search. Tech. Rep. G-2008-74, Les Cahiers du GERAD, cole Polytechnique de Montral (2008)
7. Bahl, P., Padmanabhan, V.: RADAR: an in-building RF-based user location and tracking system. In: INFOCOM 2000, Proceedings of IEEE Nineteenth Annual Joint Conference of the IEEE Computer and Communications Societies, vol. 2, pp. 775–784 (2000); doi:10.1109/INFCOM.2000.832252
8. Conn, A.R., Scheinberg, K., Vicente, L.N.: Introduction to Derivative-Free Optimization. In: MPS-SIAM Series on Optimization. SIAM, USA (2009)
9. Cricket Project: Cricket v2 User Manual. MIT Computer Science and Artificial Intelligence Lab, Cambridge, ma 02139 edn, 9-11 (2005)
10. Hightower, J., Borriello, G.: Location sensing techniques. Tech. rep., University of Washington, Department of Computer Science and Engineering, Seattle (2001)
11. Jang, J.S.R., Sun, C.T., Mizutani, E.: Neuro-fuzzy and soft computing. In: USENIX Systems Administration Conference (1997)

12. Lin, C.T., Lee, C.S.G.: Neural fuzzy systems: a neuro-fuzzy synergism to intelligent systems. Prentice-Hall, Inc., Upper Saddle River (1996)
13. Liu, H., Darabi, H., Banerjee, P., Liu, J.: Survey of wireless indoor positioning techniques and systems. IEEE Transactions on Systems, Man, and Cybernetics, Part C: Applications and Reviews 37(6), 1067–1080 (2007); doi:10.1109/TSMCC.2007.905750
14. Mestre, P., Pinto, H., Serodio, C., Monteito, J., Couto, C.: A multi-technology framework for LBS using fingerprinting. In: 35th Annual Conference of IEEE Industrial Electronics, IECON 2009, pp. 2693–2698 (2009)
15. Orr, R.J., Abowd, G.D.: The smart floor: a mechanism for natural user identification and tracking. In: CHI 2000: Extended Abstracts on Human Factors in Computing Systems, pp. 275–276. ACM, New York (2000), http://doi.acm.org/10.1145/633292.633453
16. Otsason, V., Varshavsky, A., LaMarca, A., de Lara, E.: Accurate GSM indoor location. In: Mobile Computing, Ubi Comp 2005 (2005)
17. Prasithsangaree, P., Krishnamurthy, P., Chrysanthis, P.: On indoor position location with wireless LANs. In: The 13th IEEE International Symposium on Personal, Indoor and Mobile Radio Communications, vol. 2, pp. 720–724 (2002)
18. Silva, P.M., Paralta, M., Caldeirinha, R., Rodrigues, J., Serodio, C.: Traceme - indoor real-time location system. In: 35th Annual Conference of IEEE Industrial Electronics, IECON 2009, pp. 2721–2725 (2009)
19. Want, R., Hopper, A., Veronica Falc, A., Gibbons, J.: The active badge location system. ACM Trans. Inf. Syst. 10(1), 91–102 (1992), http://doi.acm.org/10.1145/128756.128759
20. Ward, A., Jones, A., Hopper, A.: A new location technique for the active office. IEEE Personal Communications 4(5), 42–47 (1997)
21. Zadeh, L.: Fuzzy sets. Information Control 8, 338–353 (1965)

Modelling Fish Habitat Preference with a Genetic Algorithm-Optimized Takagi-Sugeno Model Based on Pairwise Comparisons

Shinji Fukuda, Willem Waegeman, Ans Mouton, and Bernard De Baets

Abstract. Species-environment relationships are used for evaluating the current status of target species and the potential impact of natural or anthropogenic changes of their habitat. Recent researches reported that the results are strongly affected by the quality of a data set used. The present study attempted to apply pairwise comparisons to modelling fish habitat preference with Takagi-Sugeno-type fuzzy habitat preference models (FHPMs) optimized by a genetic algorithm (GA). The model was compared with the result obtained from the FHPM optimized based on mean squared error (MSE). Three independent data sets were used for training and testing of these models. The FHPMs based on pairwise comparison produced variable habitat preference curves from 20 different initial conditions in the GA. This could be partially ascribed to the optimization process and the regulations assigned. This case study demonstrates applicability and limitations of pairwise comparison-based optimization in an FHPM. Future research should focus on a more flexible learning process to make a good use of the advantages of pairwise comparisons.

1 Introduction

Ecological models are abstractions of natural systems and tools for understanding complex processes and mechanisms involved. Habitat preference models are

Shinji Fukuda
Kyushu University, 6-10-1 Hakozaki, Fukuoka 812-8581, Japan
e-mail: shinji-fkd@agr.kyushu-u.ac.jp

Willem Waegeman · Bernard De Baets
Ghent University, Couple links 653, 9000 Ghent, Belgium
e-mail: {Willem.Waegeman,Bernard.DeBaets}@ugent.be

Ans Mouton
Research Institute for Nature and Forest (INBO), Kliniekstraat 25, 1070 Brussels, Belgium
e-mail: Ans.Mouton@inbo.be

B. De Baets et al. (Eds.): Eurofuse 2011, AISC 107, pp. 375–387, 2011.
springerlink.com

used in order to extract habitat preference information of a target species from observation data, or to express expert knowledge on the species. Species-environment relationships can thus be quantified and used in the decision making on plans and management options for a target ecosystem. In practice, there are two types of habitat models: univariate and multivariate models. Univariate models have been widely applied, for which the main idea consists of using a set of univariate preference functions to represent possible habitat preference of a target species. These models are used for the assessment of current status and future impacts of habitat changes in both time and space [3]. A variety of methods has been proposed and employed in practical applications, which include the habitat suitability index [3], resource selection functions [12], and other specific models such as a genetic Takagi-Sugeno fuzzy model [6, 10]. In recent years, multivariate approaches have gained more popularity as computational systems have become powerful and freely available. These include machine learning methods [14], fuzzy rule-based systems [1, 16], and statistical regression tools [2, 9]. In habitat modelling, it is often reported that the results are affected by the quality and quantity of a data set used [5, 13], which is partly because of the uncertainties inherent to observation data. The development of a sound methodology to cope with the different quality of data contributes to a better understanding and reliable assessment of target ecosystems.

Preference modelling has been one of the key topics in information sciences. Preference is used for ranking items and the ranking can be used for decision making. Recently, pairwise comparison has been gaining interest in this field, and is reported to be a sound methodology in preference learning [7, 11]. Despite the intensive works in theory, the pairwise comparison has not yet been applied to preference modelling in ecology, and it seems to be a good approach to cope with the data with uncertainties such as observation errors. It would therefore be interesting to exemplify the applicability and limitations of a pairwise comparison-based approach.

Our aim is to apply pairwise comparisons to the optimization of a Takagi-Sugeno-type fuzzy habitat preference model (FHPM) using a genetic algorithm (GA [8]). The results were compared with the previously developed FHPM optimized based on mean squared error (MSE). Three independent data sets were used for training and testing of the models. This first application of pairwise comparison provides useful information for the development of a reliable habitat assessment approach using observation data with uncertainty.

2 Methods

2.1 Data Collection

A series of field surveys focusing on Japanese medaka (*Oryziaslatipes*) was carried out in an agricultural canal in Kurume City, Fukuoka, Japan. Field surveys were conducted on three sunny days: 14 October and 5 and 9 November 2004. Two study reaches were established in the same canal: a 50-m-long study reach (1.6–2.0 m in width, 0.3% gradient) was surveyed on 14 October and 5 November, and a 30-m-long study reach (0.8–1.4 m in width, 0.3% gradient) was surveyed on 9 November. Habitat use by Japanese medaka and four physical habitat characteristics—water depth (cm, henceforth referred to as depth), current velocity (cm s^{-1}, velocity), lateral cover ratio (%, cover), and percent vegetation coverage (%, vegetation)—in the study reach were surveyed. The study reach was first mapped, then habitat use by the fish was observed, and finally the physical habitat characteristics within the reach were measured.

In the following analyses, fish distribution data are expressed as the log-transformed observed fish population density in the ith water unit ($FPD_{o,i}$; individuals per square metre), where the subscript o indicates observed, $i = 1, 2, ..., N$, denotes the water unit, and N is the total number of water units (Table 1). The size of data sets (=N) is 139 for the data set of 14 October, 130 for that of 5 November, and 86 for that of 9 November, all of which contain vectors of four habitat variables (depth, velocity, cover and vegetation) and fish population density.

Table 1 Species distribution data along with the four habitat variables of depth, velocity, cover and vegetation, each of which was observed on 14 October, 5 November, and 9 November 2004, respectively.

Date		D	V	C	VEG	$\log_{10}$(FPD+1)	presence	absence	prevalence
	maximum	65.0	61.3	50.0	100.0	1.64			
14 Oct.	mean	13.6	13.7	19.1	49.7	0.37	77	62	55.4
	minimum	2.0	1.5	0.0	0.0	0.00			
	SD	8.0	9.3	14.0	29.6	0.41			
	maximum	57.0	31.6	50.0	100.0	1.56			
5 Nov.	mean	13.6	8.8	20.6	59.5	0.35	71	59	54.6
	minimum	3.0	1.5	0.0	0.0	0.00			
	SD	8.4	5.6	14.8	31.2	0.40			
	maximum	32.0	44.7	50.0	100.0	1.88			
9 Nov.	mean	15.6	12.5	19.8	55.2	0.42	31	55	36.0
	minimum	5.0	1.5	0.0	0.0	0.00			
	SD	6.2	9.6	13.9	33.3	0.60			

SD, standard deviation; D, depth (cm); V, velocity (cm s^{-1}); C, cover (%);
VEG, vegetation (%);
FPD, observed fish population density (individuals per square metre).

2.2 *Fuzzy Habitat Preference Model*

A fuzzy habitat preference model (FHPM [10]) was employed for describing the habitat preference of the target fish. An FHPM is a 0-order Takagi-Sugeno model [15] that relates habitat variables to habitat preference by considering uncertainties such as fish behaviour and measurement errors of the habitat variables.

Simultaneous optimization of all model parameters enables an FHPM to evaluate habitat preference in an interpretable way, despite the presence of nonlinear, complex interactions between habitat variables and habitat preference.

The FHPM used here is a single-input single-output fuzzy system with four sets of if-then rules: two sets of four rules each for depth and velocity, and two sets of three rules each for cover and vegetation (14 rules in total). The input values (the four habitat variables) were expressed as symmetric triangular fuzzy numbers with centre α_l and spread σ_l, where l denotes the habitat conditions of depth, velocity, cover, and vegetation, in order to take into account the measurement errors and the spatial variance of the habitat variables. The observed value was used as the centre value α_l, and the spread σ_l was determined as follows: the spread of depth σ_{d} was 1 cm, that of velocity σ_{v} was 1 cm s^{-1}, and those of cover σ_{c} and vegetation σ_{veg} were 10% each, where the subscripts d, v, c and veg denote depth, velocity, cover and vegetation, respectively. Next, all input values were transformed by membership functions (Fig. 1) into membership values ranging from zero to one. The membership functions for depth and velocity were defined according to the ecological characteristics of Japanese medaka (Figs. 1a and 1b), and those for cover and vegetation were defined to include the available range of each of these habitat variables (Figs. 1c and 1d). A uniform partition with triangular membership functions was used, but the first and last membership functions were allowed to be trapezoidal. Each membership function has a corresponding singleton value in the consequent part. The habitat preference was calculated by taking the weighted mean of the singleton values in the consequent part, using the membership degrees as weights. Singleton values in the consequent part in the range [0, 1] were determined by a binary-coded simple GA so as to minimize an objective function described below. The present GA consisted of a population of 100 individuals, each of which has a 56-bit string (4 bits $\times$ 14 singletons). The shape of the habitat preference curves (HPCs), i.e., a set of singletons for a habitat variable, was constrained to have a unimodal or monotone form with maximum preference of one. Specifically, the model was penalized by giving large values with regard to the objective function when the above regulation was violated. Based on the objective function employed, the GA repeatedly modified the model structure (the singletons in the consequent part) using three genetic

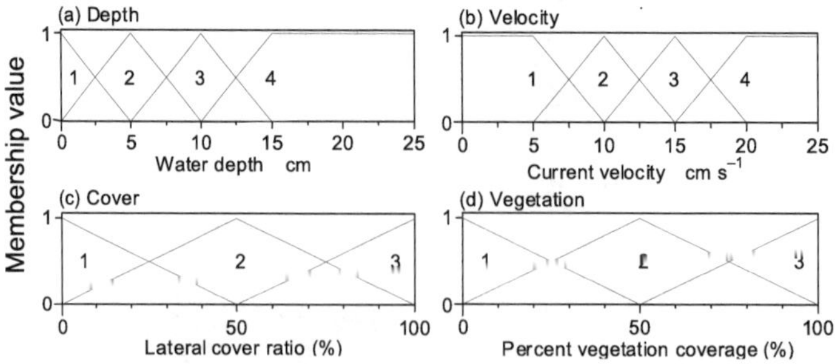

Fig. 1 Membership functions

operations: roulette wheel selection with an elitist strategy, uniform crossover and mutation at a probability of 5%. The optimal model was obtained after 2,000 iterations. The GA optimization was repeated using 20 different sets of initial conditions in order to evaluate the variablity of the model structure that resulted from the initial conditions.

2.3 Objective Function

We employed either of mean squared error (MSE) between observed and predicted fish population density or mean penalty (MP) evaluated from pairwise comparison as our objective function. The procedures to calculate each objective function are as follows.

First, the composite habitat preference according to the four habitat variables was calculated as

$$HP_i = P_{\mathrm{d},i} \times P_{\mathrm{v},i} \times P_{\mathrm{c},i} \times P_{\mathrm{veg},i} \tag{1}$$

where HP_i denotes the habitat preference in the ith water unit, and $P_{\mathrm{d},i}$, $P_{\mathrm{v},i}$, $P_{\mathrm{c},i}$, and $P_{\mathrm{veg},i}$ are the habitat preference with respect to the individual habitat variables depth, velocity, cover and vegetation, respectively. Here, the use of multiplication in Eq. (1) produces an FHPM with higher performance and consistent habitat preference curves (HPCs) [6].

To compute the MSE, fish population density of the Japanese medaka was estimated with Eq. (2):

$$FPD_{\mathrm{m},i} = \left(\frac{HP_i}{\sum_{i=1}^{N} HP_i} \right) \times \sum_{i=1}^{N} FPD_{\mathrm{o},i} \tag{2}$$

where $FPD_{\mathrm{m},i}$ is the modelled fish population density in the ith water unit, and $FPD_{\mathrm{o},i}$ is the observed fish population density. The MSE is then calculated from the observed and modelled fish population density.

The MP is an average of penalty values assigned according to a set of pairwise comparisons of composite habitat preference and observed fish population density of two different data points. That is,

$$MP = \frac{2}{N(N-1)} \sum_{\lambda_j \neq \lambda_k} penalty(\lambda_j, \lambda_k) \tag{3}$$

where λ_j and λ_k are the items (composite habitat preference and fish population density) obtained from two data points j and k. The pairwise comparison was performed as follows.

1. Calculate the composite habitat preference of data point j and k (HP_j and HP_k) using an FHPM and Eq. (1).

2. For each of data points j and k,
   2.1 Compare the habitat preference values (HP_j and HP_k), from which either of the relationships $HP_j > HP_k$, $HP_j = HP_k$, or $HP_j < HP_k$ is obtained.
   2.2 Compare the observed fish population density ($FPD_{o,j}$ and $FPD_{o,k}$), from which either of the relationships $FPD_{o,j} > FPD_{o,k}$, $FPD_{o,j} = FPD_{o,k}$, or $FPD_{o,j} < FPD_{o,k}$ is obtained.
   2.3 Assign a penalty based on the relationship obtained from 2.1 and 2.2.
   a. If the relationships are the same between the habitat preference and fish population density, then no penalty is assigned.
   b. If the relationships are the opposite between the habitat preference and fish population density, then assign a penalty of 1.
   c. If either of the relationships is equal and the absolute difference of other relationship is smaller than a predefined value[1], then assign a penalty of 0.5. Otherwise, assign a penalty of 1.
3. Compute the MP by means of Eq. (3).

2.4 Model Application and Analyses

To illustrate the difference between the FHPMs obtained from two different objective functions, these models were compared in terms of model performance and habitat preference information retrieved from them. The first data set was used for model development (training), and the remaining two data sets were used for model evaluation (testing). That is, 20 FHPMs were developed from each data set, each of which was tested using the other two different data sets. From the 20 initial conditions used in the model development, the variance of the model structures was quantified by using performance measures and the HPCs. The model performance was evaluated by the MSE, the MP and the area under receiver operating characteristics curve (AUC). The AUC is often used when evaluating species distribution models for presence-absence data [6, 14] and it is independent of the objective functions in this study. The MSE and MP were calculated using the fish population density and the AUC was calculated using the presence-absence data obtained from the fish population density data. Of these performance measures, the mean and standard deviations of the 20 FHPMs from different initial conditions were used as a measure of the predictive accuracy and of the variability of model structures, respectively. In an FHPM, an HPC can easily be obtained by providing consecutive values in the range of the corresponding membership functions (in steps of 0.1) to the FHPM and plotting the output values against the habitat variables. Specifically, the HPC shape indicates the habitat preference information retrieved from the data set used.

[1] Allowed absolute difference for fish population density was set at 0.1 and that for habitat preference was at 0.05 (about 5% of their entire range).

3 Results

The two different models (FHPMs based on pairwise comparisons and FHPMs based on MSE) produced HPCs with similar trends in preference (Figs. 2–3). The HPCs of the best models have almost the same shape with slightly different degrees of preference (solid lines in Figs. 2–3). We can, however, observe slight differences in the shapes between the two models at the water depth around 15–20 cm in the data sets of 14 October and 5 November. Slightly contradicting shapes among the 20 HPCs are found from the FHPMs based on pairwise comparison for depth in the data set of 5 November (Fig. 2(ii-a)) and for vegetation in the data set of 9 November (Fig. 2(iii-d)). In other words, some HPCs show monotone forms, but others show unimodal forms. The both FHPMs produced variable HPCs but at different habitat variables (Figs. 2–3). For instance, the HPCs obtained from the FH-PMs based on pairwise comparisons show variance in depth and vegetation (Fig. 2), whereas the HPCs from the FHPMs based on MSE show variance in cover (Fig. 3). This could be partially resulted from different penalization approaches in the model training.

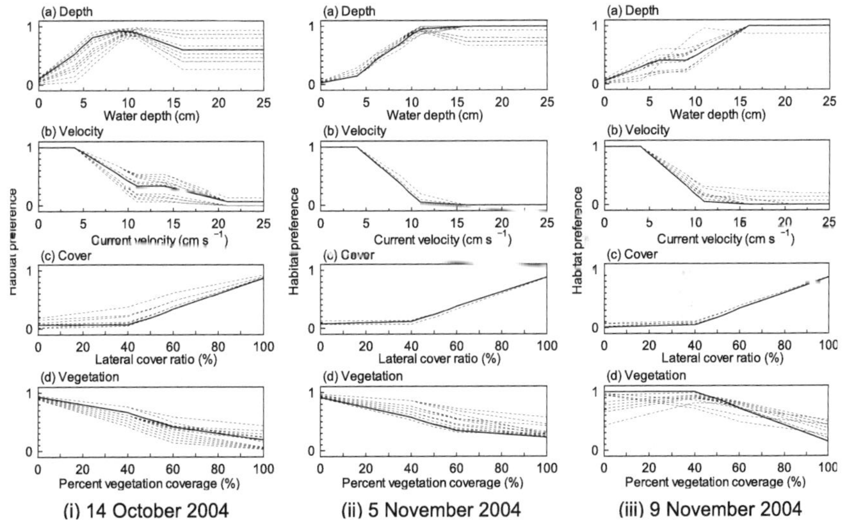

Fig. 2 Habitat preference curves obtained from fuzzy habitat preference model optimized based on pairwise comparison. Solid lines are the best curve with respect to the pairwise comparison and dotted lines are all curves obtained from 20 different initial conditions in a genetic algorithm

The performance of the two models differs by the performance measures (Table 2). The variance in HPCs seems to have some relationship with the standard deviation of all performance measures, except for the MP of FHPMs based on pairwise comparison. The FHPMs based on pairwise comparisons show better performance in MP, while the FHPMs based on MSE show better performance in

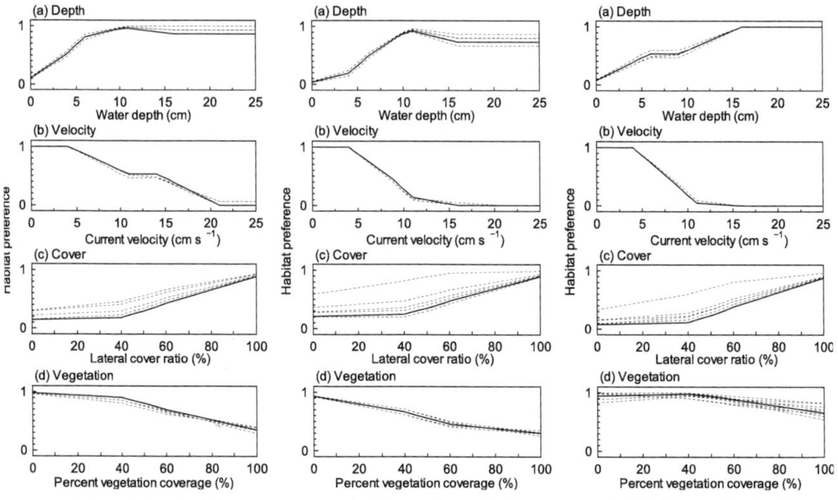

Fig. 3 Habitat preference curves obtained from fuzzy habitat preference model optimized based on mean squared error (MSE). Solid lines are the best curve with respect to the MSE and dotted lines are all curves obtained from 20 different initial conditions in a genetic algorithm

MSE and in AUC. The best FHPMs (with respect to each of the objective functions) from each data set also support the better performance of the FHPM based on MSE (Table 3). It can be seen from these results that FHPMs based on pairwise comparison failed to converge to an optimal solution.

We observe no correlation between the MP and the AUC or the MSE (Fig. 4), although there could be some correlation between these measures within a data set. Despite the differences between the data sets, the MP of FHPMs based on pairwise comparisons shows small variance within the same data set, while the AUC and MSE plots spread vertically (Fig. 4(i)). This means that AUC and MSE cannot be minimized by minimizing the MP. The MSE of FHPMs based on MSE shows a small variance, while the MP and AUC show larger variance (Fig. 4(ii)). The variance is dependent on the data set used as seen in Table 2. In contrast to the MP, the AUC and MSE show a positive relationship for both of the FHPMs (Fig. 5), of which the variance is small in the FHPMs based on MSE. This indicates that the minimization of MSE could lead to maximization of AUC through the GA optimization applied in this study.

4 Discussion

The present results demonstrated the applicability and limitations of pairwise comparison in the optimization of an FHPM, from which some possible improvements of the present FHPM can be drawn. As we observed in Fig. 4, AUC and MSE could

Table 2 Results of performance evaluation of fuzzy habitat preference models (FHPMs) optimized based on pairwise comparison (i) and FHPMs optimized based on mean squared error (MSE) (ii), of which model 1 is the FHPMs developed using the data set of 14 October, model 2 is those of 5 November, model 3 is those of 9 November 2004. Each model was tested using three data sets.

(i) pairwise comparison-based optimization				(ii) MSE-based optimization			
	Data set used				**Data set used**		
MSE	**Oct. 14**	**Nov. 5**	**Nov. 9**	**MSE**	**Oct. 14**	**Nov. 5**	**Nov. 9**
Model 1	0.305	0.172	0.441	Model 1	0.069	0.084	0.216
	±0.279	±0.124	±0.219		±0.001	±0.002	±0.006
Model 2	0.371	0.123	0.224	Model 2	0.146	0.067	0.206
	±0.044	±0.021	±0.035		±0.007	±0.000	±0.009
Model 3	1.131	0.565	0.221	Model 3	0.240	0.115	0.143
	±0.410	±0.226	±0.022		±0.042	±0.016	±0.001
AUC	**Oct. 14**	**Nov. 5**	**Nov. 9**	**AUC**	**Oct. 14**	**Nov. 5**	**Nov. 9**
Model 1	0.738	0.735	0.747	Model 1	0.875	0.818	0.893
	±0.100	±0.080	±0.100		±0.002	±0.007	±0.011
Model 2	0.701	0.817	0.892	Model 2	0.786	0.860	0.912
	±0.008	±0.007	±0.012		±0.011	±0.003	±0.005
Model 3	0.623	0.680	0.792	Model 3	0.748	0.855	0.921
	±0.028	±0.066	±0.050		±0.012	±0.003	±0.003
MP	**Oct. 14**	**Nov. 5**	**Nov. 9**	**MP**	**Oct. 14**	**Nov. 5**	**Nov. 9**
Model 1	0.229	0.289	0.317	Model 1	0.277	0.321	0.406
	±0.001	±0.004	±0.009		±0.014	±0.018	±0.027
Model 2	0.339	0.236	0.235	Model 2	0.266	0.272	0.325
	±0.001	±0.001	±0.007		±0.009	±0.012	±0.017
Model 3	0.342	0.258	0.205	Model 3	0.268	0.277	0.285
	±0.002	±0.004	±0.000		±0.011	±0.023	±0.030

not be minimized by minimizing MP, and vice versa. The insensitive response of MP would be ascribed to its qualitativeness. An MP is a quantitative measure which evaluates the pairwise relationship between observed fish population density and composite habitat preference calculated by an FHPM. Due to the qualitativeness of MP, a GA could not optimize the singleton values, from which variable HPCs were produced. Since the present GA is a binary-coded scheme, the GA could not find an optimal solution due to a discretized search space. The variance of HPCs can therefore be reduced by a real-coded GA. Fixed membership functions of the antecedent part of FHPMs could also be a cause of the insensitiveness, and thus learning habitat preference of the fish was restricted. This point could be improved by tuning the membership functions, which should, however, be limited to a certain extent in order to keep the semantics predefined from expert knowledge on the fish [16].

In contrast to the MP, the AUC positively responded to the MSE (Fig. 5). This could be because the FHPMs based on MSE could classify presence and absence of the fish by minimizing the errors between predicted and observed fish population density. For instance, an error at the point where no fish was observed but a model

Table 3 Performance of the best models from each data set using fuzzy habitat preference models (FHPMs) optimized based on pairwise comparison (i) and FHPMs optimized based on mean squared error (MSE) (ii), of which model 1 is the FHPMs developed using the data set of 14 October, model 2 is those of 5 November, model 3 is those of 9 November 2004. Each model was tested using three data sets.

(i) pairwise comparison-based optimization				(ii) MSE-based optimization			
	Data set used				Data set used		
MSE	**Oct. 14**	**Nov. 5**	**Nov. 9**	**MSE**	**Oct. 14**	**Nov. 5**	**Nov. 9**
Model 1	0.097	0.090	0.260	Model 1	0.069	0.081	0.207
Model 2	0.287	0.080	0.169	Model 2	0.137	0.066	0.196
Model 3	0.418	0.166	0.177	Model 3	0.175	0.091	0.142
AUC	**Oct. 14**	**Nov. 5**	**Nov. 9**	**AUC**	**Oct. 14**	**Nov. 5**	**Nov. 9**
Model 1	0.835	0.799	0.870	Model 1	0.879	0.829	0.911
Model 2	0.727	0.838	0.906	Model 2	0.814	0.865	0.915
Model 3	0.703	0.815	0.876	Model 3	0.773	0.862	0.926
MP	**Oct. 14**	**Nov. 5**	**Nov. 9**	**MP**	**Oct. 14**	**Nov. 5**	**Nov. 9**
Model 1	0.228	0.281	0.304	Model 1	0.257	0.296	0.368
Model 2	0.339	0.235	0.226	Model 2	0.250	0.255	0.301
Model 3	0.340	0.249	0.205	Model 3	0.259	0.258	0.256

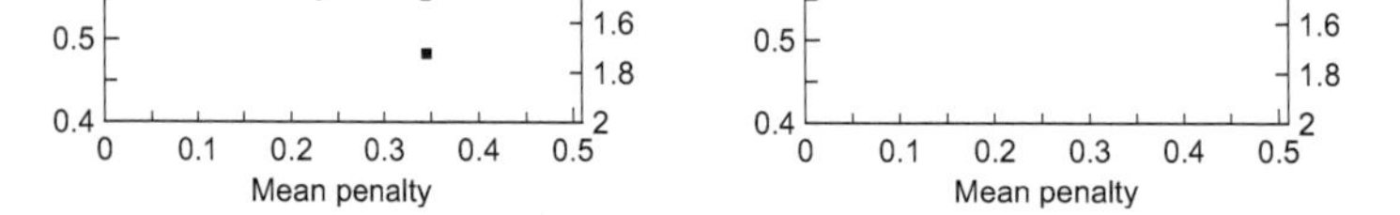

(i) Pairwise comparison-based optimization

(ii) MSE-based optimization

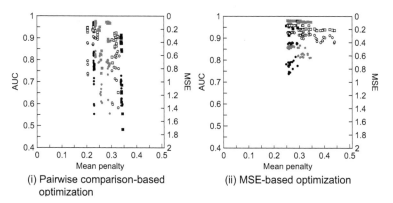

Fig. 4 Scatter diagrams between mean penalty (MP) and the area under receiver operating characteristics curve (AUC, circles) or mean squared error (MSE, squares): (i) fuzzy habitat preference models (FHPMs) optimized based on pairwise comparison and (ii) FHPMs optimized based on MSE. Black marks are the test results using the data set of 14 October, grey marks are those of 5 November, and white marks are those of 9 November 2004. Note that the axis for MSE is inverted.

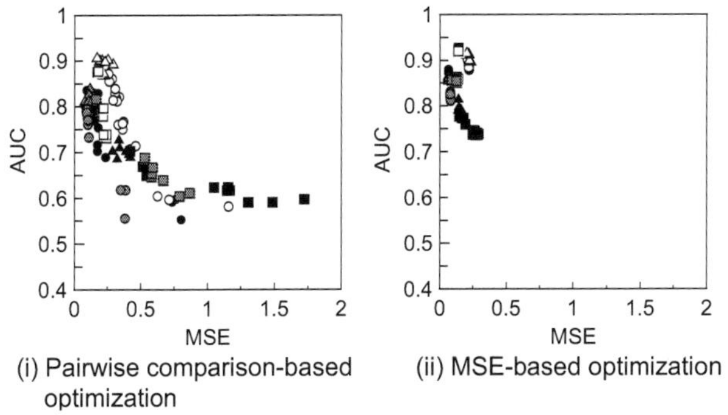

Fig. 5 Scatter diagrams between the area under receiver operating characteristics curve (AUC) and mean squared error (MSE): (i) fuzzy habitat preference models (FHPMs) optimized based on pairwise comparison and (ii) FHPMs optimized based on MSE. Circles are the results of FHPMs developed using the data set of 14 October, triangles are those of 5 November, and squares are those of 9 November. Black marks are the test results using the data set of 14 October, grey marks are those of 5 November, and white marks are those of 9 November 2004.

predicted fish presence is equal to a false positive in presence-absence analysis. Reducing the error can thus improve the model performance with regard to the AUC.

In addition to model performance, interpretation of an HPC is important in habitat modelling, from which users can learn how a target species would respond to a given condition through model output [4]. This is is an necessary process because it can further be used for designing management plans and options for a target ecosystem. The interpretability of an HPC depends on the monotonicity toward a maximal preference point and the variability of HPCs obtained. The variability of HPCs would indicate the level of uncertainty in fish habitat modelling. On the one hand, one of the variable HPCs may have a higher performance in a test data set. On the other hand, variable HPCs are not interpretable for users and may also deteriorate model performance. The variablity observed in the HPCs of FHPMs based on pairwise comparisons is ascribed to the algorithm used. In the pairwise comparison, only a qualitative relationship between modelled habitat preference and fish population density was considered, in which differences in degree of preference did not matter. Therefore, the variance in HPCs cannot be reduced by pairwise comparison if the shape is the same (even though the degrees of preference are different). In the present results, the effectiveness of pairwise comparison was not so prominent as we expected. This could be partially because the present FHPMs regulate the shape of an HPC during the optimization by a GA. If this is the case, pairwise comparison may work well when applied to habitat models without any regulation in the optimization process. Better results can be achieved by pairwise comparison using a set of individual habitat preferences for each habitat variable, instead of using composite habitat preferences as applied in this study. Different penalty assignments such as relaxation or rigidification of penalty criteria can also improve the model performance or the interpretability of the result. Further studies are necessary to clarify the mechanism of performance improvement as well as better algorithms for preference learning using pairwise comparisons, which contributes to the establishment of a sound methodology for habitat assessment.

Acknowledgements. This study was partly supported by a Grant-in-aid for Young Scientists B from the Ministry of Education, Culture, Sports, Science and Technology (MEXT), Japan. W.W. is supported as a postdoc by the Research Foundation of Flanders (FWO Vlaanderen). The authors thank Prof. E. Hüllermeier (Marburg University, Germany) whose comments helped formalize the concept of this study. S.F. thanks Prof. K. Hiramatsu (Kyushu University, Japan) for his assistance in the early phase of this study.

References

1. Adriaenssens, V., De Baets, B., Goethals, P., De Pauw, N.: Fuzzy rule-based models for decision support in ecosystem management. Sci. Total Environ. 319, 1–12 (2004)
2. Ahmadi-Nedushan, B., Hilaire, A.S., Bérubé, B., Robichaud, E., Thiémonge, N., Bobée, B.: A review of statistical methods for the evaluation of aquatic habitat suitability for instream flow assessment. River Res. Appl. 22, 503–523 (2006)
3. Bovee, K.D., Lamb, B.L., Bartholow, J.M., Stalnaker, C.B., Taylor, J., Henriksen, J.: Stream habitat analysis using the instream flow incremental methodology. U.S. Geological Survey, Biological Resources Division Information and Technology Report. USGS/BRD-1998-0004 (1998)
4. Elith, J., Graham, C.H.: Do they? How do they? Why do they differ? On finding reasons for differing performances of species distribution models. Ecography 32, 66–77 (2009)
5. Fielding, A.H., Bell, J.F.: A review of methods for the assessment of prediction errors in conservation presence/absence models. Environ. Conserv. 24(1), 38–39 (1997)
6. Fukuda, S., De Baets, B., Mouton, A.M., Waegeman, W., Nakajima, J., Mukai, T., Hiramatsu, K., Onikura, N.: Effect of model formulation on the optimization of a genetic Takagi-Sugeno fuzzy system for fish habitat suitability evaluation. Ecol. Model 222, 1401–1413 (2011)
7. Fürnkranz, J., Hüllermeier, E.: Preference Learning. Springer, Heidelberg (2010)
8. Goldberg, D.: Genetic algorithms in search, optimization, and machine learning. Addison-Wesley, Reading (1989)
9. Guisan, A., Zimmermann, N.E.: Predictive habitat distribution models in ecology. Ecol. Model 135, 147–186 (2000)
10. Hiramatsu, K., Fukuda, S., Shikasho, S.: Mathematical modeling of habitat preference of Japanese medaka for instream water environment using fuzzy inference. Trans. JSIDRE 228, 65–72 (2003) (in Japanese with English abstract)
11. Hüllermeier, E., Fürnkranz, J., Cheng, W., Brinker, K.: Label ranking by learning pairwise preferences. Artif. Intell. 172, 1897–1916 (2008)
12. Lechowicz, M.J.: The sampling characteristics of electivity indices. Oecologia (Berl.) 52, 22–30 (1982)
13. Mouton, A.M., De Baets, B., Goethals, P.L.M.: Ecological relevance of performance criteria for species distribution models. Ecol. Model. 221, 1995–2002 (2010)

14. Pino-Mejías, R., Cubiles-de-la-Vega, M.D., Anaya-Romero, M., Pascual-Acosta, A., Jordn-Lpez, A., Bellinfante-Crocci, N.: Predicting the potential habitat of oaks with data mining models and the R system. Environ. Modell Softw. 25, 826–836 (2010)
15. Takagi, T., Sugeno, M.: Fuzzy identification of systems and its appfications to modelling and comrol. IEEE Trans. Systems Man Cybernet 15, 116–132 (1985)
16. Van Broekhoven, E., Adriaenssens, V., De Baets, B.: Interpretability-preserving genetic optimization of linguistic terms in fuzzy models for fuzzy ordered classification: An ecological case study. Int. J. Approx Reasoning 44, 65–90 (2007)

Multi-feature Tracking Approach Using Dynamic Fuzzy Sets

Nuno Vieira Lopes, Pedro Couto, and Pedro Melo-Pinto

Abstract. In this paper a new tracking approach based in fuzzy concepts is introduced. The aim of this methodology is to incorporate in the proposed model the uncertainty underlying any problem of feature tracking, through the use of fuzzy sets. Several dynamic fuzzy sets are constructed according both cinematic (movement model) and non cinematic properties (image gray levels) that distinguish the feature. Meanwhile cinematic related fuzzy sets model the feature movement characteristics, the non cinematic fuzzy sets model the feature visible image related properties. The tracking task is performed through the fusion of these fuzzy models by means of a fuzzy inference engine. This way feature detection and matching steps are performed exclusively using inference rules on fuzzy sets.

1 Introduction

Object tracking plays an important role in computer vision. During the last years, extensive research has been conducted in this field and many types and applications of object tracking systems have been proposed in the literature such as automated surveillance, vehicle navigation, human computer interaction and traffic analysis [3, 5, 7].

While tracking is essential to many applications, robust tracking algorithms are still a challenge. Difficulties can arise due to noise presence in images, quick

Nuno Vieira Lopes
School of Technology and Management, Polytechnic Institute of Leiria - P-2411-901 Leiria, Portugal, and CITAB, University of Trás-os-Montes e Alto Douro,
Quinta dos Prados - P.O.Box, 1013 5001-801 Vila Real, Portugal
e-mail: nuno.lopes@ipleiria.pt

Pedro Couto · Pedro Melo-Pinto
CITAB, University of Trás-os-Montes e Alto Douro,
Quinta dos Prados - P.O.Box 1013 5001-801 Vila Real, Portugal
e-mail: {pcouto,pmelo}@utad.pt

B. De Baets et al. (Eds.): Eurofuse 2011, AISC 107, pp. 389–400, 2011.
springerlink.com

changes in lighting conditions, abrupt or complex object motion, changing appearance patterns in the object and the scene, non-rigid object structures, object-to-object and object-to-scene occlusions, camera motion and real time processing requirements. Typically, assumptions are made to constrain the tracking problem in the context of a particular application. For example, almost all tracking algorithms assume that the object motion is smooth or impose constrains on the object motion to be constant in velocity or acceleration. Multiple view image tracking or prior knowledge about objects, such as size, number or shape, can also be used to simplify this process.

In its simplest approach, the overall tracking process involves two steps: the detection of the significant objects and the trace of their trajectories. A myriad of algorithms has been developed to implement this subtasks being that each one have their strengths and weaknesses. Over the last years, extensive research has been made in this field to find an optimal tracking system, [10, 11]. Consequently many tracking techniques have been proposed in the literature. However, they are not completely accurate for all kind of scenarios and just provide good results when a certain number of assumptions are verified. This uncertainty is the key reason to study and implement new tracking approaches introducing new concepts, such as fuzzy logic, to improve the tracking process. Since the arising of fuzzy logic theory, it has been successfully applied in a large range of areas such as process control systems, automotive navigation systems, information retrieval systems and image processing. As presented beforehand, a tracking system can be seen as a multi-stage process that comprise object segmentation and object matching. Hence, fuzzy logic can appear in these two different stages. In [4, 1, 8] fuzzy object detection methods are presented while fuzzy matching approaches are introduced in [6, 9, 2].

2 Proposed Approach

The implementation of this methodology is based in some preliminary assumptions. These assumptions are commonly used in most tracking systems:

1. The feature presents smooth motion;
2. The feature has constancy of gray levels intensity;
3. For sake of simplicity, the motion between two consecutive frames can be described using a linear motion model;
4. The area occupied by the feature is small when compared with the total image area;
5. The size of the feature is somewhat preserved during the sequence.

In this approach feature brightness constancy is assumed. This situation can be described as $I(x,y,t) \approx I(x+\delta x, y+\delta y, t+\delta t)$, where δx and δy are the displacement of the local region at (x,y,t) after time δt. Nevertheless, slightly changes in illumination, camera sensor noise, among other factors that cause variations in the intensity of the feature, are tolerated.

The smoothness of the movement concerns the continuity of the feature movement. The feature movement is assumed to be continuous and, therefore, using a typical acquisition frame rate and assuming there are no occlusions or misdetections, the next position of the feature lies inside a neighbourhood of its previous position.

The size of the feature is considerably small when compared with the total image area. Assuming this, the feature can be represented as a point or by a small $A \times B$ matrix and, similar strategies to the ones used in point correspondence can be developed for feature matching.

Also, the size of the feature is considered to be approximately constant along the entire sequence. The size doesn't change drastically between frames, however, slightly changes can occur without causing the failure of the method.

2.1 Membership Functions and Fuzzy Sets

In 1965, fuzzy sets were introduced by Zadeh [12] to represent or manipulate data and information containing nonstatistical uncertainties. Let $X = \{x_1, ..., x_n\}$ be an ordinary finite non-empty set. A fuzzy set A in X is as set of ordered pairs $A = \{(x, \mu_A(x))|x \in X\}$, where $\mu_A : X \rightarrow [0,1]$ represents the degree of membership of element x with respect to the fuzzy set A.

Three fuzzy sets concerning the first three enumerated assumptions are constructed. Other fuzzy sets could be easily incorporated in the algorithm but, in order to minimise the computational resources and to increase speed and simplicity, the algorithm is constructed based only on these three fundamental and generic assumptions.

Assuming the smoothness of the feature movement, it is acceptable to suppose that the next location of the feature lies in a neighbourhood centred in its previous location. Therefore, a fuzzy set S associated with each image pixel (x,y) by means of this proximity assumption related to the feature position in the previous frame is constructed. The membership function $\mu_S(x,y) \in [0,\alpha]$ can be graphically depicted as illustrated in Fig. 1a, where the horizontal axis represents the Euclidian distance between the image pixels position and the previous location of the feature. In Fig. 1b, is depicted a pictorial description of the fuzzy set S assuming a previous position $(x,y) = (100,100)$, with $d_1 = 40$, $d_2 = 50$ and a maximum value $\alpha = 0.9$.

Two distinct zones of certainty are present in the definition of the membership function $\mu_S(x,y)$. For distances lower than d_1 the membership degree is maximum, defining a circular region centred in the feature previous position, where the new feature location is expectable with equal certainty. For distances greater than d_1 the membership degree decreases in a linear way until it reaches the zero value at distance d_2. This behavior can be explained due to the fact that, for distances greater than d_1, the certainty of finding the features becomes lower as the distance increases. The new position is not expected for distances greater than d_2 and the membership degree is zero for all these positions. This two controlling parameters d_1 and d_2 are variable during time. Both values are directly proportional to

Fig. 1 Membership function $\mu_S(x,y)$.

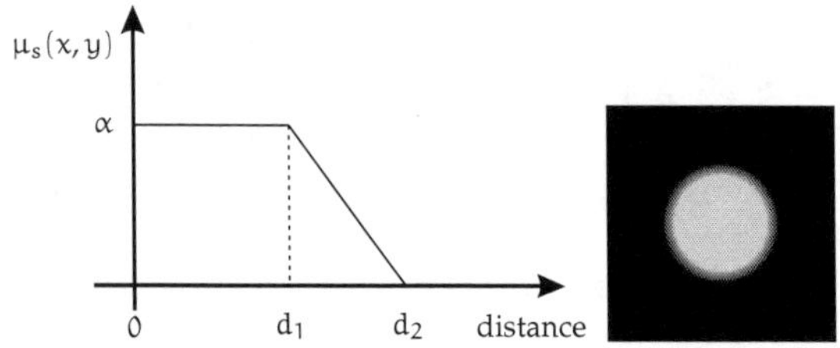

(a) Membership function $\mu_S(x,y)$ (b) Fuzzy set S.

the observed feature displacement $f_d(t)$. This displacement is based in an Euclidian distance defined as $f_d(t) = \sqrt{(\mathbf{p}(t) - \mathbf{p}(t-1))^2}$, where $\mathbf{p}(t) = (x(t), y(t))$ and $\mathbf{p}(t-1) = (x(t-1), y(t-1))$ are, respectively, the current and previous positions of the feature. To avoid abrupt changes in this parameter, a weighted sum is performed, and $\Delta d = A_f f_d(t) + (1 - A_f) f_d(t-1)$, where A_f is a constant within the interval $[0,1]$, $f_d(t-1)$ is the previous displacement and $f_d(t)$ is the current observed displacement. Then, parameters d_1 and d_2 of $\mu_S(x,y)$ are defined as

$$\begin{aligned} d_1 &= M_1 + \frac{\Delta d}{2}, \\ d_2 &= d_1 + M_2 + \frac{\Delta d}{2}, \end{aligned} \tag{1}$$

where M_1 and M_2 are two positive constants. These constants act as minimum values for these two parameters in order to deal with features denoting zero velocity. Parameter d_1 has a minimum value of M_1 and parameter d_2 has a minimum value of $d_1 + M_2$. Values M_1 and M_2 can be related with the dimensions of the feature.

The bright constancy assumed earlier ensures that the feature intensity level remains stable, or approximately stable, during the sequence. Hence, the initial gray level of the feature is considered unchangeable over time meaning that pixels denoting similar gray levels regarding the initial feature gray level are more likely to belong to the feature.

Under these conditions, a fuzzy set G is constructed in order to access the certainty of a pixel belonging to the feature in such a way that the higher the similarity in the gray levels intensity, the higher the membership degree.

Let I be an image with dimensions $M \times N$, $I(x,y)$ is the gray level of the pixel (x,y) so that $0 \le I(x,y) \le L$ and $I_f(i,j)$ an intensity matrix of dimensions $A \times B$ representing the original feature's gray levels, where $A = 2a+1$, $B = 2b+1$ and $\{a,b\} \in \mathbb{N}$.

For all (x,y) such that $a \le x \le M-a$ and $b \le y \le N-b$, the membership function $\mu_G(x,y)$ is defined as

$$\mu_G(x,y) = 1 - d_g, \tag{2}$$

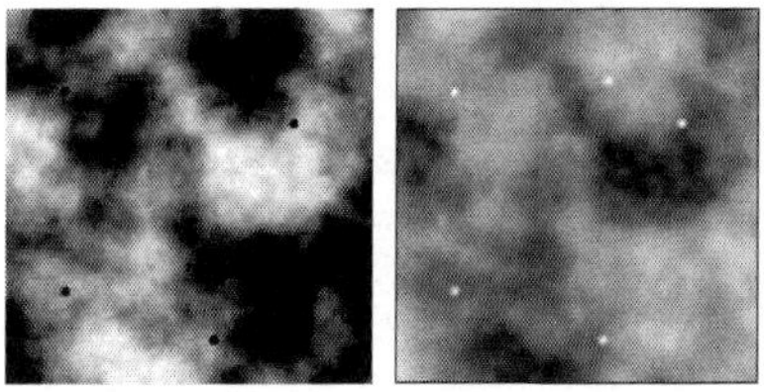

(a) Original image. (b) Fuzzy set G.

Fig. 2 Membership function $\mu_G(x,y)$

with

$$d_g = \frac{\sum_{i=-a}^{a}\sum_{j=-b}^{b}|I(x+i,y+j) - I_f(i+a,j+b)|}{A \times B \times L}. \quad (3)$$

All pixels (x,y) of the image, such that $0 \leq x \leq a \vee M - a \leq x \leq M$ and $0 \leq y \leq b \vee N - b \leq y \leq N$ have zero membership values. This set of pixels are located at the boundaries of the image and, since a and b are small positive integers, this discontinuity do not change the global performance of the method.

In Fig. 2, the output result using membership function $\mu_G(x,y)$ applied to a test image is illustrated. In order to reduce processing time and increase computational speed, this membership function is applied locally in a neighbourhood centred in the previous position of the feature.

Another membership function is constructed based on the assumption that the feature motion between two consecutive frames can be described using a linear motion model with constant acceleration. Stating this, a feature can increase its velocity between two consecutive frames. Several motion models are discussed in the literature, however, the selected motion model is a compromise between the proximity with the real motion performed by the feature and computer processing requirements. The feature can move along both the x and y axis and, therefore, the position $\mathbf{p}(t)$ can be obtained from the previous position $\mathbf{p}(t-1)$ by $\mathbf{p}(t) = \mathbf{p}(t-1) + \mathbf{v}\Delta t + \frac{1}{2}\mathbf{a}\Delta t^2$, where $\mathbf{p}(t) = [x,y]'$ is the feature position at instant t, $\mathbf{p}(t-1) = [x_0,y_0]'$ is the feature position at instant $t-1$, Δt is the elapsed time from instant $t-1$ to instant t, $\mathbf{v} = [v_x,v_y]'$ and $\mathbf{a} = [a_x,a_y]'$ are, respectively, the observed velocity and acceleration in both axis during Δt.

Kalman filter is used to estimate the state vector of the feature, i. e., its position, velocity and acceleration. Using information provided by the state vector of the feature it is possible to predict the feature position (x,y) in the next frame. This thoughts led to the development of another membership function, called $\mu_K(x,y)$. This membership function assigns a higher membership degree to pixels near the predicted location and its value decreases for locations far from this predicted point. To implement such behavior, a Gaussian shape function is used (Fig. 3a), rather than a triangular one, to ensure a swift decay in the membership values, giving more importance to locations near the predicted one.

Fig. 3 Membership function $\mu_K(x,y)$.

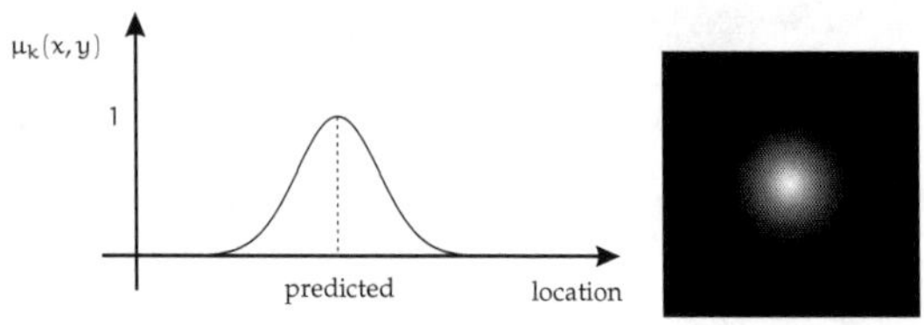

(a) Definition of $\mu_K(x,y)$. (b) Fuzzy set K.

The gaussian function shape can be changed through the standard deviation parameter σ defined as follows

$$\sigma = \frac{3M_\sigma + \Delta d}{3}, \tag{4}$$

where M_σ is a minimum value to deal with stopped features and Δd is the previous observed displacement of the feature. Then, this parameter is changed according to the velocity of the feature with higher velocities giving rise to higher standard deviation values.

This membership function $\mu_K(x,y)$ is applied in a circular neighbourhood with radius equal to 3σ. For locations whose distance to the predicted position is greater than 3σ the membership degree is zero. In Fig. 3b, is depicted the resulting fuzzy set K assuming a predicted position $(x,y) = (100,100)$ and $\sigma = 15$.

An auxiliary fuzzy set M is constructed from the fuzzy union between fuzzy sets K and S, using the maximum operator as follows:

$$\mu_M(x,y) = \vee(\mu_K(x,y), \mu_S(x,y)) \tag{5}$$

2.2 *Inference Engine*

After defining the fuzzy set M, an inference engine with the following set of fuzzy rules is constructed. The output of the engine is a fuzzy set E that will ultimately lead us to the feature position which will be the pixel (x,y) that corresponds to the highest $\mu_E(x,y)$ value.

RULE 1: **IF**, within the area defined by membership values of the fuzzy set M, such that $\mu_M(x,y) > 0, \forall(x,y)$, there is one and only one local maxima of $\mu_G(x,y) > \alpha$, **THEN**, fuzzy set E is the union between fuzzy sets S and G. This fuzzy set is constructed using the maximum operator in the following way:

$$\mu_E(x,y) = \vee(\mu_G(x,y), \mu_S(x,y)). \tag{6}$$

RULE 2: **IF**, within the area defined by membership values of the fuzzy set M such that $\mu_M(x,y) > 0, \forall(x,y)$, there are $n \geq 2$ local maxima of $\mu_G(x,y)$, located at position $(x_i,y_i), \forall i = 1,\ldots,n$ that satisfy the condition $\mu_G(x_i,y_i) > \alpha$, **THEN**, fuzzy set E is the union between fuzzy sets S and $G'_i, \forall i = 1,\ldots,n$.

The fuzzy sets G'_i are constructed using fuzzy set G in the following way:

$$\mu_{G'_i}(x,y) = \psi_i \mu_G(x,y), \forall i = 1,\ldots,n, \tag{7}$$

with

$$\psi_i = 1 - \frac{d_i}{d_{MAX}}, \forall i = 1,\ldots,n, \tag{8}$$

and

$$d_i = \sqrt{(x_i - x_K)^2 + (y_i - y_K)^2}, \forall i = 1,\ldots,n,$$
$$d_{MAX} = max\{d_i\}, \forall i = 1,\ldots,n, \tag{9}$$

where (x_K, y_K) is the location of the maximum value of fuzzy set K.
Finally, fuzzy set E is constructed using the maximum operator:

$$\mu_E(x,y) = \vee(\mu_{G'_1}(x,y),\ldots,\mu_{G'_n}(x,y),\mu_S(x,y)). \tag{10}$$

RULE 3: **IF**, within the area defined by membership values of the fuzzy set M, such that $\mu_M(x,y) > 0, \forall(x,y)$, there is no local maxima of $\mu_G(x,y) > \alpha$, **THEN**, fuzzy set E is to be equal to fuzzy set K.

$$\mu_E(x,y) = \mu_K(x,y). \tag{11}$$

The design of this inference engine and its fuzzy rules is based in human experience and previous experimental tests. People expect to find an object or a feature in its last known location or at locations in the neighbourhood, particularly when dealing with static objects or objects that denote small movement. This kind of human reasoning is modelled by fuzzy set S. When dealing with fast moving objects, people are capable to understand the feature motion pattern and consequently anticipate its next position. This thought is also valid when the object is occluded. Consequently, the fuzzy set K tends to incorporate this reasoning. According to these two behavioural attributes, the area defined by fuzzy set M, i. e., the image area where $\mu_M(x,y) > 0$, can be seen as the first area of search to locate an object or a feature. Looking for this region, if a person see an identical feature as expected, then it is plausible to consider this feature as the one that is been tracked. From RULE 1, the feature position will be the pixel, with coordinates (x,y), that denotes the maximum value of $\mu_G(x,y)$.

If multiple identical features are present in that region then, it is reasonable, based on the previous acquired motion pattern, to choose the feature near the predicted feature position (RULE 2).

In situations when the feature is not visible, the location can be only estimated by the understanding of the behavior of the motion observed until that moment (RULE 3). In this case, the feature position will be the pixel, with coordinates (x,y) with the maximum value of $\mu_K(x,y)$.

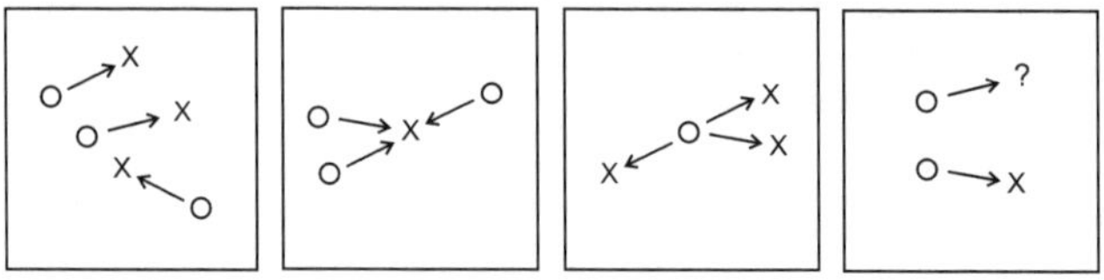

(a) (b) (c) (d)

Fig. 4 Correspondence situations in multi feature tracking.

Furthermore, if the output of the engine results from RULE 3 then, since all the $\mu_G(x,y)$ values in the considered image area are below than α, probably due to an occlusion, the membership functions of fuzzy sets S and K change in such a way that the region where $\mu_M(x,y) > 0$ becomes bigger, allowing the tracked feature to be searched in a wider area.

2.3 *Occlusion and Misdetection*

In multiple feature tracking several correspondence situations can occur. Fig. 4 depicts these situations, where $\circ$ denotes the feature position at frame $t-1$ and $\times$ denotes the feature position at frame t. The question mark (?) represents the absence of correspondence at frame t. The first situation depicted in Fig. 4a indicates that each feature is matched with a different candidate point in the next frame. When two moving features pass close each other or when one feature occludes an other, or also due to the representation of a 3D world in a 2D plane, they can appear as being just one region in the image. This situation could be seen as a merging of features or a inter feature occlusion case, Fig. 4b. The opposite situation is also considered, i. e., several united features could have different motion directions and one single region representing multiple features, could result in multiple matching points, Fig. 4c. Finally, at some instant, there are no candidate point for a feature and, without correspondence, a predicted position is assumed, Fig. 4d. All these situations must be considered in the tracking algorithm and correct procedures must be applied in each case.

In this multiple feature tracking approach it's possible to discern between inter feature occlusion and background occlusion and different actions could be taken for each case. In inter feature occlusion situation some features could have disappeared in this moment but at least one feature is present and continues to be tracked. The feature with higher confidence degree continues to be tracked normally and the occluded features suffer a slow decreasing in their confidence degree. When a background occlusion occurs, there is no candidate present in the frame and the feature suffers a higher decrease in its confidence degree. The slow decay in confidence degree when inter feature occlusion situation occurs ensures that inter occluded features continue to be tracked over several frames, giving the opportunity to some features exit from the region and be tracked.

2.4 Algorithm

According the assumptions and situations previously introduced, a new approach was developed. At the beginning, the user selects multiple features to track and each feature is associated with a confidence level. This confidence level is related with the knowledge about a feature. When the user selects the features to track at the beginning of the method, each feature has its confidence degree at the maximum value. This means that the position of each feature is known with higher certainty or confidence. When the algorithm doesn't have information about the feature, its confidence degree decreases. It is considered that situations of inter feature or background occlusion and misdetections introduce absence of information.

For all situations previously described, the algorithm performs as follows:

1. When only one feature is matched with a candidate, its confidence degree increases, Fig. 4a;
2. When several features are matched with the same candidate, the feature assigned first with that candidate see its confidence degree increasing and the other features see their confidence degree decreasing, Fig. 4b;
3. When there is no candidate to be matched with a feature, a predicted position is used and the confidence degree of this feature decreases, Fig. 4d;
4. Features with low confidence degree are removed from the list of tracked features.

All the procedures indicated previously are first applied to features with higher confidence level and this way a hierarchical matching system is performed. Situation depicted in Fig. 4c is covered by the first case since one single feature is matched with a candidate and, in this situation, the confidence degree for all the features increases. To perform the fourth case, the user introduces a minimum confidence degree and features denoting a confidence degree below this minimum value will not be tracked.

3 Results

In order to test its performance, the proposed method was applied to several microscopic sequences obtained from Bacterial Motility and Behavior group[1]. The first sequence shows a bacteria called *Serratia marcescens* moving over the surface. In this sequence, the bacteria denotes a bright color against a dark background. To illustrate the effectiveness of the approach, several bacterias are picked up and the trajectories are estimated, Fig.5. The second test sequence is an organism called *Synechococcus*, swimming in a medium. This sequence is more complex than the previous: there are more similar features in the scene, the background is not uniform and the difference of gray levels between feature and background is lower, Fig. 6. These two sequences where chosen because they present additional difficulties to the tracking process. There are features crossing each other, there are permanent

[1] `http://www.rowland.harvard.edu/labs/bacteria/index.html`

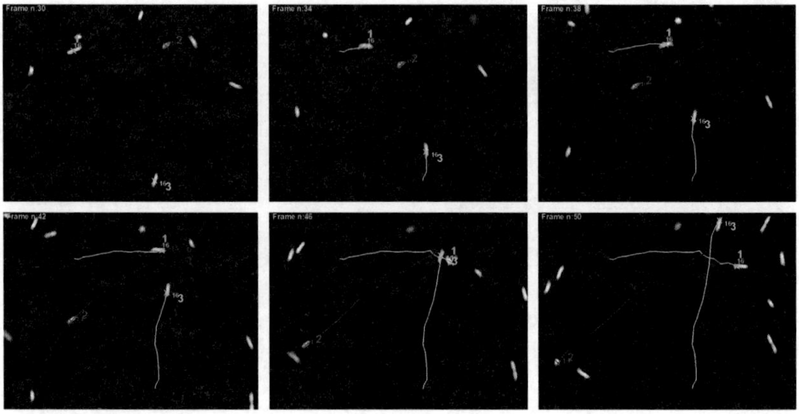

Fig. 5 Estimated trajectories between frames 30 and 50.

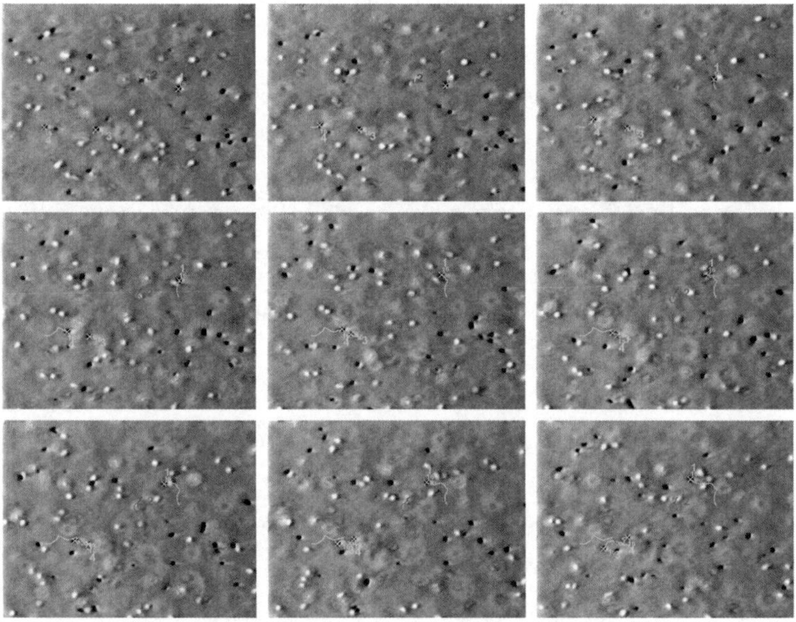

Fig. 6 Estimated trajectories between frames 40 and 70.

occlusions and the trajectories described by the features are complex. In both sequences all features are labeled with a unique identification number and their confidence level.

To implement a performance measure of the proposed approach, experimental results are compared with manually generated feature positions that represent the centre of mass of the features. Using these ground-truth positions, a tracking error measurement could be developed calculating the average and standard deviation of an Euclidian distance between experimental and ground-truth positions as presented in Table 1.

Table 1 Average distance error and standard deviation

Sequence name	Feature number	Mean	Standard deviation
Serratia marcescens	1	2.7752	1.5534
Serratia marcescens	2	1.0908	0.8948
Serratia marcescens	3	2.7443	1.5628
Synechococcus	1	0.8051	0.6809
Synechococcus	2	1.0598	0.7572
Synechococcus	3	0.8462	0.6782
Synechococcus	4	0.8032	0.6583

The first sequence generates higher error values since the moving bacterias denote the tracked feature (high gray level intensity) in all their surface and the bacterias present an elongated shape leading the method to provide experimental positions not close enough from the centre of mass. However, since the tracking process is performed correctly, these errors could be solved using morphological approaches.

4 Conclusions

In this work, a new fuzzy tracking system was introduced. With the definition of three membership functions related with cinematic and non cinematic properties of the feature and with a construction of an inference engine with three fuzzy rules the fuzzy tracking approach was implemented. Experimental tests were performed to evaluate the performance of the approach. A hierarchical implementation was constructed since features that provided more information to the method are assigned first than features with low confidence degree. The presented hierarchical matching approach for multiple feature tracking has provided encouraging results. However, these results lead us to further work with intend to improve robustness, introduce new capabilities and achieve computational efficiency over different image sequences.

References

[1] Couto, P., Pagola, M., Bustince, H., Barrenechea, E., Melo-Pinto, P.: Uncertainty in multilevel image thresholding using Atanassov's intuitionistic fuzzy sets. In: Proc. (IEEE World Congress on Computational Intelligence). IEEE International Conference on Fuzzy Systems FUZZ-IEEE 2008, pp. 330–335 (2008); doi:10.1109/FUZZY.2008.4630386

[2] Couto, P., Lopes, N.V., Bustince, H., Melo-Pinto, P.: Fuzzy dynamic model for feature tracking. In: 2010 IEEE International Conference on Fuzzy Systems (FUZZ), Barcelona, Spain, July 18-23, pp. 1–8 (2010); doi:10.1109/FUZZY.2010.5583979

[3] Hu, W., Tan, T., Wang, L., Maybank, S.: A survey on visual surveillance of object motion and behaviors. IEEE Transactions on Systems, Man, and Cybernetics, Part C: Applications and Reviews 34(3), 334–352 (2004); doi:10.1109/TSMCC.2004.829274

[4] Huang, L.K., Wang, M.J.J.: Image thresholding by minimizing the measures of fuzziness. Pattern Recognition 28(1), 41–51 (1995); doi:10.1016/0031-3203(94)E0043-K

[5] Jaimes, A., Sebe, N.: Multimodal human-computer interaction: A survey. Computer Vision and Image Understanding 108(1-2), 116 (2007) doi:10.1016/j.cviu.2006.10.019; special Issue on Vision for Human-Computer Interaction;

[6] Lazoff, H.: Target tracking using fuzzy logic association. In: Proc. IEEE International Conference on Acoustics, Speech and Signal Processing, vol. 4, pp. 2457–2460 (1998); doi:10.1109/ICASSP.1998.681648

[7] Liu, M., Wu, C., Zhang, Y.: A review of Traffic Visual Tracking technology. In: Proc. International Conference on Audio, Language and Image Processing ICALIP 2008, pp. 1016–1020 (2008); doi:10.1109/ICALIP.2008.4590198

[8] Lopes, N.V., Bustince, H., Filipe, V., Melo-Pinto, P.: Fuzziness Measure Approach to Automatic Histogram Threshold. In: Tavares, J., Jorge, N. (eds.) Computational Vision and Medical Image Processing: VipIMAGE 2007, pp. 295–299. Taylor and Francis Group, Abington (2007)

[9] Lopes, N.V., Couto, P., Bustince, H., Melo-Pinto, P.: Fuzzy Dynamic Matching Approach for Multi-Feature Tracking. In: EUROFUSE 2009, Pamplona, Spain, pp. 245–250 (2009)

[10] Moeslund, T.B., Hilton, A., Krüger, V.: A survey of advances in vision-based human motion capture and analysis. Comput. Vis. Image Underst. 104(2), 90–126 (2006); doi:10.1016/j.cviu.2006.08.002

[11] Yilmaz, A., Javed, O., Shah, M.: Object tracking: A survey. ACM Comput. Surv. 38(4), 13 (2006); doi:10.1145/1177352.1177355

[12] Zadeh, L.A.: Fuzzy Sets. Information Control 8, 338–353 (1965)

Objective Comparison of Some Edge Detectors Based on Fuzzy Morphologies

M. González-Hidalgo, S. Massanet, and A. Mir

Abstract. In this paper a comparative analysis of several edge detectors based on diverse fuzzy morphologies is performed. In addition, two different processes in order to transform a fuzzy edge image to a thin binary edge image are studied, a recently introduced unsupervised hysteresis based on the determination of a "instability zone" on the histogram and a fuzzy Atanassov's based threshold. The comparison is made according to some performance measures, such as Pratt's figure of merit and the ρ-coefficient. The goodness of the employed binarization methods is studied depending on their capability to obtain the best threshold values according to these measures.

1 Introduction

Edges are one of the most important visual clues to interpret an image. Thus, edge detection is a fundamental low-level image processing operation, which is essential to carry out several higher level operations such as image segmentation, computer vision, motion and feature analysis and recognition. Its performance is crucial for the final results of the image processing techniques. A lot of edge detection algorithms have been developed over the last decades. These different approaches vary from the classical ones ([18]) based on a set of convolution masks, to the new techniques based on fuzzy sets ([4]).

Among the fuzzy approaches, the fuzzy mathematical morphology is a generalization of the binary morphology ([19]) using concepts and techniques from the fuzzy sets theory ([1], [16]). This technique allows to obtain image structures that

M. González-Hidalgo · S. Massanet · A. Mir
Dept. of Math. and Comp. Science,
University of the Balearic Islands, 07122 Palma de Mallorca, Spain
e-mail: {manuel.gonzalez,s.massanet,arnau.mir}@uib.es

B. De Baets et al. (Eds.): Eurofuse 2011, AISC 107, pp. 401–412, 2011.
springerlink.com

are helpful in identifying region shape, boundaries, skeletons, etc. The use of fuzzy sets ensures that imprecision and uncertainty can be well handled by the fuzzy morphology. Note that a gray-scale image can be termed as a fuzzy set in the sense that is a fuzzy version of a binary image ([6, 16, 2]). The morphological operators are the basic tools of this theory. A morphological operator P transforms an image A that one want to analyse into a new image $P(A,B)$ by means of an structuring element B. The four basic morphological operators are dilation, erosion, closing and opening. In order to manipulate fuzzy sets, the use of fuzzy tools is unavoidable. Thus, conjunctors (usually continuous t-norms, but also uninorms, see [12]) and their residual implicators have been used. Recently, due to the fact that gray-scale images are not represented in practice as functions of $\mathbb{R}^n$ into $[0,1]$ because they are stored in finite matrices whose gray levels belong to a finite chain of 256 values, discrete fuzzy operators can also be used. In [10] and [8], the discrete t-norms that satisfy the algebraic properties required to a morphology in order to become a "good" one were fully determined.

Usually the fuzzy edge image is an intermediate step before its binarization. Non-maxima suppression (NMS), proposed by Canny in [5], is used as a post-processing operation along with the gradient operator for edge detection in order to obtain edges of single pixel width. After that, some binarization is carried out. In this paper, two methods will be analysed: a thresholding using Atanassov's intuitionistic fuzzy sets ([20]) and the recently introduced unsupervised hysteresis based on the determination of a "instability zone" ([15]). The obtention of a thin binary edge image is a necessary step in order to use reference images with their ground truth edge images for comparing different edge detectors through the use of some performance measures. In this paper, we want to compare with this methodology the edge detectors based on continuous t-norms, uninorms ([11]) and discrete t-norms ([9]).

The communication is organized as follows. In Section 2, definitions and properties of the fuzzy morphologies that will be used along the paper are recalled. In Section 3, the two binarization techniques are presented as well as the performance measures used in next section. In Section 4 the obtained results are presented and analysed. Finally, some conclusions and future work are pointed out.

2 Preliminaries

In this section, we will recall the main definitions and properties of the different fuzzy morphologies that will be analysed. We will use the following notation: $\mathscr{I}$ is an implicator, $\mathscr{C}$ a conjunctor, $\mathscr{N}$ a strong negation, T a t-norm, U a conjunctive uninorm with a neutral element e, $\mathscr{I}_T$ and $\mathscr{I}_U$ their respective residual implicator, and finally A a gray-scale image and B a gray-scale structuring element. An n-dimensional gray-scale image is modelled as an $\mathbb{R}^n \to [0,1]$ function and the fuzzy morphological operators are defined in the same way as De Baets in [7]:

Definition 1. *([7, 16]) The* fuzzy dilation $D_{\mathscr{C}}(A,B)$ *and* the fuzzy erosion $E_{\mathscr{I}}(A,B)$ *of A by B are the gray-scale images defined as*

$$D_{\mathscr{C}}(A,B)(y) = \sup_x \mathscr{C}(B(x-y),A(x))$$
$$E_{\mathscr{I}}(A,B)(y) = \inf_x \mathscr{I}(B(x-y),A(x)).$$

Note that as conjunctors, t-norms and uninorms can be used. For the morphology based on nilpotent t-norms, we will consider the Łukasiewicz t-norm $T_L(x,y) = \max\{0,x+y-1\}$ for all $x,y \in [0,1]$. Recall that the pair $(T_L, \mathscr{I}_{T_L})$ is the representative of the only class of t-norms (nilpotent ones) that guarantees the fulfillment of all the properties in order to have a good fuzzy mathematical morphology, including duality and generalized idempotence (see [16]). Specially, we highlight the following one that allows to build a morphological gradient useful to be used as an edge detector.

Proposition 1. *If* $B(0) = 1$ *the fuzzy dilation is extensive and the fuzzy erosion is anti-extensive,i.e.,* $E_{\mathscr{I}_T}(A,B) \subseteq A \subseteq D_T(A,B)$.

On the other hand, for the morphology based on uninorms, two types of left-continuous conjunctive uninorms and their residual implications are used. Specifically, they are defined as follows:

- *Representable uninorms*: Let $e \in (0,1)$ and let $h : [0,1] \to [-\infty,\infty]$ be a strictly increasing continuous function with $h(0) = -\infty$, $h(e) = 0$ and $h(1) = \infty$. Then

$$U_h(x,y) = \begin{cases} h^{-1}(h(x)+h(y)) & \text{if } (x,y) \notin \{(1,0),(0,1)\}, \\ 0 & \text{otherwise,} \end{cases}$$

  is a conjunctive representable uninorm with neutral element e, and its residual implicator $\mathscr{I}_{U_h}$ is given by

$$\mathscr{I}_{U_h}(x,y) = \begin{cases} h^{-1}(h(y)-h(x)) & \text{if } (x,y) \notin \{(0,0),(1,1)\}, \\ 1 & \text{otherwise.} \end{cases}$$

- A specific type of *idempotent uninorms*. Let $\mathscr{N}$ be a strong negation. The function given by

$$U^{\mathscr{N}}(x,y) = \begin{cases} \min\{x,y\} & \text{if } y \leq \mathscr{N}(x), \\ \max\{x,y\} & \text{otherwise,} \end{cases}$$

  is a conjunctive idempotent uninorm. Its residual implicator is given by

$$\mathscr{I}_{U^{\mathscr{N}}}(x,y) = \begin{cases} \min\{\mathscr{N}(x),y\} & \text{if } y < x, \\ \max\{\mathscr{N}(x),y\} & \text{if } y \geq x. \end{cases}$$

These two types of conjunctive uninorms guarantee most of the good algebraic and morphological properties associated with the morphological operators obtained from them (see [11]). An analogous result to Proposition 1 can be proved if $B(0) = e$, where e is the neutral element of the corresponding uninorm. Among the two considered classes of uninorms we will analyse the results obtained by the idempotent uninorms $U^{\mathcal{N}_C}$ and $U^{\mathcal{N}_2}$ where $\mathcal{N}_C(x) = 1 - x$ is the classical negation and $\mathcal{N}_2(x) = \sqrt{1-x^2}$. Finally, the representable uninorm U_h with $h(x) = \ln\left(\frac{x}{1-x}\right)$ is also considered.

On the other hand, a discrete framework has been recently introduced. Let L be a finite chain $L = \{0,\ldots,n\}$. The main difference is that here an N-dimensional gray-scale image is modelled by a $\mathbb{Z}^N \to L$ function instead of a fuzzy set in $\mathbb{R}^N$. The fuzzy discrete dilation and the fuzzy discrete erosion are defined in the same way as the continuous case but with the particularity that the infimum and supremum are in fact minimum and maximum. The discrete counterpart of the fuzzy operators defined in the $[0,1]$-framework is used now, in particular, discrete t-norms and their discrete residual implicators. In [10] and [8], the discrete t-norms T that have to be used in order to preserve the morphological and algebraic properties that satisfy the classical morphological operators were fully determined. Among them, a similar result to Proposition 1 hold in the discrete approach if $B(0) = n$. From the discrete t-norms that satisfy all the properties, we will take the following ones:

- $T_L(x,y) = \max\{0, x+y-n\}$, the Łukasiewicz discrete t-norm,
- the nilpotent minimum given by the following expression

$$T_{nM}(x,y) = \begin{cases} 0 & \text{if } x+y \leq n, \\ \min\{x,y\} & \text{otherwise.} \end{cases}$$

Since the Łukasiewicz discrete t-norm and its nilpotent counterpart provide almost identical results (just some values could be different due to some numerical rounding error caused by the fuzzyfication function in the continuous case), only one of them will be studied.

Consequently, all these fuzzy morphologies allow to define a *fuzzy gradient* operator that can be used in edge detection. In this case, we will use the gradient by erosion that is defined as:

$$A \setminus E_{\mathcal{I}_{\mathcal{C}}}(A,B),$$

where $\mathcal{C}$ will be a nilpotent t-norm, a discrete t-norm or a conjunctive uninorm, accordingly. This operator generates fuzzy edge images. However, in order to compare the different approaches introduced above through the use of performance measures we have to apply a thinning and a binarization to this image.

3 Non-maxima Suppression and Algorithms of Binarization

In the literature, there exist different approaches to edge detection based on fuzzy logic. These methods represent the edge image on a fuzzy way, i.e., they generate a image where each pixel value represents its membership to the edge set. However, this idea contradicts the Canny restrictions ([5]). These restrictions force a representation of the edges as binary images with edges of one pixel wide. Thus, in order to satisfy the Canny's restrictions, the fuzzy edge image has to be thinned and binarized.

First of all, the fuzzy edge image becomes a fuzzy thin edge image where all the edges have one pixel width. The fuzzy edge image will contain large values where there is a strong image gradient, but to identify edges the broad regions present in areas where the slope is large must be thinned so that only the magnitudes at those points which are local maxima remain. Non-Maxima Suppression (NMS) performs this by suppressing all values along the line of the gradient that are not peak values (see [5]). NMS has been performed using P. Kovesis' implementation in Matlab ([14]).

After that, two different binarization techniques are analysed.

- Thresholding using Atanassov's intuitionistic fuzzy sets ([20]), where restricted dissimilarity functions are used in the construction of membership functions and the ambiguity is represented by means of Atanassov's intuitionistic fuzzy sets. The dissimilarity functions $d(x,y) = \sqrt{|x^2 - y^2|}$ and the function $F(x,y) = \exp(-x^2 \cdot \ln 2)$ have been used since in [9], they constituted the best configuration according to the same performance measures used in this paper.
- An unsupervised hysteresis based on the determination of the instability zone on the histogram in order to find the low and the high thresholds (see [15]). Hysteresis allows to choose which pixels are relevant in order to be selected as edges, using their membership values. Two threshold values U_1, U_2 with $U_1 \leq U_2$ are used. All the pixels with a membership value greater than U_2 are considered as edges, while those which are lower to U_1 are discarded. Those pixels whose membership value is between the two values are selected if and only if they are connected with other pixels above U_2. The method needs some initial set of candidates for the threshold values. In this case, $\{0.01,\ldots,0.25\}$ has been introduced, the same set used in [15].

In Figure 1, the sequence of the algorithm can be observed.

3.1 Performance Measures on Edge Detection

For the comparison of the obtained results, some performance measures have been considered. These measures need, in addition to the binary thin edge image (DE) obtained, a ground truth edge image (GT) that is a binary thin edge image containing the true edges of the original image, i.e., the reference edge image. There are several performance measures on edge detection in the literature (see [17]). In this work, we will use the following measures to quantify the similarity between (DE) and (GT):

1. Pratt's figure of merit ([18]) defined as

$$FoM = \frac{1}{\max\{card\{DE\}, card\{GT\}\}} \cdot \sum_{x \in DE} \frac{1}{1 + ad^2},$$

Fig. 1 From left to right: original image and its ground truth; fuzzy edge image obtained using $U^{\mathcal{N}_C}$; fuzzy thin edge image (using NMS) and the binary thin edge image using Atanassov's thresholding with d and F (top) and using the unsupervised hysteresis (down)

where *card* is the number of edge points of the image, a is a scaling constant and d is the separation distance of an actual edge point to the ideal edge points. In our case, we considered $a = 1$ and the Euclidean distance d.

2. The ρ-coefficient ([13]), defined as

$$\rho = \frac{card(E)}{card(E) + card(E_{FN}) + card(E_{FP})},$$

where E is the set of well-detected edge pixels, E_{FN} is the set of ground truth edges missed by the edge detector and E_{FP} is the set of edge pixels detected but with no counterpart on the ground truth image. Since edges cannot always be detected at exact integer image coordinates, we consider that an edge pixel is correctly detected if a corresponding ground truth edge pixel is present in a 5×5 square neighbourhood centered at the respective pixel coordinates, as it was stated in [13].

Larger values of FoM and ρ ($0 \leq FoM, \rho \leq 1$) are indicators of better capabilities for edge detection.

4 Results and Analysis

As we have already observed, the performance measures need a dataset of images with their ground truth edge images (edges specifications) in order to compare the outputs obtained by the different algorithms. So, the images and their edge specifications from the public dataset of the University of South Florida[1] ([3]) have been used. This image dataset includes 50 natural images with their ground truth edge

[1] This image dataset can be downloaded from
`ftp://figment.csee.usf.edu/pub/ROC/edge_comparison_dataset.tar.gz`

Table 1 Results obtained

Method		Mean		St. deviation		% best-worst total		% best-worst met.	
Step 1	**Step 2**	*FoM*	ρ	*FoM*	ρ	*FoM*	ρ	*FoM*	ρ
$T_Ł$	Atan.	0.2890	0.4556	0.0970	0.1481	7-0	0-0	20-0	7-0
	Hyst.	0.3929	0.5183	0.1000	0.1311	0-0	0-7	0-20	0-33
T_{nM}	Atan.	0.2781	0.4457	0.0991	0.1526	0-40	0-33	7-40	7-47
	Hyst.	0.3886	0.5298	0.1028	0.1259	0-0	7-7	13-33	13-13
$U^{\mathcal{N}_C}$	Atan.	0.2955	0.4776	0.0911	0.1497	0-7	7-7	40-13	40-13
	Hyst.	0.4121	0.5580	0.1030	0.1377	60-13	47-7	60-13	60-13
$U^{\mathcal{N}_2}$	Atan.	0.2859	0.4574	0.0952	0.1452	7-27	13-13	20-27	27-27
	Hyst.	0.4034	0.5466	0.1010	0.1250	27-0	13-7	27-13	27-13
U_h	Atan.	0.2854	0.4551	0.0927	0.1474	0-13	13-13	13-20	20-13
	Hyst.	0.3909	0.5171	0.1009	0.1321	0-0	0-7	0-27	0-27

images. Each of the fifty images representing the domain of generic object recognition contains a single object approximately centered in the image, appearing unoccluded and set against a natural background for the object. In our experiment, we have used the first 15 images of the dataset.

First of all, the fuzzy gradient operator is applied to the original image in order to obtain the fuzzy edge image. In the displayed experiments, the following structuring element, that performs notably for all the morphologies, has been used

$$B = \begin{pmatrix} 0.5 & 1 & 0.5 \\ 1 & 1 & 1 \\ 0.5 & 1 & 0.5 \end{pmatrix}.$$

For the morphologies based on uninorms, we used the adapted structuring element eB where e is the neutral element of the uninorm and finally, in the discrete framework, its counterpart into L is used. After that, NMS is applied to the fuzzy edge image and then, Atanassov's thresholding or the unsupervised hysteresis is performed. In order to make a general analysis, Table 1 shows the mean and the standard deviation for the two performance measures using all the possible configurations. In addition, for a particular configuration, we specify the percentage of images for which this configuration is the best one or the worst one of the ten methods studied and between the methods using the same binarization method into the columns denoted by "met.". For instance, $T_Ł$ with Atanassov's thresholding is the best configuration (including the configurations with hysteresis) for 7% of the images according to *FoM* and it is the best configuration between those where Atanassov's thresholding is applied for 20 and 7% of the images according to *FoM* and ρ, respectively. Some of the used images and their best results obtained according to *FoM* and ρ are displayed in Figure 2.

Fig. 2 Original image (1st column), its ground truth edge image (2nd column) and the two best binary thin edge images according to FoM (3rd column), and ρ (4th column). From left to right and from top to down, $U^{\mathcal{N}_2}$ with Atanassov's is used in images 3, 12, with unsupervised hysteresis in image 7; T_{nM} with the unsupervised hysteresis in image 4 and $U^{\mathcal{N}_C}$ with the hysteresis is used in images 8,11.

At a first glance, it is remarkable that all the morphologies within the same binarization method provides results not far between them. Note that for example, the mean value of ρ for $U^{\mathcal{N}_C}$ with the unsupervised hysteresis (the best one) is 0.5581, while U_h with the hysteresis (the worst one) is 0.5171. However, some general comments arise from Table 1:

1. The morphology using $U^{\mathcal{N}_C}$ provides the best results (in terms of the mean value and the percentage of best results) either with the hysteresis or the Atanassov's thresholding according to both measures.
2. The other morphologies performs quite similar, although it seems that the morphologies based on uninorms outperform the ones based on t-norms.
3. The unsupervised hysteresis gives better results than the Atanassov's thresholding for all the morphologies. In general, the thresholding generates images with few and disconnected edges. On the other hand, hysteresis provides soft edge images with clear edges. This fact can be viewed in Figure 3. However, in some

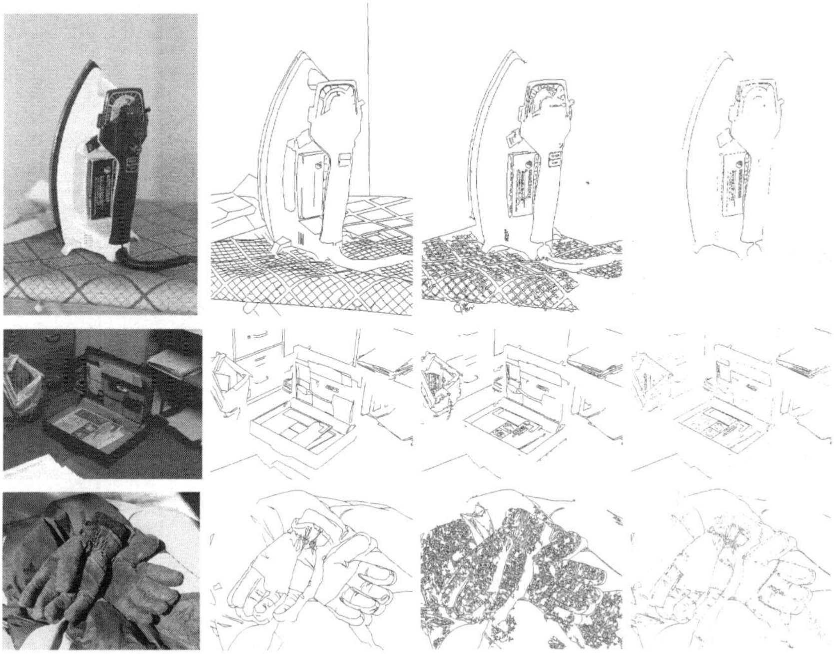

Fig. 3 Original image (1st column), its ground truth edge image (2nd column), the best result with the unsupervised hysteresis (3rd column) and the best result with Atanassov's thresholding (4th column) according to ρ. From left to right and from top to down, $U^{\mathcal{N}_2}$ is used in images 3, 12; T_{nM} in images 4,11; $U^{\mathcal{N}_C}$ in image 7 and $T_{\text{Ł}}$ in image 8.

textured images, hysteresis transforms the texture into edges while the thresholding avoids it resulting images with better performance measures values (see the gloves in Figure 3).

In order to confirm that the morphology using $U^{\mathcal{N}_C}$ with hysteresis outperforms the other ones, some statistical tests have been performed. After verifying the normality hypothesis of the differences between the results obtained with $U^{\mathcal{N}_C}$ and the other ones (except with $U^{\mathcal{N}_2}$) using a KS-test and a Shapiro test, we have made a

Table 2 p-values obtained from a t-test (or Wilconox-signed rank test) when comparing $U^{\mathcal{N}_C}$ with the other morphologies with hysteresis

Measure	$T_{\text{Ł}}$	T_{nM}	U_h
p-value considering FoM	0.03527 (0.009033)	0.03149 (0.01077)	0.03834 (0.009033)
p-value considering ρ	0.00838 (0.001007)	0.04088 (0.02063)	0.0122 (0.001678)

t-test to prove that this method provides better results. This test and even a Wilconox-signed rank test, which does not need normality hypothesis, confirm the initial conclusion. The obtained small p-values are displayed in Table 2. With respect to $U^{\mathcal{N}_2}$, we can not ensure that $U^{\mathcal{N}_C}$ is better than $U^{\mathcal{N}_2}$ statistically.

4.1 Goodness of the Binarization Methods

The binarization methods applied in this paper are two unsupervised techniques, i.e., the user has no intervention during the process. However, since we work with natural images with their corresponding ground-truth edge images and with objective performance measures, the best threshold and the best hysteresis threshold values for each measure can be determined by calculating this measure at each possible value of the threshold. It is obvious that the resulting best image will give a better (or equal) result according to the considered measure than the image obtained from the unsupervised technique. Although this idea can not be applied to any given image because we do not dispose of their ground-truth edge image in general, it can be useful in order to know the capability of the unsupervised technique to achieve the best threshold value(s) according to a given measure.

The results obtained applying all the possible values for the threshold(s) are displayed in Table 3. We have calculated these values to the fuzzy edge images obtained by T_{nM} and $U^{\mathcal{N}_C}$.

Table 3 Results obtained searching the best threshold(s)

Algorithm		Mean		% Best Technique		% of improvement	
Conjunctor	Binar. Technique	FoM	ρ	FoM	ρ	FoM	ρ
T_{nM}	Thresh.	0.4741	0.6533	40	33	70	47
	Hyst.	0.4937	0.6622	33	27	27	25
$U^{\mathcal{N}_C}$	Thresh.	0.4754	0.66	60	67	61	38
	Hyst.	0.4977	0.6697	67	73	21	20

As we can see, the mean value of the measures for the two considered morphologies are even closer now but the hysteresis gives better results than the thresholding in both morphologies. An analogous statistical study as before can be made now. Again after passing the normality test, a t-test confirms our impression with a p-value of $3.856e-05$ with $U^{\mathcal{N}_C}$. The percentage of improvement is notable if we consider all the thresholds with respect to Atanassov's thresholding. On the other hand, the unsupervised hysteresis considered in this paper achieves reasonable results and it can be used confidently in a unsupervised way. In Figure 4, note that the improvement on the measures gives also a visual improvement in the results.

Fig. 4 From top to bottom and left to right, original image and its ground-truth edge image, best result with Atanassov's thresholding, best result considering all the thresholds, best result with the unsupervised hysteresis and the best result considering all the hysteresis threshold values in the initial set $(0.01,\ldots,0.25)$ according to *FoM*. All the images are obtained with $U^{\mathcal{N}_C}$.

5 Conclusions and Future Work

In this paper, a comparative analysis of several edge detectors based on fuzzy morphologies has been carried out. The two considered performance measures, *FoM* and ρ, indicate that the edge detector from the morphology defined with the idempotent uninorm $U^{\mathcal{N}_C}$ with the unsupervised hysteresis outperforms the other edge detectors. In addition, the goodness of the two automatic binarization techniques has been studied concluding that the hysteresis based one provides results close to the best ones according to the performance measures.

As a future work, we want to compare these edge detectors with the classical ones (Canny, Sobel, Prewitt, etc.). In this way, initial experiments seem to show a big improvement of the Canny edge detector if the hysteresis is performed with the unsupervised hysteresis method used in this paper. The results outperform drastically the ones obtained using the default implementation of Canny's algorithm in Matlab or Megawave. In addition, the choice of the structuring element (size, shape, direction) in each pixel could improve the results obtained using a fixed structuring element.

Acknowledgements. This paper has been partially supported by the Spanish Grant MTM 2009-10320 with FEDER support.

References

1. Bloch, I., Maître, H.: Fuzzy mathematical morphologies: a comparative study. Pattern Recognition 28, 1341–1387 (1995)
2. Bodenhofer, U.: A unified framework of opening and closure operators with respect to arbitrary fuzzy relations. Soft Computing 7, 220–227 (2003)

3. Bowyer, K., Kranenburg, C., Dougherty, S.: Edge detector evaluation using empirical ROC curves. Computer Vision and Pattern Recognition 1, 354–359 (1999)
4. Bustince, H., Barrenechea, E., Pagola, M., Fernandez, J.: Interval-valued fuzzy sets constructed from matrices: Application to edge detection. Fuzzy Sets and Systems 160(13), 1819–1840 (2009)
5. Canny, J.: A computational approach to edge detection. IEEE Trans. Pattern Anal. Mach. Intell. 8(6), 679–698 (1986)
6. De Baets, B.: Fuzzy morphology: A logical approach. In: Ayyub, B.M., Gupta, M.M. (eds.) Uncertainty Analysis in Engineering and Science: Fuzzy Logic, Statistics, and Neural Network Approach, pp. 53–68. Kluwer Academic Publishers, Norwell (1997)
7. De Baets, B., Kerre, E., Gupta, M.: The fundamentals of fuzzy mathematical morfologies part I: basics concepts. International Journal of General Systems 23, 155–171 (1995)
8. González-Hidalgo, M., Massanet, S.: Closing and opening based on discrete t-norms. Applications to Natural Image Analysis. Accepted in EUSFLAT-LFA 2011 (2011)
9. González-Hidalgo, M., Massanet, S.: Towards an objective edge detection algorithm based on discrete t-norms. Accepted in EUSFLAT-LFA 2011 (2011)
10. González-Hidalgo, M., Massanet, S., Torrens, J.: Discrete t-norms in a fuzzy mathematical morphology: Algebraic properties and experimental results. In: Proceedings of WCCI-FUZZ-IEEE, Barcelona, Spain, pp. 1194–1201 (2010)
11. González-Hidalgo, M., Mir-Torres, A., Ruiz-Aguilera, D., Torrens, J.: Edge-images using a uninorm-based fuzzy mathematical morphology: Opening and closing. In: Tavares, J., Jorge, N. (eds.) Advances in Computational Vision and Medical Image Processing, Computational Methods in Applied Sciences. ch. 8, vol. 13, pp. 137–157. Springer, Netherlands (2009)
12. González-Hidalgo, M., Mir-Torres, A., Ruiz-Aguilera, D., Torrens, J.: Image analysis applications of morphological operators based on uninorms. In: Proceedings of the IFSA-EUSFLAT 2009 Conference, Lisbon, Portugal, pp. 630–635 (2009)
13. Grigorescu, C., Petkov, N., Westenberg, M.A.: Contour detection based on nonclassical receptive field inhibition. IEEE Transactions on Image Processing 12(7), 729–739 (2003)
14. Kovesi, P.D.: MATLAB and Octave functions for computer vision and image processing. Centre for Exploration Targeting, School of Earth and Environment, The University of Western Australia, `http://www.csse.uwa.edu.au/~pk/research/matlabfns/`
15. Medina-Carnicer, R., Muñoz-Salinas, R., Yeguas-Bolivar, E., Diaz-Mas, L.: A novel method to look for the hysteresis thresholds for the Canny edge detector. Pattern Recognition 44(6), 1201–1211 (2011)
16. Nachtegael, M., Kerre, E.: Classical and fuzzy approaches towards mathematical morphology. In: Kerre, E.E., Nachtegael, M. (eds.) Fuzzy techniques in image processing. ch. 1, vol. (52), pp. 3–57. Physica-Verlag, New York (2000)
17. Papari, G., Petkov, N.: Edge and line oriented contour detection: State of the art. Image and Vision Computing 29(2-3), 79–103 (2011)
18. Pratt, W.K.: Digital Image Processing, 4th edn. Wiley Interscience, Hoboken (2007)
19. Serra, J.: Image analysis and mathematical morphology, vol. 1, 2. Academic Press, London (1982/1988)
20. Sola, H.B., Tartas, E.B., Pagola, M., Orduna, R.: Image thresholding computation using Atanassov's intuitionistic fuzzy sets. JACIII 11(2), 187–194 (2007)

Towards a New Fuzzy Linguistic Preference Modeling Approach for Geolocation Applications

Mohammed-Amine Abchir and Isis Truck

Abstract. In many areas, fuzzy linguistic approaches have already shown their interest and successful results to express the preferences and the choices of a human. This paper focuses on the fuzzy linguistic 2-tuple representation model that is interesting and relevant when we need to express and to refer to linguistic assessments during the whole reasoning process. However, when data have a particular distribution on their axis, this model doesn't fit well the needs anymore. We propose therefore a variant version of this representation model that allow for a more realistic distribution. We also show that an operation such as an arithmetic mean is easy to implement with it and gives consistent results.

1 Introduction

Knowledge representation — and especially preference modeling — is a widely known and central problem in many fields. This is often related to preference extraction and handling that can be done through elicitation strategies. There are several ways to elicit preferences, depending on whether they are conditional or not. In the first case compete several models among which are the factorized utility models (compact models) such as the GAI (General Additive Independence) networks [1], the CP-nets (Conditional Preference networks) [2] or the LCP-nets (Linguistic Conditional Preference networks) [3]. In the second case preferences are unconditional thus don't depend on each other. The elicita-

Mohammed-Amine Abchir
Universite Paris 8, 2 rue de la Liberte, F-93526, Saint-Denis (France),
Deveryware, 43 rue Taitbout, 75009 Paris (France)
e-mail: `maa@ai.univ-paris8.fr`

Isis Truck
Universite Paris 8, 2 rue de la Liberte, F-93526, Saint-Denis (France)
e-mail: `truck@ai.univ-paris8.fr`

B. De Baets et al. (Eds.): Eurofuse 2011, AISC 107, pp. 413–424, 2011.
springerlink.com

tion can be performed through an HCI (Human Computer Interaction) [4] and NLP (Natural Language Processing) [5] in a Stimulus-Response application for instance.

In this paper, our work focuses on establishing relations between user's linguistic preferences or choices (in a dialog context) and their interpretation. Of course, such relations imply to model the linguistic preferences in a suitable representation model. Many representation models may fit with our needs, especially the classical fuzzy approach from Zadeh [6], but one model seems to be more appropriate because it enables to deal with linguistic assessments in a fuzzy way with a simple and regular representation: the fuzzy linguistic 2-tuples introduced by Herrera and Martínez [7]. Moreover, this model enables the expression of linguistic data that are unbalanced on their axis (*i.e.* data whose midterm can have *e.g.* three terms on its left and only one term on its right).

However, in some cases, the resulting fuzzy sets and associated linguistic terms obtained with this representation model do not match exactly with human preferences, especially when the data is "very unbalanced" on the axis. Therefore an intermediate representation model is needed. The aim of this paper is to present a new algorithm to represent imprecise information with this fuzzy linguistic 2-tuple model. This algorithm improves the matching of the fuzzy partitioning.

This paper is structured as follows. First, we present the context of our work that takes place in a geolocation architecture. Then, in section 3 we revise in short the fuzzy linguistic approach and the 2-tuple fuzzy linguistic representation model. Section 4 introduces a new approach that uses *linguistic hierarchies* [8] and combines some elements from both fuzzy sets and 2-tuples. We also give an illustrative example of this approach in a real world problem. We finally conclude with some remarks.

2 Industrial Context

In this section, we review our industrial context in geolocation and explain some underlying concepts such as geotracking techniques or the architecture of the application.

2.1 Geotracking

Since five or six years, geolocation has become very popular thanks to the large adoption of smartphones and navigation systems. A wide range of geotracking applications currently emerges in many areas such as logistics, security, position sharing, service recommendation...

In short, geotracking needs one or more geographic positions at regular (or irregular) intervals of the followed mobiles. This operating mode implies the collaboration of two entities: a tracking system and positioning devices.

Positioning devices can locate themselves (or be located) using different techniques. Thus, they fall into three categories:

1. Global Positioning System (GPS) devices that triangulate their positions using signals from satellites;
2. GSM devices (phones or similar) that use tracking information of cell they are connected to. This technique is known as Cell-Id;
3. WiFi (or similar) devices which location technique is similar to the one used by GSM devices.

The tracking system is responsible for collecting data coming from positioning devices (positions, battery level, the GPS/GSM signal force...). Moreover, it can trigger reactions according to given events. Regarding a device position, events can be, for example, approaching or leaving a zone of interest (in the geographical sense).

Our work takes place in a multi-partner project whose name is SALTY (for Self-Adaptive very Large disTributed sYstems). As for us, our aim is to propose solutions to automatically generate business-process applications in the geotracking context. The challenge is to be able to elicit the needs, choices and preferences of the user in order to produce a specific configuration of the generic application and to deploy this *ad hoc* application into an existing geotracking platform. The platform, called GeoHub, has been conceived by Deveryware, one of the industrial partners of SALTY project, and proposes a real-time geolocation accessible via the Internet. GeoHub acts like a middleware supporting a large set of kinds of positioning devices. Moreover, GeoHub allows Deveryware's customers to trigger high-level events (as shortly described before) and to be notified (by SMS or email) when an event happens. Thus, it can be considered as a specialized geotracking Complex Event Processor (CEP).

To help us finding which features to implement, we propose several use cases [9] that SALTY project shall address. Among them, we focus on a long distance truck tracking scenario. In this context, truck routes and destinations are planned in advance. The aim of the geotracking is to make sure that trucks reach their final destination following a predefined route and passing by intermediate waypoints. To do so, the customer (a logistician, here) needs to be notified at each step of the transportation and be alerted if any problem (*e.g.* a deviation from the planned route considered as a corridor) arises.

In the literature, fuzzy logic has of course already been applied in the geotracking context to solve a wide range of problems such as in automotive industry [10], position tracking improvement [11], etc. But most of these works focus on improving the accuracy of positioning techniques which is not the goal of our work. Inside SALTY project, our goal is to follow *approximately* a truck and notify the end-user by its passage *near* predefined waypoints, or *more or less* within the corridor.

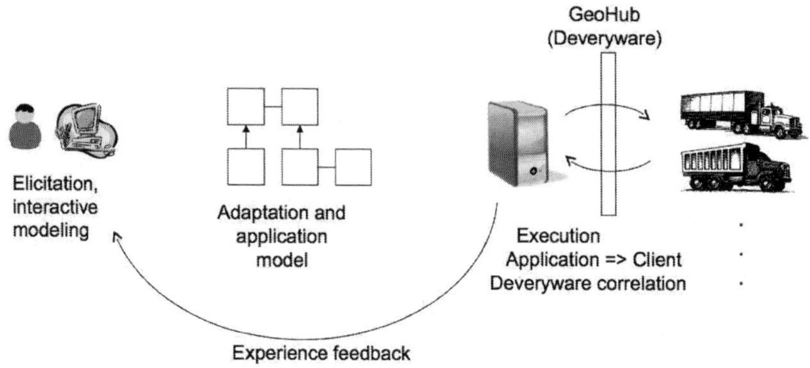

Fig. 1 Overview of the geotracking application

2.2 Architecture Overview

Deveryware offers its customers a web application called DeveryLoc that helps to configure the GeoHub. This setup is done by parameterizing predefined alerts/events types with numerical values. Overall parameters represent either distance (in meters), frequency (in seconds), device configuration or battery level (in percent), i.e. only precise values can be entered. This setup step requires from the customers to have a knowledge of a lot of technical parameters to provide the GeoHub with good enough information to reach their business objectives.

One of the goals of SALTY project is to provide an interactive tool to help non-specialist users to easily setup the GeoHub for their business applications. So this tool will be an *enhanced* DeveryLoc that will help to model at design-time the needs without having recourse to precise and technical parameters. It will, through a natural language dialog, allow customers to elicit their business-oriented objectives by using linguistic preference assessments instead of precise numerical values. Figure 1 presents an overview of the use of elicitation in business application generation.

Thus, it will be possible to indicate the positioning devices to be tracked, to setup events to be notified to setup the frequencies of position transmission, etc.

DeveryLoc, as a first evolution, will be enhanced to support linguistic terms instead of numerical ones that are less intuitive for end-users. It will allow them to build models of their applications and then configure the GeoHub consequently. These models contain all user's preferences (alert types, distances, frequencies, configuration...).

In order to catch the overall needs of the end-users, the elicitation strategy adopted by the natural language dialog will generate different models. Each kind of model is attached to a user's role in the enhanced DeveryLoc. Thus, three roles and three models can be defined:

1. an *expert role*, which consists in defining new alert types of the business application domain and their parameters (in a linguistic form when it is needed). These definitions are gathered in an *application model.*
2. an *application designer role*, which consists in defining an application using the predefined alert types. The application definition represents an *application.*
3. a *configurator role*, which consists in providing all parameters of an application: concerned devices, period of activation of alerts on the GeoHub... It is called a *configuration.*

In the *application model* (Figure 2), the *expert* defines alert types according to its domain (logistics in our example). These alert types can be used later by a business application. They contain all parameters needed by an alert, the notification triggering rules and custom data types introduced by the expert on the parameter definition. In particular, we can notice that we use *two-tuple* data types.

When linguistic parameters are chosen, their corresponding representation need to be generated and made available to the GeoHub to be used at runtime. The linguistic representation with the 2-tuples is discussed in Section 4.

In order to validate the *application models*, a Relax NG[1] grammar is used as a meta-model. Using these *application models*, the *application designer* can choose the alerts that have to be combined to generate a specific business *application.* He also provides the global parameters (that can be linguistic) of the application by choosing values for distances and frequencies for instance.

Finally, for activating an application during a period of time, a *configurator* provides the configuration parameters needed by the business application. It corresponds to the current DeveryLoc operating mode.

3 The 2-Tuple Fuzzy Linguistic Representation

In this section we review the fuzzy linguistic approach, the 2-tuple fuzzy linguistic representation model and some related works.

Due to the vagueness of their underlying aspects, many real-world problems have to be assessed by qualitative data and linguistic variables [12]. Among the various fuzzy linguistic representation models, there are two 2-tuple models: the "proportional model" and the "translation model". The first representation has been introduced by Wang and Hao in [13] and describes a model where values are defined as symbolic proportions of two successive fuzzy sets. The 2-tuple is thus composed of a pair $(\alpha l_i, (1-\alpha)l_{i+1})$, where l_i is a linguistic label represented by a fuzzy set. As this model doesn't address the fuzzy partitioning process — the authors rather focus on the

[1] REgular LAnguage for XML Next Generation.

```xml
<?xml version="1.0" encoding="iso-8859-1"?>
<geotracking-model
   uri="http://www.deveryware.com/model/TruckTracking"
   xmlns:xsd="http://www.w3.org/2001/XMLSchema">
  <alertType name="ArrivalAtWarehouseNotification"
             category="ArrivalAtADestination">
    <application-parameters>
      <param name="Delay" type="xsd:time"/>
      <param name="WarehouseLocation" type="text"/>
      <param name="DelayTolerance" type="LinguisticDistance"/>
      <param name="PositioningDevice" type="PositionDeviceType"/>
      <param name="StartTime" type="xsd:dateTime"/>
      <param name="EndTime" type="xsd:dateTime"/>
    </application-parameters>
    <trigger type="FIS">
      <param name="Pos" type="GPSValue"/>
      <fis name="triggerArrivalNotification">
        <fuzzyRule>
          <antecedent type="LinguisticDistance">
            <two-tuple linguistic-variable="Close" translation="0.0"/>
          </antecedent>
          <consequent type="TriggeringLevel">
            <two-tuple linguistic-variable="High" translation="0.25"/>
          </consequent>
        </fuzzyRule>
        ...
      </fis>
    </trigger>
  </alertType>
  <dataType name="LinguisticDistance">
    <fuzzyEnumeration>
      <label name="VeryClose"/>
      <label name="Close"/>
      ...
    </fuzzyEnumeration>
  </dataType>
  ...
</geotracking-application>
```

Fig. 2 Models captured through the enhanced DeveryLoc tool in XML format.

way to aggregate their data — it will not be further discussed in this paper. The second representation has been introduced by Herrera and Martínez in [7]. This model represents linguistic information by means of a pair (s, α), where s is a label representing a triangular fuzzy set and α is the value of the symbolic translation. The computational model developed for this representation model includes comparison, negation and aggregation operators. By default, all triangular fuzzy sets are uniformly distributed on the axis, but the targeted aspects are not usually uniform. In such cases, the representation should be enhanced with tools such as *unbalanced* linguistic term sets which are non uniformly distributed on the axis [14]. An algorithm that permits to partition data in a convenient way has been develop by Herrera and Martínez. This is explained in the next section in order to facilitate the comparison of both propositions (ours and theirs).

4 Our Proposition

As mentioned in Section 2.2, when end-users introduce linguistic parameters, the corresponding representation (i.e. their semantic values) must be generated. These representations will be used at runtime on the GeoHub.

As we are dealing with linguistic assessments which underlying aspects have not always a uniformly distributed semantic (the distribution depends on the business domain and the expert preferences), we use the 2-tuple fuzzy linguistic representation model and the *unbalanced* linguistic term sets to generate the fuzzy partitioning.

The algorithm introduced in [14] needs two inputs: the linguistic term set S (composed by the medium label denoted SC, the set of labels on its left denoted SL and the set of labels on its right denoted SR) and the density of label distribution on each side. The density can be *middle* or *extreme* according to the user's choice. For example the description of $S = \{N, L, M, AH, H, QH, VH, AT, T\}$ is $\{(2, extreme), 1, (6, extreme)\}$ with $SL = \{N, L\}$, $SC = \{M\}$ and $SR = \{AH, H, QH, VH, AT, T\}$.

However, the required inputs can be considered too demanding in a real world application because even an expert may not be able to decide the density. Moreover, only the number of labels (or terms) is taken into account in the fuzzy partitioning. Indeed, if the expert considers only three terms to qualify the weather: "frost", "snow" and "warm", the algorithm won't provide for an unbalanced linguistic term set. The only way to obtain a good distribution would be to give artificial (virtual) additional terms (such as "cold", "rather cold", etc.). Therefore, a better approach would be to ask only for the labels of the term set and their associated positions on the axis. Thus, the user preferences (the position of the labels on the axis) are fully taken into account during the fuzzy partitioning.

In our use case, trucks are supposed to stay within a corridor and the logistician has to make sure they do not deviate *too much* of this route. Typically, Deveryware's customers use five values to define the distance of the truck relative to the center of the corridor: 0m means *in the center* of the corridor, 50m means *very close to* the center, 60m means *near* the center, 80m means *far* from the center and 300m when the truck is clearly considered *out of the route*. Figure 3 shows the fuzzy partitioning that fits with reality (left of the figure) and the one obtained with the algorithm detailed in [14] (right of the figure).

The semantics of each label of the *unbalanced* term set is represented by a fuzzy membership function according to a structure that was introduced in [15] to manage multigranular linguistic information: the *linguistic hierarchy*. A linguistic hierarchy (LH) is composed of several term sets of different granularity called levels. Each level of the hierarchy is denoted $l(t, n(t))$ where t is the number of the level and $n(t)$ its granularity. Thus, a linguistic term set $S^{n(t)}$ belonging to a level t of a linguistic hierarchy LH can be denoted $S^{n(t)} = \{s_0^{n(t)}, \ldots, s_{n(t)-1}^{n(t)}\}$.

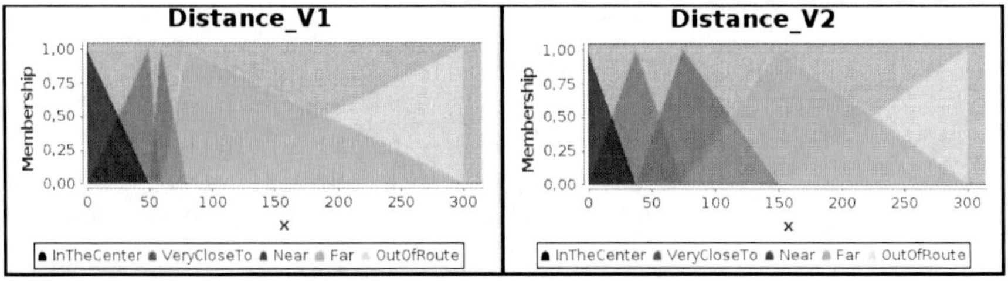

Fig. 3 The fuzzy partitioning generated for distance with two algorithms

Linguistic hierarchies have the following properties:

- levels are ordered according to their granularity;
- the linguistic term sets have an odd value granularity;
- the membership functions of the linguistic terms are triangular;
- linguistic terms are uniformly and symetrically distributed on $[0, 1]$.

The way of building the linguistic hierarchy is further described in [16].

4.1 Towards a New Partitioning

Since our goal is to operate with linguistic information introduced by end-users, the first step is to obtain their semantic representation. This process is based on the use of a linguistic hierarchy to represent each term of a given term set.

We consider an unbalanced linguistic term set S composed of pairs (s, v) where s is the label of the term and v the position of the term on the axis. Thus the term set can be denoted $S = \{(s_0, v_0), \ldots, (s_n, v_n)\}$.

The representation iterative process will assign semantics to each term according to a linguistic hierarchy LH. It implies that a label $s_i^{n(t)} \in LH$ (label number i of level t of the linguistic hierarchy) to represent a label $s_k \in S$ must be chosen. The selection of $s_i^{n(t)}$ depends on both the distance d_k between s_k and s_{k+1} and the value of v_k. The iterative assignation process is as follows:

1. Compute the distance d between s_k and s_{k+1} using v_k and v_{k+1}.
2. Starting with the level with the smallest granularity, select a level t of the hierarchy LH by choosing the first level whose step value is smaller than d. The step value of a level is the distance between two successive labels.
3. Then select the most suitable terms $s_i^{n(t)}$ and $s_{i+1}^{n(t)}$ for the assignation step. $s_i^{n(t)}$ is the term which has the nearest kernel value to v_k.
4. The representation becomes: $\underline{s_k} \leftarrow \underline{s_i^{n(t)}}$ and $\overline{s_{k+1}} \leftarrow \overline{s_{i+1}^{n(t)}}$.

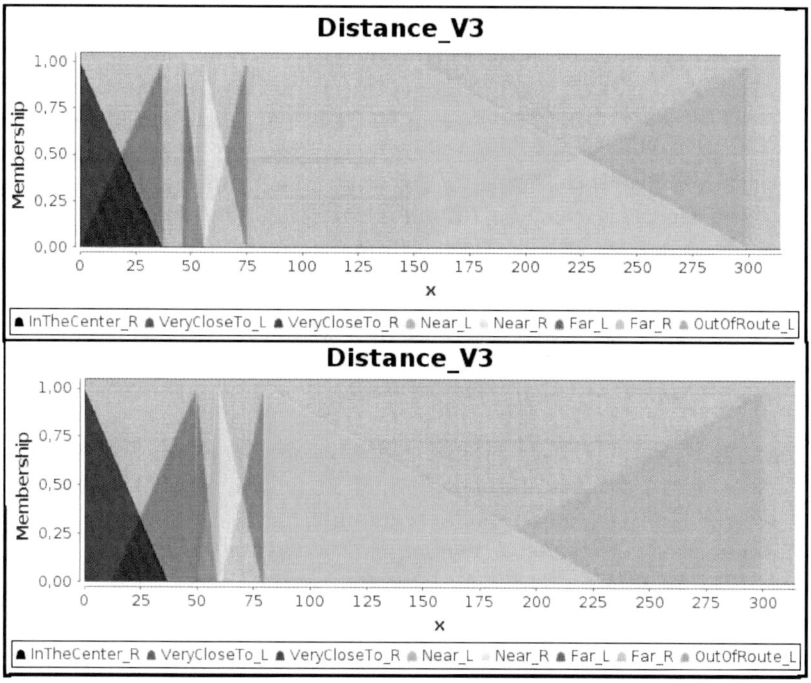

Fig. 4 The fuzzy partitioning generated for distance with our algorithm

5. As the kernel of $s_i^{n(t)}$ does not always match with v_k, a symbolic translation α_k is associated with $\underline{s_k}$ to form a linguistic 2-tuple: $(\underline{s_k}, \alpha_k)$. The same is done with $s_i^{n(t)}$
6. Move to the next pair of labels, i.e. s_{k+1} and s_{k+2}.

In order to perform the iterative assignation process, two preliminary steps must be done:

1. As linguistic hierarchies are distributed on $[0, 1]$ by default, the linguistic hierarchy LH used in the process of assignation must be scaled to match the universe of discourse of the linguistic term set S. Thus, LH is scaled using v_{max}, the maximum v value of the pairs of S. The result is a linguistic hierarchy which levels are distributed on $[0, v_{max}]$. This step is essential to guarantee the accuracy of step (3) of the assignation process.
2. By default, LH only contains a three-label level, so we automatically add as many levels as necessary to it until the obtention of a level whose step value is smaller than d_{min} (= the smallest distance between two successive terms of S). In this way, LH is refined enough to cover d_{min} in step (2) of the assignation process.

The output of the algorithm is a set of linguistic 2-tuples expressed in LH that represents a fuzzy linguistic partitioning supporting the semantics of S. Figure 4 shows the fuzzy partitioning obtained by this algorithm before (up figure) and after (bottom figure) the application of the symbolic translations (step 5 of the assignation process).

Using the step values of hierarchy levels and the distance d_k, we can easily demonstrate that this fuzzy partitioning has a minimum membership value ϵ greater than 0. Considering X the universe of discourse, x a value of this universe and f_{s_i} the membership function associated with a term s_i, this property can be denoted: $\forall x \in X, \quad f_{s_1}(x) \vee \ldots \vee f_{s_i}(x) \vee \ldots \vee f_{s_n}(x) > \epsilon > 0$

4.2 Aggregation with Our Model

As the result of our assignation process is a set of unbalanced linguistic 2-tuples expressed in LH, the unbalanced linguistic computational model introduced in [14] can be adapted to accomplish the process for linguistic information using a linguistic 2-tuple aggregation operator (weighted average, arithmetic mean, etc.). This computational model can be summarized in four steps:

1. Represent unbalanced linguistic terms in a linguistic hierarchy LH.
2. Choose a level of LH to compute this unbalanced linguistic terms.
3. Compute or aggregate them by means of the 2-tuple fuzzy linguistic computational model.
4. Express the result in the initial term set.

To illustrate the aggregation process, we consider the distance fuzzy partitioning example and suppose that we want to aggregate two values: *VeryCloseTo* and *OutOfRoute*.

According to [14], the representation of these two labels is respectively: $(s_1^9, 0)$ and $(s_2^3, 0)$. The steps of the aggregation are the following: first, both labels are represented in the same hierarchy level using the $TF_{t'}^{t}$ transformation function that transforms a linguistic label in level t to a label in level t'. Then, the 2-tuple arithmetic mean is applied to the obtained representations ($(s_1^9, 0)$ and $(s_8^9, 0)$). Finally, using the Δ and $\mathcal{LH}^{-1}$ functions the resulting representation is $(s_1^3, 0.125)$ which corresponds to $(Far, 0.125)$. Where $\mathcal{LH}^{-1}$ is the function that associates with each linguistic 2-tuple expressed in the linguistic hierarchy its respective unbalanced linguistic 2-tuple and Δ is function that associates to a value the equivalent 2-tuple.

Using our representation algorithm, both labels are represented with $(s_1^9, 12.5)/(s_5^{33}, 3.125)$ and $(s_2^3, 0)$ respectively. As we use an absolute scale, *i.e.* all levels are scaled to $[0, 300]$ on the axis, and in order to simplify the computations, the symbolic translations are left in the absolute form (instead of the normalized one where $\alpha \in [-0.5, 0.5)$). Next, we use the 2-tuple arithmetic mean to obtain the average absolute value resulting which is $(s_4^9, -12.5)$.

Finally, we use the $\mathcal{LH}^{-1}$ function to obtain $(s_1^3, 25)$ which corresponds to $(Far, 95)$.

As we can see with this simple example of an arithmetic mean aggregation, the results obtained by both algorithms are a bit different. This is due to the representation models: the "classical 2-tuple model" entails a *weak unbalanced* partitioning of the terms while our model entails a *stronger unbalanced* partitioning. For our specific context, the last results fit better reality. Thus, it seems useful to propose in the future other (more sophisticated) linguistic 2-tuples aggregation operators dedicated to our representation approach.

5 Concluding Remarks

In this paper, we have introduced an approach to deal with fuzzy linguistic preference modeling in an industrial context. The preferences are introduced by a geolocation expert using a natural language dialog interface which is not described here. The representation model and the partitioning algorithm we propose rely on the use of linguistic hierarchies and on the 2-tuple fuzzy linguistic representation model from Herrera and Martínez.

This approach allows for the representation of terms assessed in unbalanced linguistic term sets to fit better with reality by weighting the labels of these terms by their positions on the axis. It is useful when a more flexible representation is needed. We have shown that our proposition is consistent because we can compute an aggregation between our terms and obtain coherent results. In a future work, we will propose other aggregators, such as specific t-norms and t-conorms. The long-term aim is to propose a complete linguistic reasoning process with these unbalanced 2-tuples, *i.e.* an *ad hoc* generalized modus ponens.

Acknowledgements. This work is partially funded by the French National Research Agency (ANR) under grant number ANR-09-SEGI-012.

References

1. Gonzales, C., Perny, P., Queiroz, S.: GAI-Networks: Optimization, Ranking and Collective Choice in Combinatorial Domains. Foundations of computing and decision sciences 32(4), 3–24 (2008)
2. Boutilier, C., Brafman, R.I., Domshlak, C., Hoos, H.H., Poole, D.: CP-nets: A tool for representing and reasoning with conditional *Ceteris Paribus* Preference Statements. Journal of Artificial Intelligence Research 21, 135–191 (2004)
3. Châtel, P., Truck, I., Malenfant, J.: LCP-nets: A linguistic approach for non-functional preferences in a semantic SOA environment. Journal of Universal Computer Science, 198–217 (2010)
4. Booth, P.: An Introduction to Human-Computer Interaction. Lawrence Erlbaum Associates, Publishers, USA (1989)

5. Ambriola, V., Gervasi, V.: Processing natural language requirements. In: International Conference on Automated Software Engineering, p. 36. IEEE Computer Society, USA (1997)
6. Zadeh, L.: The concept of a linguistic variable and its application to approximate reasoning, i, ii and iii. IS 8 (1975)
7. Herrera, F., Martínez, L.: A 2-tuple fuzzy linguistic representation model for computing with words. IEEE Transactions on Fuzzy Systems 8(6), 746–752 (2000)
8. Herrera, F., Martínez, L.: A model based on linguistic 2-tuples for dealing with multigranularity hierarchical linguistic contexts in multiexpert decisionmaking. IEEE Transactions on Systems, Man and Cybernetics. Part B: Cybernetics (2001)
9. Melekhova, O., Abchir, M.-A., Châtel, P., Malenfant, J., Truck, I., Pappa, A.: Self-Adaptation in Geotracking Applications: Challenges, Opportunities and Models. In: The 2nd International Conference on Adaptive and Self-Adaptive Systems and Applications (ADAPTIVE 2010), pp. 68–77. IEEE, Los Alamitos (2010)
10. Cengiz, S., Vedat, T., Fevzi Baba, A.: Pneumatic motor speed control by trajectory tracking fuzzy logic controller. Sadhana 35(1), 75–86 (2010)
11. Sung, W., You, K.: Adaptive precision geolocation algorithm with multiple model uncertainties. In: Adaptive Control, In-tech., pp. 323–336 (2009)
12. Zadeh, L.A.: Quantitative fuzzy semantics. Information Sciences 3(2), 159–176 (1971)
13. Wang, J., Hao, J.: A new version of 2-tuple fuzzy linguistic representation model for computing with words. IEEE Transactions on Fuzzy Systems 14(3), 435–445 (2006)
14. Herrera, F., Herrera-viedma, E., Martínez, L.: A fuzzy linguistic methodology to deal with unbalanced linguistic term sets. IEEE Transactions on Fuzzy Systems, 354–370 (2008)
15. Cordón, O., Herrera, F., Zwir, I.: Linguistic modeling by hierarchical systems of linguistic rules. IEEE Transactions on Fuzzy Systems 10, 2–20 (2002)
16. Herrera, F., Martínez, L.: A model based on linguistic 2-tuples for dealing with multigranularity hierarchical linguistic contexts in multiexpert decisionmaking. IEEE Transactions on Systems, Man and Cybernetics. Part B: Cybernetics, 227–234 (2001)

Author Index